AF518356

Flora of Telangana

The 29th State of India

THE AUTHOR

Prof. T.Pullaiah obtained his M.Sc. (1973) and Ph.D. (1976) degrees in Botany from Andhra University. He was a Post Doctoral Fellow at Moscow State University, Russia during 1976-78. He traveled widely in Europe and USA and visited Universities and Botanical Gardens in about 17 countries. Professor Pullaiah joined Sri Krishnadevaraya University as Lecturer in 1979 and became Professor in 1993. He held several positions in the University, which include Dean, Faculty of Life Sciences, Head of the Department of Botany, Chairman, BOS in Botany, Head of the Department of Sericulture, Co-ordinator and Chairman, BOS in Biotechnology, Vice Principal and Principal, S.K.University College. He retired from active service on 31st May 2011. He was selected by UGC as UGC-BSR Faculty Fellow and is working in Sri Krishnadevaraya University. Prof. Pullaiah has published 60 books, 303 research papers and 35 popular articles. He is Principal Investigator of 20 major Research Projects totaling more than a Crore of Rupees funded by DBT, DST, CSIR, UGC, BSI, WWF, GCC etc. Under his guidance 53 students obtained their Ph.D. degrees and 35 students their M.Phil. degrees. He is recipient of Best Teacher Award from Government of Andhra Pradesh, Prof. P. Maheswari Gold Medal and Prof. G.Panigrahi Memorial Lecture of Indian Botanical Society and Prof. Y.D.Tyagi Gold medal of Indian Association for Angiosperm Taxonomy. He was Past President of Indian Association for Angiosperm Taxonomy and presently he is President of Indian Botanical Society. He was Member of Species Survival Commission of International Union for Conservation of Nature and Natural Resources (IUCN).

Flora of Telangana

The 29th State of India

Volume 2

Author
T. Pullaiah
Department of Botany
Sri Krishnadevaraya University
Anantapur – 515 003
Andhra Pradesh
India

2021
Regency Publications
A Division of
Astral International Pvt. Ltd.
New Delhi – 110 002

Flora of Telangana-*The 29th State of India*
(3 Volume Set)

Volume 1: Page 0001-0446
Volume 2: Page 0447-0892
Volume 3: Page 0893-1306

Publisher's note:
Every possible effort has been made to ensure that the information contained in this book is accurate at the time of going to press, and the publisher and author cannot accept responsibility for any errors or omissions, however caused. No responsibility for loss or damage occasioned to any person acting, or refraining from action, as a result of the material in this publication can be accepted by the editor, the publisher or the author. The Publisher is not associated with any product or vendor mentioned in the book. The contents of this work are intended to further general scientific research, understanding and discussion only. Readers should consult with a specialist where appropriate.
Every effort has been made to trace the owners of copyright material used in this book, if any. The author and the publisher will be grateful for any omission brought to their notice for acknowledgement in the future editions of the book.

Cataloging in Publication Data--DK
Courtesy: D.K. Agencies (P) Ltd. <docinfo@dkagencies.com>

Pullaiah, T., author.
Flora of Telangana : the 29th state of India / author, T. Pullaiah.
3 volumes cm
Includes bibliographical references and index.
ISBN 9789352220014 (Vol.2)
ISBN 9788189233983 (Set)

1. Plants--India--Telengana--Classification. 2. Flowers--India--Telengana--Classification. I. Title.

DDC 582.13095485 23

Published by : **Regency Publications**
A Division of
Astral International Pvt. Ltd.
– ISO 9001:2015 Certified Company –
4736/23, Ansari Road, Darya Ganj
New Delhi-110 002
Ph. 011-43549197, 23278134
E-mail: info@astralint.com Website: www.astralint.com

Laser Typesetting : **Classic Computer Services**, Delhi - 110 035

Printed at : **Replika Press Pvt. Ltd.**

Preface

Flora of Former Andhra Pradesh was published in 1997 and Additions in 2008. The book went out of print in 2010. The long awaited decision to divide the state of Andhra Pradesh coincidentally happened when I had started revising the book sitting in Missouri Botanical Garden, U.S.A. It was then the idea of separate Flora for Telangana and Andhra Pradesh began. I continued my work using the facilities of Missouri Botanical Garden Herbarium and Library. I would like to highlight the warmth and affection which I received there and was inspired by their hard work and dedication. I thank authorities of Missouri Botanical Garden, Herbarium and Library for the permission to consult the Library and Herabrium.

Out of ten districts of Telangana we already published Flora of Five districts and I have taken liberally information from these books. The contributions of the scholars in these projects who are presently well placed in various esteemed institutions include Dr. P.V. Prasanna, Prof. B. Ravi Prasada Rao, Dr. C. Prabhakar, Dr. M.Silar Mohammad and Dr. B.V.Raghava Swamy. I thank all these scholars for their contributions which enabled me to write this book.

I thank the Scientists and scholars of Botanical Survey of India, Deccan Circle, Hyderabad, especially Dr. P.Venu, Dr. P.V.Prasanna, Scientists, Dr. L.Rasingam, Dr. Shankara Rao, Alok Chorghe, P.S.Annamma, besides ministerial staff. Many a times I had stimulating discussion with them. I am confident that this centre will grow. Dr. G.V.S.Murthy, Joint Director of Botanical Survey of India, Sourthern Circle, Coimbatore was always encouraging, supporting and extending facilities at Madras Herbarium for which I am grateful to him. Dr. M.V.Ramana, Assistant Professor, Osmania University gave his thesis on Flora of Hyderabad and many a times had animated discussion on the subject. This young boy will contribute in future much to the science of Plant Taxonomy. I also thank Prof. K. Seshagiri Rao, of University of Hyderabad, for giving the details of Flora of their campus. Librarian of Kakatiya University extended Library facilities during my visit and I am grateful to them for the same. Dr.

M.Sanjappa, former Director of Botanical Survey of India was critical about my work and always wanted best from my hands and I am grateful to him for the same. May his tribe grow. Taxonomists at Calicut University helped me in nomenclature and other aspects of different groups of plants, I thank especially Prof. M.Sabu, Prof. Santhosh Nampy and Dr. A.K.Pradeep for their help in various ways.

Authorities of Sri Krishnadevaraya University were graceful enough to provide facilities for carrying out this work. I thank Prof. R.R.Venkata Raju, Prof. C.Sudhakar, Prof. B.Ravi Prasad Rao, Prof. S. Thimma Naik, Heads of Department during different periods for their support and extending the facilities in the Department. Dr. S. Sandhya Rani, the latest entrant in the Botany Department needs special mention who read the manuscript, gave suggestions for its improvement, corrected the manuscript and contributed in many ways to bring out this book. Dr. S,.Karuppusamy Assitant Professor at The Madura College and Dr. K.Sri Rama Murthy, Professor and Head, Department of Biotechnology, Montessori Mahila Kalasala, Vijayawada have also contributed much for this book.

University Grants Commission gave me BSR Faculty Fellowship which enabled me to carry out this work after superannuation and the same is gratefully acknowledged.

I thank Anil Mittal and Dinesh Arora at Astral International, New Delhi for their patience, support and encouragement to bring out his book.

I tried to make this work as perfect as possible. Had I tried for perfection I could never complete this work. I request the readers of this book to point out errors and omissions which can be rectified in future.

T. Pullaiah

Contents

RUBIACEAE

1. Trees or shrubs:
   2. Leaves whorled; ovary 5-celled .... **Hamelia**
   2. Leaves opposite; ovary 1-4-celled:
      3. Calyx lobe(s) transformed into a short leafy structure .... **Mussaenda**
      3. Calyx lobes all normal:
         4. Ovules 2 or more than 2, usually many in each cell:
            5. Flowers in globose heads:
               6. Corolla lobes imbricate; flowers ebracteolate .... **Neolamarckia**
               6. Corolla lobes valvate; flowers bracteolate:
                  7. Fruit a syncarpium .... **Morinda**
                  7. Fruit aggregated capsule:
                     8. Peduncles axillary, 2-3 together; calyx lobed; stigma clavate; capsule cuneate, not ribbed .... **Haldina**
                     8. Peduncles terminal, solitary; calyx truncate, not lobed; stigma mitiform; capsule oblong, faintly 10-ribbed .... **Mitragyna**
            5. Flowers not in globose heads:
               9. Fruit a capsule:
                  10. Corolla valvate; seeds winged .... **Hymenodictyon**
                  10. Corolla imbricate or contorted; seeds not winged .... **Wendlandia**
               9. Fruit a berry or a drupe:
                  11. Armed shrubs or trees:
                     12. Thorns often leaf-bearing........................... **Ceriscoides**
                     12. Thorns not leaf-bearing:
                        13. Berry < 1 cm across; seeds few, not embedded in pulp .... **Benkara**
                        13. Berry >2 cm across; seeds many, embedded in pulp:
                           14. Spines 1.5-2 cm long; fruit globose, smooth or slightly ribbed........... **Catunaregum**

14. Spines short; fruit ovoid, smooth .... **Tamilnadia**

11. Unarmed trees or shrubs:

15. Petiole to 0.5 cm long; lateral veins to 20 pairs; ovary one-celled .... **Gardenia**

15. Petiole to 2 cm long; lateral veins to 7 pairs; ovary 2-celled .... **Deccania**

4. Ovules single in each cell:

16. Corolla lobes contorted:

17. Stipules interpetiolar .... **Ixora**

17. Stipules intrapetiolar:

18. Flowers four-merous; style glabrous .... **Pavetta**

18. Flowers five-merous; style hairy .... **Tarenna**

16. Corolla lobes valvate:

19. Ovary 2-celled:

20. Unarmed trees; flowers 5-merous; drupe globose .... **Psydrax**

20. Armed shrubs; flowers 4-merous; drupe ellipsoid or compressed .... **Canthium**

19. Ovary more than 2-celled .... **Spermadictyon**

1. Herbs or undershrubs:

21. Flowers 5-merous .... **Dentella**

21. Flowers 4-merous or trimerous:

22. Ovules many in each cell:

23. Ovary hemispheric, seeds pitted or reticulate:

24. Calyx teeth contiguous on the capsule; capsule splitting longitudinally .... **Hedyotis**

24. Calyx teeth distant on the capsule; capsule splitting at the apex:

25. Corolla tube cylindrical, anthers and stigma included, stigma always below the anthers .... **Kohautia**

25. Corolla tube cylindrical or infundibuliform, if cylindrical, then either anthers and stigmas exserted or anthers below the stigma .... **Oldenlandia**

23. Ovary globose:

26. Seeds with ventral cavity .... **Neanotis**

26. Seeds without ventral cavity .... **Mitracarpus**

22. Ovule single in each cell:
    27. Flowers dimorphic in terminal corymbs .... **Knoxia**
    27. Flowers monomorphic in axillary fascicles .... **Spermacoce**

BENKARA Tirveng.

Benkara malabarica (Lam.) Tirveng., Taxon 32: 440. 1983. *Randia malabarica* Lam., Encycl. 3: 25. 1789; Hook.f., Fl. Brit. India 3: 111. 1880; Gamble, Fl. Madras 2: 616. 1921.

Vern.: *Pedalli*.

An armed shrub, up to 6 m high; spines short, 0.6-1cm. Leaves coriaceous, decussate or clustered, obovate-oblanceolate, 3-6 × 1.5-2.5 cm, base cuneate, margin entire, apex obtusely apiculate; petiole 0.7 cm; stipules interpetiolar, connate. Flowers in axillary, umbel like corymbose cymes, white, bracts and bracteoles free or connate, calyx turbinate, 4 mm across, lobes 5, triangular 1mm, valvate, acute, deciduous; corolla salver form, 4 cm across, tube cylindric with a ring of hairs within, lobes 5, oblong, 3 mm, recurved, obtusely acute; stamens 5, between the lobes, exserted, filaments 0.5 mm, anthers linear, 3 mm; ovary 1 mm, style 0.25 mm, stigma fusiform, 1.75 mm. Berry globose, 0.8 mm across; seeds conical or cubical, 8 per locule, not embedded in pulp.

Common in dry evergreen scrub forests of Medak district.

Fl. & Fr.: February - August.

Nathyanipally (MDK), *BR* & *CP* 11252.

Note: Tirvengadum (1983) reestablished the genus *Benkara* for the segregated taxon from *Randia*. He also discussed the reasons for reestablishing the genus and also rejection of the name *Griffithia*.

CANTHIUM Lam.

Canthium coromandelicum (Burm.f.) Alston, Handb. Fl. Ceylon 6 (Suppl.): 152. 1931. *Gmelina coromandelina* Burm.f., Fl. Indica, 132. 1768. *Canthium parviflorum* Lam., Encycl. 1: 602. 1785; Hook.f., Fl. Brit. India 3: 136. 1880. *Plectronia parviflora* (Lam.) Bedd., Fl. Sylv. S. India 124. 1872; Gamble, Fl. Madras 2: 625. 1921.

Vern.: *Balusu, Chinnabalusu*.

An armed shrub, up to 5m high, spines up to 3 cm, supra-axillary. Leaves elliptic – ovate, glabrous, 2-3 x 1-2 cm, base acute to attenuate, margin entire, apex acute, petiole 0.4 cm, stipules linear. Flowers in axillary decussate, lax cymes, greenish white; calyx cupular, 4-lobed, 1 × 2.5 mm, glabrous; corolla 5 mm across, tube 1.5 cm, villous within, lobes 4, obovate, 2.5 mm, acute, spreading or reflexed; stamens 4, (sub)sessile, anther 0.7 mm long, pointed at apex; ovary 0.7 × 1 mm, style 2.5mm, stigma capitate. Drupe globose, 1 cm across, pyrenes furrowed; seeds 1-2, brown, curved, 0.8 ××0.5 cm, rugose.

Very common in scrub forest, especially on laterite soils in Nizamabad, Medak, Ranga Reddi, Hyderabad, Karimnagar, Nalgonda and Mahabubnagar districts.

Fl. & Fr.: March - September.

Darigav RF (ADB), *GO* 4464; Biknoor (NZB), *BR* & *CPR* 7248; Mamidipally (NZB), *BR* & *KH* 9015; Rampur (MDK), *TP* & *CP* 14069; Central University campus (RR), *MSM* 15153; Rachakonda (NLG), *BVRS* 432 (NU); Near Hussain Sagar Dam (HYD), *KMS* 8025 (MH); Ramanagutta (KMM), *RR* 107994 (BSID); Jakaram (WGL), *RKP* 105263 (BSID).

CATUNAREGUM Wolf

Catunaregum spinosa (Thunb.) Tirveng., Bull. Mus. Natl. Hist. Nat. Ser. 3. Bot. 35: 13. 1978 & Taxon 27: 515. 1978. *Gardenia spinosa* Thunb., Gardenia. 16. 1780. *Randia dumetorum* (Retz.) Poiret var. *floribunda* (DC.) Gamble, Fl. Madras 616. 1921. *Gardenia dumetorum* Retz., Observ. Bot. 2: 14. 1781. *Randia brandisii* Gamble, Fl. Madras 2: 616. 1921.

Vern.: *Manga.*

Small armed shrub. Leaves elliptic – lanceolate or obovate, 1.5 – 3.5 × 1 – 1.5 cm, base acute – attenuate, margin entire, apex acute – obtuse, sparsely pubescent above, densely pubescent below, petiole short, 0.5 cm, stipules ovate, acuminate. Flowers in axillary corymbs, greenish white; calyx tube short, 1 cm, lobes 5, hairy without; corolla tubular, 6 mm, lobes 5, 4 mm, hairy without; stamens 5, attached to the base of the lobes, filament 0.5 mm, anther linear, 4.2 mm; ovary bilocular, 5 mm, ovules 2 on axile placentation, style 7 mm long, stigma bifid, 2 mm. Berry globose, hairy, 3 × 2 cm; seeds many, inserted in the pulp, white, 2 × 2 mm, curved.

Common in scrub jungles in most districts.

Fl. & Fr.: February – June.

Ankusapuram RF (ADB), *GO* 4358; Nasrullabad (NZB), *TP* & *BR* 6156; Machareddy (NZB), *BR* & *KH* 9048; Sardhana X roads (MDK), *BR* & *CP* 11620; Mohammadabad RF (RR), *MSM* & *KH* 10222; Laxmapur (MBNR), *TDCK* 13622; Amrabad (MBNR), *SKB* & *BSS* 32577; Nagarjunakonda (NLG), *BVRS* 141 (NU); Kodimial (KRN), *GVS* 20141 (MH); Khammam to Yenkur (KMM), *RR* 108545 (BSID).

CERISCOIDES Tirveng.

Ceriscoides turgida (Roxb.) Tirveng., Bull. Mus. Natl. Hist. Nat. Ser. 3. Bot. 35: 15. 1978. *Gardenia turgida* Roxb., Fl. Ind. 2: 557. 1824; Hook.f., Fl. Brit. India 3: 118. 1880; Gamble, Fl. Madras 2: 619. 1921.

Vern.: *Konda manga, Karinguva, Erri bikki, Botakari, Pedda bikki.*

Armed, deciduous tree, up to 10 m high; bark smooth, bluish grey, thick; wood white and close grained; branchlets tomentose with strong spines. Leaves thick, subsessile, chartaceous, variable, dimorphic, 1.5-2.5 × 0.8-1.5 cm, of upper branches 4-9 × 2.5 cm, hispid above, hirsute below, apex obtuse, margin entire, base attenuate; lateral nerves 8-10 pairs; stipules intra-petiolar, connate into a sheath. Flowers white, turning yellow with age, dimorphic, male flowers in fascicles; female flowers solitary, pentamerous, pedicellate; bracteoles lanceolate, free or connate. Berry globose, 5 × 4 cm, grey, turgid; seeds numerous, embedded in pulp.

Occasional, in the interior forest areas near streams in Adialabad, Nizamabad, Hyderabad and Karimnagar districts.

Fl. & Fr.: March - July.

Bheemaram nursery (ADB), *GO* 5185; Tiryani RF (ADB), *PVP* 9801; Jalalpur RF (NZB), *BR* & *KH* 9660; Mahadevpur to Kataram (KRN), *SLK* 70818 (NBG); Nellippakka RF (KMM), *RCS* 990221 (BSID).

Note: *C. curviflora* (Wall.,) Thw. is restricted to Malay peninsula and Siam. It is quite distinct from *C. ophioxyloides* Craib. Correct spelling should be *Chassalia* and not *Chasalia* Verdcourt in Kew Bull. 37: 525. 1983.

DECCANIA Tirveng.

Deccania pubescens (Roth) Tirveng., Nordic J. Bot. 3: 456. 1983. var. **candolleana** (Wight & Arn.) Tirveng. *l.c. Randia candolleana* Wight & Arn., Prodr. Fl. Ind. Orient. 39. 1834; Hook.f., Fl. Brit. India 3: 113. 1880; Gamble, Fl. Madras 2: 616. 1921. *Gardenia pubescens* Roth, Nov. Pl. Sp. 151. 1821.

Vern.: *Konda manga, Pedda manga, Jidi manga.*

Tree, up to 6 m high; bark greyish, rough; wood light brown, close grained; branchlets with elevated leaf scars. Leaves chartaceous, orbicular or ovate to obovate, 6-16.5 × 3.4 – 12.4 cm, lower surface densely tomentose than upper surface when young, base attenuate, margin entire, apex obtuse; petiole to 4 cm long. Flowers white in short axillary cymes; calyx turbinate, lobes 5; corolla 2 cm across, salver-form, tube with a ring of hairs within, lobes 5, orbicular, twisted to the left in bud; stamens 5, exserted; stigma capitate, furrowed. Berries clustered, globose, 1 × 0.5 cm, rind thin, leathery with lines on the surface; seeds, many, brown, smooth.

Common in open lands of deciduous forests in Mahabubnagar district.

Fl.: May-July. Fr.: September-October.

Appapur (MBNR), *SSR* 15544; Mallelatheertham (MBNR), *BR, SKB* & *BSS* 30713.

DENTELLA J.R. Forst. & G. Forst.

Dentella repens (L.) J.R. Forst. & G. Forst., Char. Gen. Pl. 13. 1776; Hook.f., Fl. Brit. India 3: 42. 1880; Gamble, Fl. Madras 2: 590. 1921. *Oldenlandia repens* L., Mant. Pl. 40. 1767.

Annual, prostrate, hairy herb. Leaves elliptic, 0.3 – 0.5 × 0.2 – 0.4 cm, base acute-attenuate, margin entire, apex acute-obtuse, petiole 0.1 cm long, stipules connate, scarious. Flowers axillary, solitary, white; calyx campanulate, lobes 5, equal, valvate, linear, 2 mm, acute – acuminate, hairy without; corolla campanulate, 3 mm across, lobes 5, each 3-partite, midlobe oblong, lateral lobes shorter, 4.5 mm, membranous, acute, villous within the throat; stamens 5, between the lobes, attached in the middle, included, filament 0.5mm, anthers oblong, 1mm; ovary bilocular, globular, 1.5 mm, hairy without, ovules numerous on axile placentation, stigma subulate, sessile.

Capsule globose, 3 mm, indehiscent, densely villous with coarse hairs; seeds many, minute, angular, punctuate, shiny.

All districts in moist places, especially in rice fields.

Fl. & Fr.: May-November.

Kadam dam (ADB), *GO* & *DAM* 5198; Burugidda (NZB), *BR* & *GO* 9071; Pochammaralu (MDK), *BR* & *CP* 11490; Kodimial (KRN), *GVS* 20103 (MH); Pakhal lake (WGL), *KMS* 13109 (MH); Moosi river bank (HYD), *KMS* 5977 (MH).

GARDENIA Ellis *nom. cons.*

1. Leaves petiolate, secondary nerves slender, parallel, very prominent beneath .... **G. resinifera**
1. Leaves sessile or subsessile, secondary nerves other than above:
   2. Leaves 7-30 × 2.5-18.5 cm; flowers pedicillate; calyx-lobes linear-lanceolate; stamens included; fruit to 5.5 × 5.5 mm .... **G. latifolia**
   2. Leaves 3-8.5 × 1.5-6 cm; flowers sessile; calyx lobes triangular; stamens exserted; fruit to 4 × 3 cm .... **G. gummifera**

Gardenia gummifera L.f., Suppl. Pl. 164. 1781; Hook.f., Fl. Brit. India 3: 116. 1880; Gamble, Fl. Madras 2: 618. 1921.

Vern.: *Bikki, Manchi bikki, Chinna mitta, Chit-mit.*

Deciduous tree, up to 7 m high, bark greyish brown, smooth, wood yellowish-white, close grained, branches rough and twisted. Leaves obovate-oblong, 3-8.5 × 1.5-6 cm, sessile, apex obtuse or bluntly acute, margin entire, base rounded to acute. Flowers white, change to yellow, 1-3 in axils; calyx truncate, lobes 5, puberulous without, obtuse; corolla-tube 2.5-5 cm long, white, turning yellow in the evening, lobes 5-6, obovate, obtuse; stamens 5, included, anther ca 1 cm; ovary 1-celled. Berry ellipsoid or oblong, with numerous longitudinal elevated lines, beak stout, smooth, 4 × 3 cm, seeds many, rugose.

Common in deciduous forests above 600 m on the degraded slopes in most districts.

Fl.: January- April, Fr.: May.

Gupalapatnam RF (ADB), *GO* 4403; Gannaram RF (NZB), *TP* & *BR* 6398; Manchippa (NZB), *BR* & *KH* 9027; Narsapur RF (MDK), *TP* & *CP* 11979; Dharur RF (RR), *MSM* & *TP* 11097; Kulcharam (MDK), *RG* 104137 (BSID).

Gardenia latifolia Aiton, Hort. Kew. 1: 294. 1789; Hook.f., Fl. Brit. India 3: 116. 1880; Gamble, Fl. Madras 2: 619. 1921.

Vern.: *Pedda bikki, Adavi bikki, Tharpi chettu, Pedda manga.*

Densely foliaceous, deciduous tree, up to 10 m high, bark glaucous-grey, exfoliating and leaving round depressions, wood yellowish-brown, close grained, smooth. Leaves broadly elliptic, 7-30 × 2.5 – 18.5 cm, glabrous, apex obtuse to subacute, margin entire, base rounded to narrowed into short, broad petiole; petiole 0.3 mm. Flowers in short cyme, white; calyx tubular – campanulate, tube 1.4 cm, lobes 5-11, lanceolate, 1.8 cm, acuminate; corolla 1.4 cm across, tube 6 cm, pubescent without, lobes 5-11, oblong-obovate, 3 cm, thin-fleshy; stamens 5-11, subexserted, anther 1.9 cm; ovary oblong, 0.9 cm, style 4.1 cm, stigma 2-lamellate. Berry globose, 5.5 cm across, with a crown of calyx; seeds many, rugose, 0.2cm across.

Common in deciduous forests of most districts.

Fl.: February-April, Fr.: May-November.

Mahaboobghat (ADB), *GO* & *PVP* 4264; Narsapur RF (NZB), *TP* & *CP* 11978; Anantagiri RF (RR), *MSM* & *TP* 11062; Pullayapally (MBNR), *MVS* & *DV* 38649 (BSID); Lingamaiah Kuruva, Mannanur (MBNR), *SRS* 107438 (BSID); Mahadevpur (KRN), *SLK* 70872 (NBG); Thupakulagudem (WGL), *PVS* & *NRR* 76940 (BSID); Sivampet RF (MDK), *RG* 116472 (BSID).

Gardenia resinifera Roth, Nov. Pl. Sp. 150. 1821. *G. lucida* Roxb., Fl. Ind. 2: 553. 1824; Hook.f., Fl. Brit. India 3: 115. 1880; Gamble, Fl. Madras 2: 618. 1921.

Vern.: *Karinga, Chinna karinga, Verri bikki*

Tree, up to 10 m high, bark greenish-grey, flaking off, wood yellowish white and close grained. Leaves thin-coriaceous, elliptic – oblong, 5.5-16 × 2.3 – 7 cm, bullate in the nerve axils, apex obtusely acute, margin entire, base rounded-attenuate, petiole 0.7 cm. Flowers yellow, frgrant, axillary, solitary; calyx tubuar, tube 0.7 cm, lobes 5, triangular, ca 1.5 cm long, lanceolate, puberulous without, acuminate; corolla tubular, tube 8 cm long, lobes 5, obovate, thin-fleshy; stamens 5, exserted, 9 mm long. Berry ellipsoid, 2 × 1.5 cm, with a crown of lanceolate calyx lobes; seeds rugose, whitish, many in pulp, hard.

Common in dry deciduous forests of Adilabad, Ranga Reddi, Mahabubnagar, Karimnagar and Nalgonda districts.

Fl.: March - June, Fr: July-November.

Kotapalli RF (ADB), *GO* 4521; Appapur (MBNR), *SSR* 15542; Aklaspur RF (KRN), *GVS* 22505 (MH); Rudraram RF (RR), *MSM* & *TP* 12708; Yerrakunta (MBNR), *TDCK* 16819; Kambalpally (NLG), *BVRS* 312 (NU); Poochavaram (KMM), *RCS* 98744 (BSID); Lakshmipuram (KMM), *RCS* 99005 (BSID); Thupakulagudem (WGL), *PVS* & *NRR* 76942 (BSID).

HALDINA Ridsd.

Haldina cordifolia (Roxb.) Ridsd., Blumea 24: 361. 1978. *Nauclea cordifolia* Roxb., Pl. Coromandel t. 53. 1796 & Fl. Ind. 1: 514. 1832. *Adina cordifolia* (Roxb.) Hook.f. ex Brandis, Forest Fl. N.W. India t. 33. 1874; Hook.f., Fl. Brit. India 3: 24. 1880; Gamble, Fl. Madras 2: 584. 1921.

Vern.: *Pasupu ganapa, Pasupu kadamba, Rudra ganapa, Bandari, Bandaru, Pedda kamba.*

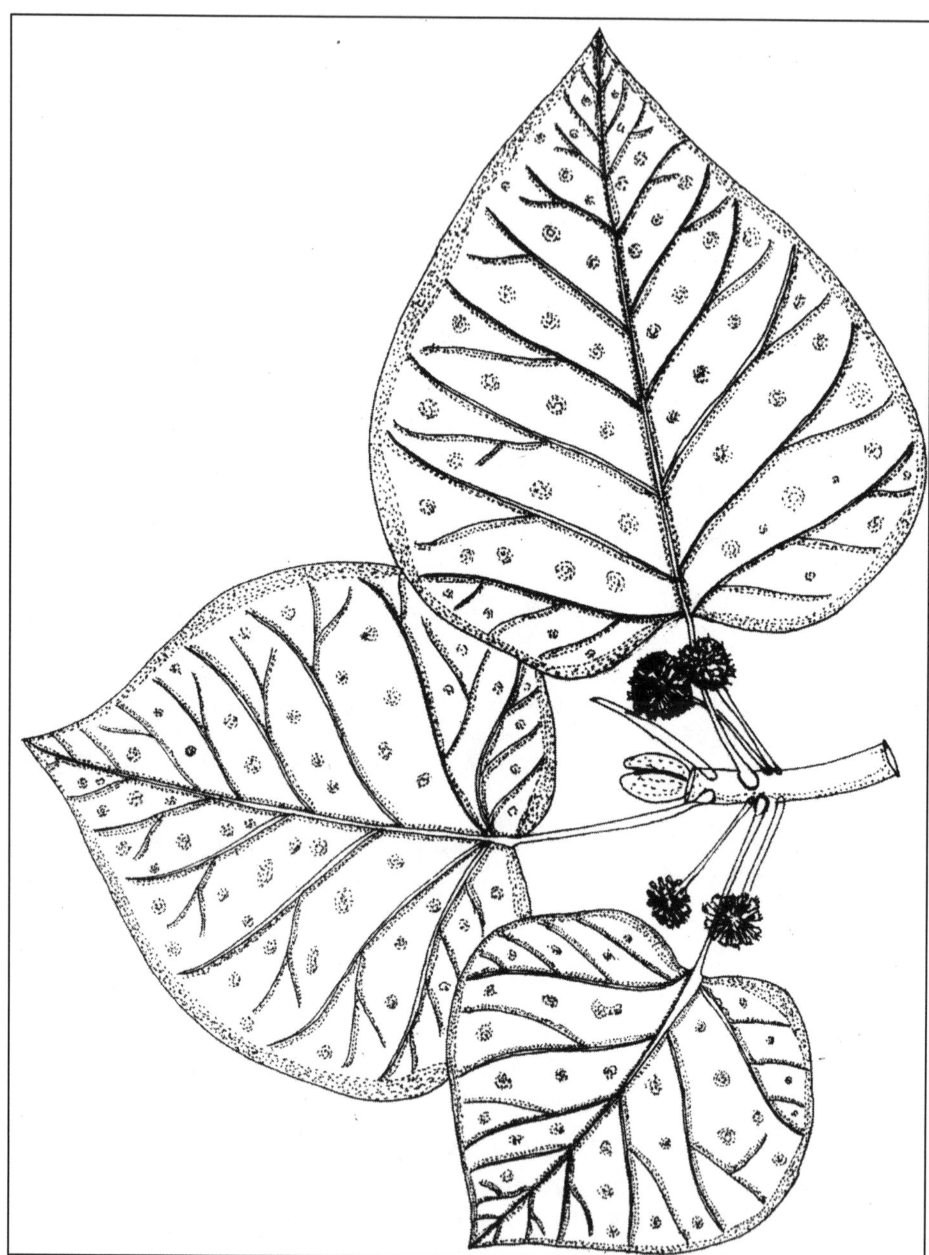

Figure 45: Haldina cordifolia (Roxb.) Ridsd.
Twig.

Large deciduous tree, up to 18 m high, crown large, spreading, trunk erect with horizontal branches, base often buttressed and fluted, bark brownish-grey, thick, furrowed, wood yellow, even-grained, fine and even textured, young branches pubescent. Leaves thick-chartaceous, orbicular-ovate, 7-22.5 × 7.5 – 25.6 cm, dark green, puberulous along nerves above, densely pubescent and pale beneath, apex shortly acuminate, margin entire, base cordate, petiole to 5.7 cm long. Flowers yellowish in axillary heads; calyx cupular, obscurely 5-angled, lobes 5, valvate, subulate, pubescent; corolla funnel-shaped, tube 4.5 mm long, pubescent without, glabrous within, lobes 5, valvate, 2 mm, oblong, recurved, acute; stamens 5, between the lobes, partly exserted, filaments 0.5 mm long, anthers oblong, hastate at base, 1 mm long, obtuse; stigma clavate. Capsule cuneate, densely downy, 5 mm, breaking into 2 follicular cocci; seeds 6, elongate, tailed above or winged, 0.3 x 0.15 mm.

Common in dry deciduous forests of all districts.

Fl.: June – September, Fr.: February-May of the following year.

Darigav RF (ADB), *GO* 4462; Jalalpur RF (NZB), *BR* & *KH* 9655; Anantasagar (RR), *MSM* & *NV* 10413; Madikonakutta – Narsapur (MDK), *KMS* 6744 (MH); Eturnagaram (WGL), *SSR* & *KSM* 18544; Near Rest House, Pakhal (WGL), *KMS* 13181 (CAL); Kambalpally (NLG), *BVRS* 316; Kudurpalli (KRN), *SLK* 70783 (NBG); Bandirevu RF (KMM), *RCS* 98934 (BSID); Umamaheswaram (MBNR), *SRS* 109139 (BSID); Veerapauram (WGL), *PVS* & *NRR* 76912 (BSID).

Note: This species has so far been treated in Floras under *Adina cordifolia* (Roxb.) Hook.f. ex Brandis. But Ridsdale (*loc.cit.)* in his revision of tribe Naucleae *s.s.* feels that the Indian taxon does not belong to the genus *Adina* Salisb., which does not occur in India and has therefore, raised the monotypic new genus *Haldina* Ridsd. to accommodate it. He differentiates the two as follows:

1. Terminal vegetative and loosely surrounded by stipules, stipules more or less free in bud, deeply bifid, some what persistent; ovules up to 4 per locule .... **Adina**
1. Terminal vegetative but flattened, stipules depressed in bud, entire, pubescent, keeled .... **Haldina**

HAMELIA Jacq.

Hamelia patens Jacq., Enum. Syst. Pl. 16. 1760; Bor & Raiz., Beautif. Ind. climbers shrubs 96. 1954.

Much-branched shrub, 3-4 m high; branches reddish– brown pubescent. Leaves elliptic to broadly oblanceolate, 6-12 x 2.5-5.5 cm, base cuneate, margins entire, apex acute or acuminate, glabrous above, pubescent beneath; petioles 1-1.5 cm long, pubescent; stipules interpetiolar, triangular, acute. Flowers sessile or subsessile, in terminal forking cymes; peduncles 1-2 cm long; calyx 1.5-2 mm long; teeth short, triangular; corolla orange – red, tube cylindric, 1.5-2.5 cm long; lobes 5, ovate. Fruits not seen.

Grown as an oranmental in gardens and house yards.

Fl. & Fr.: June- December.

HEDYOTIS L.

Hedyotis cyanantha Kurz., J. Asiat. Soc. Bengal, Pt. 2. Nat. Hist. 45(2): 136. 1876. *Hedyotis caerulea* Wight & Arn., Prodr. Fl. Ind. Orient. 412. 1834; Beddome, Icon. Pl. Ind. Or. t. 30. 1874; Hook.f., Fl. Brit. India 3 : 60. 1880. *Oldenlandia caerulea* (Wight & Arn.) Gamble, Fl. Madras 2 : 597. 1921.

Erect, scabrid herb, up to 15-20 cm high. Leaves linear to subulate,1-2 x 0.1-0.3 cm, thick herbaceous, scabrid above and below, base shortly decurrent, margins revolute, apex aristate, subsessile; stipules truncate. Flowers blue in terminal capitate clusters; calyx truncate, 1 mm long; lobes lanceolate; corolla campanulate; lobes oblong, 3 mm long. Capsule crustaceous, hardly dehiscent with persistent calyx.

Along with grasses in scrub jungles in Hyderabad district (Ramana, 2010).

Fl. & Fr.: December-March.

HYMENODICTYON Wall. *nom. cons.*

Hymenodictyon orixense (Roxb.) Mabb., Taxon 31: 66. 1982. *Cinchona orixensis* Roxb., Bot. Descr. Swietenia 21. 1793 and Med. Facts. Obs. 6: 153. 1795. *Hymenodictyon exelsum* (Roth) Wall., in Roxb., Fl. Ind. 2: 149. 1824; Hook.f., Fl. Brit. India 3: 35. 1880; Gamble, Fl. Madras 2: 589. 1921.

Deciduous tree, up to 25 m high, crown round, bark dark grey, thick, soft, corky and furrowed in old, wood white when fresh cut turning darker, young branches brown puberulous. Leaves membranous, ovate-elliptic, 13-22 × 5-12 cm, puberulous along nerves above, pubescent beneath, base acute, margin entire, apex abruptly acuminate, stipules broadly oblong, recurved, petiole to 7 cm long. Bracts foliaceous. Flowers pale green, fragrant, in terminal and axillary paniculate raceme. Capsule ellipsoid or oblong, brown, 1.6-2 × 0.3-0.8 cm, loculicidal, puberulous when young, warty, lenticelled; seeds ca. 20, broadly with flattened wings and irregularly denticulate margin.

Common in dry, mixed forest, particularly in dry rocky hills in Adilabad, Ranga Reddi, Karimnagar and Mahabubnagar districts.

Fl.: June- August, Fr.: January- March.

Anantagiri RF (RR), *TP* & *MSM* 11077; Narsapur (MDK), *RG* 106767; Sivampet RF (MDK), *RG* 116477 (BSID).

IXORA L.

1. Flowers red, scarlet or yellow .... **I. coccinea**
1. Flowers white:
   2. Calyx teeth longer than ovary .... **I. finlaysoniana**
   2. Calyx teeth equal or shorter than ovary .... **I. pavetta**

Ixora coccinea L., Sp. Pl. 110. 1753; Hook.f., Fl. Brit. India : 145. 1880; Gamble, Fl. Madras 2 : 631. 1921.

An evergreen shrub or small tree, 1-2 m high. Leaves elliptic to ovate – oblong, 5-10 x 3.2-6.3 cm, base rounded or subcordate, margins entire, apex acute; subsessile; stipules with a long rigid cuspidate point. Flowers deep red, scarlet or yellow in dense terminal corymbose cymes; calyx 2.5 mm long; teeth triangular; corolla tube 3.8 cm long; lobes elliptic-oblong. Fruits globose, 0.6 cm in diam, fleshy, crowned with the calyx teeth.

Grown in gardens and houseyards as an ornamental.

Fl. & Fr.: Almost round the year.

Ixora finlaysoniana Wall., ex G. Don, Gen. Syst. 3: 572. 1834; Gamble, Fl. Madras 2: 629. 1921.

A small tree or shrub, up to 8 m high. Leaves coriaceous, elliptic-obovate, 5-15 × 4-5 cm, base acute-cuneate, margin entire, apex acute, glabrous; petiole short, 0.6 cm; stipules sheathing, acute. Flowers in terminal corymbs, white; calyx cupular, lobes 4, lanceolate, 4 mm long, membranous, acute; corolla tubular, tube 3 × 1.5 cm, lobes 4, obovate, subacute; stamens 4, exserted; ovary 2 mm, style 3.5 cm long. Berry globose.

Cultivated as an ornamental.

Fl. & Fr.: March – September.

Ixora pavetta Andrews in Bot. Rep.t. 78. 1799. *I. arborea* Roxb. ex Sm., Cycl. 19: 5. 1811. *I. parviflora* Vahl, Symb. Bot. 3: 11.t. 52: 1794; Hook.f., Fl. Brit. India 3: 142. 1880; Gamble, Fl. Madras 2: 631. 1921.

Vern.: *Korivi.*

A small tree, up to 2.5 m high. Leaves coriaceous, elliptic-obovate, 7.5-13.5 × 3.5-6 cm, base rounded or cordate, margin entire, apex obtuse or rarely acute, glabrous, main nerves distinct; petiole very short, 0.6 cm; stipules sheathing, acuminate. Flowers in terminal branched cymes, white; calyx truncate, glabrous, tube 1 mm long, lobes 4, triangular-toothed; corolla tubular, tube 7 mm long, lobes 4, puberulous without, obtuse; stamens 4, exserted. Berry didymous, 0.7 × 0.8 cm; seeds 2, plano-convex, whitish brown, striate, 6 mm across.

Common in dry deciduous forests of all districts.

Fl. & Fr.: March – June.

Devapur (ADB), *GO* & *PVP* 4242; Laxmapur RF (NZB), *TP* & *BR* 6260; Gandhari RF (NZB), *TP* & *BR* 6329; Akkannapet (MDK), *BR* & *CP* 11562; Burugupally (MDK), *BR* & *CP* 11618; Lothuvagu (RR), *MSM* & *KH* 10291; Mahadevpur to Kataram (KRN), *SLK* 70816 (NBG); Lakshmipuram RF (KMM), *RCS* 99004 (BSID); Uskavagu, Mannanur (MBNR), *SRS* 108906 (BSID); Tadwai (WGL), *RKP* 110806; Gowraram (KMM), *PVS* 76949 (BSID).

KNOXIA L.

Knoxia sumatrensis (Retz.) DC., Prodr. 4: 569. 1830. *Spermacoce sumatrensis* Retz., Observ. Bot. 4: 23. 1786. *Knoxia mollis* R.Br. ex Wight & Arn., Prodr. Fl. Ind. Orient. 439.1834. Hook.f., Fl. Brit. India 3: 129.1880; Gamble, Fl. Madras, 2: 622.1921. *Knoxia corymbosa* auct. non. Willd. 1798; Hook.f., Fl. Brit. India 3: 128. 1880; Gamble, Fl. Madras 2: 622. 1921.

An erect herb, up to 1 m high. Leaves narrow, linear-lanceolate, 4-11 × 0.8-1.8 cm, base attenuate, margin entire, apex acute-acuminate, hairy above, more so below, especially on nerves; sessile or shortly petioled; stipule with 3-5 filiform, hairy bristles. Flowers in terminal corymbose cymes, white or bluish white; calyx cupular, lobes 4, acute; corolla tubular, tube 4.5 mm long, lobes 4, oblanceolate, acute; stamens 4, filaments 1.5 mm long. Capsule 2 mm, small.

Moist areas of the deciduous forests of Nizamabad, Khammam, Warangal, Karimnagar and Mahabubnagar districts.

Fl. & Fr.: February – December.

Bhimnagar RF (NZB), *TP* & *BR* 6212; Bhadrachalam RF (KMM), *RCS* 1043306 (BSID); Gunavaram RF (KMM); *RCS* 98937; Way to Sidhavaram (KRN), *PVS* 84077 (BSID).

KOHAUTIA Cham. & Schlecht.

Kohautia aspera (B. Heyne ex Roth) Bremek., Verh. Kon. Ned. Akad. Wetensch. Afd. Natuurk. Sect. 2. 48: 113. 1952. *Hedyotis aspera* B. Heyne ex Roth, Nov. Pl. Sp. 94. 1821. *Oldenlandia aspera* (Roth) DC., Prodr. 4: 428. 1830; Hook.f., Fl. Brit. India 3: 68. 1880; Gamble, Fl. Madras 2: 602. 1921.

Annual erect herb, up to 30 cm high. Leaves whorled, linear, 2.5-4 × 0.2-0.6 cm, pubescent, base attenuate, margin entire or undulate, apex acute, subsessile. Flowers in terminal, lax cymes, 6-14 cm, purple; calyx-tube short, 0.5 mm, lobes 5, triangular, 2mm, scabrid; corolla tubular, 3 mm across, tube narrowly cylindric, 16 mm, inflated above, lobes 5, 2.5mm, acuminate; stamens 5, filaments small, 1 mm, anther 1.75 mm; ovary 2 mm, style 8 mm, hairy, stigma bifid, 1 mm. Capsule 3 × 4 mm, 4-celled, brown; seeds black, 0.2 mm, different shapes, rugose.

Very common in waste lands and roadsides.

Fl. & Fr. : July-December.

Choutkoor (MDK), *TP* & *CP* 11916; Mohammadabad (RR), *MSM* & *NV* 10421; Mekavaripalli (RR), *MSM* & *TP* 11047; ICRISAT (MDK), *LJGV* 3900 (CAL); Rachakonda (NLG), *BVRS* 433 (NU).

MITRACARPUS Zucc.

Mitracarpus hirtus (L.) DC., Prodr. 4: 572. 1830. *M. villosus* (Sw.) DC., Prodr. 4: 572. 1830. *M. verticillatus* (Schum., & Thonn.) Vatke, Linnaea 40: 196. 1786; Sebastine & Raman, Bull. Bot. Surv. Ind. 9: 29. 1968. *Staurospermum verticillatum* Schum., & Thonn., Beskr. Guin. Pl. 73. 1827.

An erect herb, up to 35 cm high. Leaves decussate, elliptic-lanceolate, 1.6-3 × 0.6-1 cm, base attenuate, margin entire, apex acute, pubescent below, sub-sessile; stipule intrapetiolar, pectinate. Flowers in axillary and terminal capitate cymes, white. Capsule globose, 2 mm.

Rare in the forests in Khammam district.

Fl. & Fr.: November – April.

Ramavaram (KMM), *RR* 112536 (BSID).

MITRAGYNA Korth. *nom. cons.*

Mitragyna parvifolia (Roxb.) Korth., Observ. Naucl. Indic. 19. 1839; Gamble, Fl. Madras 2: 585. 1921. *Nauclea parvifolia* Roxb., Pl. Coromandel t. 52, 1796 & Fl. Ind. 1: 513. 1832. *Stephegyna parvifolia* Korth., Verh. Nat. Gesch. Ned. Bezitt. Bot. 161. 1840; Hook.f., Fl. Brit. India 3: 25. 1880.

Vern.: *Batta genupu, Pulagara chettu.*

A deciduous tree, up to 12 m high. Leaves oblong-obovate, 4.5-13 × 3.5-9 cm, base rounded, margin entire, apex acute-obtuse, glabrous; petiole 1.5-2 cm; stipules intrapetiolar, caducous. Flowers in axillary, globose heads, peduncles 4 cm, greenish yellow; calyx truncate, lobes 5, triangular; corolla salver-shaped, tube 5 mm long, lobes 5, calyculate; stamens 5, attached near the throat, exserted; ovary 2-celled, stigma mitiform. Capsule aggregated into a globose head, 1 cm.

Occasional in deciduous and moist forests in all districts.

Fl. & Fr.: March – December.

Madhavaram RF (ADB), *TP* & *GO* 5470; Rampur (ADB), *SSR* 19317; Gandhari RF (NZB), *TP* & *BR* 6335; Manchippa RF (NZB), *BR* & *KH* 9019; Narsapur forest (MDK), *CP* 11582, *KMS* 8017 (MH); Anantagiri RF (RR), *MSM* & *PRSR* 12790; Kudichintalabailu (MBNR), *TDCK* 15376; Mallelatheertham (MBNR), *BR* & *TSS* 29899; Pakhal lake (WGL), *KMS* 13183 (MH); Way from Rachepalle (KRN), *GVS* 20171 (CAL); Bollarum (NLG), *BVRS* 161 (NU).

MORINDA L.

Morinda pubescens J.E. Smith in Rees. Cyclop. 24: n. 3. 1813.var. **pubescens** Verdc., Kew Bull. 37: 543. 1983. *M. tinctoria* Roxb., Fl. Ind. 2: 197. 1824, non Noronha 1790; Hook. f., Fl. Brit. India 3: 156. 1880; Gamble, Fl. Madras 2: 652. 1921. *M. tinctoria* Roxb. var. *tomentosa* (Heyne ex Roth) Hook. f., Fl. Brit. India 3: 156. 1880. *M. tomentosa* Heyne ex Roth, Nov. Pl. Sp. 147. 1821.

Vern.: *Togara, Bandamaddi.*

A small tree, up to 4 m high. Leaves elliptic- obovate, 9-14 × 4-6 cm, base attenuate-rounded, margin undulate, apex acuminate, tomentose; petiole 1.5 cm long; stipules connate into a sheath, usually bifid. Flowers in axillary globose heads, white; calyx connate, merged in head; corolla tube 3 cm long, pubescent, lobes 5, ovate, glabrous within; stamens 5, alternate to corolla lobes; ovary inserted in syncarpous false berries. Fruit syncarpium, 2 cm diam., irregularly globose, pubescent, notched; seeds oblong-ovoid, black, 3 × 2 mm.

Common throughout the forest areas in all districts.

Fl. & Fr.: May – November.

Jaipur RF (ADB), *GO* & *PVP* 4824; Manchippa RF (NZB), *BR* & *KH* 9032; Jaipur RF (NZB), *BR* & *KH* 9658; East of Rachepalli (KRN), *GVS* 20151 (CAL); Narsapur RF (MDK), *CP* 11578, *KMS* 7977 (CAL); Mohammadabad (RR), *MSM* & *KH* 10240; Pakhal RF (WGL), *KMS* 13793 (MH); Jalupenta (MBNR), *TDCK* 15312; Pedduti (MBNR), *TDCK* 16822; Pogila (NLG), *BVRS* 339 (NU); Yetapaku (KMM), *RR* 106083 (BSID).

MUSSAENDA L.

Mussaenda frondosa L., Sp. Pl. 177. 1753.

Evergreen shrub, 1-2 m high. Leaves elliptic-oblong or ovate, 7.5-12.5 x 5-9 cm, base rounded or tapering, margins entire, apex acuminate, more or less pubescent above, pubescent on nerves beneath; petioles 0.6-2.5 cm long; stipules 6 mm long, linear to lanceolate. Flowers yellowish green outside, orange red within, in terminal cymes; calyx tube truncate, 5 angled; lobes linear-subulate; corolla tube cylindric, densely hairy within; lobes ovate-obovate. Berry globose, tip areolate, sparingly hirsute.

Grown in gardens and houseyards as an ornamental.

Fl. & Fr.: August-December.

NEANOTIS Lewis

Note: All the 14 species of *Anotis* DC. originally included in that genus by De Candolle are referable to earlier valid genera, namely *Arcytopetalum* and a few others which are distinctly natives to New World (America). None are associated with Asian genus *Anotis* auct. Therefore Walter H. Lewis (Ann. Missouri Bot. Gard. 53: 32. 1966) proposed a new generic name *Neanotis* for the related but distinct species indigenous to Asia and transferred all such species (including Indian taxa) to his new genus.

1. Glabrous herbs .... **N. lancifolia**
1. Hispid herbs .... **N. montholoni**

Neanotis lancifolia (Hook.f.) W.H. Lewis, Ann. Missouri Bot. Gard. 53: 39. 1966. *Anotis lancifolia* Hook. f., Fl. Brit. India 3: 73. 1880; Gamble, Fl. Madras 2: 605. 1921. *Hedyotis lancifolia* Dalzell, Hook. J. Bot. 2: 135. 1850. non Schumm. & Thonn. 1827. *Neanotis concanensis* Daniel & Vajravelu, J. Econ. Taxon. Bot.3: 675. 1982.

Erect, annual, glabrous herb, diffusely branched, 30 cm high, stems and branches sparsely hairy. Leaves membranous, 3-5 × 0.6-1.5 cm, ovate-lanceolate, green, sparsely on the above and paler and hairy on the nerves beneath, base acute, margin entire, apex acuminate, stipules pubescent, shortly bristly. Flowers purple in terminal and axillary cymes. Capsules compressed, 4 × 2 mm, much broader than long; seeds 2 × 1.5 mm, ellipsoid, convex on the back, pitted, black.

Rare in moist places in Adilabad and Ranga Reddi districts.

Fl. & Fr.: October – December.

Sone (ADB), *TP* & *PVP* 4029 (K & SKU); Anantagiri RF (RR), *TP* & *MSM* 11082.

Neanotis montholoni (Hook.f.) W.H. Lewis, Ann. Missouri Bot. Gard. 53: 39. 1966. *Anotis montholoni* Hook. f., Fl. Brit. India 3: 74. 1880; Gamble, Fl. Madras 2: 605. 1921.

An annual erect, hispid herb. Leaves ovate or lanceolate, 2-3 × 0.7-1.3 cm, base obtuse, margin entire, apex acute, more or less hairy on both surfaces; petiole 0.3 cm; stipules with short bristle. Flowers in terminal and axillary peduncled, many-flowered cymes, purple; peduncles hairy. Capsules clustered, sub sessile, didymous; seeds globose.

Rare in fields in Adilabad and Nizamabad districts.

Fl. & Fr.: July – September.

Mavala RF (ADB), *TP* & *GO* 5505; Neredukonda, near Nirmal (ADB), *KH* & *PVP* 5567; Gandhari (NZB), *BR* & *CPR* 9251.

NEOLAMARCKIA Bosser

Neolamarckia cadamba (Roxb.) Bosser, Bull. Mus. Natl. Hist. Nat. B. Adansonia 6 : 247. 1984. *Anthocephalus chinensis* (Lam.) A. Rich. ex Walp. Repert. 2 : 491. 1843; Ridsd., Blumea 24 : 334. 1978. *A. cadamba* (Roxb.) Miq. Fl. Ind. Bat. 2 : 135. 1856; Hook.f. Fl. Brit. India 3 : 23. 1880. *A. indicus* A. Rich., Mem. Soc. Hist. Nat. Paris 5 : 237. 1834; Gamble, Fl. Madras 2 : 583. 1921.

Large tree, up to 20 m high; branches horizontal, widely spreading, younger parts pubescent. Leaves elliptic-oblong or ovate-oblong, 12-25 x 6-10 cm, rounded or subcordate at base, margin entire, apex shortly acuminate, glabrous and shining above, pubescent beneath; petiole 2-5 cm long; stipules linear, 1.2-1.5 cm long. Flowers pale yellow in globose heads; calyx 5-6 mm long; lobes narrowly oblanceolate; corolla 6-8 mm long, tube slender. Fruit fleshy, yellow when ripe; seeds muriculate, not winged.

Planted in gardens, along roadsides as an avenue tree and near temples as a sacred tree.

Fl. & Fr.: August-November.

OLDENLANDIA L.

1. Leaves elliptic or ovate-lanceolate:
   2. Leaves > 3 cm broad, usually clustered at top of stem (*i.e.*, at base of inflorescence) due to reduced internodes and appearing verticillate .... **O. ovatifolia**
   2. Leaves < 1 cm broad, generally distributed along stem in pairs or entire stem with reduced internodes:
      3. Capsules 4-winged or angled, margins not revolute .... **O. biflora**
      3. Capsules didymous, margins revolute:
         4. Capsule indehiscent, stipules hispid .... **O. auricularia**
         4. Capsule dehiscent; stipules glabrous .... **O. pumila**

1. Leaves linear or linear-lanceolate:
    5. Flowers in exclusively terminal cymes; corolla tube slender .... **O. stricta**
    5. Flowers in terminal and axillary panicles, clustered, few-flowered cymes and umbels:
        6. Flowers white; peduncles 1-4-flowered:
            7. Capsule valves short; corolla-tube short and broad; diffuse herbs:
                8. Flowers pedicelled, usually in pairs or threes .... **O. corymbosa**
                8. Flowers sessile or sub sessile, 1 or 2 in each stem axil .... **O. diffusa**
            7. Capsule valves elongate, corolla-tube slender, erect herbs .... **O. herbacea**
        6. Flowers blue, pinkish, peduncled, 3-many-flowered:
            9. Flowers in panicles; stamens included .... **O. affinis**
            9. Flowers in capitate clusters and umbels, stamens subexserted .... **O. umbellata**

Oldenlandia affinis (Roem. & Schult.) DC., Prodr. 4: 428. 1830. *Hedyotis affinis* Roem. & Schult., Syst. Veg. 3: 194. 1818. *Oldenlandia dichotoma* (K.D.Koenig ex Roth) Hook. f., Fl. Brit. India 3: 67. 1880; Gamble, Fl. Madras 2: 601. 1921. *Hedyotis dichotoma* K.D. Koenig ex Roth, Nov. Pl. Sp. 93. 1821.

Annual, erect herb, up to 35 cm high. Leaves linear, 1-3 × 0.1-0.3 cm, base decurrent, margin entire, apex acuminate, glabrous, subsessile. Flowers in terminal and axillary dichotomous panicles, blue; calyx truncate, 2mm, lobes 4, oblong, acute; corolla tubular, tube narrowly cylindric, 1.5 mm long, lobes 4, obovate-oblanceolate, acute; stamens 4, filament 0.2 mm, anther oblong, 0.75; ovary 0.8 mm, style linear, 1.25mm, stigma bifid, 1 mm. Capsule loculicidal; seeds many, brown, ovate, hexagonal marks surrounding pit.

Occasional weed of road sides and open forests in all districts.

Fl. & Fr.: June-December.

Manchippa RF (NZB), *TP* & *BR* 6047; Gandhari RF (NZB), *BR* & *CPR* 7239; Narsapur RF (MDK), *TP* & *CP* 11974; Madikonakutta – Narsapur (MDK), *KMS* 6750 (MH); Nagarjunakonda (NLG), *BVRS* 90 (NU).

Oldenlandia auricularia (L.) K.Schumach., in Engler & Prantl, Pflanzenfam. 4 (4): 25. 1891; Gamble, Fl. Madras 2: 597. 1921. *Hedyotis auricularia* L., Sp. Pl. 101. 1753; Hook.f., Fl. Brit. India 3: 58. 1880.

Vern.: *Nelaadaviaku*.

Annual, usually trailing herb, stem glabrous, four-angled. Leaves elliptic or ovate – lanceolate, 2.5-6 × 1.5 – 2.5 cm, base rounded, margin entire, apex acute, prominently nerved, subsessile, stipules short, membranous, hispid, with few erect teeth. Flowers in axillary subsessile cyme, white; calyx 4.5 mm across, lobes 4, elongate, larger than corolla lobes, hairy without; corolla 4 mm across, lobes 4, scabrid spines within; stamens 4, filament short, 0.25 mm; stigma bifid. Capsule globose, indehiscent, 2 mm.

In forests of Khammam district in hilly areas.

Fl. & Fr.: July- December.

Mothugudem RF (KMM), *EC* 7454.

Oldenlandia biflora L., Sp. Pl. 119. 1753; Hook.f., Fl. Brit. India 3: 70. 1880; Gamble, Fl. Madras 2: 602. 1921. *O. paniculata* L., Sp. Pl. (ed.2) 2. 1767. *Thecagonum biflorum* (L.) Babu, Bull. Bot. Sruv. Ind. 11: 214. 1969 (1971).

An erect, succulent herb, up to 35 cm high. Leaves chartaceous, ovate-oblong or elliptic, 3.5-7 × 1-2 cm, base rounded, margin entire, apex acute-obtuse, glabrous; subsessile; stipules membranous, covering into a sheath. Flowers in terminal and axillary racemose panicle, white; calyx truncate, 4-angled, lobes 4, triangular, thick, acute, corolla campanulate, villous at the throat, lobes 4, obovate, acute, spreading; stamens 4, slightly exserted, anthers oblong. Capsule loculicidal, 0.3 × 0.3 cm.

Occasional in open habitats along with grasses inWarangal and Hyderabad districts.

Fl. & Fr.: September – February.

Pakhal lake (WGL), *KMS* 13167 (CAL).

Oldenlandia corymbosa L., Sp. Pl. 119. 1753; Hook.f., Fl. Brit. India 3: 64. 1880; Gamble, Fl. Madras 2: 600. 1921. *Hedyotis corymbosa* (L.) Lam., Encycl. 1: 272. 1792.

Annual, erect, scaberulous herb. Leaves linear-lanceolate, 1.5-3 × 0.7-1.2 cm, base decurrent, margin undulate, apex acute, scabrid, prominently nerved. Flowers white, in axillary, peduncled cyme, usually in pairs; calyx tube short, triangular, 2 mm across, lobes 4, 0.2 mm, scabrid; corolla tube short, 1 mm, lobes 4, 1.5 mm; stamens 4, filament short, 0.5 mm, anther 0.75 mm; ovary 1 mm, style slender. 1.25 mm, stigma bifid, 0.5.mm. Capsule loculicidal, 3 × 2 mm, with persistent calyx, 4-celled; seeds to 0.2 mm, ovate, blackish, rugose, dotted.

Common in forests and in cultivated fields in all districts.

Fl. & Fr.: June-January.

Ankusapuram (ADB), *GO* 4319; Laxmapur (NZB), *TP* & *BR* 6279; Choutkoor (MDK), *TP* & *CP* 11944; Mohammadabad RF (RR), *MSM* & *KH* 10496; Nagarjunakonda (NLG), *BVRS* 104 (NU); Pakhal lake (WGL), *KMS* 11614 (MH); Narsapur (MDK), *KMS* 6621 (MH); Hamsanpally (MDK), *RG* 116445 (BSID); Kazipet (WGL), *RKP* 105228 (BSID); Ramavaram RF (KMM), *RR* 112623 (BSID).

Oldenlandia diffusa (Willd.) Roxb., Fl. Ind. 1: 444. 1820; Hook. f., Fl. Brit. India 3: 65. 1880; Gamble, Fl. Madras 2: 601. 1921. *Hedyotis diffusa* Willd., Sp. Pl. 1: 566. 1798.

Annual, diffuse herb, up to 25 cm high. Leaves decussate, chartaceous, linear, 2-4 × 0.2-0.4 cm, base decurrent, margin undulate, apex acute-apiculate, 1-nerved, subsessile. Flowers in axillary pairs or solitary, white, pedicel up to 3 mm; calyx subtruncate, puberulous without, lobes 4, lanceolate, 1.5 mm long, ciliate, aristate; corolla tubular, 6 mm across, lobes 4, oblong, united in the middle, glabrous on both sides, acute; ovary 1.2 mm long, stigma linearly bifid, hairy. Capsule loculicidal, 3 x 3 mm with persistent calyx lobes, acute.

Occasional in moist places in Adilabad, Warangal and Hyderabad districts.

Fl. & Fr.: July-November.

Ankusapuram (ADB), *GO* 4316; Mallurgutta (WGL), *RKP* 110962 (BSID).

Oldenlandia herbacea (L.) Roxb., Fl. Ind. 1: 445. 1820; Gamble, Fl. Madras 2: 601. 1921. *Hedyotis herbacea* L., Sp. Pl. 102. 1753. *Oldenlandia heynii* R. Br. ex G. Don, Gen. Syst. 3: 531. 1834; Hook.f., Fl. Brit. India 3: 65. 1880.

Annual, erect herb, up to 30 cm high. Leaves opposite, decussate, linear-lanceolate, 2-4.5 × 0.2-0.4 cm, base decurrent, margin entire, apex acute, glabrous, turning black when dry, sessile. Flowers axillary, solitary, some times 2 in each axil, white; calyx tubular, 0.5 mm, lobes 4, lanceolate, 1.25 mm, acute; corolla 2 mm across, tube 3 mm, lobes 4, ovate, 2.25 mm, acute; stamens 4, included, filament 0.5 mm, anther oblong, 0.75 mm; ovary 1.25 mm, style 3 mm, stigma bifid, 1.5 mm, hairy. Capsule loculicidal, 2.5 × 2 mm, crown protruded; seeds reticulate, blackish.

Common in most districts both in plains and hills.

Fl. & Fr.: June-December.

Bijjur RF (ADB), *GO* & *PVP* 4981; Annaram (NZB), *BR* & *CPR* 7169; Narsapur tank (MDK), *TP* & *CP* 12003; Narayanpur RF (MDK), *BR* & *CP* 11274; Madikonakutta, Narsapur (MDK), *KMS* 6749 (CAL); Mohammadabad (RR), *MSM* 15150; Amrabad RF (MBNR), *BSS* & *SKB* 32529; Near Mysamma temple (MBNR), *SRS* 109083 (BSID), Nagarjunakonda (NLG), *BVRS* 95 (NU).

Oldenlandia ovatifolia (Cav.) DC., Prodr. 4: 427. 1820. *Hedyotis ovatifolia* Cav., Ic. 6: 52. 1801. *Thecagonum ovatifolium* (Cav.) Babu, Bull. Bot. Surv. Ind. 11: 214. 1969 (1971). *Oldenlandia nudicaulis* Roth, Nov. Pl. Sp. 95. 1821; Hook.f., Fl. Brit. India 3: 70. 1880; Gamble, Fl. Madras 2: 602. 1921. *Hedyotis nudicaulis* (Roth) Wight & Arn., Prodr. Fl. Ind. Orient. 416. 1834.

An erect, annual herb, up to 10 cm high. Leaves 4 in a whorl, radical, chartaceous, broadly ovate, 3-6 × 2.5-3 cm, base rounded, margin entire, apex acute, sparsely pubescent below; sessile; stipules pectinate. Flowers in terminal, long peduncled, trichotomously branched cymes; calyx truncate, lobes 4, triangular, acute; corolla white, campanulate, tube 0.5 mm long, lobes 4, obovate, acute; stamens 4, exserted. Capsule 4 x 5 mm, hemispheric.

Rare in moist deciduous forests and stream banks of Karimnagar and Adilabad districts.

Fl. & Fr.: September – February.

Vempalli RF (ADB), *GO* 5500; Aklaspur (KRN), *GVS* 25612 (MH).

Oldenlandia pumila (L.f.) DC., Prodr. 4 : 425. 1830. *Hedyotis pumila* L.f., Suppl. Pl. 119. 1781. *Oldenlandia crystallina* Roxb., Fl. Ind. 1: 443. 1820; Hook.f., Fl. Brit. India 3: 65. 1880; Gamble, Fl. Madras 2: 601. 1921.

Annual, diffuse herb. Leaves elliptic-oblong, 0.7-1.5 × 0.3-0.5 cm, base decurrent, margin entire, apex acute, glabrous, sessile-subsessile. Flowers in terminal and axillary, 1-2-flowered cyme, white; calyx 4-lobed, up to 0.1 cm long, deltoid, acute, margins hispidulous; corolla up to 0.2 cm long, lobes 4, ovate, acute, pilose within; stamens 4, adnate to the top of corolla-tube, alternating with the lobes. Capsule 0.3 × 0.2 cm, glabrous.

Occasional in marshy localities in Hyderabad district (Ramana, 2010).

Fl. & Fr.: September - December.

Oldenlandia stricta L., Mant. Pl. 2: 200. 1771; Hook.f., Fl. Brit. India 3: 68. 1880; Gamble, Fl. Madras 2: 602. 1921. *Hedyotis stricta* (L.) Rao & Hemadri, Indian Forester 99: 378. 1973, non Smith 1811, nec Wall., 1820. *H. graminifolia* L.f., var. *graminifolia* Deb & Dutta, Taxon 32: 285. 1983. *H. graminifolia* L.f., var. *shuteri* (Hook.f.) Deb & Dutta, Taxon 32: 285. 1983. *Olednlandia shuteri* Hook.f., Fl. Brit. India 3: 69. 1880; Gamble, Fl. Madras 2: 602. 1921. *Hedyotis shuteri* (Hook.f.) Rao & Hemadri, Indian Forester 99: 378. 1973.

Annual, erect, glabrous herb, up to 35 cm high. Leaves chartaceous, linear-lanceolate, 1.5-4 × 0.2-0.5 cm, base decurrent, margin undulate, apex acute, 1-nerved, subsessile. Flowers purplish white, in terminal, unilateral, lax, elongated cyme; calyx tubular, lobes 4, 2.5mm, acute; corolla 3 mm across, campanulate, 4 mm, lobes 4, oblong, 6 mm, acute; stamens 4, exserted, filament 2.5 mm, anther linear, 3 mm; ovary 3 mm, style 6 mm, stigma linearly bifid, 2 mm. Capsule loculicidal, 5 × 3 mm, crown protruded.

Occasional, on alluvial soil, often in fields along rivulets in Karimnagar district (Naqvi, 2001).

Fl. & Fr.: July - December.

Oldenlandia umbellata L., Sp. Pl. 119. 1753; Hook.f., Fl. Brit. India 3: 66. 1880; Gamble, Fl. Madras 2: 601. 1921. *Hedyotis puberula* (G. Don) Arn., Pug., Pl. Ind. Or. 18: 342. 1836. *Oldenlandia puberula* G. Don, Gen. Syst. 3: 530. 1834. *Hedyotis umbellata* (L.) Lam. Tabl. Encycl. 1: 272. 1792, non Wall., 1788. *Oldenlandia wightii* Hook. f., Fl. Brit. India 3: 66. 1880; Gamble, Fl. Madras 2: 601. 1921. *Hedyotis wightii* (Hook.f.) K.K.K. Nair, J. Bombay Nat. Hist. Soc.79: 453. 1982.

Vern.: *Chiriveru*.

Annual or perennial herb, up to 15 cm high. Leaves decussate, appearing clustered, linear, 1.5 × 0.3 cm, base decurrent, margin undulate, apex acute-apiculate, scabrous, (sub) sessile. Flowers in terminal or in upper axils, capitate clusters, single or paired, white; calyx truncate, 2 mm across, lobes 4, lanceolate, 3 mm, puberulous without; corolla 5 mm across, campanulate, lobes 4, lanceolate, 4 mm, acute, incurved; stamens 4, subexserted, filament 2 mm, anther oblong, 1.75 mm; ovary 2 mm, style linear, 5.25 mm, hairy, stigma bifid. Capsule subglobose, loculicidal, 1 × 2 mm, crown protruded; seeds brown, hexagonal, pitted.

Occasional weed of waste lands, road sides and open forests in all districts.

Fl. & Fr.: August – January.

Sirpur (ADB), *GO* & *PVP* 4917; Sattenapalle (ADB), *TP* & *PVP* 4094; Kubgall RF (NZB), *TP* & *BR* 6203; Yancha (NZB), *BR* & *PSPB* 9587; Nagulabanda (MDK), *CP* 11325; East Narsapur (MDK), *KMS* 6699 (MH); Nagarjunasagar (NLG), *KMS* 9740 (MH); Karimnagar rest house (KRN), *GVS* 62318 (MH); Pakhal RF (WGL), *KMS* 13162 (CAL); Central University (RR), *MSM* 15147; Vatavarlapally (MBNR), *VBH* 84966 (BSID); Sathiyampet RF (KMM), *RR* 112642 (BSID); Pillalamarri (NLG), *BVRS* 47 (NU).

PAEDERIA L. *nom.cons.*

Paederia foetida L., Mant. Pl. 1: 52. 1767; Hook.f., Fl. Brit. India3: 195. 1880.

A slender, twining shrub, foetid when bruished. Leaves opposite, elliptic-ovate or oblong-ovate, 6-11 × 2.5-5 cm, base rounded or cordate, margin entire, apex acuminate or acute, glabrous; petiole 6 cm; stipule intrapetiolar, short. Flowers in terminal and axillary trichotomously branched, panicled cyme, greenish purple; panicles puberulous, 11 cm long; bract linear, 2.5 mm, hairy; bracteole linear, 2 mm, hairy. Fruit compressed, 7 × 5 mm.

Rare in moist deciduous forests in Warangal district.

Fl. & Fr.: August – December.

Mallur (WGL), *CSR* 1634 (KU).

PAVETTA L.

Pavetta indica L., Sp. Pl. 110. 1753; Hook.f., Fl. Brit. India 3: 150. 1880; Gamble, Fl. Madras 2: 633. 1921. var. **indica.**

Vern.: *Papatakammi.*

A small tree, up to 3 m high. Leaves chartaceous, obovate-oblong or oblanceolate, 7-20 × 6-10 cm, base acute-rounded, margin entire, apex acute, glabrescent; petiole 1 cm long; stipules connate. Flowers in terminal or axillary trichotomously branched corymbose 12 cm long cymes, white; peduncle 7 cm; calyx-tube 2 mm long, lobes 4, minute; corolla tube 1 cm long, lobes 4, obovate, recurved, pubescent within, 4 × 1.5 mm; stamens 4, alternate to the corolla lobes, anthers extruded, 4 mm long; ovary inferior, 1 mm across, style 2 cm long, stigma 4 mm, bifid, swollen. Berry fleshy, globose, glabrous, 0.6 × 0.4 cm, crown 1.5 mm across; seeds 4 in 2 locules, brown, 4 × 2 mm.

Common in deciduous forests of Adilabad, Mahabubnagar and Karimnagar districts.

Fl. & Fr.: May –November.

Appapur (MBNR), *TDCK* 13634; Kodimial (KRN), *GVS* 20122 (MH); Parnasala RF (KMM), *RCS* 98766 (BSID); Yetapaka (KMM), *RR* 106079.

Key to Varieties

1. Leaves glabrescent .... **P. indica** var. **indica**
1. Leaves tomentose .... **P. indica** var. **tomentosa**

Pavetta indica L. var. **tomentosa** (Roxb. ex Sm.) Hook.f., Fl. Brit. India 3: 150. 1880; Gamble, Fl. Madras 2: 683. 1921. *P. tomentosa* Roxb. ex Sm. in Rees, Cyclop. 26: n.2. 1819; Gamble, Fl. Madras 3: 1879. 1935.

A small tree, up to 3 m high. Leaves elliptic-oblong, 8-12 × 4.5- 7 cm, base acute-rounded, margin entire, apex acute, softly pubescent above, tomentose below; petiole 1.5 cm long; stipules connate. Flowers in terminal, trichotomously branched corymbose cymes, white; calyx tomentose without; corolla tomentose within, tube 1 cm, lobes 5; stamens 5, exserted; style hairy. Fruit a drupaceous berry, globose or subdidymous.

Common in forests of Medak, Ranga Reddi, Nalgonda and Mahabubnagar districts.

Fl. & Fr.: May – February.

Narsapur RF (MDK), *TP* & *CP* 14033; Anantagiri RF (RR), *TP* & *MSM* 12791.

PSYDRAX Gaertn.

1. Flowers in cymes; fruit globose .... **P. dicoccos**
1. Flowers in umbels; fruit obovoid .... **P. umbellata**

Psydrax dicoccos Gaertn., Fruct. Sem. Pl. 1: 125. 1788. var. **dicoccos.** *Canthium dicoccum* (Gaertn.) Merr., Philipp. J. Sci. 35: 8. 1928. *Plectronia didyma* (Gaertn.) Kurz., For. Fl. Burma 2: 35. 1877; Gamble, Fl. Madras 2: 624. 1921. *Canthium didymium* Gaertn., Fruct. 3: 94. t. 26. 1805; Hook.f., Fl. Brit. India 3: 132. 1880.

Ver.: *Alla balusu.*

An unarmed tree, up to 3 m high; branches smooth. Leaves very coriaceous, ovate-lanceolate or suborbicular, 4-12 × 1.5- 6 cm, base acute or cordate, margin entire, apex usually obtusely caudate – acuminate, petiole 0.5 cm. Flowers 5-merous, in axillary short-peduncled cyme, greenish white; calyx 5-6-toothed, 2 × 1.5 cm, pubescent without; corolla cream, 6 mm across, tube 2 mm, villous within, lobes 5-6, oblong, reflexed, 2 mm long, acute; stamens 5-6, filament 1.5 mm, anthers 1.0 mm. Drupe globose.

In dry evergreen forests and open forests, often on river banks in Mahabubnagar district.

Fl. & Fr.: December -May.

Appapur (MBNR), *TDCK* 13634; Umamaheswaram (MBNR), *SRS* 109145 (BSID).

Psydrax umbellata (Wight) Bridson, Kew Bull. 48: 762. 1993. *Canthium umbellatum* Wight, Icon. Pl. Ind. Orient. t. 1034. 1845; Hook.f., Fl. Brit. India 3: 132. 1880. *C. dicoccum* (Gaertn.) Teijsm & Binn. var. *umbellatum* (Wight) Sant. & Merch., Bull. Bot. Surv. India 3: 107. 1961. *Plectronia didyma* (Gaertn.) Kurz. var. *umbellata* (Wight) Gamble, Fl. Madras 2: 624. 1921.

Vern.: *Alli, Balusu, Nalla balusu, Nakkina, Nakkeri, Niralli.*

Densely foliaceous, evergreen, unarmed tree, up to 10 m; wood pale yellow; young branches four-angled, glabrous, stipules lanceolate. Leaves coriaceous, elliptic-obovate, 7-15 × 3-5 cm, glabrous, base acute-attenuate, margin entire, apex acuminate; petiole to 1 cm. Flowers whitish in axillary, pedunculate umbels; calyx tube very short, 2 mm across, lobes 4, 0.5 mm; corolla tube short, 3 mm, villous within, lobes 4, oblong, 4 mm, acute; stamens 4, attached at the throat of the lobes, filament 1 mm, anther mucronate, 2.25 mm; ovary 2 mm, style 5 mm, stigma fusiform, 2.5 mm. Drupe obovoid, 0.6 × 0.7 cm, brown, 2-seeded; seeds obovate, granular, 4 × 3 mm.

Occasional in deciduous forests in Adialabad and Mahabubnagar districts.

Fl. : December-February; Fr.: April-May.

Gupalapatnam (ADB), *GO* 4407; Uskavagu, Mannanur (MBNR), *SRS* 107434, 108909 (BSID).

SPERMACOCE L.

1. Stem winged .... **S. latifolia**
1. Stem not winged:
   2. Leaves linear-lanceolate .... **S. pusilla**
   2. Leaves elliptic or ovate-obovate:
      3. Stem terete with densely pilose hairs; leaves highly flexuous; corolla-tube equaling the lobes .... **S. hispida**
      3. Stem 4-angled, with minutely appressed hairs; leaves rarely flexuous; corolla-tube 3-4 times longer than lobes, occasionally equal .... **S. articularis**

Spermacoce articularis L.f., Suppl. Pl. 119. 1782; Sivarajan & Vasu, Taxon 35: 363-67. 1986. *Borreria articularis* (L.f.) F.N. Williams, Bull. Herb Boissier, Ser. 2. 5 : 956. 1905.

Vern.: *Madanak.*

An annual, erect herb, up to 25 cm high, stem 4-angled, glabrous. Leaves opposite, decussate, elliptic-obovate, 1.5-4 × 1-2 cm, base acute-attenuate, margin entire, apex acute, subsessile; stipule connate with the base of leaves into a sheath, pectinate. Flowers in axillary and terminal fascicles, pinkish white; calyx tube 4 mm, lobes 4, linear-oblong, 2-4 mm, hairy, persistent; corolla infundibuliform, lobes 4, gamopetalous, throat of corolla hairy; stamens 4, ca 1 mm; ovary 2-locular, 1 ovule per locule, style 6 mm, stigma bifid. Capsule globose, hispid, 0.5 × 0.2 cm, with a crown of calyx; seeds narrow, oblong, testa reticulate.

Common weed in waste places of the low forests in all districts.

Fl. & Fr.: June – February.

Vempalli (ADB), *GO* & *PVP* 4741; Manchippa RF (NZB), *TP* & *BR* 6018; Kamareddy (NZB), *BR* 9273; Amrabad RF (MBNR), *BSS* & *SKB* 32579; Aklaspur (KRN), *GVS* 2039(MH); Pandivagu (MDK), *BR* & *CP* 11204; Way to Pagarikutta – Narsapur (MDK), *KMS* 6612 (MH); Kusumasamudram RF (RR), *MSM* & *NV* 10340; Pakhal RF (WGL), *KMS* 11598 (MH); Vijayapuri (NLG), *KMS* 9948 (MH); Rangapur (WGL), *PVS* 76958 (BSID); Yelandu (KMM), *RR* 105942 (BSID).

Spermacoce hispida L., Sp. Pl. 102. 1753; Hook.f., Fl. Brit. India 3: 200. 1880. *Borreria hispida* (L.) K. Schum., in Engl. & Prantl, Pflanzenfam. 4(4): 144. 1891; Gamble, Fl. Madras 2: 654. 1921.

An annual, erect, hispid herb, up to 26 cm high; stem terete. Leaves ovate-obovate, 2-3.5 × 1-1.8 cm, acute-attenuate at base, undulate margin, acute at apex; subsessile; stipules connate with the base of leaves into a sheath, pectinate. Flowers in few-flowered axillary fascicles, pink; calyx oblong-globose, hispid without, lobes 4, equal, obovate, acute-apiculate; corolla tubular, tube 2.5 mm long, with a ring of hair below within, lobes 4, oblong, acute; stamens 4, exserted. Capsule hispid, 0.5 × 0.3 cm; seeds broadly ellipsoid, coarse, closely pitted, black.

Common in dry lands in Warangal. Medak, Khammam, Mahabubnagar and Nalgonda districts.

Fl. & Fr.: January –February.

Appapur (MBNR), *BR* & *BSS* 30658; Jakaram (WGL), *RKP* 105297; Medak to Bodhan (MDK), *RG* 113533 (BSID); Ramavaram (KMM), *RR* 112578 (BSID); Nalgonda (NLG), *KMS* 9848 (CAL); Nagarjunasagar (NLG), *BVRS* 103 (NU).

Note: The taxonomy of *Spermacoce hispida* – *articularis* complex has been quite perplexing. The two taxa originally described as two different species of *Spermacoce* by Linnaeus and his son respectively have been subsequently united because of overlapping variation in superficial leaf and floral characters and treated as a highly polymorphic species. Authors who preferred its generic position under *Spermacoce* adopted the epithet *hispida* for the combined taxon while those who treated it under *Borreria* preferred *articularis*. Studies by Sivarajan & *al.* (1986) have shown that the two are distinct species. We are also of the opinion that the two are distinct species.

Spermacoce latifolia Aubl., Hist. Pl. Guiane 55. 1775. *Borreria latifolia* (Aubl.) K. Schum., Mart. Fl. Bras. 6(6): 61. 1888.

An annual diffuse herb, stem 4-angled, angle prominently winged and hispid. Leaves broadly ovate or elliptic, 2-3 × 0.5 – 1 cm, base cuneate, margin entire, apex acute, scabrid on both sides, impressly veined; subsessile; stipules with papillate bristles. Flowers in axillary few-flowered cymes, pink. Capsule densely pubescent, 0.5 × 0.3 cm; seeds ellipsoid, testa reticulate-rugose.

Rare in Warangal district.

Fl. & Fr.: November – March.

Pasra (WGL), *RKP* 10811, 10813 (BSID).

Spermacoce ocymoides Burm.f., Fl. India 34: t. 13. F.1. 1768.

Pubescent, erect herbs, up to 50 cm tall. Leaves decussate, elliptic-oblong, 3.5 × 1.5 cm, sparsely scabrous, nerves 3-4 pairs, base cuneate to attenuate, apex acute, subsessile; stipules 1 × 0.8 cm. Flowers in axillary fascicles, about 40-50 per node, white, 2 mm across, 4-merous. Capsule truncate, 1.5 × 1 mm, dehiscing transversely.

Occasional weed in Warangal district (C.S. Reddy, 2001).

Fl. & Fr.: July – December.

Hanamkonda (WGL), *CSR* 475 (KU).

Spermacoce pusilla Wall., in Roxb., Fl. Ind. 1: 379. 1820. *Borreria pusilla* (Wall.,) DC. Prodr. 4: 453. 1830. *B. stricta* (L.f.) K. Schumach., in Engl. & Prantl, Pflanzenfam. 4(4): 143. 1891; Gamble, Fl. Madras 2: 654. 1921. *Spermacoce stricta* L.f., Suppl. 120. 1781; Hook.f., Fl. Brit. India 3: 200. 1880.

An annual, erect herb, up to 35 cm high. Leaves linear-lanceolate, 2.5 × 0.4-0.8 cm, base attenuate, margin entire, apex acute, glabrous; sessile; stipules broad, truncate with marginal bristles. Flowers in axillary fascicles, pink; calyx truncate, villous without, lobes 4, linear, acute; corolla tubular, tube narrowly cylindric, 1.5 mm long, lobes 4, oblanceolate, acute; stamens 4, exserted. Capsule oblong, 3.5 cm; seeds 2 mm long, longitudinally striated.

Common in foot hills of Nizamabad, Medak, Ranga Reddi, Khammam, Warangal, Nalgonda and Mahabubnagar districts.

Fl. & Fr.: August – November.

Manchippa RF (NZB), *TP* & *BR* 6061; Borgaum (NZB), *BR* 9280; Narsapur RF (MDK), *TP* & *CP* 11967; way to Pagarikutta – Narsapur (MDK), *KMS* 6613 (MH); Kanmankalva RF (RR), *MSM* 10465; Durachipalli (NLG), *BVRS* 38 (NU); Nellippakka RF (KMM), *RCS* 99082 (BSID); Lawal beat, Pasra (WGL), *RKP* 108152 (BSID).

Note: Sivarajan & *al.* (1987) erected a new species *Spermacoce ramanii* listing out several differences. We tried to differentiate them in our specimens and found many intergrading characters. Further when one character agrees, the other character is contradictory. Hence we keep all our specimens in *Spermacoce pusilla* only.

SPERMADICTYON Roxb.

Spermadictyon suaveolens Roxb., Pl. Coromandel t. 236. 1815. *Hamiltonia suaveolens* Roxb., Fl. Ind. 2: 223. 1824; Hook.f., Fl. Brit. India 3: 197. 1880; Gamble, Fl. Madras 2: 650. 1921.

A small tree, up to 4 m high. Leaves elliptic-ovate, 8-13.5 × 3.5 – 7 cm, base acute, margin entire, apex obtuse-acute; petiole 1.5 – 3 cm long; stipule interpetiolar, connate, acute. Flowers in terminal panicles, violet pink, scented; calyx truncate, lobes 4, triangular, acute; corolla tubular, tube narrowly cylindric, 1.7 mm long, puberulous without, lobes 4, ovate, acute; stamens 4, included. Capsule ellipsoid; seeds 3-quetrous.

Rare in dry deciduous forests of Hyderabad and Mahabubnagar districts.

Fl. & Fr.: November – March.

Nilagiri view Point (MBNR), *SRS* 109706 (BSID).

TAMILNADIA Triveng.

Tamilnadia uliginosa (Retz.) Triveng. & Sastre, Mauritius Inst. Bull. 8(4): 85. 1979. *Gardenia uliginosa* Retz., Observ. Bot. 2: 14. 1781. *Randia uliginosa* (Retz.) DC., Prodr. 4: 386. 1830; Hook.f., Fl. Brit. India 3: 110. 1880; Gamble, Fl. Madras 2: 615. 1921.

An often armed, medium sized tree, up to 4 m high. Leaves obovate-oblanceolate, 6-16.5 × 2.5-6 cm, base narrowed, margin entire, apex acute-obtuse, glabrous, shining, main nerves 8-10 pairs; petiole 1.5 cm; stipule intrapetiolar, deciduous. Flowers axillary, solitary, white; calyx truncate, lobes 4-5, thick, triangular, acute; corolla campanulate, tube 1.5 cm long, lobes 4-5, obovate, pubescent without, obtuse; stamens 5, anthers linear; ovary 2-celled, style stout, stigma thick. Berry ellipsoid, 6 x 4 cm; seeds smooth, compressed.

All open dry forests in wet places and on black cotton soils in Adilabad, Warangal and Mahabubnagar districts.

Fl. & Fr.: June – February.

Way to Kunthala water falls (ADB), *GO* & *DAM* 5113; Eturnagaram (WGL), *SSR* & *KSM* 18535; Pakhal lake (WGL), *KMS* 13189 (MH); Farhabad to Vatavarlapally (MBNR), *SRS* 104589 (BSID); Bureddypalli bavi (MBNR), *VBH* 84998 (BSID).

TARENNA Gaertn.

Tarenna asiatica (L.) Kuntze ex K. Schumach., Bot. Tidsskr. 24: 332. 1902. *Chomelia asiatica* (L.) Kuntze, Revis. Gen. Pl. 1: 278. 1891; Gamble, Fl. Madras 2: 613. 1921. *Rondeletia asiatica* L., Sp. Pl. 172. 1753. *Webera corymbosa* Willd., Sp.Pl. 1: 1224. 1798; Hook.f., Fl. Brit. India 3: 102. 1880.

Vern.: *Komari, Papidi, Konda papidi, Komi.*

Evergreen shrub to small tree, bark dark grey, fairly rough; wood light brown, hard, close grained; branchlets glabrous. Leaves coriaceous, oblong or elliptic-oblong, 6-13 × 2-5 cm, glabrous, black when dry, apex acute-acuminate, margin entire, base acute or rounded, petiole to 1.8 cm long; stipules with broad persistent triangular base, cuspidate blade. Flowers white, fragrant, to 1.5 cm across, in terminal, corymbose cymes; calyx campanulate, lobes 5, 1 mm, puberulous without; corolla salver form, tube 2 mm, villous within, lobes 5, oblong, 3.5 mm, reflexed; stamens 5, between the lobes, exserted, filament 0.5 mm, anther linear, 3 mm; ovary bilocular, 1 mm, style columnar, 4 mm, stigma fusiform. Berry black, globose, 6 mm dia. with 2 mm long crown of calyx lobes; seeds 6-8.

Common in forests in Adilabad, Warangal, Khammam and Mahabubnagar districts.

Fl.: April-July, Fr.: October-December.

Jaipur RF (ADB), *GO & PVP* 4949; Mannanur (MBNR), *TDCK* 13638; Umamaheswaram (MBNR), *SRS* 109128 (BSID); Lawal beat, Pasra (WGL), *RKP* 108169 (BSID); Rollapadu (KMM), *RR* 108021 (BSID); Jallada –Kothagudem (KMM), *RR* 106088 (BSID).

WENDLANDIA DC. *nom. cons.*

1. Anther exserted .... **W. heynei**
1. Anther inserted:
    2. Calyx teeth shorter than the tube .... **W. tinctoria**
    2. Calyx teeth as long as or longer than the tube .... **W. gamblei**

Wendlandia gamblei Cowan, Notes Royal Bot. Gard. Edinb. 16: 271. 1932.

A small tree. Leaves coriaceous, elliptic-oblong, 6-10 × 3-5.5 cm, base cuneate, margin entire, apex acute – shortly acuminate. Flowers creamish yellow in terminal and axillary panicle.

In deciduous forests in Nallamalais of Mahabubnagar district.

Fl. & Fr.: March – November.

Gundem beat, Mannanur (MBNR), *SRS* 109186; Farahabad to Pullaechelma (MBNR), *SRS* 109198.

Wendlandia heynei (Roem. & Schult.) Santapau & Merchant, Bull. Bot. Surv. India 3: 110. 1961. *Roendeletia heynei* Roem. & Schult., Syst. Veg. 5: 234. 1819. *Wendlandia exserta* (Roxb.) DC., Prodr. 4: 411. 1830; Hook.f., Fl. Brit. India 3: 37. 1880; Gamble, Fl. Madras 2: 587. 1921.

A small, deciduous tree, 4-6 m high. Leaves ovate-lanceolate, 4-11 × 0.6-4 cm, slightly pubescent and dull above, densely greyish tomentose beneath, base acute to attenuate, margin entire, apex acute. Flowers in terminal or axillary panicles, white, sweet scented; bract ca 1.5 mm; calyx glabrous with 5 short teeth; corolla salviform, 5-lobed, lobes 2 mm, tube 4 mm; anthers 5, partially exserted; ovary globose, style 5 mm, stigma bifid. Capsule pubescent, crowned by the erect, lanceolate sepals.

Rare in the open deciduous forests in Mahabubnagar district.

Fl. & Fr.: March – May.

Way to Farahabad (MBNR), *GS* 18280.

Wendlandia tinctoria DC., Prodr. 4: 411. 1830; Hook.f., Fl. Brit. India 3: 38. 1880; Gamble, Fl. Madras 2: 587. 1921 ssp. **tinctoria**

A shrub, up to 2 m high. Leaves elliptic-obovate, 6.5 - 15 × 3.5-5.5 cm, acute at base, entire margin, shortly acuminate at apex, glabrous above, pubescent near vein below, prominently nerved; petiole 2 cm long; stipules ovate, cuspidate. Flowers in terminal dense panicles, white. Capsule globose, 0.3 × 0.2 cm, loculicidal, 2-valved.

Occasional in forests in Mahabubnagar district.

Fl. & Fr.: February – August.

Way to Farahabad (MBNR), *SSR* 15534; Pullayapally (MBNR), *MVS* & *DV* 38646; Appapur (MBNR), *VSL*, *DV* & *BR* 39278.

ASTERACEAE (*nom. alter.* COMPOSITAE)

1. Flowes all ligulate; sap milky; pappus fine, capillary or silky .... **LACTUCEAE**
    2. Capitula solitary on leafless scape; scapigerous herbs .... **Taraxacum**
    2. Capitula several on branching leafy stem; non scapigerous:
        3. Achenes distinctly beaked; beak slender .... **Lactuca**
        3. Achenes not beaked:
            4. Achenes narrow, truncate at both ends .... **Launaea**
            4. Achenes compressed .... **Sonchus**
1. Flowers all tubular or the outer only ligulate; sap watery; pappus other than silky:
    5. Involucral bracts uniseriate .... **SENECIONEAE**
        6. Plants succulent .... **Kleinia**
        6. Plants not succulent:
            7. Heads homogamous .... **Emilia**
            7. Heads heterogamous .... **Senecio**
    5. Involucral bracts more than one seriate:
        8. Heads homogamous, not rayed; pappus strictly present:
            9. Leaves opposite, rarely upper ones alternate .... **EUPATORIEAE**
                10. Pappus of very short scales:
                    11. Involucral bracts narrow; achenes 5-angled; pappus scales paleaceous .... **Ageratum**
                    11. Involucral bracts broad; achenes 3-ribbed; pappus scales clavate .... **Adenostemma**
                10. Pappus of long hairs .... **Chromolaena**
            9. Leaves alternate:
                12. Anther cells tailed or auricled or mucronate; style branches elongated .... **VERNONIEAE**
                    13. Herbs rhizomatous; heads subtended by leafy bracts .... **Elephantopus**
                    13. Herbs or shrubs, not rhizomatous; heads not subtended by leafy bracts .... **Vernonia**
                12. Anther cells cleft at base or subentire; style branches not elongated:

14. Style branches thickened towards base .... **CYNAREAE**

15. Involucral bracts spinescent:

16. Heads compound; leaves spinescent; deeply lobed .... **Echinops**

16. Heads simple; leaves not spinscent; dentate or lyrately lobed .... **Oligochaeta**

15. Involucral bracts aristate-acuminate or acuminate:

17. Heads 4-5-flowered, in compound corymbs .... **Goniocaulon**

17. Heads many-flowered, usually solitary or few .... **Tricholepis**

14. Style branches not thickened towards base .... **MUTISIEAE : Dicoma**

8. Heads heterogamous, rayed or not (except **Caesulia** and **Lagascea);** pappus present or absent:

18. Receptacle epaleaceous:

19. Leaves opposite .... **HELENIEAE : Flaveria**

19. Leaves usually alternate:

20. Anthers base obtuse .... **ASTEREAE**

21. Pappus of many slender hairs, setaceous:

22. Female florets ligulate (atleast marginal ones), ligule exceeds the pappus .... **Erigeron**

22. Female florets not ligulate, if present do not exceed the pappus .... **Conyza**

21. Pappus absent or a lacineate cup:

23. Receptacle convex; achene glandular .... **Grangea**

23. Receptacle cup like; achenes glabrous ...... **Cyathocline**

20. Anther base sagittate or tailed .... **INULEAE**

24. Heads aggregated into dense globose or ovoid clusters:

25. Herbs aromatic; stems winged or atleast leaves decurrent; florets pink .... **Sphaeranthus**

25. Herbs not aromatic; stems not winged; florets white .... **Blepharispermum**

24. Heads not aggregated into clusters:

26. Stems winged (except *Laggera aurita*):

27. Plants diffuse or prostrate herbs; leaves linear-oblong, pappus absent; bisexual florets minute and sterile .... **Epaltes**

27. Plants erect herbs; leaves elliptic-oblanceolate; pappus present; bisexual florets fertile .... **Laggera**

26. Stems not winged:

28. Plants marshy herbs; heads solitary, white .... **Caesulia**

28. Plants other than marshy; heads in racemes:

29. Heads rayed:

30. Pappus biseriate, outer row of scales forming cup, inner of setaceous free hairs .... **Pulicaria**

30. Pappus uniseriate, setaceous hairs free .... **Pentanema**

29. Heads not rayed:

31. Involucral bracts broad, scarious .... **Gnaphalium**

31. Involucral bracts other than above:

32. Stem woody; involucral bracts broad, rigid .... **Pluchea**

32. Stem herbaceous; involucral bracts narrow, slender .... **Blumea**

18. Receptacle paleaceous:

33. Leaves opposite, all cauline:

34. Inner involucral bracts partly or completely enclose the achenes of ray florets; pappus or ray and disc florets absent; bisexual florets sterile, ray florets fertile .... **Acanthospermum**

34. Inner involucral bracts do not enclose the achenes of ray florets, almost free from achenes; pappus of either ray and disc florets present or rarely absent; bisexual florets fertile, ray florets usually sterile:

35. Heads one-flowered, clustered together or aggregated in compound fashion, subtended by leafy bracts; ligules absent .... **Lagascea**

35. Heads separate, not aggregated in compound fashion, not subtended by leafy bracts; ligules present:

36. Leaves and phyllaries with oil glands .... **Tagetes**

36. Leaves and phyllaries lacking oil glands:
    37. Pappus of awns:
        38. Achenes dimorphic, of ray winged, of bisexual florets not winged; involucral bracts rigid; pappus glabrous .... **Synedrella**
        38. Achenes monomorphic, not winged; involucral bracts soft; pappus awns otherwise:
            39. Pappus awns plumose; achenes densely silky .... **Tridax**
            39. Pappus awns retrorsely barbellate; achenes hispid or glabrous .... **Bidens**
    37. Pappus of scales, bristles, or absent:
        40. Receptacle conical:
            41. Style branches truncate; heads rather large; achenes usually ciliate along margins .... **Spilanthes**
            41. Style branches acute; heads small; achenes not ciliate along margins:
                42. Pappus present .... **Galinsoga**
                42. Pappus absent .... **Guizotia**
        40. Receptacle flat, or slightly convex:
            43. Bisexual florets completely enclosed by palea; achenes curved and narrowly beaked at apex .... **Sclerocarpus**
            43. Bisexual florets free from palea; achenes straight, truncate to obtuse at apex:
                44. Stems erect; heads on conspicuous strong penduncles; ligules yellow:
                    45. Heads less than 5 cm in diam .... **Wedelia**
                    45. Heads more than 5 cm in diam .... **Helianthus**

44. Stems often decumbent; head subsessile to sessile or with weak peduncle, white:
    46. Bisexual florets yellow and red .... **Gaillardia**
    46. Bisexual florets white:
        47. Palea broad, dentate at apex .... **Blainvillea**
        47. Palea narrow, acute to gland tipped .... **Eclipta**

33. Leaves alternate, radical and cauline or cauline only:
    48. Prostrate, decumbent or scapigerous low herbs:
        49. Pappus of awns .... **Glossocardia**
        49. Pappus absent .... **Chrysanthellun**
    48. Erect herbs to sub-shrubs:
        50. Heads unisexual; female heads eligulate, 2-flowered, covered by hooked spines; anemophilous .... **Xanthium**
        50. Heads bisexual; female florets ligulate; entomophilous .... **Parthenium**

ACANTHOSPERMUM Shrank

Acanthospermum hispidum DC., Prodr. 5:522. 1836; Gamble, Fl. Madras 2: 704.1921; Chowdhery in Hajra & *al.,* Fl. India 12 : 361. 1995.

Erect herb, to 40 cm high, stem striated, sparsely pubescent below, white serecious in younger parts; branches in dichotomies. Leaves opposite, simple, sessile, one veined from base and veins raised on lower surface; sparsely pubescent on both surfaces, obovate, spathulate or elliptic, 3-8 x 1-3 cm, base attenuate, margin entire to dentate, ciliate, apex obtuse to apiculate. Heads solitary, axillary in forks, yellow, 1 cm, heterogamous, rayed, subsessile. Involucral bracts biseriate, not exceed the florets at maturity; outer 5, elliptic to oblong-elliptic, 4-6 mm, basally united, 3-veined, pubescent on dorsal side, margin ciliate with long multicellular hairs, acute to sub-acute, rarely bidentate; inner ones connate, 4-6 mm, enveloping the ovary of ray florets, compressed, prickly, apically 2-awned, awns 3-5 mm. Receptacle slightly convex, 2-3 mm. Ray florets 8, tube minute, glabrous, ligule elliptic, 1.5 mm, 3-veined, glandular and hairy on dorsal side, 3-lobed, lobes ovate to oblong, 0.3 mm, obtuse; style 1.5 mm, branches 0.7 mm, often recurved, flat, obtuse, papillate. Bisexual florets 7, tube 1 mm, cylindrical, pubescent with multicellular hairs; limb 1 mm, 5-lobed, lobes ovate-elliptic, 0.8 – 1 mm, pubescent with hairs and glands, rarely midveined, subacute; stamens 5, exserted to the level of lobes, anthers oblong, 1 mm, hood ovate, incurved, base sagittate, auricles obtuse; style not exserted, 2 mm, branches broad, obtuse, papillate; pappus absent. Achene of ray obovate, 2 mm, glabrous, enclosed with echinate inner phyllaries; of bisexual floret cylindrical, 1 mm, glandular, sterile, apex truncate.

Abundant weed in foot hills and plains in all districts.

Fl. & Fr.: Throughout the year.

Bijjur (ADB), *GO* & *PVP* 4873; Mosra – Malkapur Road (NZB), *TP* & *BR* 6091; Pilutla RF (MDK), *BR* & *CP* 11271; Nagulabanda (MDK), *CP* 11354; Kusumasamudram RF (RR), *MSM* & *KH* 11025; Nagarjunasagar (NLG), *BVRS* 50 (NU); Bhupatipur (KRN), *GVS* 20174 (MH).

ADENOSTEMMA J.R.Forst. & J.G.A.Forst.

Adenostemma lavenia (L.) Kuntze, Revis. Gen. Pl. 1: 304. 1891; Uniyal in Hajra & *al.*, Fl. India 12: 347. 1995.

Erect herbs up to 1m high. Leaves simple, opposite, broadly ovate, 2-17 × 1-10 cm, sparsely puberulous, base truncate – attenuate, margin serrate – dentate, apex acute. Heads a few, terminal, white, homogamous, not rayed. Involucral bracts sub 2-seriate, free or sometimes connate at base, glabrous or hairy on back, ciliate on margins. Achenes compressed, oblanceolate, 3 – ribbed, slightly curved, minutely glandular, pappus scales clavate.

Rare along the streams at high elevations in Mahabubnagar district.

Fl. & Fr. September – January.

AGERATUM L.

Ageratum conyzoides L., Sp. Pl. 839. 1753; C.B. Clarke in Hook.f., Fl. Brit. India 3: 243. 1881; Gamble, Fl. Madras 2: 672. 1921; Uniyal in Hajra & *al.*, Fl. India 12: 348. 1995.

Erect, aromatic herb, 30-80 cm high, stem terete, branched, pilose. Leaves simple, opposite or the upper alternate, ovate, 2-3 × 1-5 cm, base acute to attenuate, margin dentate, apex acuminate, hirsute, 3-veined from base, surface with spreading hairs and sessile glandular below; petiole to 5 cm. Heads 4-6 in terminal corymbs, 3-4 mm, blue or white, bracteate, homogamous, not rayed; pedicel to 1 cm, glandular. Involucral bracts 18-24, biseriate, subequal, not exceed the florets. Receptacle convex, 1-2 mm across, glabrous. Florets 50-60, corolla 1.5-2 mm. tube glandular, 5-lobed, acute; stamens 5, anthers 0.6-0.8 mm, hood ovate, obtuse, basally auricled; style 2-2.5 mm, exserted to 1 mm, branches elongate, obtuse, papillate. Pappus of 5 scales, lanceolate, 2 mm, exceeding the florets, united at base, form a ring. Achenes 2 mm, black, 5-angled, glabrous or scabrid on angles, pappus scales paleaceous.

Moist places in all districts.

Fl. & Fr.: July – March.

Surzapur (ADB), *TP* & *PVP* 4111; Mosra – Malkapur road (NZB), *TP* & *BR* 6095; Narsapur (MDK), *BR* & *CP* 11232; Narsingi (MDK), *BR* & *CP* 11536; Kanmankalva RF (RR), *MSM* 10467; Irlapenta (MBNR), *BSS* & *SKB* 32386; Kambalpally (NLG), *BVRS* 323 (NU); Kodimial (KRN), *GVS* 21839 (MH); Pakhal (WGL), *KMS* 11604 (MH); Yenkur (KMM), *RR* 108561 (BSID); Umamaheswaram (MBNR), *SRS* 109136 (BSID).

BIDENS L.

1. Achenes curved or twisted .... **B. cynapifolia**
1. Achenes straight:
   2. Ligule white .... **B. pilosa**
   2. Ligule yellow .... **B. bipinnata**

Bidens bipinnata L., Sp. Pl. 832. 1753; Chowdhery in Hajra & *al.*, Fl. India 12: 367. 1995. *Bidens pilosa* L. var. *bipinnata* (L.) Hook.f., Fl. Brit. India 3: 309. 1881.

Erect herb to 1 m high, stems quadrangular, striated, internodes to 30 cm, glabrous. Leaves opposite, petiolate, sparsely pilose, lower one bipinnately compound, uppermost one pinnate, to 12 cm long, bipinnate one with 2-3 lateral secondaries, lateral segments 1-2 pairs, terminal one 2-3-lobed, segments one-veined from the base, glabrous or sparsely pilose on both surfaces. Heads solitary, terminal on divaricate branches, yellow, heterogamous, rayed. Involucral bracts around 21, subequal. Receptacle 4-5 mm across, flat to slightly concave, paleaceous. Ray florets 5; corolla 6.5 mm, tube 1.5 mm, glabrous, ligule 5 mm, strongly 5-nerved, sessile, glandular and sparsely hairy along veins on dorsal side, retuse or minutely 3-lobed. Bisexual florets around 40; corolla 4.2 mm, tube 1.2 mm, cylindrical, glabrous, 5-lobed, lobes triangular, acute; stamens 5, black, anthers 1.5 mm, hood ovate, acute, base sagittate, auricles free, obtuse; style 5 mm, branches 1mm, elongate, hairy on dorsal side, acute. Pappus barbellate, awns 3 mm. Achenes of ray 5 mm, linear, sterile, sparsely hispid at apex, 4-ribbed, colorless.

Open forests near and around streams in Adilabad, Medak, Karimnagar, Hyderabad, Ranga Reddi and Nalgonda districts.

Fl. & Fr.: November – January.

Bhimaram RF (ADB), *GO* 4496; Narsapur forest nursery (MDK), *BR* & *CP* 11222; Mohammadabad RF (RR), *MSM* 10444; Rachakonda (NLG), *BVRS* 434 (NU).

Bidens cynapiifolia Kunth, Nov. Gen. & Sp. Pl. 4: 235. 1820. *B. bipinnata* L. var. *cynapiifolia* (Kunth) Gomeg, Ann. Hist. Nat. Marid 19: 275. 1890.

Erect herb, to 1 m high, stem quadrangular, striated, glabrous. Leaves uni or bipinnately compound, opposite, petiolate, lateral segments 1-3 pairs, prominently veined, glabrous on upper surface, acute to subacuminate. Heads 1-3, terminal, heterogamous, rayed. Involucral bracts around 19, biseriate. Receptacle 3-4 mm across, paleaceous. Ray florets 3, white, 5 mm, 5-veined, sparsely hairy on dorsal side, bidentate at apex. Bisexual florets 3.5 mm, tube 1.5 mm, glabrous, limb 2 mm, 5-lobed; stamens 4, slightly exserted, anthers linear-oblong, 1.5 mm, hood ovate, subacute, base sagittately auricled; style 3.5 mm, exserted, branches slender, hairy; ovary oblong, 2 mm, 4-ribbed, glabrous. Pappus 3-4, retrorsely barbellate awns, 1.5-2.5 mm. Achenes of ray sterile, of bisexual florets fertile, 9-13, curved in upper region or twisted with 2 spirals, glabrous or hispidulous in a few marginal ones, 4-ribbed.

Around and near the streams in Nizamabad district.

Fl. & Fr.: October – January.

Manchippa (NZB), *TP* & *BR* 6033, *CPR* 14499. .

Bidens pilosa L., Sp. Pl. 832. 1753.

Key to Varieties

1. Ray florets present, heads longer than wide .... var. **radiata** f. **indivisa**
1. Ray florets absent, heads wider than long .... var. **pilosa**

Bidens pilosa L. var. **pilosa** Sherff., Field. Mus. Nat. Hist. Bot. 16:412. 1937; Chowdhery in Hajra & *al*., Fl. India 12: 373. 1995. *Bidens chinensis* auct. non Willd. Hook.f., Fl. Brit. India 3: 309. 1881.

Erect herb, to 50 cm high, stem terete, glabrous. Leaves opposite, trifoliolate, rarely upper ones simple, petiole to 1 cm. Heads solitary or two, terminal, yellow, homogamous. Involucral bracts biseriate, not exceed the florets, outer 8, equal in length, 3-veined, glabrous, margin ciliate. Receptacle paleaceous. Bisexual florets yellow, tube glabrous, 1 mm, cylindrical, limb glandular pubescent, 5-lobed, lobes ovate, margin glandular; stamens 5, anthers black, linear, 2 mm, exserted, hood ovate, acute, base sagittate, auricles subacute; style exserted, 4 mm, branches papillate Achenes monomorphic, linear, 8-10 mm, tuberculate-hispid, 4-ribbed.

In open forest areas of the hill tops in Medak, Hyderabad and Mahabubnagar districts.

Fl. & Fr.: November – February.

Narsapur forest nursery (MDK), *BR* 11222; Ranjhole (MDK), *CP* 11676.

Bidens pilosa L. var. **radiata** Sch.-Bip. forma **indivisa** Sherff, Bot. Gaz. 88: 297. 1929 & Field Mus. Nat. Hist. Bot. 439. 1937.

Annual erect herb, up to 50 cm high, stem striated. Leaves simple, ovate-lanceolate, 1.5-4 × 0.5-2 cm, base obtuse to attenuate, margin serrate, apex sub-acuminate. Heads 2-3, terminal, heterogamous, yellow. Achenes obovate-cuneate, 8-ribbed.

Rare in open places in Nizamabad district.

Fl. & Fr.: October – January.

Manjeera Barrage (NZB), *CP* & *BR* 14087.

BLAINVILLEA Cass.

Blainvillea acmella (L.) Philipson, Blumea 6 : 350. 1950; Chowdhery in Hajra & *al*., Fl. India 12: 377. 1995. *Verbesina acmella* L., Sp. Pl. 901. 1953. *Blainvillea latifolia* (L.f.) DC. in Wight, Contrib. Bot. Ind.17. 1834. & Prodr. 5: 492. 1836; Hook.f., Fl. Brit. India 3: 305. 1881. *Eclipta latifolia* L.f., Suppl. 378. 1781. *Blainvillea rhomboidea* Cass., Dict. Sci. Nat. 29: 494. 1823; Gamble, Fl. Madras 2: 706. 1921.

Erect herb, to 1 m high, stem terete, striated, divaricately branched. Leaves simple, opposite, alternate in upper parts, petiole to 2 cm, hispid, 3-veined, scabrid above, hispid below, ovate or elliptic-lanceolate, 2-10 × 1-6 cm, base cuneate, margin coarsely serrate, apex acute or acuminate. Heads solitary or 3, terminal, 6 mm, heterogamous, rayed. Involucral bracts 8, biseriate, ovate, 3-6 mm, 3-5-veined, ciliate marginally. Ray florets 5; corolla 2 mm, tube 1.2 mm, glabrous, ligule 0.8 mm, 2-3-lobed, lobes obtuse, glabrous; style 2 mm, branches broad, subacute to obtuse. Bisexual florets 12; corolla 1.5 mm, tube 0.6 mm, glabrous, limb 1 mm, 4-lobed, lobes ovate, 0.5 mm, pubescent with colleters and sessile glands on dorsal side, acute; stamens 5, black, anthers oblong-elliptic, 0.5 mm, hood broad, obtuse, base auricled, auricle obtuse; style 1.5 mm, branches 0.3-0.4 mm, papillate on dorsal side, acute. Pappus of two unequal short bristles, hispid. Achenes obovate, 3-5 mm, dimorphic, of ray triquetrous, of disc compressed or biconvex, black, whitish pubescent, glabrous and shiny, wrinkled or tuberculated.

Wet places in low elevations in most of the districts.

Fl. & Fr.: October – April.

Ankusapuram (ADB), *GO* 4331; Manchippa RF (NZB), *TP* & *BR* 6014; Mombojipally (NZB), *BR* & *KH* 6427; Narsapur RF (MDK), *KMS* 6692; Mohammadabad RF (RR), *MSM* & *NV* 10428; Nagarjunasagar (NLG), *BVRS* 521 (NU).

BLEPHARISPERMUM Wight

Blepharispermum subsessile DC. in Wight, Contrib. Bot. Ind. 11. 1834; Hook. f., Fl. Brit. India 3: 276. 1881; Pant in Hajra & *al.*, Fl. India 13: 43. 1995.

Erect herb with many branches from woody root stock, to 60 cm high, stems striated, solid, glabrous. Leaves simple, alternate, petiole 2-3 cm, one-veined, glaucous beneath, glabrous upper, narrowly or broadly elliptic, 2.5-8.5 × 1-4 cm, cuneate or obtusely attenuate at base, entire, acute or obtusely apiculate. Heads many, white, 8-10 mm, aggregated into globose clusters. Bract of common receptacle absent, leafy when present; bracts of individual heads 2-3 scales, axial scale lanceolate, 9 mm, strongly one-veined, glabrous. Female florets 2, tubular, 3 mm, glandular, 3-4-lobed, lobes ovate, obtuse; style 3.5 mm, exserted, branches broad, recurved, glabrous. Bisexual florets 4, 4.5 mm, glandular above the middle, 5-lobed, lobes lanceolate, 1.8 mm, subacute; stamens 5, anthers 3 mm, linear-oblong, hood obtuse, sagittately auricled; style 5 mm, branches 1 mm, recurved, glabrous. Pappus of unequal bristles. Achenes obovoid, 3.5 mm, black, of female larger than those of bisexual ones, trigonous, densely ciliate along the margins, slightly obcordate at apex.

In the open forests in Medak district.

Fl. & Fr.: November – January.

Narsapur (MDK), *TP* & *CP* 11980; Kalcharam (MDK), *TP* & *CP* 14079.

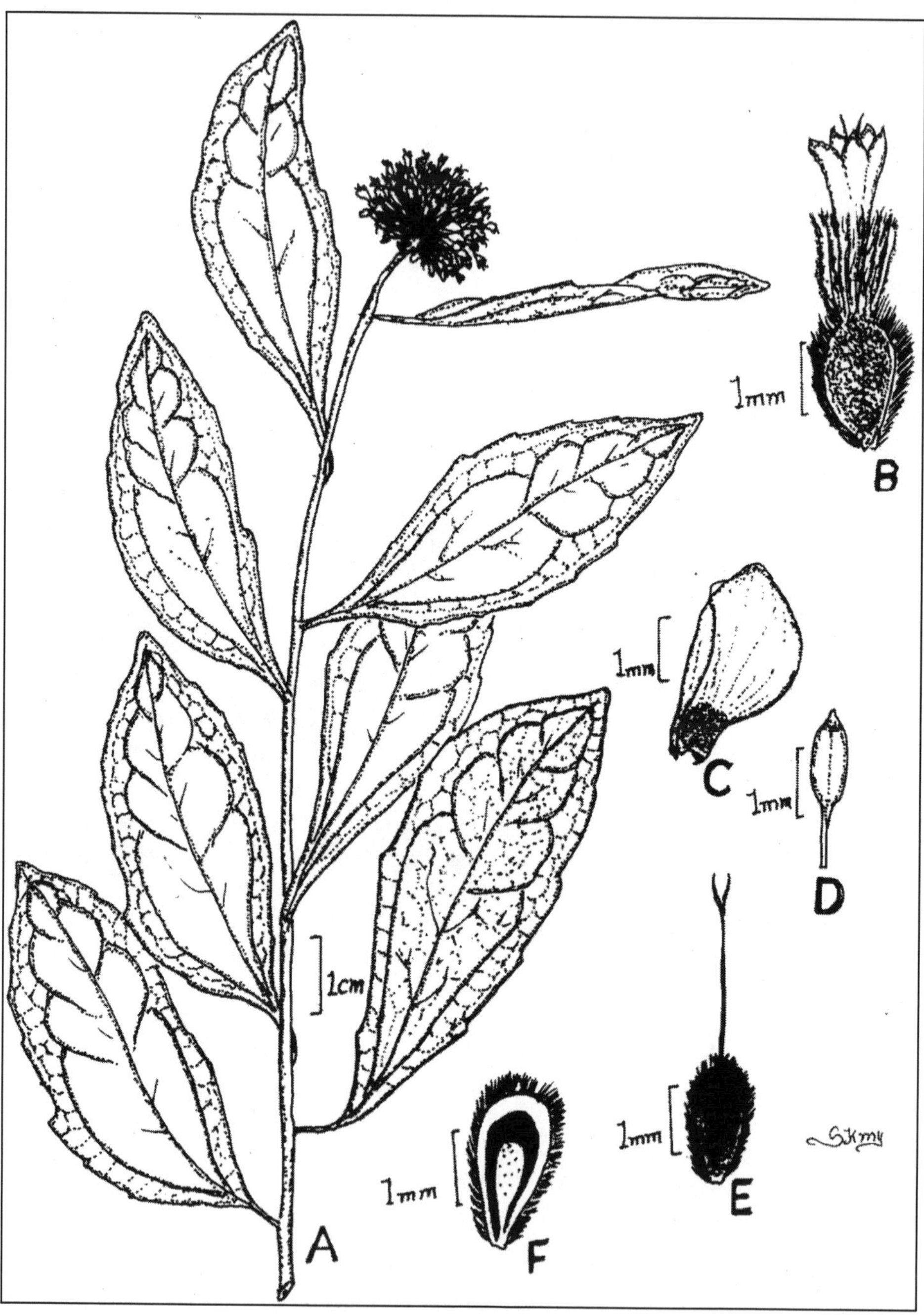

Figure 46: Blepharispermum subsessile DC.

A. Twig, B. Floret, C. Petal, D. Stamen, E. Pistil, F. Ovary l.s.

BLUMEA DC *nom. cons*

1. Leaf margins toothed; anthers abortive:
   2. Plants prostrate; bisexual florets multicellular hairy except on tube .... **B. oxyodonta**
   2. Plants erect; bisexual florets multicellular hairy on lobes and tubes .... **B. eriantha**
1. Leaf margins otherwise; anthers fertile:
   3. Heads solitary or few:
      4. Uppermost leaves opposite or sub-opposite .... **B. bifoliata**
      4. Uppermost leaves alternate .... **B. obiliqua**
   3. Heads variously paniculate or clustered:
      5. Receptacle pubescent or pilose .... **B. laciniata**
      5. Receptacle glabrous or fimbriate:
         6. Plant parts densely silky sericeous:
            7. Bisexual florets pubescent with colleters; achenes ribbed .... **B. hieracifolia**
            7. Bisexual florets pubescent with multicellular hairs; achenes striated .... **B. lacera**
         6. Plant parts variously pubescent or glabrous:
            8. Bisexual florets colourless with pink tinge .... **B. mollis**
            8. Bisexual florets yellow:
               9. Plant parts with glandular indumentum ...... **B. membracacea**
               9. Plant parts eglandular .... **B. virens**

Blumea bifoliata (L.) DC. in Wight, Contrib. Bot. Ind. 15. 1834; Hook.f., Fl. Brit. India 3: 261. 1881; Gamble, Fl. Madras 2: 686. 1921; Kumar in Hajra & *al.*, Fl. India 13: 120. 1995. *Conyza bifoliata* L., Sp. Pl. 862. 1753.

Erect or ascending herb, 10-40 cm high, stems simple or branched from base, villous with multicellular hairs and stipitate glands. Leaves simple, lower alternate, uppermost opposite or sub-opposite, sessile, one-veined from base, villous and glandular on both surfaces, margin serrate, apex acute or apiculate. Heads solitary, axillary and terminal, yellow, 6-7 mm, heterogamous, not rayed. Involucral bracts 5, biseriate, equal in length to the florets. Female florets about 140, tubular, 3 mm, filiform, glabrous, 3-lobed; style 5 mm, exceeding the lobes, branches linear, obtuse. Bisexual florets 10-13, 4-5 mm, glabrous below, 5-lobed, lobes ovate-triangular, pubescent with unicellular hairs and colleters on dorsal side, apex acute; stamens 5, anthers 2 mm, hood rounded at apex, base sagittately tailed; style 5 mm, slightly exserted, branches linear, obtuse. Pappus 4 mm, white, uniseriate, setaceous hairs, not exceeding the florets. Achene 1 mm, oblong, hairy.

Wet places of the forests in Nizamabad, Hyderabad, Mahabubnagar and Nalgonda districts.

Fl. & Fr.: November – March.

Salura (NZB), *BR* 7278; Erloo to Mahabubnagar (MBNR), *SRS* 104516 (BSID); Rachakonda (NLG), *BVRS* 392 (NU).

Blumea eriantha DC. in Wight, Contrib. Bot. Ind. 15. 1834; Hook.f., Fl. Brit. India 3: 266. 1881; Gamble, Fl. Madras 2: 686. 1921; Kumar in Hajra & *al.*, Fl. India 13: 122. 1995 .

Erect herb, strongly aromatic, 30-100 cm high, stem brown, striated, puberulous to densely silky in younger parts, globose woolly buds clustered at the base and on roots. Leaves alternate, simple, sessile, one veined, raised on both surfaces, densely silky, silvery and shiny in younger ones, obovate, base attenuate, margin spinulose dentate, apex acute or obtusely apiculate. Heads terminal or axillary, cymosely paniculate or solitary, yellow, 6 mm, heterogamous, not rayed. Involucral bracts 50, lanceolate, pubescent with multicellular hairs and glands on dorsal side, acute to sub-acuminate. Female florets around 150, 3.5 mm, tubular, filiform, 3-lobed, pubescent on lobes, occasionally hairy at base; style not exserted, branches linear, obtuse. Bisexual florets around 60, 4 mm, pubescent with multicellular hairs, often with colleters on lobes and at base, 5-lobed, lobes ovate, sub-acute; stamens 5, anthers shrinked and abortive, 1.5 mm, hood oblong, narrow, base sagittately tailed; style 2-3 mm, branches minutely papillate, obtuse. Pappus uniseriate, setaceous with hairs, equal to the florets. Achene oblong to elliptic-oblong, 0.5 mm, pubescent.

Rare clumps in the forests in Adilabad, Nizamabad, Ranga Reddi and Warangal districts.

Fl. & Fr.: February – May.

Sathermott (ADB), *TP* & *PVP* 4107; Utnoor (ADB), *GO* & *PVP* 4604; Gandhari RF (NZB), *TP* & *BR* 6338; Sirikonda RF (NZB), *BR* 9697; Dharmapur (RR), *MSM* & *KH* 10547; Pakhal (WGL), *ANH* 15938 (MH).

Blumea hieracifolia (D. Don) DC. in Wight, Contrib. Bot. Ind. 15. 1834. var. **macrostachya** (DC.) Hook.f., Fl. Brit. India 3: 263. 1881; Gamble, Fl. Madras 2: 687. 1921; Kumar in Hajra & *al.*, Fl. India 13: 126. 1995. *B. macrostachya* DC., Prodr. 5: 442. 1836.

Simple or sparingly branched, usually scapigerous herbs. Basal leaves in rosette, oblanceolate or spathulate, regularly shortly dentate, stem leaves usually smaller and more oblong, often very white-woolly. Heads in close clusters, terminal to the stem or its branches. Achenes ribbed, pubescent at first, then glabrous.

Rare along streams in Ranga Reddi district.

Fl. & Fr.: Aughust – January.

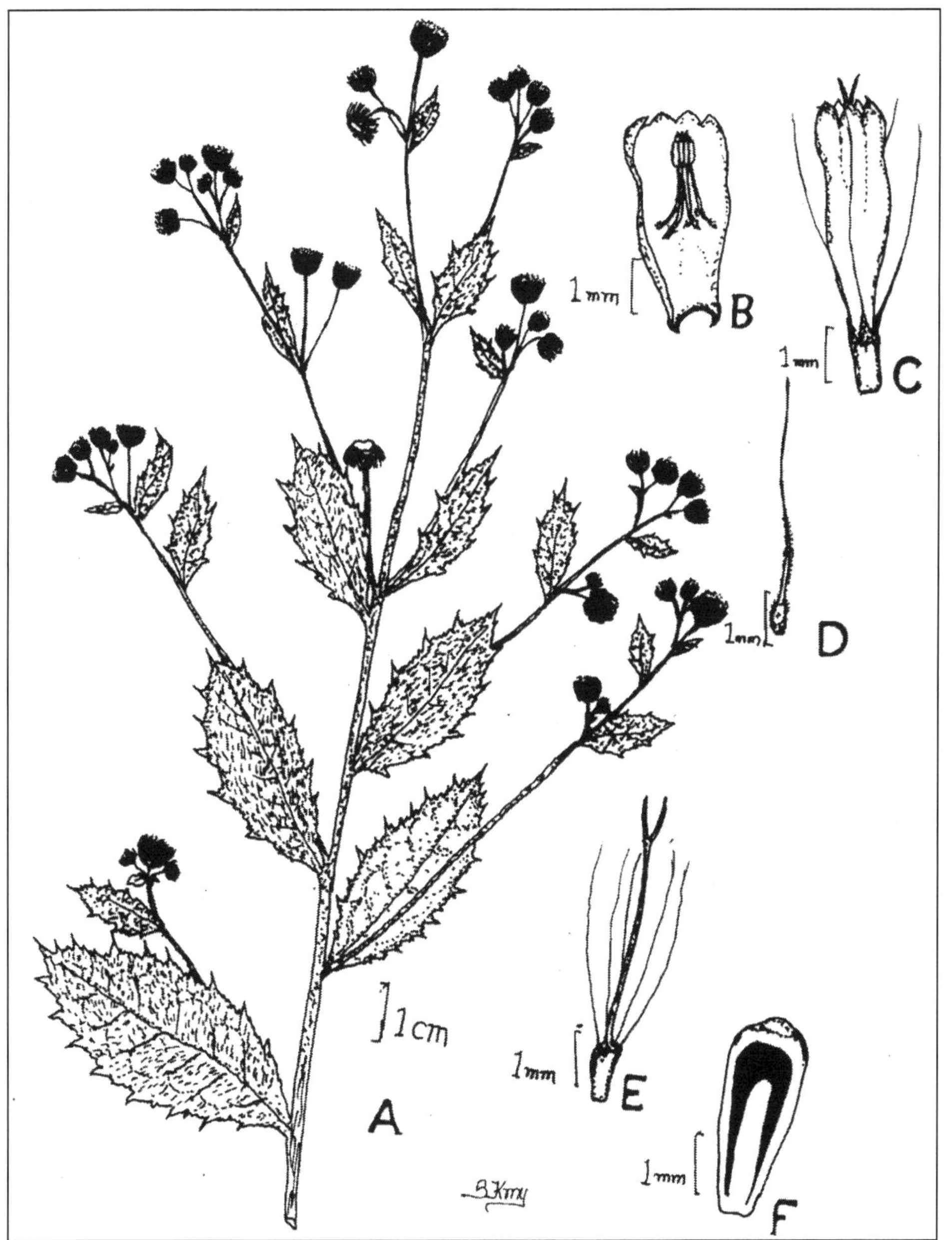

Figure 47: Blumea eriantha DC.

A. Twig, B. Corolla open, C. Floret, D. Pistil, E. Pistil & pappus, F. Ovary l.s.

Blumea lacera (Burm.f.) DC. in Wight Contrib. Bot. Ind. 14. 1834; Hook.f., Fl. Brit. India 3: 263. 1881; Gamble, Fl. Madras 2: 687. 1921; Kumar in Hajra & *al.*, Fl. India 13: 128. 1995. *Conyza lacera* Burm. f., Fl. Ind. 180. t. 59. f.1. 1768.

Erect aromatic herb, to 1m high, stem striate, pubescent with white hairs and glands. Leaves alternate, simple, sessile, one-veined, densely white silky and glandular on both surfaces, obovate to elliptic, base attenuate and clasped to stem, margin dentate with sharp teeth, apex obtusely apiculate or acute. Heads in cymose clusters, terminal and upper axillary, yellow, 6 mm, heterogamous, not rayed. Involucral bracts around 50, 3-seriate, equal, exceeding the florets. Female florets 90-100, tubular, filiform, 3-4 mm, glabrous, except few uni or multicellular hairs on lobes, 3-lobed; style 5 mm, exserted, branches linear, obtuse. Bisexual florets around 14, 4.5 mm, gradually expanding, multicellular hairy at the base of tube; 5-lobed, lobes ovate, acute, pubescent; stamens 5, anthers 1.5 mm, oblong, hood ovate, base sagittately tailed; style 4.5 mm, slightly exserted, bulged at base, branches linear, minutely hirsute on dorsal side. Pappus uniseriate, setaceous hairs, 4 mm, not exceed the florets. Achene oblong, 0.5-0.7 mm, striated, hairy.

Occasional in cultivated fields in Warangal and Nizamabad districts.

Fl. & Fr.: December – April.

Armoor (NZB), *BR* 7295.

Blumea laciniata (Roxb.) DC., Prodr. 5: 436. 1836; Hook.f., Fl. Brit. India 3: 264. 1881; Gamble, Fl. Madras 2: 688. 1921; Kumar in Hajra & *al.*, Fl. India 13: 128. 1995. *Conyza laciniata* Roxb., Fl. Ind. 3: 42. 1832.

Erect herb, to 1m high, stem striate, brown, densely glandular, sparsely pilose. Leaves alternate, simple, sessile, one-veined, coverd with hirsute hairs, oblong or ovate, deeply lobed, base attenuate and clasped to stem, margin dentate, apex acute. Heads in terminal or axillary panicles, yellow, 6 mm, heterogamous, not rayed. Involucral bracts around 60, 3-4 seriate, equal, exceeding the florets, acute. Female florets 100, tubular, filiform, 3.2 mm, 3-lobed, glabrous, style 2.4 mm, not exserted, branches linear, obtuse. Bisexual florets around 18, 4 mm, tubular, 5-lobed, lobes ovate-triangular, pubescent with unicellular hairs; stamens 5, inserted, anthers 2 mm, oblong, hood ovate, base sagittately tailed; style 2.2 mm, inserted, branches linear, obtuse. Pappus uniseriate, setaceous hairs, 3.2 mm, not exceed the florets. Achene oblong, 0.8-1 mm, oblong, ribbed, hairy.

Edges of streams and rocky slopes in Warangal (C.S. Reddy, 2001) and Hyderabad districts (Rajagopal, 1973).

Fl. & Fr.: December – April.

Blumea membranacea DC., Prodr. 5: 440. 1936; Hook.f., Fl. Brit. India 3: 265. 1881. var. **membranacea**; R. R. Rao & *al.*, Fl. Ind. Enum.-Ast. 17. 1988; Kumar in Hajra & *al.*, Fl. India 13: 135. 1995.

Erect, aromatic herb, up to 100 cm high, stem striated, sparsely pubescent with soft spreading hairs and glands. Leaves alternate, simple, sessile, one-veined, sparsely velutinous on both surfaces, elliptic-oblanceolate, 1.5-11 × 0.6-3.5 cm, base attenuate, margin dentate, apex acute to short acuminte. Heads in terminal or axillary, lax panicles, 5-6 mm, yellow, heterogamous, not rayed. Involucral bracts around 50, exceeding the florets, completely reflexed at maturity, rigid, 4-seriate, sparsely pubescent and glandular, apex acute. Receptacle 2 mm, flat, glabrous. Female florets 50-80, 4 mm, glabrous, 3-lobed; style exserted, branches linear, obtuse. Bisexual florets 3-5; corolla 4 mm, tube glabrous, 5-lobed, lobes ovate, sub-acute, pubescent with unicellular hairs and colleters; stamens 5, anthers oblong, 2 mm, hood ovate, truncate to obtuse, base sagittately tailed; style 4 mm, slightly exserted, branches linear, obtuse, hirsute on dorsal side. Pappus of uniseriate, setaceous, white hairs, not deciduous. Achene oblong to elliptic-oblong, 1 mm, ribbed.

Waste places and road sides in the forests.

Fl. & Fr.: December – March.

Blumea mollis (D. Don) Merr., Philipp. Jour. Sci. (Bot.) 5: 395. 1910; Kumar in Hajra & *al.*, Fl. India 13: 135. 1995 . *Erigeron molle* D. Don, Prodr. Fl. Nep. 172. 1825. *Blumea wightiana* DC. in Wight, Contrib. Bot. Ind. 14. 1834; Hook.f., Fl. Brit. India 3: 261. 1881; Gamble, Fl. Madras 2:686. 1921. *B. neilgherriensis* Hook.f., Fl. Brit. India 3: 262. 1881; Gamble, Fl. Madras 2: 686. 1921.

Erect, aromatic herb, to 1 m high, stem striated, covered with white pilose hairs and stipitate glands. Leaves alternate, simple, upper sessile and lower petiolate, petiole 2 cm, pilose and glandular, one-veined, elliptic or obovate, 2-11 × 0.6-4.5 cm, base attenuate to acute, margin dentate or double dentate, entire in uppermost leaves, apex acute or obtusely apiculate. Heads many, in terminal panicles and racemes in the upper axils, 5 mm, purple, heterogamous, not rayed. Involucral bracts 50, 3-seriate, not exceeding the florets, entirely reflexed at maturity. Female florets 50-85, tubular, 3 mm, filiform, colourless, 2-3-lobed, glabrous; style exserted or inserted, branches linear, obtuse. Bisexual florets 4-15, 3.5 mm, glabrous except in lobes, 5-lobed, lobes ovate, papillate along margins, sub-acute, pubescent with colleters on dorsal side; stamens 5, anthers 2 mm, hood oblong, truncate at apex, base sagittately tailed; style 3-3.5 mm, usually not exserted, branches linear, obtuse. Pappus of uniseriate, setaceous hairs, not exceed the florets, white. Achene 0.5 – 0.7 mm, oblong, striated, hairy.

Along the edges of streams, wet places and road sides of the forests in Adilabad, Warangal, Karimnagar and Nalgonda districts .

Fl. & Fr.: July – January.

Alampalli (ADB), *TP* & *PVP* 4167; Kunthala water falls (ADB), *GO* & *DAM* 5130; Bhupatipur (KRN), *GVS* 22293 (MH); Kambalpally (NLG), *BVRS* 609 (NU); on the way to Bhadrachalam RF (KMM), *RCS* 98962 (BSID); Karkalpadu, Amanagal RF (MBNR), *SRS* 104552 (BSID); Pakhal (WGL), *RKP* 108266 (BSID).

Blumea obliqua (L.) Druce, in Rep. Bot. Exch. Club. Brit. Isles 4: 609. 1916. var. **obliqua** Randeria, Blumea 10: 286. 1960. *Erigeron obliqum* L., Mant. 2: 573. 1771. *Blumea amplectens* DC. in Wight, Contrib. Bot. Ind. 13. 1834; Hook.f., Fl. Brit. India

3: 260. 1881; Gamble, Fl. Madras 2: 686. 1921; Kumar in Hajra & *al*., Fl. India 13: 137. 1995.

Erect herb, to 40 cm high, stem striated, covered with pilose hairs and colleters. Leaves simple, alternate, one-veined, pubescent on both surfaces, ovate-oblong, 2-8 × 0.5-2.5 cm, base semiamplexicaul to auriculate, apex acute. Heads terminal, solitary or two, pinkish, 5 mm, heterogamous, not rayed. Involucral bracts around 90, 3-seriate, equal to the florets, pubescent, tip acute. Receptacle 3 mm, slightly convex, areolate, glabrous. Female florets 350-400, 4.5 mm, glabrous, 3-lobed; style slightly exserted, branches linear, obtuse. Bisexual florets around 40, 5 mm, gradually expanding, glabrous on tube, 5-lobed, lobes ovate, triangular, sub-acute, multicellular hairs on lobes; stamens 5, anthers oblong, 2.5 mm, hood obtuse, base sagittately tailed; style 5.5 mm, slightly exserted, branches obtuse at apex, papillate on dorsal side. Pappus of uniseriate, setaceous, white hairs, 4-5 mm, equaling the florets. Achene oblong, 0.5 mm, hairy.

Usually near wet fields in Adilabad, Nizamabad, Warangal, Medak and Mahabubnagar districts

Fl. & Fr.: October – February.

Sone (ADB), *TP* & *PVP* 4062; Mallipet (NZB), *BR* & *CPR* 9237; Poosanpalle (MDK), *CP* 11958; Gangapur (MDK), *TP* & *CP* 11958; Edigonipally (MBNR), *MV* & *BR* 40704; Khammam hills (KMM), *RR* 113807; Gangaram (WGL), *RKP* 110949 (BSID).

Blumea oxyodonta DC. in Wight, Contrib. Bot. Ind. 15. 1834; Hook.f., Fl. Brit. India 3: 266. 1861; Gamble, Fl. Madras 2: 686. 1921; Kumar in Hajra & *al*., Fl. India 13: 139. 1995.

Diffuse herb, to 30 cm, stems many from base, spreading, procumbent or ascending, villous with white long glandular hairs. Leaves radical and clustered at the base, densely silky white and glandular hairs on both surfaces, basal leaves obovate, 5 ×× 1.5 cm, base attenuate; upper ones elliptic, base rounded, 1 × 0.5 cm, margin spinous dentate, apex acute. Heads terminal and axillary, to 6 cm, heterogamous, not rayed. Involucral bracts 36, 3-seriate, slightly exceeding the florets, completely reflexed at maturity, green or violet brown, villous on both surfaces, acute or acuminate. Receptacle 2 mm, flat, alveolate, glabrous. Female florets 80, filiform, 3.2 mm, tubular, glabrous on tube, 3-lobed, lobe sub-acute, papillate along the margins; style exserted, to 0.5 mm, branches linear, obtuse. Bisexual florets 20, 4 mm, tube and limb 3.2 mm, gradually expanding from base, glabrous, tube 5-lobed, lobes ovate, 0.5 mm, sub-acute, papillate along the margins on ventral side; stamens 5, anthers abortive or not developed, hood obtuse, base fused with connectives, auricles not distinct; style 3.2 mm, exserted, to 0.5 mm, papillate above the middle, branches linear, obtuse, minutely hirsute on dorsal side. Pappus of very deciduous, setaceous, white hairs, exceeding the florets. Achene oblong, 0.8-1 mm, pubescent.

Near streams in Adilabad, Warangal, Medak and Hyderabad districts.

Fl. & Fr.: February – June.

Pochera (ADB), *CPR* 13229; Erloo to Koilkonda (MBNR), *SRS* 104513 (BSID); Hamsanpally RF (MDK), *RG* 116428 (BSID).

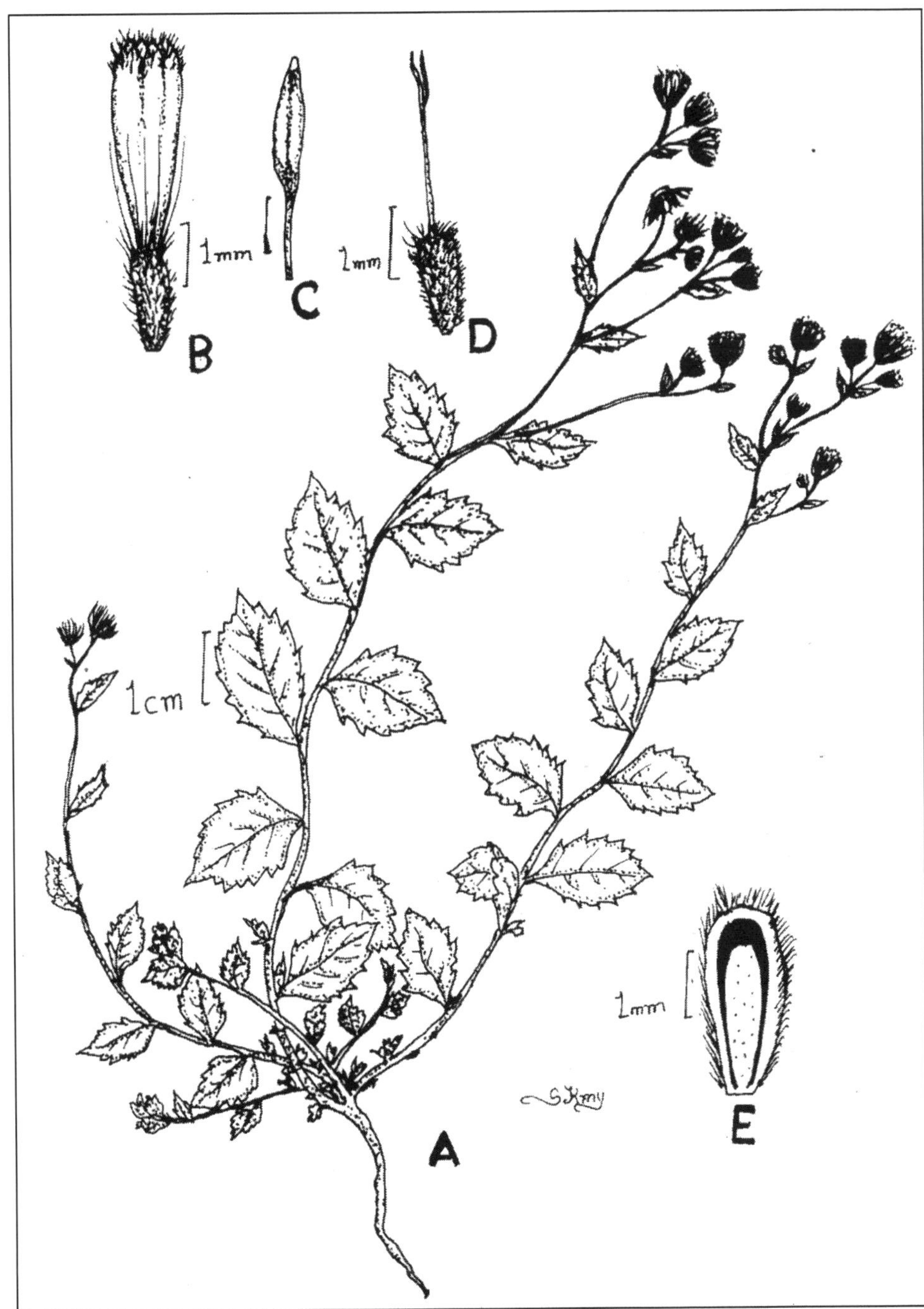

Figure 48: Blumea oxyodonta DC.

A. Twig, B. Floret, C. Pistil, D. Stamen, E. Ovary l.s.

Blumea virens Wall. ex DC. in Wight Contrib. Bot. Ind. 14. 1834; Hook.f., Fl. Brit. India 3: 264. 1881; Gamble, Fl. Madras 2: 687. 1921; Kumar in Hajra & *al*., Fl. India 13: 143. 1995.

Erect herb, simple or bushy, to 1.5 m high, stem striated, glabrous or sparsely pilose with white hairs especially in younger parts. Leaves simple, alternate, thin, lower ones petioled to 2 cm, uppermost sessile, one-veined, lower leaves lyrately lobed, upper ones ovate to oblanceolate, base attenuate, margin serrate, apex acute to apiculate. Heads axillary and terminal in lax panicles, 4 mm, yellow, heterogamous, not rayed. Involucral bracts around 40, 4-5-seriate, exceeding the florets, entirely reflexed at maturity. Receptacle 2.5 mm across, alveolate, glabrous. Female florets around 70, filiform, tubular, 4 mm, glabrous, 3-lobed; style inserted, 2.5 mm, branches linear, obtuse. Bisexual florets around 10, 4 mm, 5-lobed, lobes ovate, 0.5 mm, sub-acute, pubescent with unicellular hairs and colleters on dorsal side; stamens 5, anthers oblong-elliptic, 2 mm, hood ovate, obtuse, base sagittately tailed; style inserted, 3 mm, branches linear, obtuse, minutely hirsute on dorsal side. Pappus of uniseriate, setaceous, white hairs, equal to the florets. Achene 0.8 mm, oblong or elliptic-oblong, pale brown, ribbed, hairy.

Road sides and waste palces of the forests.

Fl. & Fr.: November – March.

Gundaram RF (NZB), *BR* & *KH* 9681; Rachakonda (NLG), *BVRS* 435 (NU); Pakhal (WGL), *ANH* 15928 (MH).

CAESULIA Roxb.

Caesulia axillaris Roxb., Pl. Coromandel 1: 64. t. 93. 1759; Hook. f., Fl. Brit. India 3: 291. 1881; Gamble, Fl. Madras 2: 702. 1921; Pant in Hajra & *al*., Fl. India 13: 2. 1995.

Erect herb, to 40 cm high, branches from base, stem terete, glabrous. Leaves simple, alternate, sessile, one-veined, glabrous, linear-lanceolate, 2-14 × 0.3-1 cm, base amplexicaul with obtuse auricles, margin coarsely serrate, apex acute to acuminate. Heads solitary, axillary, white, 0.1-1 × 1-1.5 cm, homogamous, not rayed, sessile. Involucral bracts 2, 2.5 cm, leafy, base auricular, scarious, margins serrate. Receptacle flat or slightly concave, 4 mm across, naked in the centre, paleaceous. Florets 130-140, corolla 3 mm, glabrous, tube 1.2 mm, limb 1.8 mm, deeply lobed towards base, 5-lobed, lobes ovate-lanceolate, obtuse; stamens 5, black, anthers linear-oblong, 2 mm, exserted, hood elongated, obtuse, base sagittately auricled; style 2 mm, branches broad, very short, obtuse. Pappus of biseriate scales, ovate, 1.5 mm, folded, uninerved, apiculate. Achene obovate, 3.5 mm, compressed, flat, black, broadly winged, notched at apex.

Common in marshy areas in all districts.

Fl. & Fr.: August – January.

Sone (ADB), *TP* & *PVP* 4018; Kadthala (ADB), *GO* & *DAM* 5011; Kammarpally (NZB), *TP* & *BR* 6237; Gandhari RF (NZB), *BR* & *CP* 7243; Poosanpalle (MDK), *TP* & *CP* 11952; Mohammadabad RF (RR), *MSM* 10493; Kodimial (KRN), *GVS* 21844 (MH);

Pakhal (WGL), *KMS* 11694 (MH); Parnasala RF (KMM), *RCS* 98768 (BSID); Chownapur (MDK), *RG* 106754 (BSID).

CHROMOLAENA DC.

Chromolaena odorata (L.) R.M. King & H. Rob., Phytologia 20: 204. 1970; Hook.f., Fl. Brit. India 3: 244. 1881. *Eupatorium odoratum* L., Syst. Nat. ed.10. 1205. 1759; Uniyal in Hajra & *al.*, Fl. India 12: 354. 1995.

Erect herb to subshrub, 1-2 m tall, stem terete, striated, glabrous in lower parts, sparsely to densely pubescent in upper parts, glandular when young, much branched. Leaves simple, opposite, petioled to 1.5 cm, pilose and glandular, suprabasally 3-veined, ovate, 3.5-9 × 1.4-5.5 cm, cuneate at base, sparsely serrate-dentate, teeth apiculate, acute to sub-acuminate. Heads in terminal and axillary corymbs, pink, 1-3 cm, bracteate, homogamous, not rayed. Involucral bracts around 28-32, 5-seriate, not exceed the florets, pubescent with pilose, margin scarious, apex obtuse, ciliate. Florets 26-32, 1.1 cm including style exertion, limb 5.5 mm, glabrous, 5-lobed, lobes ovate, 0.6 mm, papillate, sub-acute; stamens 5, anthers 2.5 mm, inserted, hood ovate, obtuse, base obtusely truncate; style 1.1 cm, exserted, branches 7 mm, flat, expanded, muricate along margins, obtuse. Pappus 38, scabrid, white, uniseriate hairs, not exceed the florets. Achene cuneate-oblanceolate, 4 mm, black, 5-ribbed, ribs scabrid.

Occasional in forest clearing areas in Khammam district.

Fl. & Fr. December – May.

Kinnerasani Wild Life Sanctuary (KMM), *RR* 112678 (BSID).

CHRYSANTHELLUM L. C. Rich.

Chrysanthellum americanum (L.) Vatke, Abh. Naturwiss. Vereine. Bremen 9: 122. 1885; Chowdhery in Hajra & *al.*, Fl. India 12: 379. 1995. *Artemisia americana* L., Sp. Pl. 895. 1753. *Chrysanthellum indicum* DC., Prodr. 5: 631. 1836; Hook.f., Fl. Brit. India 3: 310. 1881.

Erect or diffuse herb, 10-20 cm high, stem 4-ribbed, angled, sparsely hispid or glabrous, simple or dichotomously branched. Leaves bipinnately compound, alternate, petiolate to 2 cm, with narrow hyaline wing, uppermost sessile, whitish beneath, margin entire, apex sharply apiculate. Heads solitary, axillary or terminal, yellow, 3-4 mm, heterogamous, rayed. Involucral bracts 8, biseriate, not exceed the florets, brown, ovate-elliptic, 3.2 mm, glabrous. Receptacle 1.5-2 mm across, palaeaceous, glabrous. Ray florets around 9, 2.2 mm, tube 0.2 mm, cylindrical, pubescent with colleters, limb 0.4 mm, elliptic, 2-veined, 2-lobed at apex, lobes minute; style 1.5 mm, exserted, to 0.8 mm, branches short, to 0.4 mm, flat, obtuse. Bisexual florets around 18, 2 mm, pubescent with colleters towards base, 4-lobed, lobes 0.5 mm, triangular-ovate, thick along margins, glabrous or rarely papillate, obtuse; stamens 4, anthers 0.7 mm, connective free at base, filaments fused, hood ovate, obtuse or truncate, base broadly obtuse; style 2 mm, brown, exserted, to 0.5 mm, branches 1 mm, flat, broad and thick brown at base, papillate on dorsal side. Pappus absent. Achenes dimorphic; marginal ones oblong, 3 mm, whitish brown, 4-ribbed,

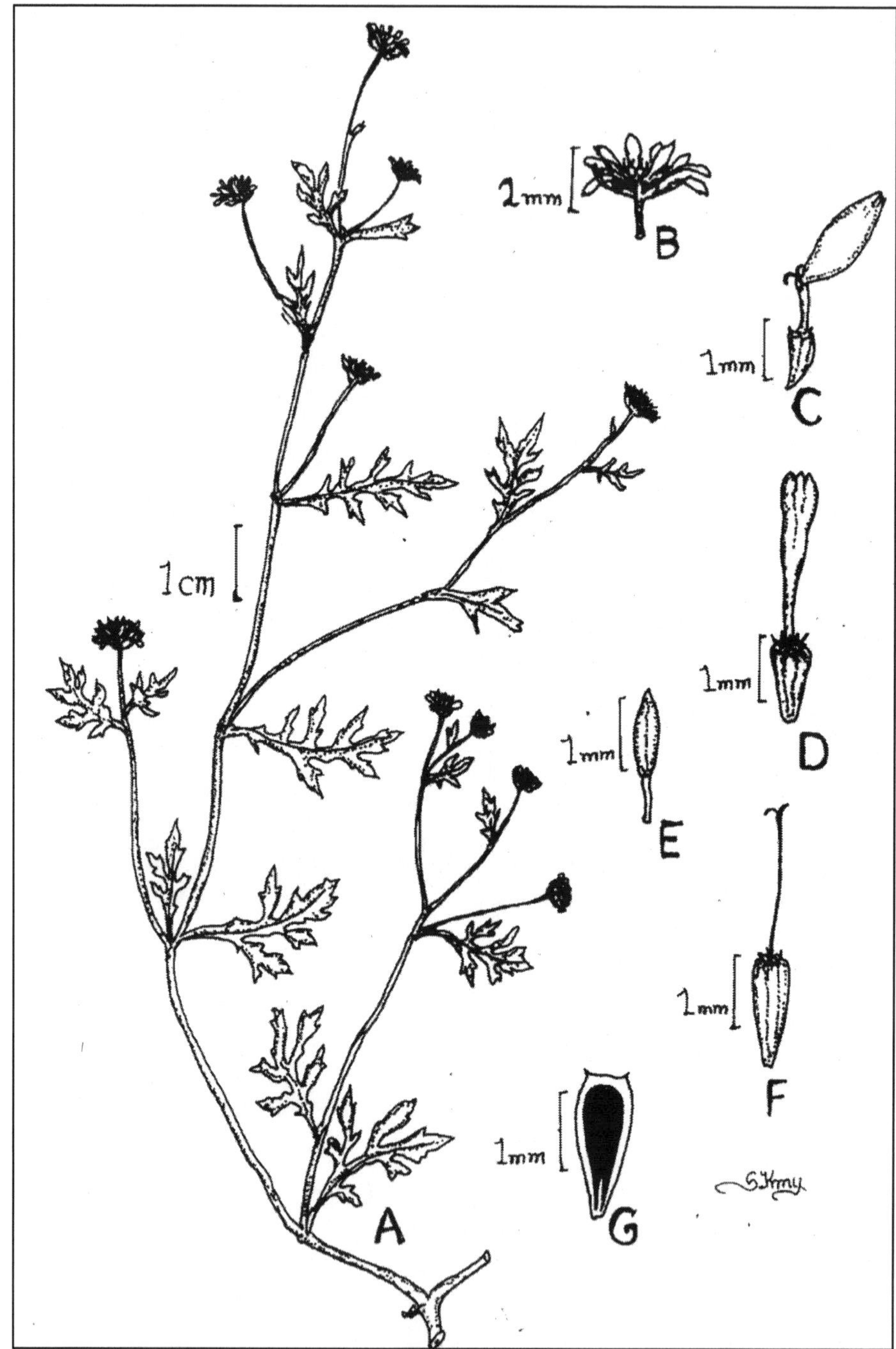

Figure 49: Chrysanthellum americanum (L.) Vatke.

A. Twig, B. Head, C. Ray floret, D. Disc floret, E. Stamen, F. Pistil, G. Ovary l.s.

middle ribs once striated, margins tubercled; inner ones elliptic, 3 mm, convex on one side and concave on another side, 2-ribbed, winged, wings yellow-white.

Rare in wet palces of the forests.

Fl & Fr.: September - December.

Gandhari (NZB), *BR* & *CPR* 7222; Pagarikutta – Narsapur (MDK), *KMS* 6632 (MH); Anantagiri RF (RR), *TP* & *MSM* 11083.

CONYZA Less.

Conyza japonica (Thunb.) Less. ex DC., Prodr. 5: 382. 1836; Hook.f., Fl. Brit. India 3: 258. 1881; Gamble, Fl. Madras 2: 682. 1921; Hajra in Hajra & *al.*, Fl. India 12: 105. 1995. *Erigeron japonicum* Thunb., Fl. Jap. 312. 1784.

Erect herbs, up to 40 cm high. Vegetative parts pubescent with white multicellular hairs and colleters. Leaves simple, alternate, rarely form a rosette at base, obovate or oblong – elliptic, 2-6 × 1- 2.5 cm, base attenuate, auricled in upper ones, margin serrately dentate, apex apiculate. Heads a few in terminal corymbs, yellow, heterogamous, not rayed. Involucral bracts lanceolate, scarious margined, pubescent. Outer florets female, fertile, 2-many-seriate. Disc florets bisexual, fertile, few, tubular, 5-toothed; anther bases obtuse, entire; style arms of female flattened, usually slender. Achenes elliptic or obovate, muricate.

Rare on open hill slopes at high elevations in Medak and Mahabubnagar districts.

Fl. & Fr.: November - February.

Medak (MDK), *RG* 104097, 104094 (BSID).

CYATHOCLINE Cass.

1. Leaves cauline only; anther base obtuse .... **C. purpurea**
1. Leaves both radical and cauline; anther base sagittate .... **C. manilaliana**

Cyathocline manilaliana Prabhakar Raju & Venkataraju, Rheedea 9: 151. 1999.

Erect, aromatic herb, up to 40 cm high, stem simple, often branched from base, ultimate branches in dichotomies, pubescent with hairs and glands, axillary buds densely covered by persistent silky hairs. Leaves simple, radical and cauline, cauline ones alternate, sessile, puberulous and glandular on both surfaces, coriaceous, upper most ovate, lower spathulate or oblong, 1-13 × 0.4-3 cm, lyrately or deeply lobed, lobes triangular, margin dentate, apex acute. Heads few, in compound corymbs, terminal, pink purple, heterogamous, not rayed, involucral bracts 2-seriate, subacuminate, margins not scarious, anthers 2-3.5 mm long, anther base deeply sagittate with sharp auricles. Achenes of female florets oblong or elliptic, occasionally narrowly rostrate at both ends, glabrous, of disc minute, around 0.2 mm, sterile.

Rare in Pochera fields in Adilabad district.

Fl.: January – March; Fr.: February – April.

Pochera (ADB), *RVR* & *CPR* 13230 (Holotype – MH).

Note: This species has not been listed in Flora of India (Hajra & *al.*, 1995).

Cyathocline purpurea (Buch.-Ham. ex D. Don) Kuntze, Revis. Gen. Pl. 333. 1891; Hajra in Hajra & *al.*, Fl. India 12: 111. 1995. *Tanacetum purpureum* Buch.-Ham. ex D. Don, Prodr. Fl. Nepal 181. 1925. *Cyathocline lyrata* Cass. in Ann. Sci. Nat. Ser.1. 17: 420. 1829; Hook.f., Fl. Brit. India 3: 246. 1881; Gamble, Fl. Madras 2: 679. 1921.

Erect, aromatic herb, 40 cm high, stem glabrous, pilose in younger parts. Leaves simple, alternate, sessile, chartaceous, one-veined, sub-glabrous to pilose on both surfaces, lyrately-deeply lobed, 2-14 × 0.6-2 cm, lobes ovate to triangular, base auricled, margin dentate, apex acute. Heads panicled, terminal, pink-purple, small, 2 mm, heterogamous, not rayed. Involucral bracts 20, biseriate, pilose, sessile glandular on dorsal side, linear, acuminate. Female florets around 150, tubular, filiform, corolla 1.2 mm, tube 0.8 mm, glands clustered at base, glabrous above, limb 0.4 mm, 3-lobed, lobes ovate; style 1 mm, branches short, sub-acute. Bisexual florets 8-15, corolla 2 mm, glandular, 5-lobed, lobes ovate, 0.5 mm, acute; stamens 5, anthers linear oblong, 1 mm, hood ovate, glandular on dorsal side, obtuse, base auricled; style 2 mm, papillate above the middle, branches linear, 0.8 mm, acute. Pappus absent. Achene of female florets oblong or elliptic, glabrous, of bisexual florets 0.2 mm, sterile.

In moist places of the forests of Adilabad and Nizamabad districts.

Fl. & Fr.: December – April.

Alampalli (ADB), *TP* & *PVP* 4164; Potchera (ADB), *GO* & *DAM* 5079, 5080, *CPR* 13231; Nagaram (NZB), *BR* 9708.

DICOMA Cass.

Dicoma tomentosa Cass., Bull.-Sci. Soc. Philom. Paris 1818; 12. 1818; Hook.f., Fl. Brit. India 3 : 387. 1881; Gamble, Fl. Madras 2: 728. 1921; Hajra in Hajra & *al.*, Fl. India 13: 171. 1995.

Erect or diffuse herb, up to 30 cm high, stem clothed with white tomentum. Leaves simple, alternate, sessile, sparsely tomentose on upper side, densely white cottony beneath, obovate or linear, 1-5 × 0.3-0.7 cm, base attenuate, serrately denticulate, apiculate at apex. Heads solitary, axillary, homogamous, not rayed. Involucral bracts 24-34, 4-5 seriate, linear-lanceolate, 1.2-1.8 cm long, ending in a spine point. Receptacle 3-4 mm, pitted, fimbriate. Florets 9-11, tube 2.5 mm, cylindrical, pubescent, deeply lobed, lobes 5, oblong-lanceolate, 3 mm, papillate; stamens 5, anthers oblong, 4 mm, hood elongate, subacute, base sagittately tailed; style 7 mm, cylindrical, branches fused except at tip, sparsely hairy. Pappus 3-seriate, united at base, equal to florets, scarious. Achene obovate, brown, 3 mm, 10-ribbed, densely clothed with silky hairs developed from the base.

Common in scrub forests in Medak and Nalgonda districts.

Fl. & Fr.: November – February.

Choutkoor (MDK), *TP* & *CP* 11942; Nagarjunasagar (NLG), *BVRS* 522 (NU)

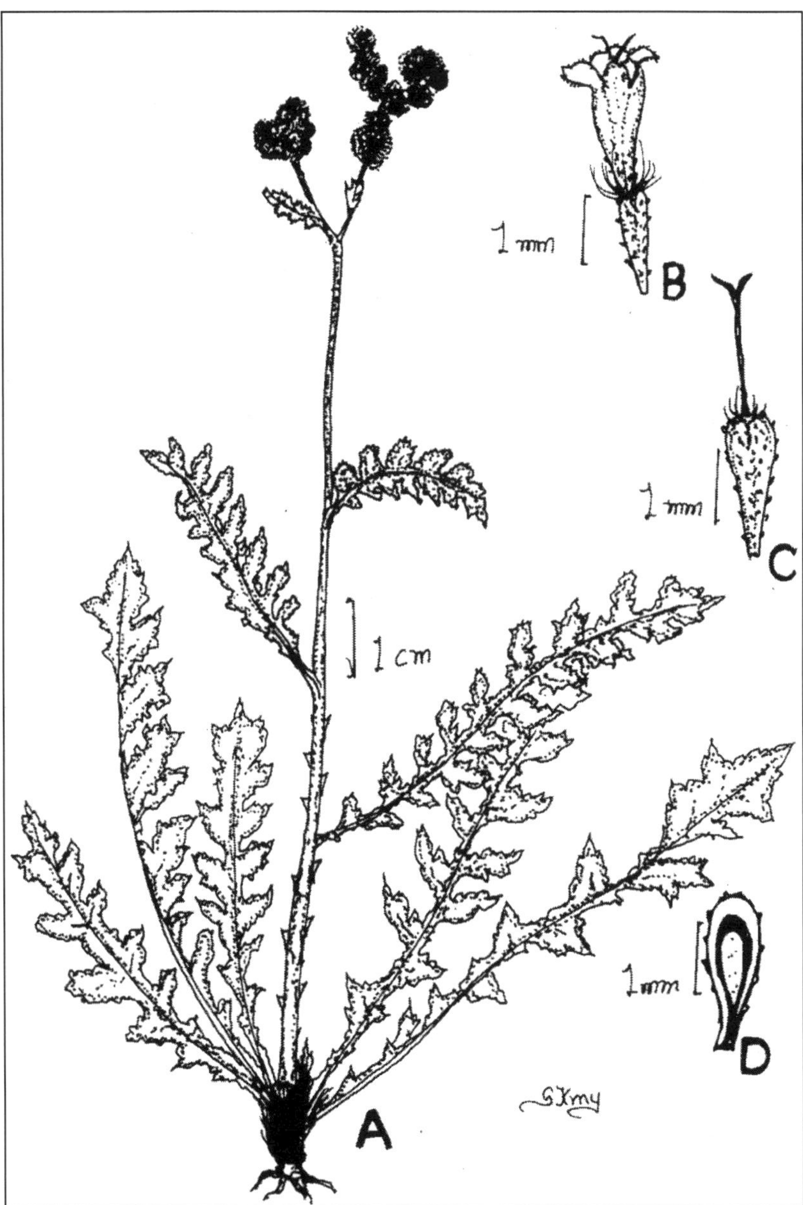

Figure 50: Cyathocline purpurea (Buch.-Ham. ex D.Don) Kuntze
A. Habit, B. Floret, C. Pistil, D. Ovary l.s.

ECHINOPS L.

Echinops echinatus Roxb., Fl. Ind. 3: 447. 1832; Hook.f., Fl. Brit. India 3: 358. 1880; Gamble, Fl. Madras 2: 724. 1921; Hajra in Hajra & *al.*, Fl. India 12: 178. 1995.

Erect herb, up to 50 cm high, stem striated, glabrous. Leaves simple, alternate, sessile, hispidulous above, densely white tomentose below, 3-13 × 1.5- 5 cm, lobed, lobes ovate, spinescent, base sub-amplexicaul. Heads many, collected into globose clusters, terminal on branches, white tomentose, homogamous, not rayed. Involucral bracts many, multiseriate, unequal, linear-lanceolate, glabrous, spinescent. Receptacle narrow, glabrous. Florets 12 mm long, tube 6 mm, glabrous, corolla lobes deeply divided, lanceolate, acute; stamens 5, anthers oblong, 4 mm, hood 1 mm, oblong, obtuse, base sagittately tailed; style 14 mm, branches 2.5 mm, flat, glabrous, obtuse. Pappus 1-1.5 mm, uniseriate, short scales, connate at base, persistent. Achene obovate, 5 mm, ribbed, densely silky.

In open forests at low altitudes in most districts.

Fl. & Fr.: August – December.

Sone (ADB), *TP* & *PVP* 4067; Gandhari (NZB), *TP* & *BR* 6351; Mombojipally (MDK), *BR* & *CP* 11452; Lankalakunta (RR), *MSM* 10269; Mannanur (MBNR), *TDCK* 13601; Amrabad (MBNR), *BSS* & *SKB* 32528; Azampur (NLG), *BVRS* 496; Kodimial (KRN), *GVS* 21828 (MH).

ECLIPTA L.

Eclipta prostrata (L.) L., Mant. Pl. 2: 286. 1771; Chowdhery in Hajra & *al.*, Fl. India 12: 381. 1995. *Verbesina prostrata* L., Sp. Pl. 902. 1753. *Eclipta alba* (L.) Hassk., Pl. Jav. Rar. 528. 1848; Hook.f., Fl. Brit. India 3: 304. 1880; Gamble, Fl. Madras 2: 705. 1921. *Verbesina alba* L., Sp.Pl. 902. 1753.

Erect herb, to 40 cm high, stem terete, striated, sparsely hirsute, branched. Leaves simple, opposite, sessile, scabrid on both surfaces, elliptic or elliptic-lanceolate, 3-6 × 1.5 cm, base obtuse, margin entire, apex acute. Heads 3-6, terminal or axillary, white, 5 x 8 mm, heterogamous, rayed. Involucral bracts 10-12, 2-3-seriate, herbaceous, margin ciliate, acute. Receptacle 3-4 mm across, convex, paleaceous. Ray florets around 60, 2-3 seriate, corolla 2-2.5 mm, tube 0.8 mm, glabrous, ligulate. Bisexual florets around 25, corolla 1.5 mm, tube 0.4 – 0.5 mm, glabrous, minutely hispid, 5-lobed, acute; stamens 5, black, anthers linear, 0.8 mm, hood absent or very short and obtuse, base auriculate; style 1.2 mm, branches short, papillate, acute. Pappus of 2 very short scales, connate. Achene obovate to oblong, 2 mm, triquetrous, tuberculate or wrinkled, margined, hairy at apex.

Moist areas of the low altitudes in most districts.

Fl. & Fr.: Throughout the year.

Jannaram (ADB), *TP* & *PVP* 4196; Mosra (NZB), *TP* & *BR* 6082; Sirnapally RF (MDK), *BR* & *GO* 9083; Patancheru (MDK), *CP* 11690; Kusumasamudram RF (RR), *MSM* & *NV* 10350; Osmangunta (NLG), *BVRS* 284; Kodimial (KRN), *GVS* 20083 (MH); On the way to Wyra from Khammam (KMM), *RR* 108047 (BSID); Khammam to Dhaniyaigudem (KMM), *RR* 108524 (BSID).

Note: *Verbesina prostrata, V. alba* and *V. pseudoacmella* were the three combinations made by Linnaeus (in his Species Plantarum 1753 on page 902) for three different taxa which in course of time have been worked out to be conspecific. The problem then was as to which of the three should be selected as the basionymn. Linnaeus himself in 1771 (*loc. cit.*) reduced the three into two species merging *V. alba* and *V. prostrata* into one – *Eclipta prostrata*. In the light of Article 57 of the Code the priority is to be given to the combination chosen by the author who fuses the two taxa and as such *Eclipta prostrata* (L.) L. is the valid name for this taxon and the combination *E. alba* made by Hassk. (1848 *loc. cit.*) on the base of *Verbesina alba* L. (1753) does not stand since Linnaeus himself in 1771 (*loc. cit.*) has preferred the other epithet (see Santapau in JBNHS 54: 475. 1957 & in BUBSI 3: 15. 1961).

ELEPHANTOPUS L.

Elephantopus scaber L., Sp. Pl. 814. 1753; Hook.f., Fl. Brit. India 3: 242. 1881; Gamble, Fl. Madras 2: 676. 1921; Uniyal in Hajra & *al.*, Fl. India 13: 333. 1995.

Erect, scapigerous herb, to 40 cm high, stem strigose, dichotomously branched above. Leaves radical and cauline, simple, sessile, often petioled to 3 mm, hirsute, elliptic- obovate, the upper leaves obovate to lanceolate, 3-25 × 1-6 cm, base cuneate-attenuate, margin crenulate-serrate, apex obtuse or acute. Heads homogamous, not rayed, pink, sessile, aggregated, forming a large flat-topped terminal inflorescence, subtended by 3 common leafy bracts. Involucral bracts 8, biseriate, scarious, acute. Florets 4 in each head, corolla 7.5 mm, glabrous, 5-lobed, lobes obtuse; stamens 5, anthers 1.5 mm, hood ovate, sub-acute, base sagittately auricled; style 7 mm, not exserted, branches 1 mm, broad, obtuse. Pappus 5, uniseriate, scabrid, 5.5 m, broad at base. Achenes oblong, 2.5 mm, truncate at apex, narrowed at base, nearly 10-ribbed.

Rare as undergrowth in forests in Warangal and Nizamabad districts.

Fl. & Fr.: October – January.

Mudheli RF (NZB), *BR* & *CPR* 7209.

EMILIA Cass.

Emilia sonchifolia (L.) DC. in Wight, Contrib. Bot. Ind. 24. 1834; Hook.f., Fl. Brit. India 3: 336. 1881; Gamble, Fl. Madras 2: 716. 1921; Mathur in Hajra & *al.*, Fl. India 13: 212. 1995. *Cacalia sonchifolia* L., Sp. Pl. 833. 1753.

Erect herb, 40 cm high, stem terete, pilose in lower parts, glabrous in upper parts. Leaves simple, lower leaves lyrately lobed, radical and cauline, 5-11 × 1.5 – 3.5 cm, cauline one alternate, petioled to 2.5 cm, sessile in upper leaves, glabrous to pilose. Heads in loose corymbs, terminal, pink, 6-8 mm, homogamous, not rayed. Involucral bracts 8, uniseriate, oblong, 8 mm, pilose, scarious, acute or papillate. Receptacle 2 mm across, flat to slightly concave, alveolate. Florets ca 45, corolla glabrous, tube 6 mm, 5-lobed, acute and papillate at apex; stamens 5, inserted, anther linear, 1 mm, hood obovate, acute, base obtuse; pappus 7 mm, copious, white silky, setose hairs, uniseriate, persistent; ovary oblong, 2 mm, ribbed and hairy.

In moist places of the forests in all districts.

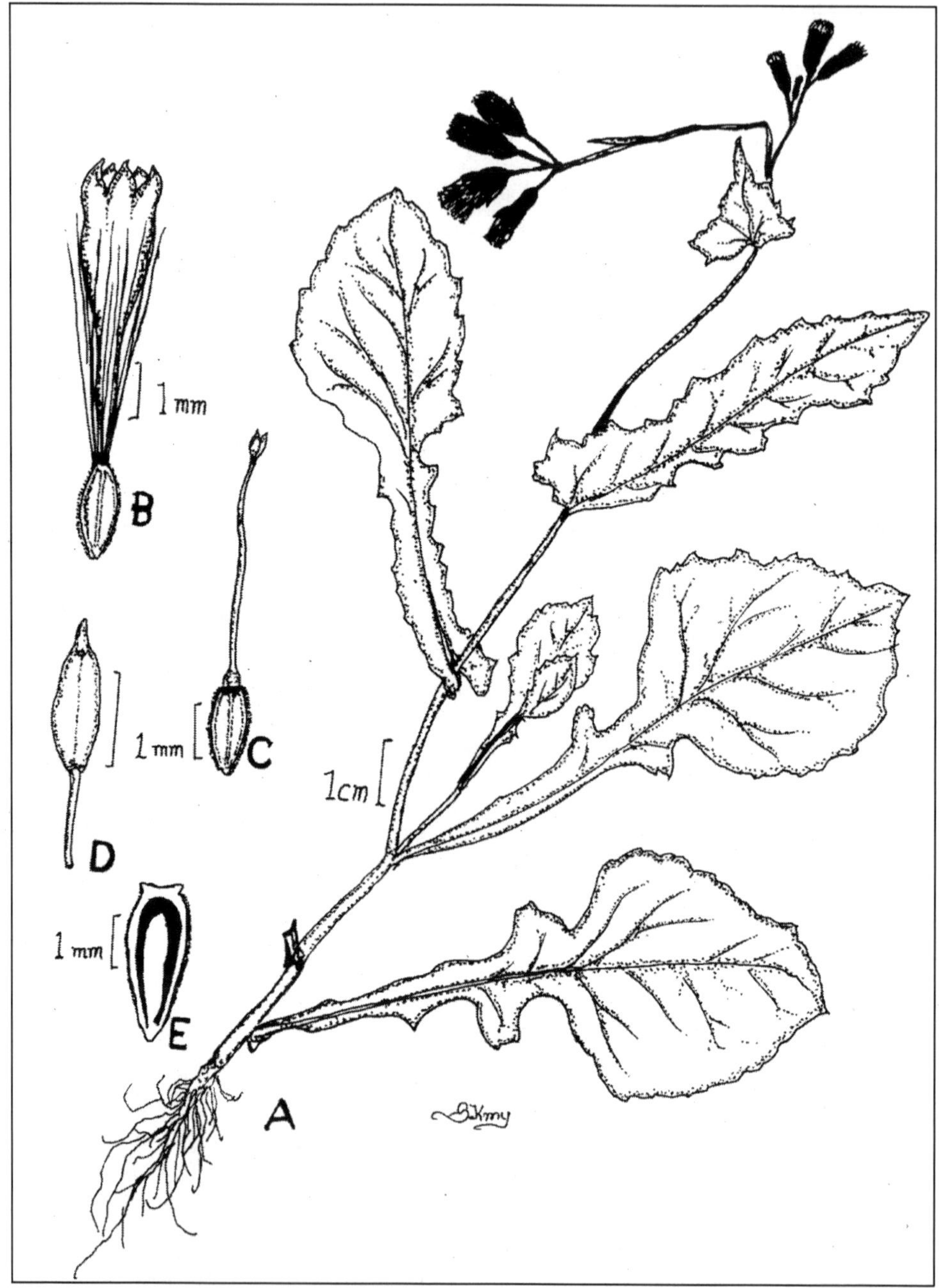

Figure 51: Emilia sonchifolia (L.) DC.

A. Habit, B. Floret, C. Pistil, D. Stamen, E. Ovary l.s.

Fl. & Fr.: July – March.

Kadam dam bank (ADB), *TP* & *PVP* 4040; Ibrahimpet RF (NZB), *TP* & *BR* 6128; Nagulabanda (MDK), *CP* 11356; Anantasagar RF (RR), *MSM* & *KH* 11050; Kodimial (KRN), *GVS* 21832 (MH); Ramavaram RF (KMM), *RR* 112627; Lawal beat, Pasra range (WGL), *RKP* 10818.

EPALTES Cass.

1. Heads 0.3-0.4 cm across; involucres hemispherical; receptacle naked; achenes ellipsoid .... **E.pygmaea**
1. Heads 0.7 cm across; involucres broadly campanulate; receptacle pitted; achenes oblong .... **E. divaricata**

Epaltes divaricata (L.) Cass., Bull. Sci. Soc. Philom. Paris 139. 1818; Hook.f., Fl. Brit. India 3: 274. 1881; Gamble, Fl. Madras 2: 691. 1921; Kumar in Hajra & *al.*, Fl. India 13: 147. 1995. *Ethulia divaricata* L., Pl. Mant. 1: 110. 1767.

Diffuse herb, up to 15 cm high, stem winged, glabrous, divaricately branched. Leaves simple, alternate, sessile, glabrous, linear-oblong to subspathulate, 1.5-4 × 0.2-0.4 cm, base decurrent on stem, margin distantly denticulate, apex acute or apiculate. Heads solitary, terminal, pink, 5 mm across, heterogamous, not rayed. Involucral bracts around 40, 4-seriate, unequal, lanceolate, acuminate. Female florets glabrous, 3-4-lobed, lobes short, obtuse; style 2 mm, exserted, branches linear. Bisexual florets glabrous, 5-lobed, lobes ovate-triangular, glandular on dorsal side, acute; stamens 5, anthers linear, hood truncate, base sagittately auricled; style 3.2 mm, exserted, spathulate at apex. Pappus absent. Achenes of female floret obovoid, brown, minutely tuberculate, striated or ribbed, of bisexual florets minute, sterile.

Near streams and wet places of the forests of Warangal, Karimnagar and Mahabubnagar districts.

Fl. & Fr.: November – April.

Pakhal (WGL), *KMS* 11674 (MH); Aklaspur (KRN), *GVS* 22498 (MH).

Epaltes pygmaea DC., Prodr. 5: 461. 1836; Hook.f., Fl. Brit. India 3: 274. 1881; Gamble, Fl. Madras 2: 691.1921; Kumar in Hajra & *al.*, Fl. India 13:147. 1995.

Annual herb, up to 12 cm high, stems minutely winged. Leaves linear – lanceolate, 1-2 × 0.2 – 0.5 cm, base cuneate, margin entire, apex obtuse. Flowers purple in terminal heads, violet, solitary or few, 0.3-0.4 cm across, peduncled. Involucral bracts multiseriate, broad, acute or obtuse, dry, rigid. Receptacle naked, flat. Ray florets multiseriate, female fertile; corolla filiform, 2-3-toothed, much shorter than style. Disc florets bisexual, tubular, sterile; corolla 3-5-fid.. Achenes minute, ellipsoid, smooth, brownca 0.4 mm long.

Reported from Central University campus in Ranga Reddi district (Seshagiri Rao, 2012).

Fl. & Fr.: April.

ERIGERON L.

Erigeron sublyratus DC. in Wight, Contrib. Bot. Ind. 9. 1834; Hajra in Hajra & *al.*, Fl. India 12: 125. 1995. *E. asteroides* Roxb., Fl. Ind. ed. 2. 3: 432. 1832; Hook. f., Fl. Brit. India 3: 254. 1881; Gamble, Fl. Madras 2: 681. 1921.

Erect herb, up to 50 cm high, stem striated, glandular white hirsute, unbranched. Leaves simple, alternate, sessile, glandular, sparsely to densely pubescent on upper surface, densely hirsute along veins beneath, obovate, 2-7 × 3-2 cm, distantly serrate, teeth apiculate. Heads solitary or two, heterogamous, rayed. Involucral bracts unequal, lanceolate, glandular on dorsal side, margin scarious, acuminate. Receptacle convex, fimbriate in centre. Florets trimorphic, outer female ligulate, middle tubular and inner ones bisexual. Pappus of uniseriate, setaceous hairs, form ring at base, not exceeding the florets. Achenes 4-ribbed in bisexual florets, obovoid, rarely ribbed in female florets, densely hairy with short wire like hairs, hairs forked at apex.

Rare in moist forest areas in Khammam and Mahabubnagar districts.

Fl. & Fr.: January – April.

Rollapadu (KMM), *RR* 108043.

FLAVERIA Juss.

Flaveria trinervia (Spreng.) C.Mohr. in Contrib. U.S. Nat. Herb. 6: 810. 1901; Chowdhery in Hajra & *al.*, Fl.India 12: 386. 1995. *Oedera trinervia* Spreng., Bot. Gart. Halle 63. 1800. *Flaveria australasica* Hook. in T.L.Mitch. J. Exped. Trop. Australia 117. 1848; Gamble, Fl. Madras 2: 711. 1921.

Annual erect, glabrous herbs, to 40 cm high. Leaves opposite, elliptic, 1.5-4.5 × 0.5-1.5 cm, base cuneate or attenuate, margin denticulate or distantly sharply serrulate, apex acute. Heads one-flowered, aggregated in cymes, subtended by floral leaves, solitary, terminal or in the forks of dichotomies, yellow, heterogamous. Involucre campanulate; involucral bracts 2-4, outer linear-lanceolate, 3-4 mm, inner oblong, 4-4.5 mm, folded, entire or minutely dentate at apex. Ray florets whitish yellow, short, up to 1.5 mm long, oblique or suborbicular, almost enclosed in the bracts. Disc florets 2-2.5 mm long; corolla campanulate, tube pubescent, lobes acute; anthers 1 mm long; style exserted, 2 mm long, stigma truncate. Achenes obovate, ribbed.

Occasional in waste places, especially in moist places in most districts.

Fl. & Fr.: September – March.

Indervelli (ADB), *GO* & *DAM* 5414; Uppalwai (NZB), *BR* 9567; Mohammadabad RF (RR), *MSM* & *NV* 10431; Kodimial (KRN), *GVS* 25670 (MH); Pillutla (MDK), *RG* 106786 (BSID); Erloo to Mahabubnagar (MBNR), *SRS* 104507.

GAILLARDIA Foug.

1. Annual herbs; corolla often red .... **G. pulchella**
1. Perennial herbs or under shrubs; corolla often yellow with or without spots .... **G. aristata**

Gaillardia aristata Pursh, Fl. Am. Sept. 2 : 573. 1814; Bailey, Man. Cult. Pl. 1014.1949.

Erect, hirsute herb, 30-75 cm high; stem much–branched, terete. Leaves variable, lanceolate, oblong or spathulate, 5-15 x 1-2 cm, entire or sinuate– pinnatifid, acute, gland –dotted beneath; petioles short or obscure. Heads 7-10 cm across, solitary on long peduncles. Involucral bracts 2-3-seriate, linear lanceolate, 0.5-1 cm long, green, acute, hispid. Marginal florets pure yellow or with yellow and red mixed colour. Central florets purple. Achenes oblong, densely villous all over. Pappus of 5-10, awned scales.

Cultivated in gardens as an hedge and bedding ornamental.

Fl. & Fr.: November-February.

Gaillardia pulchella Foug., Mem. Acad. Sci. Paris 1786 : 5. 1788; Bailey, Man. Cult. Pl. 1014. 1949; Chowdhery in Hajra & *al.*, Fl.India 12: 430. 1995.

Erect, annual softly pubescent herb, up to 30-40 cm high. Leaves lower pinnatifid, upper entire, oblong-lanceolate or spathulate, 5-10 cm long, pubescent. Heads 3-5 cm across, solitary on long peduncles; involucral bracts 2-3-seriate, linear lanceolate, 0.5-1 cm long, green, acute, hispid; marginal florets variously colored; central florets yellow. Achens oblong, densely hairy. Pappus of scale like bristles.

Cultivated in gardens as an hedge and bedding ornamental.

Fl. & Fr.: November-March.

GALINSOGA Ruiz & Pavon

Galinsoga parviflora Cav., Icon. & Descr. Pl. 3: 41. t. 281. 1796; Hook.f., Fl. Brit. India 3: 311. 1881; Gamble, Fl. Madras 2: 711. 1921; Chowdhery in Hajra & *al.*, Fl. India 12: 388. 1995.

Erect herb, up to 40 cm high, stem striated, glabrous. Leaves simple, opposite, petiolate, glabrous or sparsely puberulous on both the surfaces, 1.5-2.5 × 0.5 cm, base attenuate, margin entire to distantly serrate, apex acute. Heads solitary or 2-3, axillary or terminal, yellow, heterogamous, rayed. Involucral bracts 8, biseriate, margin scarious to ciliate. Receptacle conic, paleaceous. Ray florets 4-5, corolla 5 mm, tube 0.8 mm, densely hairy, 2 or 3-lobed. Bisexual florets 22-25, corolla 1 mm, tube 0.3-0.4 mm, densely hairy, 5-lobed; stamens 5, anthers linear, hood ovate, base auricled. Pappus in ray florets absent. Bisexual florets paleaceous scales, united to form a ring at base, ciliate along the margins, apex obtusely ciliate. Achenes dimorphic, of ray compressed, obovate, 1.5 mm, broad at apex, triquetrous, whitish black, pubescent towards apex, of disc obovate-cuneate, 1 mm, black, ribbed, hairy.

In moist places in Nizamabad district.

Fl. & Fr.: October – January.

Kamareddy (NZB), *BR* & *CPR* 7159.

GLOSSOCARDIA Cass.

1. Awns retrorsely barbellate; involucre hemispherical .... **G. bidens**
1. Awns glabrous; involucre funnel form .... **G. bosvallea**

Glossocardia bidens (Retz.) Veldkamp, Blumea 35: 468. 1991. *Zinnia bidens* Retz., Observ. 5: 28. 1788. *Glossogyne bidens* (Retz.) Alston in Trimen Handb. Fl. Ceylon 6: 168. 1931; Chowdhery in Hajra & *al.*, Fl. India 12: 393. 1995. *Glossogyne pinnatifida* DC. in Wight, Contrib. Bot. Ind. 19. 1934; Hook.f., Fl. Brit. India 3: 310. 1881; Gamble, Fl. Madras 2: 710. 1921.

Scapigerous herb, flowering stems 30-40 cm high, erect, ramose, almost leafless. Leaves radical, pinnatisect, petiolate, to 3 cm long, lobes-lanceolate, 0.5-3 × 0.1-0.3 cm, apex acute, cauline one scale like. Heads heterogamous, solitary, terminal, yellow, rayed. Involucral bracts 2-3-seriate, linear-oblong, obtuse or acute. Receptacle flat, paleaceous. Ray florets 12, 3-lobed, ligulate. Bisexual florets 4-5-lobed, lobes ovate, obtuse. Pappus of two slender retrorsely barbellate awns, 2-3 mm. Achenes linear-oblong, 6-7 mm, compressed, brown, 5-ribbed.

Rare in open forest areas in Mahabubnagar district.

Fl. & Fr.: October – May.

Glossocardia bosvallea (L.f.) DC. in Wight, Contrib. Bot. Ind. 19. 1834; Gamble, Fl. Madras 2: 709. 1921; Chowdhery in Hajra & *al.*, Fl. India 12: 391. 1995. *Verbesina bosvallea* L.f., Suppl. Pl. 379. 1781. *Glossocardia linearifolia* Cass., Dict. Sci. Nat. 19: 62. 1821; Hook. f., Fl. Brit. India 3: 308. 1881.

Diffuse or ascending herb, up to 20 cm high, stems many from the base, ribbed, glabrous to sparsely puberulous. Leaves bipinnate, in rosette at base and alternate on stems, sessile when young, petiolate in old ones, to 3 mm, broad at base, decurrent on stem, entire, apiculate. Heads solitary, terminal or axillary, leaf-opposed, yellow, heterogamous, rayed. Involucral bracts 5, biseriate, sparsely ciliate, acute. Receptacle narrow, flat, glabrous in centre, paleaceous towards margin. Ray florets only one in number, corolla 2-3-lobed, glabrous. Bisexual florets 5, corolla 3-4-lobed, glabrous, obtuse; stamens 3-4, anthers linear-oblong, hood short, obtuse to retuse. Pappus 2-3, black hairs. Achenes linear-elliptic, black, densely villous with yellow hairs.

In waste places of the forests.

Fl. & Fr.: October – December.

Wankidi RF (ADB), *TP* & *GO* 5483; Manchippa RF (NZB), *TP* & *BR* 6022; Choutkoor (MDK), *TP* & *CP* 11933; Anantagiri RF (RR), *TP* & *MSM* 11073; North Nagarjunakonda valley (NLG), *KTH* 9847 (CAL); Nagarjunasagar (NLG), *BVRS* 54 (NU); Koriguttalu (KMM), *RR* 106039 (BSID); Jakaram (WGL), *RKP* 105300 (BSID).

GNAPHALIUM L.

1. Heads in leafy spikes:
   2. Leaves in a basal rosette .... **G. coarctatum**
   2. Leaves not in a basal rosette
      3. Pappus hairs coherent at base to from a ring .... **G. pensylvanicum**
      3. Pappus hairs free .... **G. polycaulon**
1. Heads in leafless corymbose clusters .... **G. luteo-album**

Gnaphalium coarctatum Willd., Sp. Pl. 3: 1886. 1804; Pant in Hajra & *al.*, Fl. India 13: 85. 1995. *G. spicatum* Lam., Encycl. 2: 757. 1786 (non Mill. 1768).

An erect, annual herb, up to 20 cm high, stems simple or sparingly branched at base, decumbent, appressed greyish-white tomentose. Leaves in a basal rosette, 2-4 × 0.4-1 cm, smaller, distant above, spathulate, discolorous, glabrous or sparsely araneose on the upper surface, appressed greyish-white tomentose on lower surface. Capitula heterogamous, 2-3 mm, campanulate, numerous in axillary clusters in slightly interrupted spicate inflorescence. Phyllaries 3-seriate, outer bracts ovate-elliptic, 1.4-2 mm long, inner ones oblong-lanceolate, equaling the florets, tips golden brown. Florets 2-2.5 mm long, tips whitish with purplish tinge. Disc florets bisexual, 2 or 3, petals 5-toothed. Ray florets female, many, petals 3-4, toothed. Achenes 0.4-0.6 mm, oblong-elliptic, minutely glandular, pappus whitish, 2.3-2.6 mm long, bases coherent in smooth ring, deciduous.

Along river banks in Warangal district (C.S. Reddy and V.S. Raju, 1999)

Fl. & Fr.: March – August.

A South American species, now adventives in some parts of Old World.

Gnapahalium luteo-album L., Sp. Pl. 851. 1753 subsp. **luteo-album**; Pant in Hajra & *al.*, Fl. India 13: 89. 1995.

Erect herb, up to 30 cm high, stem densely woolly, unbranched. Leaves simple, sessile, radical and cauline, cauline ones alternate, densely white woolly, margin entire, apex acute. Heads in leafless corymbs, terminal, yellow, heterogamous, not rayed. Involucral bracts biseriate, broadly scarious, white woolly. Receptacle 2 mm across, glabrous, flat. Female florets around 80, 2 mm, tube glabrous, 3-lobed, glandular on dorsal side. Bisexual florets 4-8, corolla 2 mm, 5-lobed, lobes ovate, acute, glandular on dorsal side; stamens 5, anthers linear, 1 mm, hood ovate, acute, base sagittately tailed. Pappus uniseriate, slender, setaceous white, deciduous hairs. Achenes elliptic, 0.5 mm, dark orange coloured, 3-ribbed, 3-angled, papillate.

Open places in Medak district

Fl. & Fr.: August – November.

Rusthempet RF (MDK), *RG* 96250.

Gnaphalium pensylvanicum Willd., Enum. Hort. Berol. 867. 1809: Pant in Hajra & *al.*, Fl. India 13: 89. 1995. *G. purpureum* auct.non L.; Hook. f., Fl. Brit. India 3: 289. 1881.

Erect herb, up to 50 cm high. Leaves simple, alternate, spathulate, 1.5-8.5 × 0.5-1.5 cm sub-amplexicaul, sparsely woolly above, densely white – appressed below, flat, entire, obtusely apiculate. Heads 5-10 in axillary and terminal leafy clusters, golden yellow, densely white- woolly. Achenes oblong – elliptic, yellowish towards margins, brown, papillate.

Rare along sandy localities of Godavari river bed in Warangal district (C.S. Reddy, 2001).

Fl. & Fr.: November – January.

Mangapet (WGL), *CSR* 1420 (KU).

Gnaphalium polycaulon Pers., Syn. Pl. 2: 421. 1807; Kumar in Hajra & *al.*, Fl. India 13: 91. 1995. *G. multicauline* Willd., Sp. Pl. 3: 188. 1803. *G. pluricaule* Poiret in Lam., Encycl. Suppl. 2 : 807. 1812. *G. indicum* auct. pl. (non L. 1753) : Hook.f., Fl. Brit. India 3: 209. 1881; Gamble, Fl. Madras 2: 698. 1921.

Prostrate or decumbent herb, up to 20 cm high, stem appressed white woolly. Leaves simple, alternate, sessile, sparsely tomentose, densely woolly below, oblanceolate to spathulate, 2-6 × 0.3-1 cm, margin entire, apex acute. Heads in spikes, terminal or axillary, yellow-white, heterogamous, not rayed. Involucral bracts 16, biseriate, oblong-elliptic, glandular woolly. Receptacle flat, 1 mm across, glabrous. Female florets around 80, corolla 1.5 mm, tube glabrous, minutely lobed. Bisexual florets 5, corolla 1.5 mm, tube glabrous, 5-lobed; stamens 5, anthers linear, hood ovate, base sagittately tailed. Pappus 10, uniseriate, free, slender, white hairs, exceeding the florets. Achenes oblong, 0.3-0.6 mm, papillate.

In wet places of the forests in Adilabad, Nizamabad, Medak, Ranga Reddi, Khammam, Nalgonda and Mahabubnagar districts.

Fl. & Fr.: November- February.

Satnella (ADB), *GO* & *PVP* 4217; Laxmapur (NZB), *TP* & *BR* 6275; Nagaram (NZB), *BR* 9713; Narsapur tank (MDK), *TP* & *CP* 12004; Rangammagudem RF (RR), *MSM* & *KH* 10578; Bhongir fort (NLG), *BVRS* 554 (NU); Dimmagudem (KMM), *RCS* 99066; Appapur (MBNR), *BSS* & *SKB* 30589; Pullaechelma (MBNR), *SRS* 109713 (BSID); Farahabad to Vatavarlapalli (MBNR), *SRS* 104590A.

GONIOCAULON Cass.

Goniocaulon glabrum Cass., Bull. Sci. Philom. 1917. 34. 1817; Hook.f., Fl. Brit. India 3: 377. 1881; Gamble, Fl. Madras 2: 725. 1921; Hajra in Hajra & *al.*, Fl. India 12: 181. 1995.

Erect herb, up to 1 m high, stem fistular, angular, sparsely gland dotted. Leaves alternate, simple, sessile, gland dotted, obovate, 9 × 3.5 cm, base cuneate, clasped to stem, margin distantly serrate, apex obtuse or apiculate. Heads 4-5-flowered, in terminal, compound corymbs, pink, homogamous, not rayed. Involucral bracts 24, 6-

7-seriate, glabrous, acute. Receptacle obconic, 1 mm across, narrow, paleaceous. Florets usually 4, rarely 5-6, corolla 1.2 mm, 5-lobed, glabrous; stamens 5, anthers 6 mm, hood ovate, base deeply sagittate, filaments hairy. Pappus of scales, in different heights, oblanceolate to spathulate, formed into a ring at base, yellowish, obtuse at apex. Achenes cylindrical, oblong, 20-ribbed, glabrous, slightly constricted and toothed at apex, basal areola horizontal.

Rare in the low altitude forests in Adilabad, Nizamabad, Warangal, Karimnagar and Mahabubnagar districts.

Fl. & Fr.: December - May.

Itkial (ADB), *TP* & *PVP* 4130; Nirmal – Potchera (ADB), *CPR* 13225; Gandhari (NZB), *TP* & *BR* 6343; Aklaspur (KRN), *GVS* 22535 (MH).

GRANGEA Adans.

Grangea maderaspatana (L.) Poiret in Lam., Encycl. Suppl. 2: 825. 1812; Hook.f., Fl. Brit. India 3: 247. 1881; Gamble, Fl. Madras 2: 680. 1921; Hajra in Hajra & *al.*, Fl. India 12: 127. 1995. *Artemisia maderaspatana* L., Sp. Pl. ed. 2: 1190. 1763.

Vern.: *Nemalipadam.*

Prostrate herb, up to 30 cm high, striated, whitish pubescent and glandular. Leaves radical and cauline, cauline alternate, lyrate-pinnatifid, sessile, densely whitish pubescent, oblanceolate or oblong in outline, to 7 × 2.5 cm, apex apiculate. Heads solitary or two, terminal, yellow, heterogamous, not rayed. Involucral bracts 3-seriate, pubescent. Receptacle convex, fleshy, glabrous. Female florets marginal, 2-3-seriate, corolla 1.5 mm, tube 0.6 mm, 3-lobed, glandular. Bisexual florets around 70, corolla 1.5 mm, tube 0.5 mm, glandular, 5-lobed; stamens 5, anthers linear, hood ovate, acute, base obtuse. Pappus a laciniate cup. Achene obovate, 1.5 mm, green, ribbed, glandular and hairy, stalked.

In moist places and edges of the marshy areas. Fl. & Fr.: August – February.

Potchera water falls (ADB), *GO* & *DAM* 5083; Laxmapur (NZB), *TP* & *BR* 6278; Narsapur RF (MDK), *KMS* 7965 (MH); Pocharam Wild Life Sanctuary (MDK), *RG* 116420 (BSID); Venkatammacheruvu (RR), *MSM* & *KH* 10268; Amrabad (MBNR), *BSS* & *SKB* 32543; Dimmagudem (KMM), *RCS* 99065 (BSID); Eturnagaram (WGL), *RKP* 110831 (BSID); Nagarjunasagar (NLG), *BVRS* 129 (NU).

GUIZOTIA Cass. *nom. cons.*

Guizotia abyssinica (L.f.) Cass. in Dict. Sci. Nat. 59: 248. 1829; Hook.f., Fl. Brit. India 3: 308. 1881; Gamble, Fl. Madras 2: 708. 1921; Chowdhery in Hajra & *al.*, Fl. India 12: 395. 1995. *Polymia abyssinica* L.f., Sp. Pl. Suppl. 83. 1781.

Vern.: *Valiselu.* English: Nigerseed.

Erect herb, up to 1 m high, stem terete striated, sparsely puberulous. Leaves simple, opposite, sessile, glabrous, elliptic to elliptic-lanceolate, 5-7 × 1-1.5 cm, base sub-amplexicaul, margin distantly serrulate, apex subacuminate and gland tipped. Heads 3-5, subcorymbose, terminal, yellow, heterogamous, rayed. Involucral bracts 5, ovate-elliptic to elliptic, pubescent with pilose, margin ciliate, acute. Receptacle

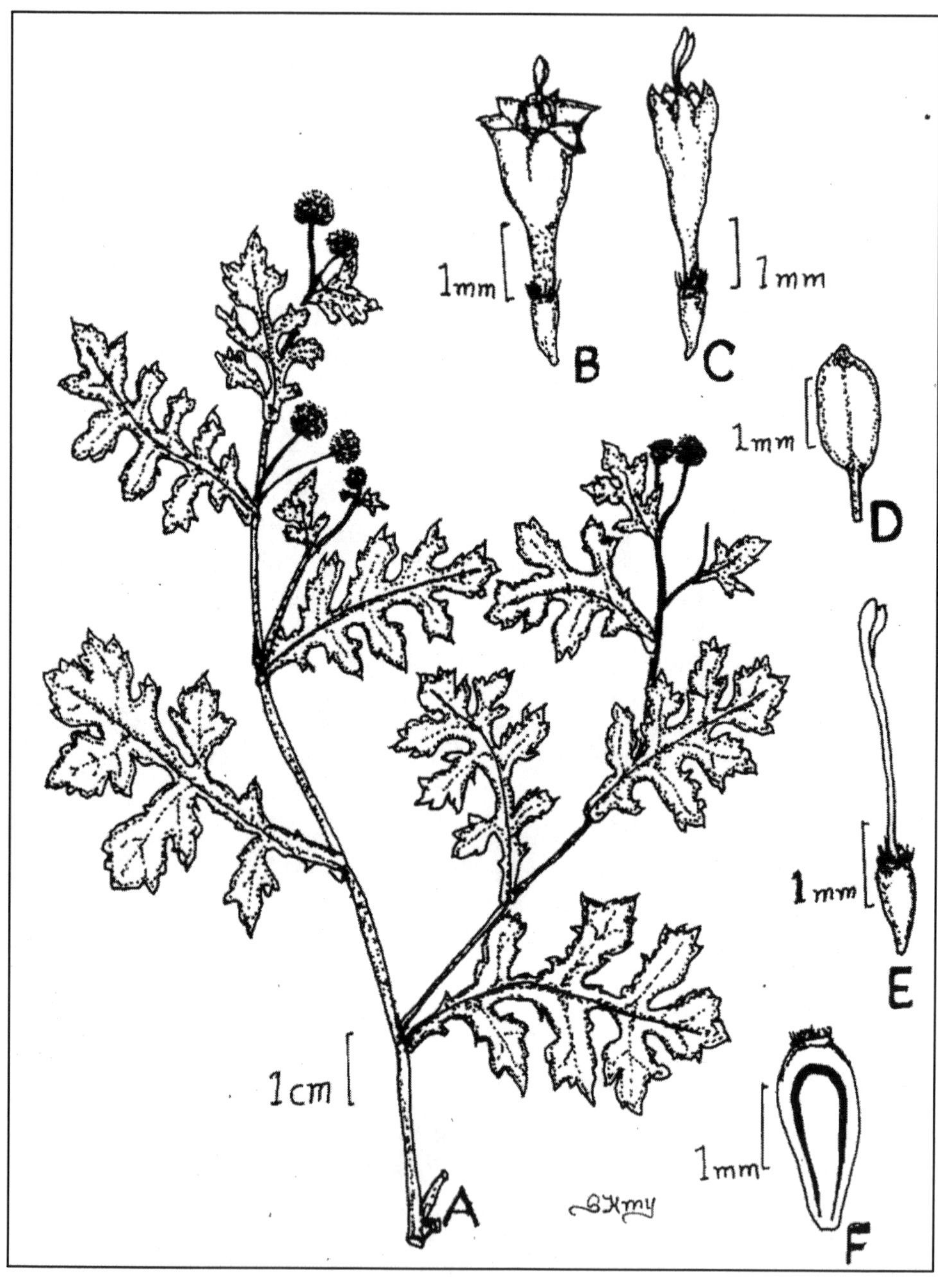

Figure 52: Grangea maderaspatana (L.) Poiret

A. Twig, B&C. Florets, D. Stamen, E. Pistil, F. Ovary l.s.

narrow, 2 mm across, concave, paleaceous. Ray florets 8, 1.7 cm, tube 2 mm, densely hairy, ligulate, 3-lobed. Bisexual florets around 30, corolla cylindric, 5.5 mm long, densely hairy, glabrous above, 5-lobed; stamens 5, black, anthers 2.5 mm, hood ovate-triangualr, base sagittate; ovary oblong to narrowly oblanceolate, 4 mm, triquetrous, glabrous, densely hairy at apex.

Rare in forests clearing areas.

Fl & Fr.: October – December.

Kadthala (ADB), *GO* & *DAM* 4998.

HELIANTHUS L.

Helianthus annuus L., Sp. Pl. 904. 1763; C.B. Clarke, Comp. Ind. 139. 1876.

Vern.: *Podduthirugudu chettu.* English: Sunflower.

Erect, simple or rarely sparingly branched, robust, annual undershrub, to 1.5 m high. Leaves broadly ovate to ovate-elliptic, 10-15 x 7-10 cm, thick–chartaceous, sparsely scabrous, subcordate to truncate at base, acute or shortly acuminate at apex, serrate; petiole to 8 cm long. Heads yellow, radiate, 8-25 cm across, solitary, nodding; phyllaries thick, broadly lanceolate, acuminate, minutely dentate. Achenes 4-angled, 4-5 mm long. Pappus scales 1.5-3 mm long.

Cultivated for its oil seeds.

Fl. & Fr.: March–July.

KLEINIA P. Miller

Kleinia grandiflora (Wall. ex DC.) N. Rani in Matthew, Fl. Tamilnadu Carnatic 2: 801. 1983; Mathur in Hajra & *al.*, Fl. India 13: 226. 1995. *Notonia grandiflora* Wall., ex DC., Arch. Bot. (Paris) 2: 518. 1833; Hook. f., Fl. Brit. India 3: 337. 1881; Gamble, Fl. Madras 2: 717. 1921.

Erect herb to sub-shrub, 1-1.5 m high, stem cylindrical, whitish green, glabrous, marked with prominent leaf scars. Leaves simple, alternate, crowded at terminal, sessile, fleshy, glabrous, spathulate or elliptic, 3-12 × 1.5-5.5 cm, base attenuate, margin entire, apex obtuse. Heads in corymbs, terminal, white, homogamous, not rayed. Involucral bracts 13, uniseriate, lanceolate, glabrous. Receptacle flat or slightly convex, glabrous. Florets around 70, corolla 2.5 mm, glabrous, tube 1.5 cm, 4-5-lobed, lobes ovate to ovate-elliptic, brownish dotted on dorsal side, margins papillate, obtuse; stamens 5, not exserted, anthers linear, hood lanceolate, base obtuse; pappus around 80, slender, white, uiseriate, setaceous hairs, deciduous; ovary oblong, 5 mm, brown, glabrous, ribbed, dilated at apex.

Occasionally cultivated as ornamental.

Fl. & Fr.: February – July.

Mannanaur RF (MBNR), *TDCK* 15368.

LACTUCA L.

Lactuca runcinata DC. in Wight, Contrib. Bot. Ind. 26. 1834; Hook.f., Fl. Brit. India 3: 403. 1881; Gamble, Fl. Madras 2: 731. 1921; Mamgain & R.R.Rao in Hajra & *al.*, Fl. India 12: 284. 1995.

Erect herb, up to 1 m high, stem terete, striated, glabrous. Leaves simple, chiefly radical and cauline, sessile, radical ones lyrately lobed, cauline ones lanceolate to linear, 2-30 × 0.4-13 cm, base clasped to stem, margin spinulosely ciliated, apex obtuse to acute. Heads solitary or fascicled, on racemose branches, homogamous, rayed. Involucral bracts around 20, biseriate, scarious margins, acute. Receptacle slightly concave, glabrous. Florets 24, corolla 12 mm, 5-lobed; stamens 5, anthers linear, 2 mm, hood ovate, obtuse, base sagittately auricled. Pappus 6 mm, copious, white, silky, persistent. Achenes 4 mm, obovate-cuneate, black, ribbed, tuberculated.

Rare on the road sides of the forests.

Fl. & Fr.: March – May.

Gaarlapadu RF (MBNR), *SRS* 109087 (BSID).

LAGASCEA Cav. *nom. cons.*

Lagascea mollis Cav. in Anal. Cienc. Nat. 6: 332. 1803; Hook.f., Fl. Brit. India 3; 302. 1881; Gamble, Fl. Madras 2: 703. 1921; Chowdhery in Hajra & *al.*, Fl. India 12: 397. 1995.

Erect herbs, up to 50 cm high, stem pale, slender, striate, pubescent. Leaves simple, opposite, petiolate, densely pubescent, ovate, 1-4 × 0.6-2.5 cm, base obtuse, margin coarsely serrate, apex acute or acuminate. Heads one-flowered, aggregated into terminal compound capitulum. Involucral bracts 5, connate at base, 5 mm, densely puberulous; corolla 5 mm, tube glabrous, 5-lobed, sparsely hairy; stamens 5, brown, anther 2 mm, hood ovate, base sagittate. Pappus absent. Achenes obovate, 2 mm, glabrous, apex hairy.

In wet places of the forests in all districts.

Fl. & Fr.: September – May.

Surzapur (ADB), *TP* & *PVP* 4112; Manchippa (NZB), *TP* & *BR* 6055; Choutkoor (MDK), *TP* & *CP* 11903; Rangammagudem (RR), *MSM* & *KH* 10313; Farahabad (MBNR), *TDCK* 15332; Kambalpally (NLG), *BVRS* 523 (NU); Gaarlapaadu RF (MBNR), *SRS* 109095; Pakhal (WGL), *KMS* 11571 (MH); Kazipet (WGL), *RKP* 105210 (BSID); Kodimial (KRN), *GVS* 20117 (MH).

LAGGERA C. H. Schultz. Bip. ex K.H.E. Koch

1. Stem winged .... **L. alata**
1. Stem not winged .... **L. aurita**

Laggera alata (D.Don) Sch.-Bip. ex Oliver, Trans. Linn. Soc. 29: 95. 1973; Hook. f., Fl. Brit. India 3: 271. 1881; Gamble, Fl. Madras 2: 689. 1921; Kumar in Hajra & *al.*, Fl. India 13: 148. 1995. *Erigeron alatum* D.Don, Prodr. Fl. Nepal 171. 1825.

Erect herb, up to 1 m high, stem striated, puberulous, winged, wings entire. Leaves simple, alternate, sessile, glandular and puberulous, elliptic to oblanceolate, 2.5-10 × 0.5- 3 cm, base decurrent on to stem, margin denticulate, apex acute. Heads solitary, terminal or axillary, racemosely arranged on flowering stem, purple, heterogamous, not rayed. Involucral bracts multiseriate, foliaceous, glandular hairy. Receptacle slightly concave, glabrous. Female florets around 300, corolla tubular, 4 mm, glabrous, 3-lobed. Bisexual florets 14-15, corolla 5 mm, glabrous, 5-lobed; stamens 5, not exserted, anthers linear, hood ovate, base sagittate. Pappus 10-18, slender, setose hairs, uniseriate, free, whitish. Achenes oblong, 1 mm, ribbed, hairy.

Occasional on hill slopes in Mahabubnagar district.

Fl. & Fr.: November – February.

Pullaechelma, Farahabad (MBNR), *SRS* 109159 (BSID).

Laggera aurita (L.f.) Benth. ex C.B. Clarke, Compos. Ind. 92. 1876; Hook.f., Fl. Brit. India 3: 271. 1881; Gamble, Fl. Madras 2: 689. 1921; Kumar in Hajra & *al.*, Fl. India 13: 150. 1995. *Conyza aurita* L.f., Suppl. 367. 1781.

Erect aromatic herb, up to 80 cm high, stem striated, solid, pubescent with villous hairs. Leaves simple, alternate, sessile, villous on both sides, elliptic-oblanceolate, 1.5-8 × 1-2.5 cm, base auriculate and decurrent with 2-3 linear lobes on each side, margin dentate, apex acute. Heads 3-8, in corymbs, terminal on shoot and axillary branches, pinkish tinged, heterogamous, not rayed. Involucral bracts multiseriate, unequal, lanceolate, pubescent, acuminate. Receptacle flat, slightly concave, glabrous. Female florets glabrous, 3-lobed. Bisexual florets cylindrical, 5-lobed, pubescent; stamens 5, anthers 2 mm, hood ovate, base sagittate. Pappus uniseriate, slender, setose hairs, deciduous. Achenes oblong, 1 mm, brown, strongly ribbed, hairy throughout.

In waste places of the forests in Adilabad, Nalgonda, Hyderabad, Karimnagar and Mahabubnagar districts.

Fl. & Fr.: September – February.

Sone (ADB), *TP* & *PVP* 4062; Nirmal (ADB), *CPR* 13236; Molachintapally (MBNR), *TDCK* 15307; Gaarlapaadu RF (MBNR), *SRS* 109094; Krishna river bank (NLG), *KMS* 9809 (MH); Aklaspur (KRN), *GVS* 22524 (MH).

LAUNAEA Cass.

1. Leaves with white cartilaginous teeth .... **L. procumbens**
1. Leaves without cartilaginous teeth .... **L. sarmentosa**

Launaea procumbens (Roxb.) Ramayya & Rajagopal, Kew Bull. 23: 465. 1969; Mamgain & R.R. Rao in Hajra & *al.*, Fl. India 12: 309. 1995. *Prenanthes procumbens* Roxb., Fl. Ind. 3: 404. 1824. *Launaea nudicaulis* auct. non. (L.) Hook.f. sensu stricto. Hook.f., Fl. Brit. India 3: 416. 1881; Gamble, Fl. Madras 2: 733. 1921.

Rhizomatous herb, much branched from base, stem terete, glabrous. Leaves in rosette at base and cauline, cauline ones alternate, sessile, glabrous, oblanceolate-oblong, lyrately lobed or pinnatifid, 3-12 × 1-3.5 cm, base obtuse or attenuate, margin cartilaginous denticulate, apex acute. Heads many, clustered at nodes, yellow, homogamous, rayed. Involucral bracts 18, triseriate, ovate-elliptic, glabrous. Receptacle glabrous, 1.5 mm across. Florets 20, corolla 13 mm, tube 5 mm, tubular, glabrous below, sparsely hairy above, ray oblong, 5-lobed; stamens 5, anthers linear, hood ovate, base sagittate. Achenes pale yellow, polymorphous, 2.5-4 mm long; outer slightly curved, truncate at base; inner columnar, very thickly ribbed, truncate at both ends, smooth or obtusely rugose, much shorter than pappus. Pappus copious, white, slender, silky, forming a ring at base.

In sandy areas and waste places of the forests in Adialabad, Nizamabad, Medak, Karimnagar and Mahabubnagar districts.

Fl. & Fr.: January – October.

Darigv (ADB), *GO* 4576 (MH); Kadam dam (ADB), *GO* & *BR* 9414; Aklaspur (KRN), *GVS* 22534 (MH); Rudroor (NZB), *BR* & *KH* 9637; Pochammaralu (MDK), *BR* & *CP* 11486; Gaarlapaadu RF (MBNR), *SRS* 109099.

Launaea sarmentosa (Willd.) Sch. Bipp. ex Kuntze, Revis. Gen. Pl. 1: 350. 1891; Mamgain & R.R. Rao in Hajra & *al.*, Fl. India 12: 310. 1995. *Prenanthes sarmentosa* Willd., Phytographia 10.t. 6.f.2. 1794 & Sp. Pl. 3: 1540. 1803. *Launaea pinnatifida* Cass. Ann. Sci. Nat. (Paris), 23: 85. 1831; Hook.f., Fl. Brit. India 3: 416. 1881; Gamble, Fl. Madras 2: 833. 1921.

Prostrate herb, to 40 cm long, rooting at every node. Leaves spathulate, 1.5-6.5 × 0.5-2 cm, base cuneate-attenuate, margin denticulate or lobed, apex obtusely apiculate or acute. Heads solitary at nodes, yellow, homogamous. Achenes oblong, obtusely 4-gonous.

Seshagiri Rao (2012) reported from Central University caampus in Rangaa Reddi district.

Fl. & Fr.: December – February.

OLIGOCHAETA (DC.) K.H.E. Koch

Oligochaeta ramosa (Roxb.) Wagenitz, Verroffent. Geobot. Inst. 37. 323. 1962. *Cardus ramosus* Roxb., Fl. Ind. 3: 407. 1832. *Volutarella divaricata* Benth. in Benth. & Hook. f., Gen. Pl. 2: 476. 1873; Hook. f., Fl. Brit. India 3: 383. 1881; Gamble, Fl. Madras 2: 727. 1921. *Amberboa ramosa* (Roxb.) Jeffrey, Scientist 3: 29. 1958; R.R.Rao in Hajra & *al.*, Fl. India 12: 153. 1995.

Erect herb, 30 cm high, dichotomously branched, branches striated, puberulous. Leaves simple or lyrate, sessile, powdery and gland dotted, obovate to spathulate, 2-10 × 0.4 -4 cm, base attenuate, margin distantly apiculate, apex acute. Heads solitary, terminal, spinescent, pink, homogamous, not rayed. Involucral bracts 5-6-seriate, ovate-elliptic, white tomentose on dorsal side, apex spinescent. Receptacle flat, paleaceous, glabrous. Florets around 25, corolla 15 mm, glabrous, tube cylindrical, 7 mm, 5-lobed; stamens 5, anthers 2 mm, hood elongate on dorsal side, filaments free. Pappus multiseriate, in different heights. Achenes 2-4 mm, angled, glabrous, basal areole oblique.

Occasional in waste lands in most districts.

Fl. & Fr.: December – January.

Potchera (ADB), *GO* & *DAM* 5050; Gowraram (NZB), *TP* & *BR* 6383; Choutkoor (MDK), *TP* & *CP* 11943; Kusumasamudram RF (RR), *MSM* & *NV* 10352; Rechapalli (KRN), *GVS* 22268 (MH); Eturnagaram (WGL), *RKP* 110829 (BSID); Perantapally (KMM), *RCS* 98994 (BSID).

PARTHENIUM L.

Parthenium hysterophorus L., Sp. Pl. 988. 1753; Chowdhery in Hajra & *al.*, Fl. India 12: 403. 1995.

Erect herb, up to 1.5 m high, stem striated or angular, puberulous to appressed pilose or scabrid. Leaves pinnatifid, alternate, when young rosette at base, whitish pilose, densely hirsute, obovate, spathulate or oblong in outline, 2-15 × 0.5-7 cm, base truncate, decurrent on petiole, lobes acute; petiole to 4 cm. Heads in panicles, terminal, white, 3 mm, heterogamous, rayed. Involucral bracts 5, biseriate, ovate-elliptic, puberulous, acute or obtuse. Receptacle convex, conic, paleaceous. Ray florets 5, corolla 1 mm, glabrous, 2-lobed. Bisexual florets 60, corolla 1 mm, glabrous, 4-lobed; stamens 4, anthers linear-oblong, 1 mm, inserted, hood ovate, base sagittate. Pappus of two scales in ray florets, absent in bisexual florets. Achenes of ray obovate, 2 mm, black, triquetrous, sparsely tuberculated, of bisexual florets oblanceolate, 1 mm, glabrous, sterile.

Common in waste places and roadsides in all districts.

Fl. & Fr.: Throughout the year.

Sone (ADB), *TP* & *PVP* 4035; Mosra (NZB), *TP* & *BR* 6070; Narsapur (MDK), *BR* 11219; Mohammadabad RF (RR), *MSM* & *KH* 10236; Azampur (NLG), *BVRS* 495 (NU); Kudurpalli (KRN), *SLK* 70756 (NBG); Erloo to Koilkonda road (MBNR), *SRS* 104512 (BSID); Mannanur (MBNR), *SRS* 109727 (BSID).

PENTANEMA Cass.

Pentanema indicum (L.) Ling, Acta Phyt. Sin. 10: 179. 1965; Kumar in Hajra & *al.*, Fl. India 13: 28. 1995. var. **indicum.** *Inula indica* L., Sp. Pl. ed. 2. 1937. *Vicoa auriculata* Cass., Ann. Sci. Nat. Ser. 1. 17: 418. 1829; Hook.f., Fl. Brit. India 3: 247. 1881. *V. indica* (L.) DC. in Wight, Contrib. Bot. Ind. 10. 1834; Gamble, Fl. Madras 2: 493. 1921.

Erect herb, up to 1 m high, stem striate, reddish, sparsely puberulous. Leaves simple, alternate, sessile, scabrid on upper surface, densely pilose to powdery tomentose on lower surface, lanceolate, linear, oblong-lanceolate or narrowly elliptic-lanceolate, 1.5-6 × 0.5-2 cm, base cordate, amplexicaul and auricled, margin entire, apex acute. Heads few, paniculate, terminal, yellow, heterogamous, rayed. Involucral bracts 3-seriate, unequal, puberulous, acute. Receptacle convex, fimbriate. Ray florets around 27, tube cylindrical, glabrous, ligulate, 2-lipped, yellowish glandular, 3-lobed. Bisexual florets around 150, corolla 3.8 mm, glabrous, 5-lobed, glandular on dorsal side; stamens 5, anthers linear, hood ovate, base sagittate. Pappus absent in ray

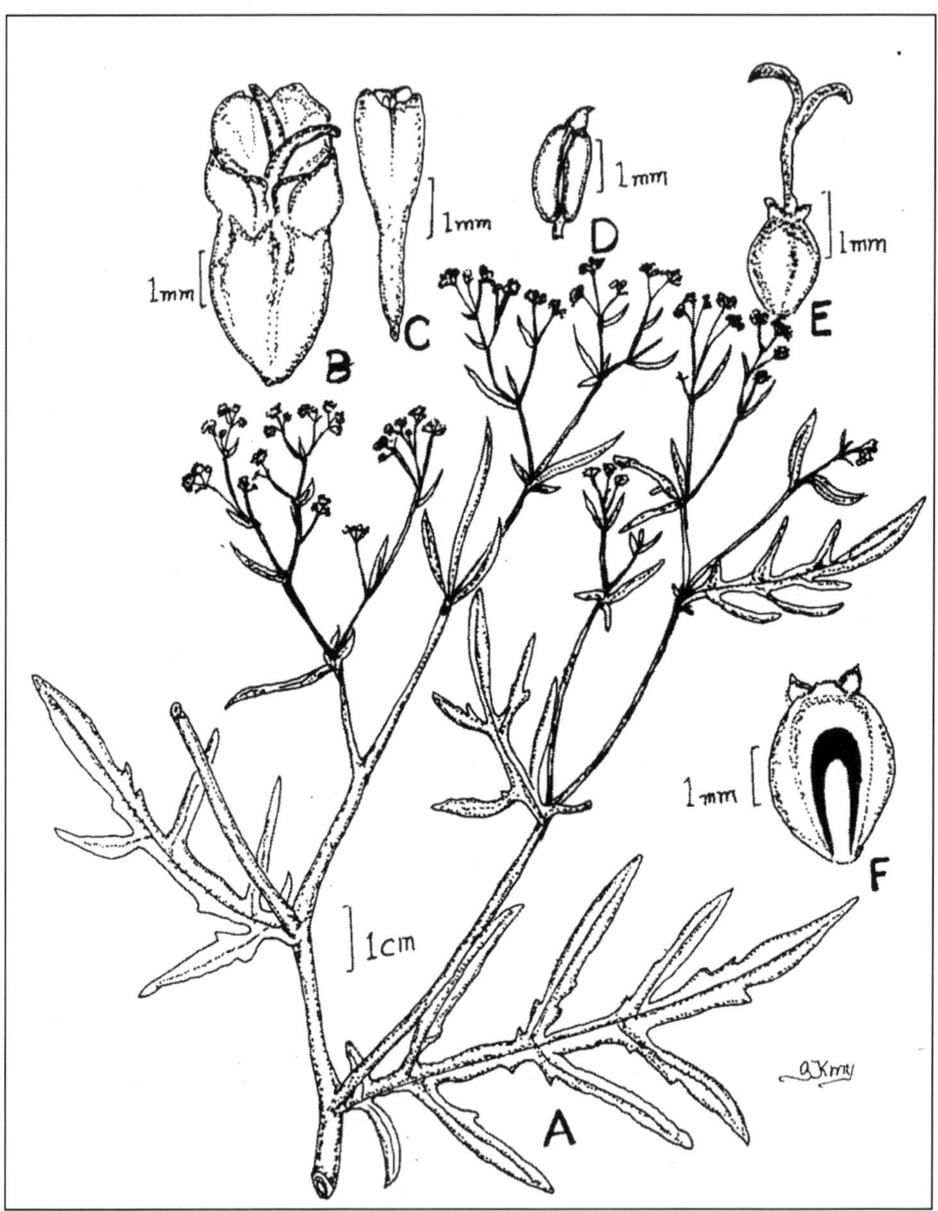

Figure 53: Parthenium hysterophorus L.

A. Twig, B. Ray floret, C. Stamen, E. Pistil, F. Ovary l.s.

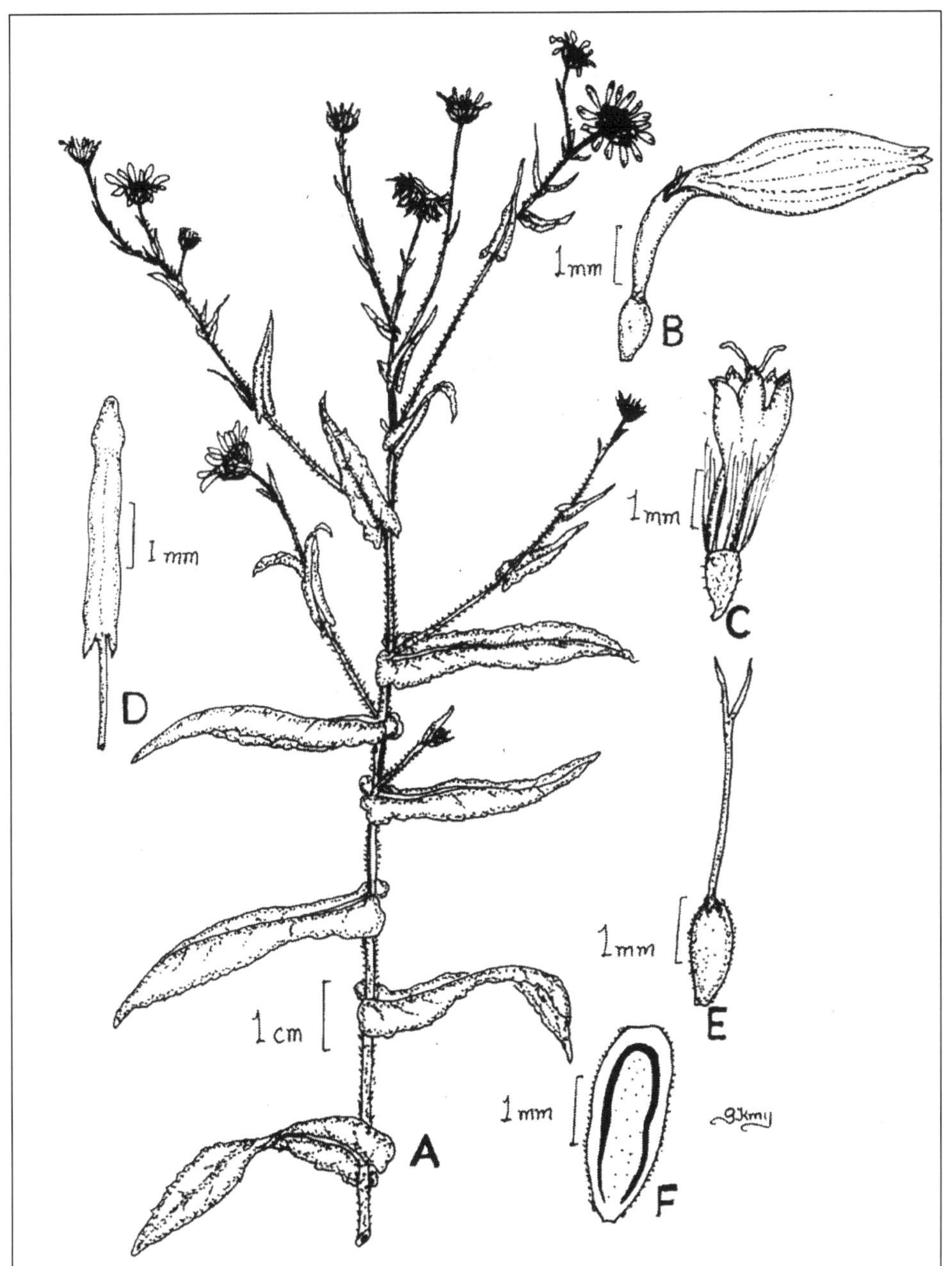

Figure 54: Pentanema indicum (L.) Ling

A. Twig, B. Ray floret, C. Disc floret, D. Stamen, E. Pistil, F. Ovary l.s.

florets, uniseriate in bisexual florets. Achenes to 1 mm, glabrous in ray florets, hairy and striated in bisexual florets.

In open places and edges of the forests in most districts.

Fl. & Fr.: September – December.

Sone (ADB), *TP* & *PVP* 4037; Vempalli RF (ADB), *GO* 4438; Nasarullabad RF (NZB), *TP* & *BR* 6157; Narsapur RF (MDK), *RG* 113470 (BSID); Pakhal (WGL), *KMS* 11532 (MH); Kodimial (KRN), *GVS* 21874 (MH); Sathiyampet to Penugadappa (KMM), *RR* 112647 (BSID); Nilgiri view poit (MBNR), *SRS* 109714 (BSID).

PLUCHEA Cass.

Pluchea tomentosa DC. in Wight, Contrib. Bot. Ind. 16. 1834; Hook.f., Fl. Brit. India 3: 272. 1881; Gamble, Fl. Madras 2: 689. 1921; Kumar in Hajra & *al*., Fl. India 13: 155. 1995.

Erect, woody sub-shrub, 1.2 m high, stems striated, puberulous. Leaves simple, alternate, sessile, whitish tomentose below, glandular, elliptic or ovate-elliptic, 1.5-3.5 × 1-2.5 cm, base auricled, margin serrate, apex acute to apiculate. Heads in compound corymbs, terminal and axillary, 6 mm, heterogamous, not rayed. Involucral bracts multiseriate, pinkish tinged, rigid, sparsely glandular, ovate, acute. Receptacle 2.5 mm across, slightly convex, fimbriate. Female florets filiform, tubular, glabrous, 3-lobed, exserted or inserted. Bisexual florets glabrous, 5-lobed; stamens 5, anthers 2.5 mm, exserted, hood ovate, base sagittate. Pappus hairy, uniseriate, dry white, equal to those of bisexual florets, longer than the female florets. Achenes oblong, ribbed, hairy.

Rare in the stream banks of the forests in Ranga Reddi district

Fl. & Fr.: October – January.

Lingampalle (RR), *MSM* 12743.

PULICARIA Gaertn.

1. Outer involucral bracts slender, linear-lanceolate .... **P. foliolosa**
1. Outer involucral bracts herbaceous, spathulate, obovate or linear-oblong:
    2. Herbs diffuse to erect, to 30 cm high, peduncle not thickened below the head .... **P. angustifolia**
    2. Herbs erect, 50-60 cm high, peduncle thickened below the head .... **P. wightiana**

Pulicaria angustifolia DC., Prodr. 5. 479. 1836; Hook.f., Fl. Brit. India 3: 299. 1881; Gamble, Fl. Madras 2: 702. 1921; Kumar in Hajra & *al*., Fl. India 13: 33. 1995.

Diffuse erect herb, 15-30 cm high, stems terete, puberulous. Leaves simple, alternate, sessile, puberulous on both surfaces, oblanceolate, subspathulate or oblong, 1-2.5 × 0.3-0.8 cm, base obtuse, margin serrate, apex acute. Heads solitary, terminal, yellow, heterogamous, rayed. Involucral bracts 3-4-seriate, equal in height, outer linear-oblong, foliaceous, glandular and hairy, acute or apiculate. Receptacle 4 mm across, alveolate, slightly convex. Female florets 16, tube 10 mm, pubescent, ligule oblong, 3-lobed. Bisexual florets 150, some of them sterile, tube reduced, anthers not developed. Fertile florets 6.5 mm, sparsely pubescent, 5-lobed; stamens 5, anthers linear-oblong, 2.5 mm, hood acute, base sagittate. Pappus biseriate, outer short fused scales formed into a cup, inner free, setaceous, deciduous hairs. Achenes oblong, ellipsoid, ribbed, hairy throughout.

Rarely in scrub forests in Karimnagar, Nalgonda and Mahabubnagar districts.

Fl. & Fr.: October – February.

Nagarjunasagar (NLG), *KMS* 9756 (CAL); Kodimial (KRN), *GVS* 21874 (MH); Thaskapally (MBNR), *SRS* 109074 (BSID).

Pulicaria foliolosa DC., Prodr. 5. 480. 1836; Hook.f., Fl. Brit. India 3: 298. 1881; Kumar in Hajra & *al*., Fl. India 13: 35. 1995.

Erect herb, up to 40 cm high, stems striated, pilose glandular; root stock woody. Leaves simple, alternate, sessile, pilose glandular on both surfaces, spathulate or oblong, 2-7.5 × 0.2-1.5 cm, base obtuse, sub-auricled, margin entire to obscurely serrate, apex apiculate. Heads solitary, terminal and axillary, yellow, heterogamous, rayed. Involucral bracts 3-seriate, equalling the florets, linear-lanceolate, glandular and hairy, acute or apiculate. Receptacle 5 mm across, pitted, slightly convex. Female florets 180, tube 3.1 mm, glabrous, ligule oblong, 3-lobed. Bisexual florets 260, 3.1 mm, glabrous, 5-lobed; stamens 5, anthers linear-oblong, 2.5 mm, hood narrowly ovate, base sagittate. Pappus biseriate, outer scales fused into a cup, inner free, setaceous, deciduous hairs. Achenes ellipsoid, brown, pubescent.

In moist places of the forests in Medak district.

Fl. & Fr.: October – January.

Pochammaralu (MDK), *BR* & *CP* 11485.

Pulicaria wightiana (DC.) C.B. Clarke, Compos. Ind. 178. 1876; Hook.f., Fl. Brit. India 3: 298. 1881; Gamble, Fl. Madras 2: 702. 1921; Kumar in Hajra & *al*., Fl. India 13: 40. 1995. *Poloa wightiana* DC. in Guill. Arch. Bot. 2: 515. 1833.

Erect herb, 50-60 cm high, stem striated, puberulous. Leaves alternate, simple, sessile, puberulous on both surfaces, oblong, 1-4 × 0.3-0.8 cm, base sub-auricled, margin serrate to entire, apex obtuse, or obtusely apiculate. Heads solitary, terminal, yellow, 2.5 cm across, heterogamous, rayed. Involucral bracts 3-seriate, subequal, outer spathulate or obovate, 7.5 mm, herbaceous, slightly exceed the inner, pilose and glandular on dorsal side, acute. Receptacle slightly convex. Female florets around 30, 12 mm, exceed the head to 5-6 mm, tube pubescent, ligule oblong, 3-lobed. Bisexual florets around 300, 4 mm, 5-lobed, each lobe 2-5-dentate, acute; stamens 5, anthers exserted, oblong, hood ovate, base sagittate. Pappus biseriate, outer short scales formed into a cup, inner free, white, setaceous hairs. Achenes oblong, 1-1.2 mm, striated, brown, truncate at apex, pubescent.

In waste places and open forests in most districts.

Fl. & Fr.: October – January.

Manchippa RF (NZB), *TP* & *BR* 6017; Pandivagu (MDK), *BR* & *CP* 11207; Thoniganla RF (MDK), *RG* 112484 (BSID); Mohammadabad (RR), *MSM* 10437; Umamaheswaram (MBNR), *GS* 18209; Irlapenta (MBNR), *BSS* & *SKB* 32378 (BSID); Nagarjunakonda valley (NLG), *KTH* 9740 (CAL); Kambalpally (NLG), *BVRS* 202 (NU),

SCLEROCARPUS Jacq.

Sclerocarpus africanus Jacq., Icon. Pl. Rar. 1 : 17, t. 176. 1781; Gamble, Fl. Madras 2: 706. 1921; Chowdhery in Hajra & *al.*, Fl. India 12: 405. 1995.

Erect herb, up to 80 cm high, stem terete, hirsute, divaricately branched. Leaves simple, alternate, petiole to 1 cm, sparsely hirsute on upper surface, densely sericious on lower surface, elliptic-ovate, 1.5-6 × 0.5-3 cm, base attenuate, margin distantly serrate, apex acute. Heads solitary, terminal, yellow, heterogamous, rayed. Involucral bracts 4-5, leafy, innermost subtending the ray florets, oblong-oblanceoalte, with foliaceous tip, hirsute, acute or apiculate. Receptacle narrow, convex, paleaceous. Female florets 3, corolla 5 mm, tube 2 mm, pubescent, ligule ovate, 3 mm, hirsute, 2-lobed, acute or obtuse. Bisexual florets 12, corolla 5 mm, tube 2 mm, hairy towards apex, 5-lobed, lobes lanceolate, hairy; stamens 5, black, anthers linear-oblong, hood narrow, obtuse, base sagittate. Pappus absent. Achenes of female florets elliptic or oblong-elliptic, 4 mm, sterile, colourless, hairy along the margins, striated, of bisexual florets obovate-cuneate, 6 mm, black, striated, glabrous, slightly curved and beaked at apex, acute.

Rare in the open forests and stream banks in Nizamabd, Karimnagar and Medak districts.

Fl. & Fr.: September – December.

Bhimnagar (NZB), *BR* 6220; Kondagutta (KRN), *GVS* 25686 (MH); way to Peddakutta - Narsapur (MDK), *KMS* 6760 (MH).

SENECIO L.

Senecio tenuifolius Burm. f., Fl. Ind. 181. t. 60. f.4. 1768; Hook.f., Fl. Brit. India 3: 345. 1881; Gamble, Fl. Madras 2: 722. 1921; Mathur in Hajra & *al.*, Fl. India 13: 278. 1995.

Erect herb, up to 30 cm high, stem striated, glabrous. Leaves pinnatified, upper ones deeply lobed to base, alternate, sessile, glabrous on both surfaces, 1-1.5 × 0.1-0.8 cm, lobes subacute to acute. Heads solitary or two, terminal, rarely axillary, yellow, 8 mm, heterogamous, rayed. Involucral bracts 12-13, uniseriate, elliptic-oblanceoalte, 4 mm, reflexed at maturity, acute. Receptacle 1.5 mm across, glabrous. Female florets 6, corolla 4 mm, tube short, hairy, elliptic, 3-lobed. Bisexual florets 20-32, corolla 2.5 mm, tube 1 mm, glabrous, rarely pubescent, 5-lobed; stamens 5, anthers 1.2 mm, hood ovate, obtuse, base sagittate. Pappus 1-2 mm, scabrid, uniseriate, yellowish free hairs. Achenes obovate, 2 mm, ribbed, hairy between ribs.

Occasional in forests in Ranga Reddi district.

Fl. & Fr.: October – December.

Kusumasamudram RF (RR), *MSM* & *NV* 10365.

SONCHUS L.

1. Leaf auricles obtuse:
   2. Plants not rhizomatous; achenes winged .... **S. asper**
   2. Plants rhizomatous; achenes not winged .... **S. wightianus**
1. Leaf auricles at least in upper ones acute .... **S. oleraceus**

Sonchus asper (L.) Hill., Herb. Brit. 1: 47. 1769; Hook.f., Fl. Brit. India 3: 414. 1881; Mamgain & R.R.Rao in Hajra & *al.*, Fl. India 12: 318. 1995; Venkata Raju & Prabhakar Raju, J. Econ. Taxon. Bot. 24: 98. 2000.

Erect herb, up to 1.5 m high, usually unbranhced, glabrous below, glandular above, fistular. Leaves cauline, sessile, lower ones spathulate, upper ones oblong or oblong-lanceolate, margin spiny, apex acute to acuminate. Heads 3-8 in umbels or cymes, terminal, bracteate, yellow, homogamous, rayed, peduncle to 6 cm, densely glandular, involucral bracts 4-seriate, florets exceeding invlucral bracts. Achene oblong to oblong-ellipsoid, flat, 3-ribbed on each face, glabrous, pappus heterogenous, outer simple, slender, copius, silky, forming into a ring at base, inner of 15-20 setaceous deciduous hairs.

Rare in rice fields in Adilabad district.

Fl. & Fr.: February – April.

Nirmal (ADB), *CPR* 13232.

Sonchus oleraceus L., Sp. Pl. 794. 1753; Hook.f., Fl. Brit. India 3 : 414. 1881; Gamble, Fl. Madras 2: 732. 1921; Mamgain & R.R.Rao in Hajra & *al.*, Fl. India 12: 321. 1995.

Erect herb, up to 50 cm high, stem fistular, striated, glabrous below, glandular when young. Leaves cauline and radical, sessile, cauline ones alternate, glabrous on both surfaces, upper ones lanceolate, lower radical ones lobed, lateral lobe lanceolate, terminal lobe hastate, 7-14 × 1.2 -7.5 cm, base amplexicaul, margin sharply denticulate, apex acuminate. Heads cymose in upper axils or cymosely umbellate and terminal, yellow, whitish tomentose at base when young, homogamous, rayed. Involucral bracts 28, 3-4-seriate, minutely setose hairs, scarious along the margins, apex obtuse. Receptacle convex, 7 mm across, glabrous. Florets pubescent, 5-lobed, lobes obtuse; stamens 5, exserted, hood blackish, obtuse, base sagittate. Pappus heterogenous, white, silky, slender hairs. Achenes towards the periphery or sterile, towards centre fertile, obovoid, 3 mm, flat, 4-ribbed, muricate on ribs.

Rare on the road sides of the forests of Adilabad, Nizamabad, Medak, Ranga Reddi, Hyderabad and Mahabubnagar districts.

Fl. & Fr.: October – February.

Kadthala (ADB), *GO* & *DAM* 5021; Laxmapur (NZB), *TP* & *BR* 6291; Devampally (NZB), *BR* & *KH* 9633; Mombojipally (MDK), *BR* & *CP* 11631; Rangammagudem (RR), *MSM* & *KH* 10314; Erloo to Mahabubnagar (MBNR), *SRS* 104510 (BSID).

Sonchus wightianus DC., Prodr. 7: 187. 1838; Hook.f., Fl. Brit. India 3: 414. 1881; Gamble, Fl. Madras 2: 732. 1921; Mamgain & R.R.Rao in Hajra & *al.*, Fl. India 12: 321. 1995.

Rhizomatous erect herb, up to 80 cm high, usually unbranched, stem fistular, green, striated; glabrous below, pubescent with stalked glands in young parts. Leaves rosette at base when young, alternate at maturity, sessile, glabrous, oblong, spathulate or lanceolate in outline, 6 – 34 × 1 – 7.5 cm, lyrate or lobed, base auricled, margin denticulately spinulose. Heads few, homogamous, cymosely paniculate, terminal, deep yellow-orange, peduncle to 5 cm, involucral bracts 40, 4-seriate. Achenes brown, fertile or sterile, 4 mm slightly narrowed towards apex, truncate at both ends, ribbed, muricate, pappus heterogenous, 8 mm, outer slender, silky, copious, formed into a ring at base.

Very rare along the roadsides in Medak district.

Fl. & Fr.: November – December.

SPHAERANTHUS L.

Sphaeranthus indicus L., Sp. Pl. 927. 1753; Hook.f., Fl. Brit. India 3: 275. 1881; Gamble, Fl. Madras 2: 692. 1921; Kumar in Hajra & *al.*, Fl. India 13: 160. 1995.

Diffuse or suberect, aromatic herb, up to 30 cm high, stems glandular, spreading, whitish pilose, winged, wings denticulate. Leaves simple, alternate, sessile, glandular and whitish pilose, obovate or spathulate, 1-2.5 × 0.5-1 cm, base narrowed, subamplexicaul, margin sharply serrate-dentate, apex spinulosely acute or mucronate. Heads many, sessile, aggregated to form a compound head; solitary, terminal, pink-purple, spherical, glandular-pubescent. Common receptacle 4 mm, globose. Involucral bracts 11, lanceolate, glabrous, ciliate along the margins, acute. Female florets 9-10, corolla 2 mm, tubular, sparsely glandular, 3-lobed. Bisexual florets 2, 3 mm, glandular at base, globose, 5-lobed; stamens 5, anthers 1.5 mm, hood ovate, base sagittate. Pappus absent. Achenes of female florets elliptic-oblong, 1 mm, hairy, often 3-angled. Achenes of bisexual florets oblong, 1 mm, cylindrical, glabrous, rarely glandular.

Common in water stagnant areas of the forests, waste lands and road sides.

Fl. & Fr.: November – April.

Sone (ADB), *TP* & *PVP* 4060; Yellareddy (NZB), *TP* & *BR* 6299; Medak (MDK), *BR* & *CP* 11418; Mohammadabad RF (RR), *MSM* & *KH* 10205; Mannanur (MBNR), *TDCK* 13610; Pakhal (WGL), *KMS* 11592 (MH); Eturnagaram (WGL), *RKP* 110830 (BSID); Kodimial (KRN), *GVS* 21835 (MH); Koida (KMM), *PVS* 84043 (BSID); near Mysamma temple (MBNR), *SRS* 104471 (BSID); Nelikal (NLG), *BVRS* 460 (NU).

SPILANTHES Jacq.

1. Heads rayed; achenes without pappus:
    2. Involucral bracts uniseriate .... **S. uliginosa**
    2. Involucral bracts biseriate .... **S. calva**
1. Heads discoid:
    3. Heads white .... **S. radicans**
    3. Heads yellow .... **S. paniculata**

Spilanthes calva DC. in Wight, Contrib. Bot. Ind. 19. 1834; Chowdhery in Hajra & *al.*, Fl. India 12: 409. 1995. *Acmella repens* (Watt.) L.C.Rich. in Pers. Syn. Pl. 2: 473. 1807; Venkata Raju & Prabhakar Raju, Rheedea 9: 41. 1999; Pullaiah & Karuppusamy, Fl. Andhra Pradesh 5: 29. 2008.

Diffuse pubescent herb with soft hairs on lower parts, rooting at lower nodes. Leaves simple, opposite, lanceolate, 3.5 – 4.5 × 1.2-2 cm, pilose along veins, base obtuse, margin distantly serrate, rarely entire, apex acute. Heads solitary, terminal, appears axillary due to further growth of axillary branches, heterogamous, rayed. Involucral bracts biseriate, elliptic-lanceolate, 5 mm long, pubescent. Florets bisexual; corolla campanulate, tubular, tube 1.5 mm long, lobes 5, ovate, reflexed; stamens 5, subexserted, anther-base sagittate. Achenes dimorphic, compressed, epappose.

Rare in marshy areas in Warangal district.

Fl. & Fr.: October – January.

Mahabubabad (WGL), *RKP* 111507 (BSID).

Spilanthes paniculata Wall., ex DC., Prodr. 5: 624. 1836; Chowdhery in Hajra & *al.*, Fl. India 12: 410. 1995.

Erect or diffuse herb, up to 2 cm high, stem striated, sparsely pilose. Leaves simple, opposite, petiolate to 6 mm, sparsely pilose on both surfaces, ovate, 1.5 × 1 cm, base cuneate, margin distantly serrate, apex acute. Heads 2 or 3, terminal or axillary, yellow, homogamous, not rayed. Involucral bracts 10, biseriate, ovate, hairy, margin scarious, acute. Florets 80, 2 mm, glabrous, 4-5-lobed, lobes ovate, subacute; stamens 4-5, anthers oblong, hood ovate, base sagittately tiled. Pappus of two unequal bristles. Achenes dimorphic, marginal trigonous, inner ones laterally compressed, flat, black, hairy throughout.

Wet places in the forests.

Fl. & Fr.; November – January.

Potchera water falls (ADB), *GO* & *DAM* 5088.

Spilanthes radicans Jacq., Collect. Bot. Chem. Hist. Nat. 11(3): 1714. 1804; Chowdhery in Hajra & *al.*, Fl. India 12: 412. 1995.

Erect to decumbent herb, to 1 m long, rooting at lower nodes, stems striated, fistular, glabrous in lower parts, sparsely pilose when young, branches trichotomous, rarely dichotomous.

Leaves simple, opposite, sparsely pilose or glabrous on both surfaces, ovate to ovate-lanceolate, 2-5 × 0.8 – 2 cm, base obtusely cuneate-attenuate, margin entire or distantly serrate and ciliate, apex acute. Heads many, terminal and axillary, white, homogamous, not rayed, involucral bracts 6-8, 2-seriate. Achenes dimorphic, marginal ones trigonous, inner ones laterally compressed, biconvex, black, obovate, hairy towards apex, pappus of two subequal bristles

Rare weed of moist fields in Warangal district (C.S.Reddy, 2001).

Fl. & Fr.: October – January.

Hanamkonda (WGL), *CSR* 1273 (KU).

Spilanthes uliginosa Sw., Nov. Sen. Pl. Scu. Prod. Descr. Veg. Ind. Occ. 110. 1788; Chowdhery in Hajra & *al*., Fl. India 12: 412. 1995. *Acmella uliginosa* (Sw.) Cass., Dict. Sci. Nat. ed. 2.24: 331.1822.

Erect herb, 25 – 40 cm high, stem striated, fistular, sparsely pilose on internodes and densely so at nodes, usually unbranched. Leaves simple, opposite, petiolate, winged with attenuate leaf base, margin dentate or serrate, apex obtuse. Heads solitary or few, terminal or in upper axils, yellow, heterogamous, rayed, involucral bracts 5, uniseriate. Achenes dimorphic, marginal ones triquetrous, inner ones flat to biconvex, oblong to oblong – cuneate, black, glabrous, pappus absent.

Rare near streams in wet places in Medak district.

Fl. & Fr.: October – January.

Gummadidala (MDK), *BR* & *CP* 14129.

SYNEDRELLA Gaertn.

Synedrella nodiflora (L.) Gaertn., Fruct. Sem. Pl. 2: 456. t. 171. 1791; Hook.f., Fl. Brit. India 3: 308. 1881; Gamble, Fl. Madras 2: 708. 1921; Chowdhery in Hajra & *al*., Fl. India 12: 413. 1995. *Verbesina nodiflora* L., Amoen. Acad. 4: 290. 1787.

Erect herb, to 1 m high, stem striated, glabrous. Leaves simple, petiolate, to 1 cm long, scabrous on both surfaces, ovate-elliptic, base cuneate, margin serrate to serrulate, apex acute. Heads solitary or many in clusters, heterogamous, rayed. Involucral bracts 4-6, biseriate, elliptic-lanceolate, pilose-scarious, ciliate margins, apex acute. Receptacle narrow, paleaceous. Female florets 5; corolla 4 mm, tube 2.2 mm, glabrous, 2-3-lobed. Bisexual florets 10, corolla 3 mm, tube 1.5 mm, glabrous, 5-lobed; stamens 5, black, anthers elliptic, hood elliptic-obtuse, base sagittately tailed. Pappus of 2 short awns. Achenes of ray obovate to elliptic, winged, of bisexual florets obovate-cuneate, 3-ribbed, puberulous.

Common near the wet palces in Hyderabad district.

Fl. & Fr.: October – January.

TAGETUS L.

1. Heads 3-4 cm across .... **T. patula**
1. Heads 6-8 cm across .... **T. erecta**

Tagetes erecta L., Sp. Pl. 887. 1753; Grierson in Dassan. & Fosb., Rev. Handb. Fl. Ceylon 1 : 234. 1980.

Vern.: *Banthi.*

Erect, simple or sparingly branched, strong-scented, annual herb, 15-60 cm high; stems glabrous or puberulent, ribbed. Leaves 5-20 cm long; leaflets or segments opposite or alternate, lanceolate-oblong, 1-5 x 0.2-1 cm, margin serrate to subentire, apex acute to acuminate, glabrous or puberulent near base, with few marginal glands near base of teeth. Heads solitary, 6-8 cm across; involucre campanulate, 1.5-2 cm long; ray-florets 5-8 yellow; disc florets 100-200, yellow. Achenes blackish, angled, pubescent on angles, 7-9 mm long.

Cultivated in gardens and houseyards as an ornamental.

Fl. & Fr.: October – December.

Tagetes patula L., Sp. Pl. 887. 1753.

Vern.: *Banthi.*

Erect, strongly aromatic, much-branched, annual herb, to 50 cm high. Leaves pinnatisect; leaflets or segments narrowly lanceolate–oblong, acutely serrate, with numerous scattered glands. Heads cymose, sometimes solitary, 3-4 cm across; peduncles 3-15 cm long, scarcely thickened at top; involucre 1-1.2 cm long with 5-7 triangular, acute teeth; ligule or ray-florets yellow with brownish red markings, 1-1.5 cm long, emarginated at apex. Pappus of 1-2 long awned scales and 2-3 shorter blunt ones.

Grown in gardens as an ornamental.

Fl. & Fr.: September-December.

TARAXACUM Weber

Taraxacum javanicum Soest, Wentia 10: 56. f 27. 1963; Mamgain & Rao in Hajra & *al.*, Fl. India 12: 258. 1995. *T. officinale* auct. non Weber; Hook.f., Fl. Brit. India 3: 401. 1881; Gamble, Fl. Madras 2: 731. 1921.

Perennial herb with rhizomatous rootstock and milky. Leaves all radical, sessile, oblanceolate, glabrous, 6 – 18 × 1.5 – 4 cm, irregularly pinnatifid, lobes linear or triangular, acute, toothed, scapes 10 – 15 cm long. Heads ligulate, solitary on leafless scape, glabrous, outer involucral bracts reflexed, spreading, inner ones erect, lingules yellow. Achenes 4 mm long, glabrous, flattened, ribbed, narrowed at base and spinulose at apex, pappus copious, soft, whitish, 5 mm long.

Rare weed of human habitations in Khammam district (K.N. Reddy & *al.*, 2000).

Fl. & Fr.: June – December.

Bhadrachalam (KMM), *KNR* 111 (KU).

TRICHOLEPIS DC.

Tricholepis radicans DC., Prodr. 6 : 564. 1838; Hook.f., Fl. Brit. India 3 : 381. 1881; Gamble, Fl. Madras 2 : 726. 1921; Hajra in Hajra & *al.*, Fl. India 12: 223. 1995.

Ascending herb, up to 40 cm high, stems angular, ribbed, scabrid, glabrous. Leaves simple, sometimes in rosette at base, alternate, sessile, decurrent along stem, gland dotted on upper surface, 1-5 × 0.1-2 cm, base clasped to the stem, sub auricled, margins narrowly recurved, serrate-dentate, apex apiculate. Heads solitary, terminal, purple, homogamous, not rayed. Involucral bracts numerous, multiseriate, margin serrate, acuminate tip. Receptacle fleshy, flat, bristly acicular. Florets 30; corolla 12 mm, tube 7 mm, glabrous, 5-lobed, lobes lanceolate, acute; stamens 5, anthers 8 mm, hood lanceolate, base sagittate. Pappus stiff paleaceous hairs, many seriate. Achenes oblong, brown, striated, glabrous, lateral and hollowed.

Common in foot hills and scrub forests in Adilabad, Ranga Reddi, Karimnagar, Hyderabad and Mahabubnagar districts.

Fl. & Fr.: September – December.

Naskal (RR), *MSM* 15115; Kodimial (KRN),.*GVS* 21870 (MH).

TRIDAX L.

Tridax procumbens L., Sp. Pl. 900. 1753; Hook.f., Fl. Brit. India 3: 311. 1881; Gamble, Fl. Madras 2: 711. 1921; Chowdhery in Hajra & *al.*, Fl. India 12: 418. 1995.

Procumbent herb, to 30 cm, stem terete, scabrid. Leaves simple, opposite, petiolate, to 1 cm, scabrid and gland dotted, ovate-lanceolate, 1.5-5 × 0.8-3 cm, base cuneate, margin incised dentate or lobed, apex acute. Heads solitary, terminal, yellow, heterogamous, rayed. Involucral bracts 10, biseriate, herbaceous, densely hirsute. Ray florets 5, corolla 9 mm, densely pubescent, 2-3-lobed. Bisexual florets around 40, corolla 8 mm, pubescent, 5-lobed; stamens 5, anthers 3 mm, exserted, hood triangular-ovate, acute, base sagittate. Pappus multiseriate, paleaceous bristles, unequal, aristate-acuminate, plumose, persistent. Achenes obconical, 2 mm, truncate at apex, ribbed, densely white silky.

Common in waste places, road sides and open forests in all districts.

Fl. & Fr.: Throughout the year.

Sone (ADB), *TP* & *PVP* 4049; Kammarpally (NZB), *TP* & *BR* 6196; Sangareddy (MDK), *BR* & *CP* 11288; Vikarabad (RR), *MSM* 19238; Pakhal (WGL), *KMS* 11669; Kodimial (KRN), *GVS* 20039 (MH); Khammam to Dhaniyaigudem (KMM), *RR* 108530 (BSID).

VERNONIA Schreb.

1. Herbs; stems hebaceous:
    2. Plant parts glabrous .... **V. cinerea**
    2. Plant parts pubescent; leaves whitish below .... **V. albicans**
1. Sub-shrubs; stem woody .... **V. anthelmintica**

Vernonia albicans DC. in Wight, Contrib. Bot. Ind. 6. 1834; Gamble, Fl. Madras 2: 473. 1921; Uniyal in Hajra & *al.*, Fl. India 13: 351. 1995.

Erect herb, up to 1 m high, stem terete, striated, solid, tomentose. Leaves simple, alternate, petiolate, to 2-5 mm, sparsely pilose, yellow glandular on upper surface, sessile, yellow glandular beneath, ovate, elliptic-lanceoalte, 2-3 × 0.4-2 cm, base cuneate, apex acute or apiculate. Heads in corymbs, terminal, purple, homogamous, not rayed. Involucral bracts 50, 4-5-seriate, linear-lanceolate, rigid, pubescent, sub-acuminate. Receptacle slightly convex, flat, alveolate. Florets around 45; corolla glabrous, deeply 5-lobed, lobes acute at apex; stamens 5, anthers 2 mm, inserted, yellow, hood triangular, ovate, acute, base sagittate. Pappus biseriate, short, persistent scales, united at base, ciliate at apex. Achenes oblong or obovate, brown, ribbed, densely white hairy along the ribs and glandular between the ribs and at base.

Occasional in waste lands in Adilabad, Ranga Reddi, Hyderabad, Nalgonda and Mahabubnagar districts.

Fl. & Fr.: October – January.

Sone (ADB), *TP* & *PVP* 4039; Central University campus (RR), *MSM* 15144; Gacchibowli (RR), *MSM* 15199; Amrabad (MBNR), *BSS* & *SKB* 32544; Gaarlapadu (MBNR), *SRS* 109101 (BSID); Kambalpally (NLG), *BVRS* 201 (NU).

Vernonia anthelmintica (L.) Willd., Sp. Pl. 3: 1634. 1803; Hook.f., Fl. Brit. India 3: 236. 1881; Uniyal in Hajra & *al.*, Fl. India 13: 357. 1995. *Conyza anthelmintica* L., Sp. Pl. 861. 1753. *Centratherum anthelminticum* (L.) Kuntze, Revis. Gen. Pl. 1: 30. 1891; Gamble, Fl. Madras 2: 667. 1921.

Subshrubs, up to 1.5 m high, stem woody, terete, striated, pilose or sessile glandular. Leaves simple, alternate, sessile, pilose on upper surface, densely glandular on lower surface, elliptic or elliptic-lanceolate, 6-20 × 1.5-9 cm, base cuneate-attenuate, margin serrate or serrate-dentate, apex acuminate. Heads subcorymbose or solitary, terminal, pink, homogamous, not rayed. Involucral bracts 36, 3-4-seriate, herbaceous, puberulous on dorsal side, spathulate, obtuse. Receptacle flat, glabrous. Florets around 40; corolla 13 mm, tube 6 mm, narrow, glabrous, 5-lobed, lobes ovate-lanceolate, glandular at tip, acute; stamens 5, anthers linear, hood ovate, obtuse, base sagittate. Pappus biseriate, persistent scales, free above, deciduous. Achenes oblong or narrowly obovate, 8-10-ribbed, blackish, pubescent.

Rare in waste places and open forests Medak, Mahabubnagar and Hyderabad districts.

Fl. & Fr.: October – January.

Borapur (MBNR), *BSS* & *SKB* 32359; Mallelatheertham (MBNR), *TSS* & *BR* 29849; Narsapur (MDK), *KMS* 6783 (MH, BSID).

Vernonia cinerea (L.) Less., Linnaea 4: 291. 1829; Hook.f., Fl. Brit. India 3: 233. 1881; Gamble, Fl. Madras 2: 676. 1921; Uniyal in Hajra & *al.*, Fl. India 13: 367. 1995.

Erect herb, up to 100 cm high, stem striated, puberulous. Leaves simple, alternate, subsessile, puberulous on both surfaces, broadly elliptic or lanceolate, 1.5-8 × 4-5 cm, base cuneate-attenuate, margin entire, apex apiculate or acute. Heads in corymbs, terminal, pink, homogamous, not rayed. Involucral bracts 4-seriate, unequal, linear-lanceolate, pubescent, margin ciliate, sharply acuminate. Receptacle slightly convex, flat, alveolate, naked. Florets around 24, corolla 3 mm, tube 2 mm, glabrous, shortly lobed, glandular on dorsal side; stamens 5, anthers 0.5 mm, hood ovate, base sagittate. Pappus biseriate, free scales, white, deciduous. Achenes obovate, densely hairy.

Common in open forests and waste places in all districts.

Fl. & Fr.: August – April.

Manchippa (NZB), *TP* & *BR* 6214; Patancheru (RR), *BR* & *CP* 11313; Mohammadabad RF (RR), *MSM* & *NV* 10417; Appapur (MBNR), *TDCK* 16815; Molachintapally (MBNR), *TDCK* 15309; Kambalpally (NLG), *BVRS* 610 (NU); Bhupatipur (KRN), *GVS* 22294 (MH); Edrallapalli (KMM), *RCS* 98749 (BSID); Khammam hills (KMM), *RR* 113809 (BSID).

WEDELIA Jacq.

1. Leaves subsessile, spathulate-lanceolate, margin entire or serrate .... **W. chinensis**
1. Leaves distinctly petioled, ovate, acuminate, serrate .... **W. urticaefolia**

Wedelia chinensis (Osbeck.) Merr., Philipp. J. Sci. 12: 111. 1917; Chowdhery in Hajra & *al*., Fl. India 12: 424. 1995. *Solidago chinensis* Osbeck. in Dagb. Ostind. Tesa. 241. 1757. *W. calendulacea* (L.) Less., Syn. Gen. Comp. 222. 1832; Hook.f., Fl. Brit. India 3: 306. 1881.

Procumbent herb, rooting at the lower nodes. Leaves simple, opposite, spathulate-oblanceolate, acute, trinerved, attenuate, margins flat or slightly inrolled, serrate or entire, 2-6 × 0.5-1.5 cm, shortly appressed hairs on both surfaces. Heads yellow, solitary, terminal, heterogamous, rayed. Involucral bracts biseriate, acute or obtuse at apex, appressed pubescent.

Ray florets few, 3-lobed. Disc florets cylindric, limb 2.5 mm long, glabrous 5-lobed. Achenes dark brown, rugulose, glabrous. Pappus a minute irregularly margined hairs.

Seshagiri Rao (2012) reported from Central University campus in Ranga Reddi district.

Fl. & Fr.: Throughout the year.

Wedelia urticaefolia DC. in Wight, Contrib. Bot. Ind. 18. 1834; Hook. f., Fl. Brit. India 3: 306. 1881; Gamble, Fl. Madras 2: 707. 1921; Chowdhery in Hajra & *al*., Fl. India 12: 426. 1995.

Erect herb, to 60 m high, stem terete, striated, solid, hispid. Leaves simple, opposite, petiolate, to 5 mm, hispid to scabrid on both surfaces, ovate, 3-6 × 1-3 cm, base rounded, margin coarsely serrate, apex acute or acuminate. Heads solitary, terminal, yellow, heterogamous, rayed. Involucral bracts 8, biseriate, oblong or elliptic-lanceolate, foliaceous, green, hirsute, apex apiculate or acuminate. Ray florets 10, corolla hairy, lobes acute. Bisexual florets 15, corolla 8 mm long, glabrous, 5-lobed; stamens 5, black, anthes 3 mm, exserted, oblong, hood ovate, base sagittate. Pappus absent in ray florets, short cup-shaped in bisexual florets. Ovary of ray florets obovate, 3 mm, truncate and hairy at apex, scabrid along the margins, of bisexual florets 2.5 mm, 3-4-ribbed, scabrid along ribs, truncate and hairy at apex.

Occasional in moist places in Ranga Reddi district.

Fl. & Fr.: October – January.

Anantagiri RF (RR), *TP* & *MSM* 11055.

XANTHIUM L.

1. Fruiting involucre pale-green, covered with hooked bristles, devoid of glands or hairs; apical beaks divergent .... **X. indicum**
1. Fruiting invlucre greyish-green to red brown, covered with hairs and glands; apical beaks straight or incurved .... **X. pungens**

Xanthium indicum Koenig in Roxb., Fl. Ind. 3: 601. 1832; Chowdhery in Hajra & *al.*, Fl. India 12: 427. 1995. *Xanthium strumarium* L., Sp. Pl. 987. 1753; Hook.f., Fl. Brit. India 3: 303. 1881; Gamble, Fl. Madras 2: 73. 1921; Prabkar Raju & R.R.Venkata Raju in Pullaiah & Moulli, Fl. Andhra Pradesh 2: 542. 1997.

Sub-shrub, up to 1.5 m high, stem brown, hispidulous. Leaves simple, alternate, petiolate, to 7 cm, hispidulous, prominently 3-veined from base, yellow glandular on lower surface, broadly obovate, 3-10 × 1.5- 10 cm, lobed, base rounded or cordate, margin serrate to dentate, apex acuminate. Heads monoecious, cyme in axils or clustered at nodes on a racemose axis, each cyme has one male and female at base. Male heads 5-6 mm, white. Involucral bracts 10, biseriate, 3 mm, scarious, pubescent, acute. Receptacle cylindrical, 3 mm, elongate, paleaceous. Florets around 60, corolla 3 mm long, campanulate, hairy and glandular, 5-lobed, lobes acute; stamens 5, black, free, exserted, anthers oblong, hood incurved, base auriculate. Female heads solitary, ovoid, to 1 cm, 2-flowered. Involucral bracts biseriate, outer 8, free, linear-lanceolate, pubescent, acute; inner connate, covered with uncinate spines, spine 5 mm, spreading, hispidulous at base, hooked, ending in two stout rostra at the apex, erect or slightly incurved, hispidulous, corolla absent; ovary ovoid, 5mm, brown, glabrous. Pappus absent. Achenes elliptic, to 1.8 cm, greyish-black, flat on one surface, tapering to a fine point at the apex.

Forest edges in low altitudes and marshy areas in most districts.

Fl. & Fr.: September – March.

Indervelli (ADB), *GO* & *PVP* 4531; Mosra (NZB), *TP* & *BR* 6065; Patancheru (MDK), *BR* & *CP* 11320; Gacchibowli (RR), *MSM* 12774; Sarlapally (MBNR), *TDCK* 15324; Borapuram (MBNR), *BSS* & *SKB* 32358; Dhaniyaigudem (KMM), *RR* 108098 (BSID); Cherukupally (NLG), *BVRS* 664 (NU); Pakhal (WGL), *KMS* 11665 (MH); Kodimial (KRN), *GVS* 21842 (MH).

Xanthium pungens Wallroth, Beitr. Bot. 1(2): 227 & 231-232. 1844; Maheshwari, Curr. Sci. 42(5): 327-328. 1973; Chowdhery in Hajra & *al.*, Fl. India 12: 429. 1995.

Undershrubs, up to 1 m high; stems angular, ribbed, sparsely hispid with short, thick hairs. Leaves cordate, entire or 3-5-lobed, 6-10 × 10-12 cm, minutely hispid on both surfaces, base cuneate, margin irregularly, coarsely dentate with acute teeth, apex acute; petiole 4.5 -5.5 cm long. Female involucres yellowish-green turning red, oblong or ovoid, glabrescent or subglabrous, with minute hairs and glands, involucral spines slender, straight, usually 4 mm long, apex straight or hooked. Fruiting heads dark brown, 15-20 × 5-9 mm, bearing 2 terminal, straight or incurved beaks with hooked or straight apices.

A noxious weed, occurs in waste places and by the roadsides.

Fl. & Fr.: October – February.

Musthyal (WGL), *CSR* 1081 (KU); Pakhal (WGL), *KMS* 11665 (MH).

Note: It is the most common *Xanthium* in the state.

YOUNGIA Cass.

Youngia japonica (L.) DC., Prodr. 7: 194. 1838; Mamgain & R.R.Rao in Hajra & *al.*, Fl. India 12: 329. 1995. *Prenanthes japonica* L., Mant. Pl. 107. 1767; Hook.f., Fl. Brit. India 3: 395. 1881; Gamble, Fl. Madras 2:730. 1921.

Scapigerous herb, flowering branches erect, up to 40 cm, stems densely pilose towards base. Leaves in rosette at base, lower ones lobed, simple, spathulate to oblong – lanceolate, 2 – 10 × 0.5 – 4 cm, sparsely puberulous, margin denticulate or entire, apex acute. Heads in terminal panicles, yellow, homogamous, rayed. Achenes fusiform, straight, slightly curved, 10-ribbed, scabrid along ribs, apex dilate to form a cup like ring on which pappus arises.

Rare on damp shady localities in Warangal district (C.S. Reddy & *al.*, 1999; C.S. Reddy, 2001).

Fl. & Fr.: November – January.

Waddepally (WGL), *CSR* & *VSR* 1013 (KU).

CAMPANULACEAE

1. Flowers in dense spikes; stamens attached at the base of corolla ..... **Sphenoclea**
1. Flowers not in spikes; stamens free from corolla .... **Wahlenbergia**

SPHENOCLEA Gaertn. *nom. cons.*

Sphenoclea zeylanica Gaertn., Fruct. Sem. Pl. 1: 113 t. 24. f. 5. 1788; Hook.f., Fl. Brit. India 3: 438. 1881; Gamble, Fl. Madras 2: 739. 1921.

An erect, stout annual herb, up to 30 cm high, stem hollow, root fibrous. Leaves alternate, lanceolate, 2.5-7.5 × 1-2 cm, glabrous, glaucous, acute at both ends, margin entire, sessile or subsessile. Flowers 1.2 mm across, in long-peduncled, terminal or leaf-opposed 0.8–3 cm long spikes, calyx tube adnate to the ovary, hemispheric, lobes 5, ovate, connate below on the line of dehiscence of the capsule and falling off with the lid; corolla white or greenish yellow, campanulate, 5-lobed, lobes valvate; stamens 5, on the base of corolla, filaments dilated at base, anthers short, free; style short, stigma hardly lobed. Capsule 3–4 mm across, depressed globose, membranous, circumscissile; seeds many, minute, sparsely verrucose.

Rare, in marshy places – rice fields, margins of ponds etc in Khammam and Warangal districts.

Fl.: December – March.

On the way to Wyra from Khammam (KMM), *RR* 108033 (BSID).

WAHLENBERGIA Schrader ex Roth *nom. cons.*

Wahlenbergia erecta (Roth ex Roem. & Schult.) Moel. & Tuyn in Steenis, Fl. Males. Ser. 1. 6: 113. 1960. *Dentella erecta* Roth ex Roem. & Schult., Syst. Veg. 5: 1819-20. *Cephalostigma schimperi* Hook.f., Fl. Brit. India 3: 428. 1881; Gamble, Fl. Madras 2: 738. 1921.

Annual erect branched herb, up to 40 cm high, stems often flexuose, angular, hispid hairy. Leaves elliptic-oblong to oblong, gradually smaller and narrower, to 2.8 × 1 cm, chartaceous, sparsely hairy above, hispid hairy beneath on the midrib, base narrowed or rounded, margin undulate-denticulate, apex acute or apiculate. Flower solitary, terminal and axillary, aggregated into a leafy racemes; pedicel to 1.5 cm, filiform, hairy; bracts 2, lanceolate, calyx lobes 5, unequal, linear, partly adnate to ovary; corolla white or pale blue, 6 mm across, lobes 5, as long as calyx, linear, acute; stamens 5, alternate to corolla lobes, free from the corolla lobes, filaments membranous, broad at the base; ovary inferior to superior, stigma bifid. Capsule loculicidal, 3.3 x 2.4 mm, apex cone-shaped, 3-valved, hispidulous, many-seeded; seeds shining, brown, ellipsoid

Frequent in open places on laterite and rocky soil in Nizamabad district.

Fl. & Fr.: September – December.

Mudheli RF (NZB), *BR* & *KH* 7139; Pillutla RF (MDK), *BR* 11260.

LOBELIACEAE

LOBELIA L.

1. All anthers bearded at apex; seeds 3-gonous .... **L. alsinoides**
1. Two anterior anthers bearded at apex; seeds ellipsoid .... **L. heyneana**

Lobelia alsinoides Lam., Encycl. 3: 588. 1792. *L. trigona* Roxb., Fl. Ind. 2: 111. 1824. Hook. f., Fl. Brit. India 3: 423. 1881; Gamble, Fl. Madras 2: 736. 1921.

Erect or ascending herb, up to 25 cm high, stems 3-angled and winged. Leaves alternate, broadly ovate, upper leaves ovate-oblong, 1-1.4 × 1-1.2 cm, glabrous, base obtuse or rounded or slightly cordate or narrowed into a short petiole or sometimes sessile, margin crenate-dentate, apex acute. Axillary flowers solitary, terminal ones forming a lax raceme; bracts linear, to 3 mm; pedicel to 1.8 cm long; calyx lobes 5, blue, irregular with 2 dorsal lobes, 1 ventral lobe and 2 lateral (trifid), lobes broadly ovate; stamens 5, alternate with corolla lobes, filaments 5 mm long, almost free in the lower half, anthers basifixed, unequal, bearded at apex; ovary inferior, style bifid. Capsule 4 mm long, obconical, glabrous; seed 3-gonous, brown, 0.5 mm long.

Common in wet places through out the state.

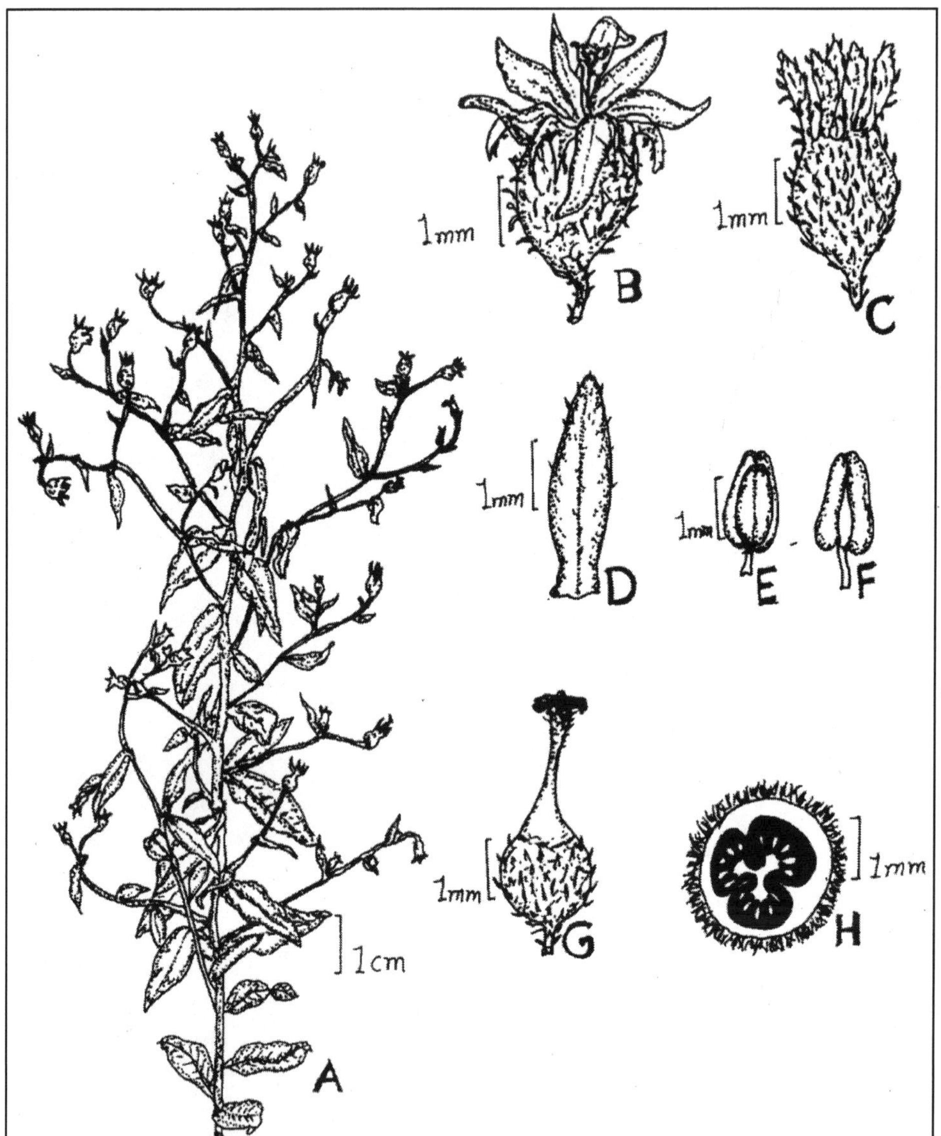

Figure 55: Wahlenbergia erecta (Roth ex Roem. & Schult.) Moel & Tuyn.

A. Twig, B. Flower-entire, C. Calyx with ovary, D. Petal, E&F. Stamens, G. Pistil, H. Ovary c.s.

Fl & Fr.: May - August.

Jannaram (ADB), *TP* & *PVP* 4184; Ibrahimpet (NZB), *TP* & *BR* 6131; Annaram RF (NZB), *BR* 7189; Narsapur RF (MDK), *BR* & *CP* 14113; Mohammadabad RF (RR), *MSM* & *KH* 10600; Anantsagar RF (RR), *MSM* & *KH* 11011.

Lobelia heyneana Roem. & Schult. Syst. Veg. 5: 50. 1819. *L. trialata* Buch.-Ham. ex D. Don, Prodr. Fl. Nepal 157. 1825; Hook.f., Fl. Brit. India 3: 425. 1881; Gamble, Fl. Madras 2: 736. 1921.

Erect or suberect herb, up to 20 cm high, stem glabrous, 3-winged, 3-angled. Leaves shortly petioled, ovate-elliptic, to elliptic-lanceolate upwards, 1-3 × 0.5-1.8 cm, base decurrent, margin serrate-dentate, serratures sometimes apiculate, suddenly narrowed into the petiole, to 0.5 cm, apex obtuse or subacute. Flowers axillary, solitary, often forming terminal raceme; pedicel to 1.8 cm long, angled; bracteoles linear; calyx lobes 5, unequal, linear, 2.5 to 2.8 mm long; corolla bluish-purple, trifid, 0.3 cm; stamens 5, alternate with corolla lobes, 2 anterior anthers bearded at apex; style 2 mm, glabrous, stigma capitate. Capsule obconic, 5 x 3 mm; seeds numerous, ellipsoid, smooth, brown, 0.4 mm long.

Occasional on laterite soil in moist and shady places in Adilabad district.

Fl. & Fr.: July – December.

Birsaipet (ADB), *GO* & *PVP*4612.

PLUMBAGINACEAE

PLUMBAGO L.

Plumbago zeylanica L., Sp. Pl. 151. 1753; Hook.f., Fl. Brit. India 3: 480. 1882; Gamble, Fl. Madras 2: 744. 1921.

Vern.: *Tella chitramulam.* English: The Ceylon Lead Wort.

Perennial undershrub, 60 cm – 1 m high, branchlets scandent, striate, stem 4-angled, glabrous. Leaves alternate, thin, 3.5–7.5 × 2.5–4.5 cm, elliptic-ovate, base cuneate or attenuate, margin entire, apex acute, apiculate; petiole to 1 cm long, auriculate at base; stipules auriculate. Spiciform racemes 6-9 cm long, rachis striate, glandular hairy; calyx tubular, to 1.2 cm, 10-ribbed, covered with stalked glands, 5-toothed; corolla white, salver-form, tube narrowly cylindric, lobes 5, spreading, obovate; stamens 5, free, filaments to 2 cm long, anthers exserted, parallel; style filiform, stigma 5-fid. Capsule membranous, enclosed in persistent calyx, 5-lobed, lobes deeply channeled on the back; seed one, dark brown or black.

Common in open places, scrub forests and among bushes in most of the districts.

Flowers with a peak during December – March. Fr.: Throughout the year.

Vempalli RF (ADB), *GO* 4454; Birsaipet (ADB), *GO* & *PVP* 4591; Pocharam (NZB), *BR* 9501; Gnanapur fields (MDK), *CP* 11784; Medak (MDK), *RG* 113507 (BSID); Kommireddipalli (RR), *MSM* 10326; Hajipur hills (MBNR), *TDCK* 13629; Amangal to Shadnagar (MBNR), *SRS* 104582 (BSID); Kambalpally (NLG), *BVRS* 213 (NU).

PRIMULACEAE

ANAGALLIS L.

1. Corolla blue; filaments hairy; leaves opposite, whorled upwards .... **A. arvensis**
1. Corolla white; filaments glabrous; leaves opposite or subopposite, upper ones alternate .... **A. pumila**

Anagallis arvensis L., Sp. Pl. 148. 1753; Hook.f., Fl. Brit. India 3: 506. 1882; Gamble, Fl. Madras 2: 747. 1921.

Erect or ascending herbs, to 50 cm high, much branched, stems 4-angled. Leaves opposite, whorled upwards, elliptic-ovate, oblong-ovate or lanceolate, 1-2 x 0.5-0.8 cm, membranous, apex acute, subsessile. Flowers solitary, axillary, pedicel to 2 cm long; calyx lobes 5, 4 mm, midrib prominent, apex shortly acuminate; corolla blue, lobes 5, rotate, glandular hairy, obtuse; stamens 5, filaments hairy, adnate to the corolla, free from each other; ovary 7 mm, style 2 mm. Capsule globose, 4-6.2 mm diameter; seed flattened, subtrigonous.

A weed of cultivated fields.

Fl. & Fr.: Jan.-Mar.

Rooplathanda (NZB), *TP* & *BR* 6399; Armoor fields (NZB), *BR* 7289.

Anagallis pumila Sw., Prodr. Veg. Ind. Occ. 40. 1788; Gamble, Fl. Madras 2: 747. 1921. *Centunculus tenellus* Duby in DC., Prodr. 8: 72. 1844; Hook.f., Fl. Brit. India 3: 506. 1882.

Erect, slender, little branched herb, 4-12 cm high, stems glabrous, angular, somewhat winged. Leaves opposite or subopposite, upper ones alternate, small, 7 × 4 mm, ovate, obovate, oblanceolate, base narrowed, margin entire, apex acute or obtuse, apiculate. Flowers axillary, solitary; pedicels 5 mm long, filiform; calyx lobes 5, unequal, narrowly lanceolate, acuminate. 2.8 mm long; corolla white, divided nearly to the base, 2.5 mm long, lanceolate, apiculate; stamens 5, adnate to the base of the corolla lobes, opposite to the lobes, filaments glabrous, slender, short; ovary glandular. Capsule subglobose, seeds red, small, trigonous, solid.

Occasional on open moist places among grasses in Hyderabad district.

Fl. & Fr.: October - December.

Thumallakunta (HYD), *TR* 701 (HH).

SAPOTACEAE

1. Calyx lobes 4 .... **Madhuca**
1. Calyx lobes 6-8:
   2. Secondary nerves few, distant; stamens 8 .... **Mimusops**
   2. Secondary nerves numerous, close; stamens 6 .... **Manilkara**

MADHUCA Ham. ex Gmel.

Madhuca longifolia (J. Koenig) Macbr., Contr. Grey Herb. (n.s) (53). 17. 1918. var. **longifolia.** *Bassia longifolia* J. Koenig ex L., Mant. Pl. 563. 1771; Hook.f., Fl. Brit. India 3: 544. 1882; Gamble, Fl. Madras 2: 763. 1921.

Vern.: *Pedda ippa, Natu ippa*

Deciduous tree, 15-20 m high; bark dark yellowish-grey to brown, scaly; wood red, close grained, coarse and even textured; branchlets brownish-yellow, glabrous. Leaves chartaceous, oblanceolate, 5.5-8 × 2-3 cm glabrous, green above, paler beneath, base shortly narrowed, margin entire, apex acute; secondary nerves 10-12 pairs, raised above; petiole to 2 cm. Flowers creamish-white, long pedicellate umbels. Berry ellipsoid, glabrous, 4 x 2 cm; seed 1, compressed, light brown, smooth, shining with a fairly thick and hard testa.

Occasional, in open forests in Warangal, Hyderabad and Mahabubnagar districts.

Fl.: Mar. Fr.: May-June.

Pullajalamola (MBNR), *VBH* 84980 (BSID); Uskavagu, Mannanur (MBNR), *SRS* 108913 (BSID); near Pakhal lake (WGL), *RKP* 108293 (BSID).

Key to Varieties

1. Leaves elliptic or elliptic-oblong; anther connectives apiculate; pedicel stout; berry ovoid-oblong, 1-4-seeded .... var. **latifolia**
1. Leaves oblanceolate; anther connectives toothed; pedicel slender; berry ellipsoid, 1- seeded .... var. **longifolia**

Madhuca longifolia (J.Koenig) Macbr. var. **latifolia** A. Chev., Rev. Int. Bot. Appl. Agric. Trop. 23: 149. 1943. *Bassia latifolia* Roxb., Pl. Coromandel t. 19. 1795 & Fl. Ind. 2: 526. 1832; Hook.f., Fl. Brit. India 3: 544. 1882; Gamble, Fl. Madras 2: 763. 1921. *Madhuca indica* J. F. Gmel., Syst. Nat. 2: 799. 1791.

Vern.: *Ippa, Ippi, Madhookamu, Ippa chettu.*

Deciduous tree, up to 18 m high; trunk short; bark thick, dark with vertical cracks, the inner bark red, milky; wood reddish-brown, straight or shallowly interlocked grained, medium to fine textured; branchlets glabrous, spreading. Leaves clustered near the ends of branches, chartaceous, thick, elliptic or elliptic-oblong, 8-20 × 4-10 cm, glabrescent, base rounded or acute, margin entire, apex acute or shortly acuminate; secondary nerves ca. 12 pairs, 1.6 cm apart; petiole to 3.5 cm long. Flowers cream, dense in fascicles on drooping, densely tomentose pedicels; calyx lobes 4, obovaate, 7-9 mm, tomentose; corolla 6-12-lobed; stamens 16 in 2 or 3 whorls, anthers lanceolate, connectives apiculate; ovary villous, 6-celled, style 2.5 mm, stigma obtuse. Berry ovoid-oblong, 4 x 2 cm, fleshy, greenish; seeds 1-4, brown, 1-3 cm long, polished, slightly broad.

Common, in mixed deciduous forests. It prefers a sandy soil but also grows on shallow dry soils. It is a drought hardy species, strong light demander and coppices well.

Fl.: February-April; Fr.: June-July.

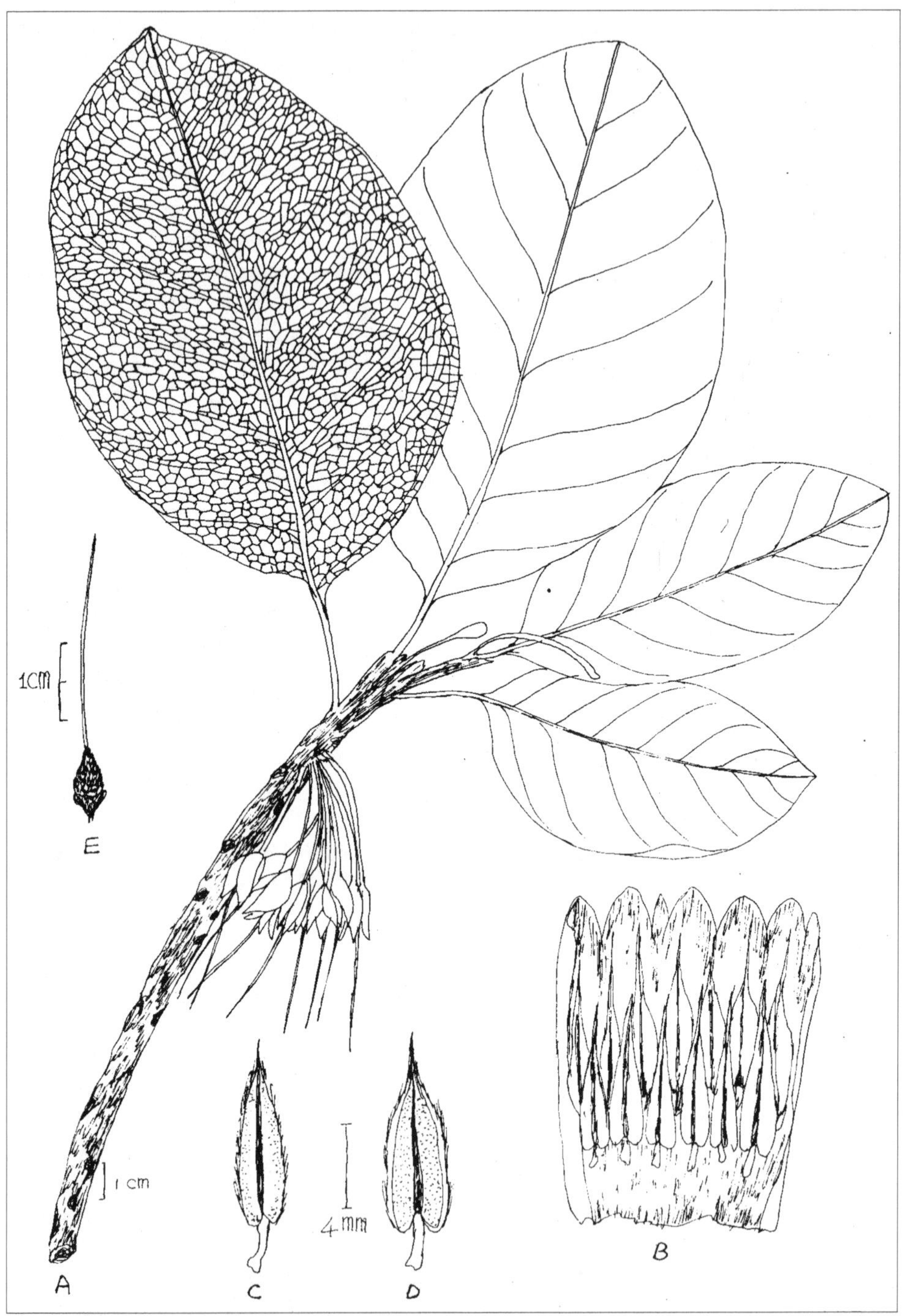

Figure 56: Madhuca longifolia (J.Koenig) Macbr. var. **latifolia** A. Chev., (Syn.: *Madhuca indica* J.F. Gmel.)

A. Twig, B. Calyx lobes, C. Stamens, D. Pistil, E. Berries.

Macodi (ADB), *PVP* & *GO* 21901; Sundergiri (ADB), *GO* & *PVP* 4236; Lingaram (NZB), *BR* & *GO* 9062; Narsapur RF (MDK), *CP* 11587; Gangapur RF (MDK), *BR* & *CP* 11427; Mohammadabad (RR), *MSM* 10224; Appapur (MBNR), *TDCK* 13662; Eturunagaram (WGL), *SSR* 18541; Rajakonda (NLG), *BVRS* 395 (NU).

Note: Fresh seed has a high percentage of fertility, but the seed quickly loses its viability if kept and is much subject to insect and fungus attack. The tree suffers greatly from the attacks of *Loranthus*.

MANILKARA Adans. emend. Gilly *nom. cons.*

1. Leaf apex acute or retuse; flowers solitary; berry 3-7-seeded .... **M. zapota**
1. Leaf apex obtuse or emarginated; flowers in fascicles; berry 1- seeded:
    2. Leaves broadly elliptic; flower pedicel to 4.5 cm long; berry globose, depressed .... **M. roxburghiana**
    2. Leaves broadly oblong; flower pedicel to 0.8 cm long; berry ellipsoid .... **M. hexandra**

Manilkara hexandra (Roxb.) Dubard., Ann. Inst. Bot. Geol. Colon. Marseille (Ser.3)23. 9. f. 2. 1915. *Mimusops hexandra* Roxb., Pl. Coromandel t. 15. 1795 & Fl. Ind. 2: 238; Hook. f., Fl. Brit. India 3: 549.1882; Gamble, Fl. Madras 2: 766. 1921.

Ver.: *Puttapala, Pala nemmi, Manchipala, Pala chettu, Kanupala.*

Evergreen tree, up to 15 m high, with a shady crown; trunk erect; bark blackish-grey, crimson inside, deeply furrowed; wood red and close grained. Leaves clustered at the ends of branches, coriaceous, obovate or oblong, 4-11.3 × 4-6.2 cm, glabrous, dark green above, and paler beneath, base obtuse or acute, margin entire, apex obtuse or emarginated; secondary nerves 20-25 pairs, parallel, impressed above, raised below; petiole to 2 cm. Flowers whitish, 3-6 in leaf axils, calyx-segments 6, in two whorls, up to 4 x 3 mm, tawny tomentose without, slightly pubescent within at top, margin slightly scarious, ovate; corolla 9 mm across, cream coloured, glabrous, lobes 18, broadly elliptic, clawed; stamens glabrous, filaments long, anthers sagittate, staminodes alternate; ovary 6-lobed, style subulate. Berry reddish-yellow, ellipsoid, slightly curved, oblong, 0.5-1.2 x 0.5-0.8 cm, smooth; seed 1, rarely 2, ovoid, reddish brown, 1-1.5 cm long, and shining.

Common, in dry evergreen forests in Adilabad, Nalgonda and Hyderabad districts. It prefers sandy loam and loamy soils; in unfavourable conditions (dry situations) it becomes stunted and shrubby like. It is a light demander.

Fl.: November-January; Fr.: April-July.

Sothermat (ADB), *TP* & *PVP* 4108; Inchapalli RF (ADB), *GO* & *DAM* 5151; Farhabad (MBNR), *TDCK* 15343; Saleswaram (MBNR), *SKB* & *BSS* 30723; Ramavaram (KMM), *RR* 112553; Edrallapalli (KMM), *RCS* 98763 (BSID); Rangapur (WGL), *PVS* & *NRR* 76964 (BSID); Kambalpally (NLG), *BVRS* 340 (NU).

Manilkara roxburghiana (Wight) Dubard, Ann. Inst. Bot.-Geol. Col. Marseille (Ser.3) 23: 10. f. 2. 1915. *Mimusops roxburghiana* Wight, Icon. Pl. Ind. Orient. 4: t. 1588. 1850; Hook. f., Fl. Brit. India 3: 548. 1882; Gamble, Fl. Madras 2: 766.1921.

Vern.: *Renga, Magula.*

Large evergreen tree; wood reddish-brown; branchlets puberulous. Leaves broadly-elliptic, 4-9 × 2.8-5.8 cm, glabrous, brownish above, paler beneath when dry, base obtuse, margin entire, apex emarginate or obtuse; secondary nerves 14 pairs, not very close, inconspicuous; petiole to 3.5 cm long. Flowers white, 3-4 in axillary fascicles. Berry globose, 1cm long.

Rare, in dry hills in Mahabubnagar district.

Fl. & Fr.: March-December.

Uskavagu, Mannanur (MBNR), *SRS* 108905 (BSID); near Mallelatheertham (MBNR), *SRS* 108980 (BSID).

Manilkara zapota (L.) P. Royen, Blumea 7: 410. 1953. *Achras zapota* L., Sp. Pl. 1190.1753. *A. sapota* L., Sp. Pl. (ed. 2) 470. 1762; Hook.f., Fl. Brit. India 3: 534. 1882; Gamble, Fl. Madras 2: 759. 1921.

Vern.: *Sapota, Seema ippa;.* Eng.: Sapota.

Evergreen tree, up to 12 m high; bark greyish, smooth, scaly; branchlets puberulous. Leaves crowded at the ends of branchlets, elliptic-oblong, 7.5-11 × 2.3-3.9 cm, glabrous, base acute, margin entire, apex acute or retuse; secondary nerves 15-20 pairs, parallel; petiole 2-5 cm. Flowers white, solitary, axillary, pedicellate; calyx rusty tomentose, calyx lobes 3+3; corolla lobes 6, contorted, stamens 6, opposite to corolla lobes, inserted at its base, staminodes alternating with the petaloid stamens, ovary 10-12-celled. Berry brown, globose, 3-5 cm dia., fleshy, rough; seeds 3-7, black, glossy, oblong, compressed, 1-1.2 cm long.

Commonly cultivated for its fruits in plains and lower slopes.

Fl.: February-April; Fr.: May. Fruits edible.

Farhabad (MBNR), *TDCK* 16816.

MIMUSOPS L.

Mimusops elengi L., Sp. Pl. 349. 1753; Hook.f., Fl. Brit. India 3: 548.1882; Gamble, Fl. Madras 2: 765. 1921.

Vern.: *Pogada, Bogada, Pagada, Vakulamu, Kesara*; Eng.: Bullet wood.

Evergreen handsome tree, up to 10 m high; crown hemispherical; trunk cylindrical; bark dark grey, deeply fissured; wood dark red, close-grained, shallowly inter-locked, even and very fine textured; branchlets glabrous. Leaves subcoriaceous, elliptic-ovate or elliptic-oblong, 4.5-15 × 4-7.5 cm, glabrous, base acute, margin entire, apex acute or acuminate; petiole to 2.5 cm. Flowers white, fragrant, 2-6 in fascicles; calyx 9 mm long, tomentose without, lobes 8 in two whorls, ovate-lanceolate, inner whorl narrower than outer, with a median longitudinal groove on the dorsal surface; corolla 1.4 cm across, lobes 8, dorsal petaloid appendages 16, slightly connate at base into pairs forming outer series, stamens 8, staminodes ovate-lanceolate. Berry orange-yellow, ovoid, fleshy, 1.5 cm across; seed solitary, ovoid, compressed, brown, shining, 1.5-2.3 × 1-1.3 cm, smooth with hard testa and soft whitish albumen.

It is often cultivated.

Fl.: February-April; Fr.: Following February-July.

Mannanur (MBNR), *SRS* 107451 (BSID); Medak (MDK), *RG* 113518 (BSID).

EBENACEAE

DIOSPYROS L.

1. Flowers 3-merous; ovary 3-loculed; fruiting calyx not accrescent .... **D. vera**
1. Flowers 4-5-merous; ovary 4-10-loculed, fruiting-calyx accrescent (except *D. chloroxylon*) .... **Diospyros**
   2. Leaves tomentose or pubescent on lower or both surfaces:
      3. Petiole to 5 mm; female flowers sessile; calyx lobes 4 .... **D. chloroxylon**
      3. Petiole to 1.5 cm long; female flowers pedicellate; calyx lobes 5:
         4. Leaves elliptic; pedicel stout; corolla tubular; calyx thick to 8 mm; stamens 10-12; pericarp crustaceous .... **D. melanoxylon**
         4. Leaves broadly ovate; pedicel not stout; corolla not tubular; calyx thin, to 3 mm; stamens 16; pericarp leathery .... **D. exsculpta**
   2. Leaves glabrous on both surfaces
      5. Male flowers in cymes.-Leaves black when dry .... **D. sylvatica**
      5. Male flowers in umbels or fascicles .... **D. montana**

Diospyros chloroxylon Roxb., Pl. Coromandel t. 4: 1795 & Fl. Ind. 2: 538. 1832; C.B. Clarke in Hook.f., Fl. Brit. India 3: 560. 1882; Gamble, Fl. Madras 2: 776.1921.

Vern.: *Nela ulimiri, Ullinda, Ellinda, Thorika, Illinta.*

Evergreen tree, up to 10 m high; sometimes spinescent; bark dark brown, deeply cracked, peeling off in rectangular small scales; wood yellowish-grey; branchlets rusty tomentose. Leaves thinly-coriaceous, elliptic or elliptic-oblong, 1.6-5.6 × 0.8-3 cm, slight pubescent above, fulvous hairy beneath, apex and base acute or obtuse; nerves inconspicuous; petiole to 5 mm. Male flowers small in axillary clusters; female ones solitary, sessile; calyx lobes 4, to 4 mm across, orbicular, densely tomentose without; corolla white, lobes 4, with a tuft of hairs at the back of each corolla-lobe; stamens 16, glabrous, staminodes 8. Berry brown, globose, 1 cm across, smooth, shining, fruiting calyx enlarged, submembranous, rusty hairy, flat; seeds 2-3.

Common, in dry deciduous forests in most districts. It produces root suckers, prefers laterite and sand stone hills.

Fl.: March- June. Fr.: January-April.

Sattenapalli RF (ADB), *TP* & *PVP* 4070; Vempalli RF (ADB), *GO* & *PVP* 4793; Laxmapur RF (NZB), *TP* & *BR* 6264; Gandhari RF (NZB), *TP* & *BR* 6341; Ghanapur RF (NZB), *BR* & *KH* 7118; Narsapur RF (MDK), *TP* & *CP* 11986; Pillutla RF (MDK), *BR* 11266; Gangapur RF (MDK), *BR* & *CP* 11437; Mohammadabad RF (RR), *MSM* & *KH* 10247; Yerragondupenta (MBNR), *TDCK* 13697; Eturunagaram (WGL), *SSR* 18538; Poochavaram RF (KMM), *RCS* 98731 (BSID).

Diospyros exsculpta Buch.-Ham., Trans. Linn. Soc. London 15: 110. 1827 excl. syn.; non Dalzell 1861; Sant. & Wagh, Bull. Bot. Surv. India 5: 106. 1963. *D. tomentosa* Roxb., Fl. Ind. 2: 532. 1832; C.B. Clarke in Hook.f., Fl. Brit. India 3: 564. 1882; Gamble, Fl. Madras 2: 772. 1921.

Medium-sized tree; bark greyish black, exfoliating in rectangular scales; wood reddish-brown, fairly straight or curly grained, fine and even textured; young branchlets densely rusty tomentose. Leaves broadly ovate, 7.5-15 × 5-8 cm, base rounded, margin entire, apex obtuse or slightly acute; nerves and reticulations conspicuous above; petiole short. Flowers white, 4-5 merous; male flowers in short cymes; female flowers solitary. Berry globose, 2.5-3.7 cm dia., glabrous; seeds 1.2 cm long, shortly oblong, slightly compressed.

Occasional in Karimnagar district in deciduous forests.

Fl.: April- May, Fr.: October- December.

Rechapalli (KRN), *GVS* 22256 (MH).

Diospyros melanoxylon Roxb., Pl. Coromandel t. 46. 1795 & Fl. Ind. 2: 530. 1832; C.B.Clarke in Hook.f., Fl. Brit. India 3: 564. 1882; Gamble, Fl. Madras 2: 772. 1921.

Vern.: *Tuniki, Tumiki, Beedi aku, Tellagada, Tuki, Manchi tumiki*. Eng.: Ebony, Tendu.

Deciduous tree, up to 18 m high; bark dark grey or black, exfoliating in rectangular scales; wood reddish-brown with an irregular black heart wood; young branches tawny tomentose. Leaves coriaceous, broadly elliptic, 6.5-30 × 3.2-14 cm, glabrescent above, pubescent beneath, base acute or rounded, margin entire, apex subacute or obtuse; secondary nerves 8-10 pairs, prominent on both sides; petiole to 1cm. Male flowers 3-12 in 2.5 cm long, tawny-tomentose cymes; calyx lobes 5, ovate, to 9 mm, thick, tomentose; corolla white, 1 mm across, tubular, thin-fleshy, tube 1 cm long, appressed-hairy without, lobes 5, ovate, reflexed, stamens 10-12, filaments to 4 mm long. Female flowers solitary, pedicel stout; calyx lobes 5, ovate, margin folded; ovary globose, 4 mm across, tomentose, style 3 mm long; staminodes 5. Berry globose, 3-4.5 cm across, yellow when ripe, smooth, supported by the thick calyx with wavy reflexed margins; seeds 4-6, compressed, oblong, brown, 5 mm long, shining with a wrinkled testa.

Common in dry, mixed deciduous forests in all districts. It readily grows on poor denuded soil, hot dry hill slopes in moist valleys. It prefers loamy soil with dominant clay fraction and loose porous soil. It is a good coppicer and it also reproduces by root suckers.

Neelvai RF (ADB), *GO* & *DAM* 5164; Mallampet RF (ADB), *TP* & *GO* 5446; Manchippa RF (NZB), *BR* & *KH* 9025; Jalalpur RF (NZB), *BR* & *KH* 9657; Narsapur RF (MDK), *BR* & *CP* 11412; Kusumasamudram (RR), *MSM* & *KH* 11035; Eturunagaram (WGL), *SSR* 18530; Saleswaram (MBNR), *SKB* & *BSS* 30728; Ramavaram (KMM), *RR* 112548 (BSID); Cherukupally (NLG), *BVRS* 175 (NU).

Diospyros montana Roxb., Pl. Coromandel t. 45. 1795 & Fl. Ind. 2.538. 1832; C.B. Clarke in Hook.f., Fl. Brit. India 3: 555. 1882; Gamble, Fl. Madras 2: 776. 1921. *D. cordifolia* Roxb., Pl. Coromandel t. 49. 1795; Gamble, Fl. Madras 2: 776.1921. *D. montana* Roxb. var. *cordifolia* (Roxb.) C.B.Clarke in Hook. f., Fl. Brit. Inida 3: 555. 1882.

Vern.: *Erragath, Erra goda, Jagadagondi, Muchi tunki.*

Semi-evergreen tree, up to 8 m high, often spiny; bark greyish-black, peeling off in thin flakes; wood grey, tinged with yellow or brown with darker patches, and fine grained. Leaves thick-coriaceous, obovate or elliptic, 5-12 × 2-6 cm, glabrous, apex acute, margin entire, base obtuse; secondary nerves 10 pairs, faint; petiole 7 mm long. Flowers white; male flowers in umbels, peduncle 5 mm; calyx lobes 4, ovate, 2.5 mm, thick, ciliate, obtuse, basally connate; corolla greenish, 7 mm across, tubular-campanulate, tube 5 mm long, lobes 4, ovate; stamens 16, in 8 pairs, inner the shorter, filaments 1.5 mm long; pistillode conical; female flowers solitary, axillary, drooping on short pedicels, pedicel to 0.4 cm long; calyx and corolla the same as in male flowers, ovary globose, style 1.5 mm long, stigma capitate; staminodes 4. Berry yellow turning black, globose, 2.5-3 cm in dia., glaucous, pendulous; fruiting calyx thick, reflexed; seeds 5, brown, compressed.

Common in deciduous forests in most districts.

Fl.: March-May. Fr.: December-February.

Sirpur RF (ADB), *GO* 4388; Rampur (ADB), *SSR* 19318; Chedmal RF (NZB), *BR* & *CPR* 9254; Mohammadabad RF (RR), *MSM* & *KH* 10225; Pakhal (WGL), *KMS* 13178 (MH); Nilgiri view point (MBNR), *SRS* 109712 (BSID).

Diospyros sylvatica Roxb., Pl. Coromandel t. 47. 1795; C.B.Clarke in Hook.f., Fl. Brit. India 3: 559. 1882; Gamble, Fl. Madras 2: 773.1921.

Vern.: *Pasara gatha, Tella gatha, Gatha.*

Evergreen tree, up to 10 m high; bark black, smooth, thin; wood grey with black streaks and irregular patches; branchlets glabrous. Leaves coriaceous or membranous, elliptic, 3.5-9 × 2.5-3.5 cm, glabrous, shining above and paler beneath, black when dry, base acute or obtuse, margin entire, apex acuminate; petiole to 1 cm. Flowers white; male flowers few in 1.2 cm long, axillary cymes; Female flowers solitary, axillary. Berry globose, 1.3-1.6 cm in dia., yellowish, smooth; fruiting calyx enlarged, spreading, flat, foliaceous; seeds 2-8, oblong, 1 cm long, rugose, shining, brown.

Occasional, near streams and hill slopes in the moist deciduous forests in Mahabubnagar district.

Fl.: April-May; Fr.: January-March.

Mallelatheertham (MBNR), *SRS* 108996 (BSID).

Diospyros vera (Lour) A.Chev., Cat. Pl. Jard. Bot. Saigon 31. 1919. *Maba buxifolia* (Rottb.) A.L. Juss., Ann. Mus. Natl. Hist. Nat. 5: 418. 1804; C.B. Clarke in Hook.f., Fl. Brit. India 3: 551. 1882; Gamble, Fl. Madras 2: 768.1921. *Pisonia buxifolia* Rottb., Nyesaml. Kongel Danske Viden S.K. Selsk. Skr. 2: 536. t. 4. f. 2. 1783. *M. ebenus* Wight, Icon. Pl. Ind. Orient. t.t. 1228 & 1229. 1848.

Vern.: *Alli, Tella alli, Chinna-ullingi, Uti, Yerruti, Pisinki.*

Shrub to tree, up to 8 m high; bark grey to black, rough; wood grey with darker streaks, close-grained, branchlets glabrous. Leaves coriaceous, obovate-spathulate, 1.5 × 0.6-1.2 cm, glabrous, apex obtuse or emarginate, margin entire, base acute or attenuate; secondary nerves not distinct; petiole to 2 mm. Flowers white, to 4.5 mm long, 3-merous, bisexual in fascicles. Berry purplish black, oblong, 5 mm across, crowned by calyx; seed one.

Occasional in Adilabad district in deciduous forests.

Fl.: April- May. Fr.: June-July.

Mallelatheertham (MBNR), *BR* & *SKB* 30711.

OLEACEAE

1. Erect or climbing shrubs; fruit didymous berry .... **Jasminum**
1. Trees; fruit loculicidally 2-valved capsule:
    2. Capsule orbicular, compressed, thin, erect; seeds not winged; leaves rough above; flowers white-orange .... **Nyctanthes**
    2. Capsule obovoid, woody, pendulous; seeds winged; leaves smooth; flowers yellowish-brown .... **Schrebera**

JASMINUM L.

1. Leaves simple:
    2. Calyx pubescent:
        3. Calyx-lobes minute, deltoid .... **J. arborescens**
        3. Calyx-lobes linear-subulate .... **J. sambac**
    2. Calyx glabrous or nearly so:
        4. Calyx-lobes less than 1 cm long .... **J. angustifolium**
        4. Calyx-lobes more than 1 cm long .... **J. cuspidatum**
1. Leaves compound:
    5. Cymes and flowers sessile, fulvous-tomentose .... **J. brevilobum**
    5. Cymes and flowers stalked, glabrous to pubescent:
        6. Leaves with the two lateral leaflets very small and often wanting; leaflets pubescent .... **J. auriculatum**
        6. Leaves with all the three leaflets present, the lateral ones little smaller than the terminal .... **J. flexile**

Jasminum angustifolium (L.) Willd., Sp. Pl. (ed. 4). 1: 36. 1797; C.B.Clarke in Hook.f., Fl. Brit. India 3: 598. 1882; Gamble, Fl. Madras 2: 790. 1923. *Nyctanthes angustifolium* L., Sp. Pl. 6. 1753. *Jasminum ritchei* C.B.Clarke var. *purpurea* C.B.Clarke in Hook.f., Fl. Brit. India 3: 598. 1882. *p. p.*

Vern.: *Adavimalle, Chirumalle*; Eng.: Wild jasmine.

Scandent shrub, branchlets glabrous. Leaves simple, elliptic-ovate, 1.5-4 × 0.5-1.5 cm, base obtuse or rounded, apex acute, glabrous. Flowers in 1-3-flowered terminal cymes, white, up to 3 cm across, pedicels up to 1.2 cm long; calyx glabrous, lobes 5-6, linear, to 5 mm long; corolla tube up to 1.2 cm long, lobes up to 9, each 1.5 × 0.2 cm, lanceolate; stamens 2. Fruit ellipsoid or globose, 8 mm across.

Occasional in forests of Mahabubnagar and Hyderabad districts.

Fl. & Fr. May.

Farahabad (MBNR), *TDCK* 15344.

Jasminum arborescens Roxb., Fl. Ind. 1: 94. 1820; C.B.Clarke in Hook.f., Fl. Brit. India 3: 594. 1882; Gamble, Fl. Madras 2: 789. 1923. *J. roxburghianum* Wall. ex C.B.Clarke in Hook.f., Fl. Brit. India 3: 595. 1882; Gamble, Fl. Madras 2: 789. 1923.

Vern.: *Garuda malli, Erri malli, Tivva malli, Naga malli, Malletiga, Adavimalle.*

Villous, climbing shrub, branchlets softly pubescent. Leaves elliptic-ovate or ovate, 4-9 × 3-7 cm, softly tomentose on both sides, base rounded, apex acute, main nerves 7 pairs, petiole 1.2 cm long, densely pubescent. Flowers white in lax trichotomous, densely villous, many-flowered cymes, fragrant, pedicels 0.5-1.2 cm long, calyx pubescent, lobes linear, to 2.5 mm long, corolla tube up to 1.2 cm long, lobes 10-12, linear or narrow-oblong. Berry solitary, globose or ellipsoid, black.

Rare in forests in Medak district, cultivated elsewhere.

Fl. & Fr.: December – May.

Narsapur (MDK), *KMS* 8019 (CAL, MH).

Jasminum auriculatum Vahl, Symb. Bot. 3: 1. 1794; C.B.Clarke in Hook.f., Fl. Brit. India 3: 600. 1882; Gamble, Fl. Madras 2: 790. 1923.

Vern.: *Malletiga, Adavi teega malli, Sanna jaji, Vasanthi.*

Scandent shrub, branchlets densely pubescent. Leaves mostly 3-foliolate, the lateral leaflets when present very much reduced, often wanting, terminal leaflet 1.8-5 × 1-3 cm, broadly ovate, base narrowed or rounded, margin entire, apex obtuse or acute. Flowers white, in dense corymbose cymes; calyx-lobes 5, pubescent without; corolla tube 0.8-1.2 cm long, lobes 5-7. Berry globose, black.

Common in forests in all districts. Often cultivated.

Fl. & Fr.: August – December.

Jaipur (ADB), *KH* 5558, *PVP* 9427; Mamidipally RF (NZB), *BR* & *KH* 6434; Gangapur RF (MDK), *TP* & CP 12099; Edupayalu RF (MDK), *RG* 113424 (BSID);

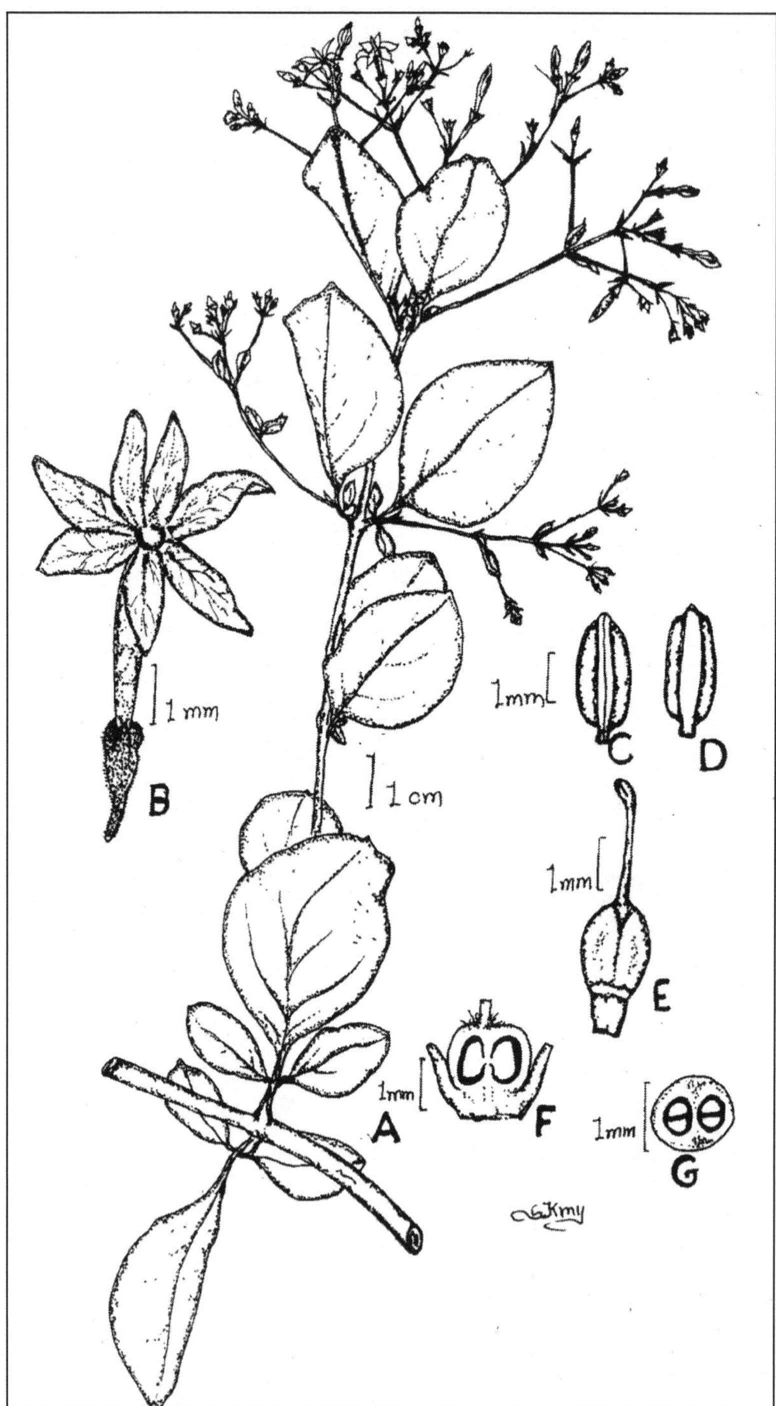

Figure 57: Jasminum auriculatum Vahl.

A. Twig, B. Flower-entire, C&D. Stamens, E. Pistil, F. Ovary l.s., G. Ovary c.s.

Kusumasamudram RF (RR), *MSM* & *NV* 10356; Mannanur (MBNR), *TDCK* 16802; Kodimial (KRN), *GVS* 20080 (MH); Bhadrachalam (KMM), *RCS* 104328, 98973 (BSID); Kambalpally (NLG), *BVRS* 341 (NU).

Jasminum brevilobum A.DC. in DC. Prodr. 8: 307. 1844; C.B.Clarke in Hook.f., Fl. Brit. India 3: 600. 1882; Gamble, Fl. Madras 2: 790. 1923.

Vern.: *Jazimalli.*

A large climbing shrub with spirally twisted stem, tawny pubescent. Leaves deltoid-ovate, 2.5-5.5 × 1.5-4 cm, chartaceous, base truncate, apex mucronate, petiole to 0.5 cm long, with two small, lateral leaflets. Flowers in terminal corymboid cymes, sessile; bract 3 mm, bracteole 1.5 mm; calyx lobes 5, triangular, 3 mm thick, puberulous; corolla 2 cm across, tube 1.7 cm long, lobes 5, obovate, glabrous on both sides; stamens 2, included, filaments 1.5 mm long, anthers apiculate, ovary 2 mm, tomentose above, style 7 mm long. Berry globose, 0.8 cm across.

Cultivated ornamental.

Fl. & Fr.: July-August.

Jasminum cuspidatum Rottl., Neue Schriften Ges. Naturf. Freunde Berlin 4: 192. 1803. *J. rigidum* Zenker, Pl. Ind. 5. t.6. 1835; C.B.Clarke in Hook.f., Fl. Brit. India 3: 598. 1882; Gamble, Fl. Madras 2: 790. 1923.

Vern.: *Adavi malli, Pedda malli, Naga malli.*

Dense rigid shrub, rarely scandent, stems glabrous, young parts pubescent. Leaves elliptic or obovate, 1.8-5.8 × 0.6-1.5 cm, coriaceous, shining, acute at both ends, petiole to 4 mm long, nerves prominent. Flowers white, fragrant, in 1-5-flowered, corymbose, terminal and axillary cymes; calyx lobes 5, unequal, glabrous, lobes more than 1 cm long; corolla-tube up to 2.5 cm long, lobes 6, stamens 2. Berry oblong, globose, bilobed, shining, black.

Common near streams in deciduous forests in Ranga Reddi and Mahabubnagar districts.

Fl. & Fr.: January – May.

Pullaechelma (MBNR), *SRS* 109200 (BSID).

Jasminum flexile Vahl, Symb. Bot. 3: 1. 1794; C.B.Clarke in Hook.f. Fl. Brit. India 3: 601. 1882; Gamble, Fl. Madras 2: 791. 1923. *J. azoricum* sesnu Burm.f., Fl. Ind. 6: 1768 *p.p.* non L.

Vern.: *Jazi.* Eng.: Azores jasmine.

Straggling shrub. Leaves trifoliolate, leaflets lanceolate, glabrous on both sides, dark green, 5-7 × 2-3 cm, margin entire, apex caudate. Cymes compound, flowers many, large, white, fragrant; calyx lobes five, small, imbricate, lanceolate, 0.5 mm long, acute apex; corolla 5-lobed, lobes lanceolate, spreading, white with slightly pinkish, corolla-tube long, 2-2.5 cm; stamens 2, inserted. Berry green, 1 cm wide, black when ripe.

Cultivated in gardens .

Fl. & Fr.: November – March.

Jasminum sambac (L.) Aiton, Hort. Kew. 1. 1: 8. 1789; C.B.Clarke in Hook.f., Fl. Brit. India 3: 591. 1882; Gamble, Fl. Madras 2: 788. 1923. *Nyctanthes sambac* L., Sp. Pl. 6. 1752.

Vern.: *Gundu malle, Boddu malle, Malli;* Eng.: Arabian jasmine.

Erect or some times small climbing shrub, branches sarmentose. Leaves simple, broadly ovate or elliptic, 3-8 × 2-5 cm, chartaceous, shining, pubescent along nerves, base rounded or truncate, apex acute to acuminate, nerves 6 pairs, petiole 0.5 cm long, hairy. Flowers white, fragrant, in 3-flowered terminal cymes; calyx-tube 1.8 mm, pubescent, lobes 5-7, linear-subulate, 6-10 mm long, corolla-tube 1.5 cm, lobes 6. Berry globose, violet when ripe.

Found near villages as an escape from cultivation.

Fl. & Fr.: March-August.

Macha Reddy (NZB), *BR* & *KH* 5545.

NYCTANTHES L.

Nyctanthes arbor-tristis L., Sp. Pl. 6. 1753; C.B.Clarke in Hook.f., Fl. Brit. India 3: 603. 1882; Gamble, Fl. Madras 2: 792. 1923.

Vern.: *Parijathamu, Pagada malli, Krishti.*

Small tree, 4-6 m high, bark light brown, roungh, branchlets 4-angular, often pendent. Leaves coriaceous, ovate to elliptic, 7-14 × 3-7.2 cm, scabrous with short bulbous hairs above, pubescent beneath, base obtuse or rounded, margin entire, apex short to long acuminate, secondary nerves 3-5 pairs, prominent beneath, petiole to 7 mm. Flowers white-orange, very fragrant, to 2.5 cm across in trichotomous, axillary and terminal cymes. Capsule brown, elliptic-oblong or orbicular, compressed, 2-valved, 0.7-1.6 × 0.8-1.8 cm, glabrous, epicarp thick, seeds 2.

Common in dry deciduous forests and often cultivated in gardens.

Fl. & Fr.: Throughout the year.

Vempalli (ADB), *GO* 4433; Ibrahimpet (NZB), *TP* & *BR* 6135; Kondagutta (KRN), *GVS* 25688 (MH); Narsapur RF (MDK), *TP* & *CP* 11973; Pegarikutta east (MDK), *KMS* 6081 (MH); Kothapalli (RR), *MSM* & *KH* 10446.

Notes: *Nyctanthes arbor-tristis* was assigned to family Oleaceae till Airy Shaw (in Kew Bull. 272. 1952) transferred it to the family Verbenaceae under new subfamily Nyctanthoideae. Stant (Kew Bull. 273-276. 1952) pointed out several morphological and anatomical features different from those in Oleaceae that prove Verbenaceous affinity of this plant, but nothing conclusive can be said at present, since the finding of other authors give weight to the arguments on either sides. Crete (1963) on embryological evidence and Das & *al.* (1966) on chemotaxonomic evidence establish a close relationship between *Nyctanthes* and

Oleaceae. Kapil and Vani (1966) support de Jussieu (1789) treatment of assigning *Nyctanthes* to Oleaceae. Kundu (1966), however on anatomical and palynological evidences suggests that *Nyctanthes* may belong neither to Oleaceae nor to Verbenaceae.

SCHREBERA Roxb. *nom. cons.*

Schrebera swietenioides Roxb., Pl. Coromandel t. 101. 1799 & Fl. Ind. 1: 109. 1832; C.B.Clarke in Hook.f., Fl. Brit. India 3: 604. 1882; Gamble, Fl. Madras 2: 792. 1923.

Vern.: *Mokkem, Makkam, Ishuga rashi, Magalingam, Mokkalapa*; Eng.: Weaver's bean tree.

Deciduous tree, up to 20 m high, crown globose, bark pale grey, nearly smooth exfoliating in thin flakes, lenticellate. Leaves imparipinnate, to 25 cm long, petiole 7 cm long, 7-9-foliolate, leaflets coriaceous, ovate to ovate-lanceolate, 2.5-8.5 × 1-3.5 cm, glabrous above, villous beneath, base often unequal sided, margin entire, apex acute-acuminate, secondary nerves 5-9 pairs, petiolule to 3 mm. Flowers yellowish-brown, fragrant at night, numerous in 8 cm long, trichotomous cymes. Capsule woody, obovoid or pear-shaped, to 6 x 2.5 cm, pendulous, light brown, very hard and woody, dotted with scabrous specks, loculicidal, 2-valved; seeds 20-30, flat, winged, 2-3.7 cm.

Common, in the hills of dry deciduous forests of Karimnagar and Mahabubnagar districts.

Fl.: April – May. Fr.: Through out the year, dehiscing by March.

Farahabad (MBNR), *MVS* & *MB* 38022; Rechapalli (KRN), *GVS* 22247 (MH).

SALVADORACEAE

1. Armed shrub; fruit white when ripe .... **Azima**
1. Unarmed tree; fruit red when ripe .... **Salvadora**

AZIMA Lam.

Azima tetracantha Lam., Encycl. 1: 343. 1783; Hook.f., Fl. Brit. India 3: 620. 1882; Gamble, Fl. Madras 2: 800. 1923.

Vern.: *Tella uppili, Mursangu, Mundra kampa, Puttu podha, Uppara podha.*

Scrambling armed shrub, up to 5 m high, spines straight, paired in the leaf axils, 1.5-2.8 cm long, branchlets quadrangular, pale pubescent. Leaves elliptic-oblong to lanceolate, 3-6 × 2-4 cm, glabrous, greyish-green, margin entire, tip acute with spine, secondary nerves 1-3; petiole 3-5 mm long. Flowers fascicled, zygomorphic, dioecious; calyx 4-toothed, bipartite, unequal, imbricate; corolla cream-coloured, 3 mm wide, lobes 4, free; stamens free from the corolla, 4; pistillode conical, ovary bilocular, staminodes 4. Berry 2-seeded, white coloured when ripe, 7.5-12 mm diam.

Coomon in scrub forests in foot hills of Nalgonda district.

Fl. & Fr.: January – July.

Peddvura (NLG), *BVRS* 505 (NU).

SALVADORA L.

Salvadora persica L., Sp. Pl. 122. 1753; Hook.f., Fl. Brit. India 3: 619. 1882; Gamble, Fl. Madras 2: 799 1923. var. **wightiana** (Thw.) Verdc., Kew Bull. 19. 150. 1964. *S. wightiana* Planchon ex Thwaites, Enum. Pl. Zeyl. 190. 1860.

Vern.: *Ghunia, Gonni, Wara gogu, Gonni, Chekkara chettu;* Eng.: Tooth brush tree, Salt bush, Meshwak.

Large straggling evergreen shrub or small tree, 8-10 m high, crown compact, trunk short, branches drooping, bark grey, deeply cracked, branchlets glabrous, pale yellowish. Leaves thick-chartaceous, ovate-lanceolate or elliptic-oblong, 3.2 – 6.2 × 1.7-2.8 cm, glabrous, base rounded or obtuse, margin entire, apex obtuse or subacute, secondary nerves 3-5 pairs, slightly raised beneath; petiole 0.7-2 cm. Flowers yellowish-green, 3 mm across in axillary and terminal copious panicles; calyx campanulate, 4-lobed, lobes rounded; corolla cream, 3 mm across, shortly tubular, lobes 4, recurved, 1.5 mm, membranous, emarginated; stamens 4, inserted at the middle of the corolla tube. Drupe red, globose, 3-4 mm diam., smooth, juicy; seed 1, globose, 2 mm across.

Common in open places, especially in saline soils and black cotton soils. Planted in Mosques and Muslim burial grounds

Fl. & Fr.: January-May. .

Rudroor (NZB), *BR* & *KH* 9636.

APOCYNACEAE

1. Armed shrubs; fruit berry .... **Carissa**
1. Unarmed shrubs, trees, lianas or twining or erect herbs; fruit follicle, drupe or capsule:
    2. Corona present:
        3. Corona petaloid .... **Nerium**
        3. Corona scale like .... **Wrightia**
    2. Corona absent:
        4. Corolla lobes overlapping to right:
            5. Small trees:
                6. Branchlets with irregular peel off flakes; flowers in axillary cymes .... **Holarrhena**
                6. Branchlets with marked leaf scar; flowers in terminal cymes .... **Cerbera**
            5. Lianas:
                7. Fruit capsule .... **Allamanda**
                7. Fruit follicles:
                    8. Tender parts and inflorescence glabrous .... **Anodendron**

8. Tender parts and inflorescence pubescent:
    9. Branchlets, flowers not rusty pubescent; calyx divided to below the middle; corolla lobes longer than the tube .... **Aganosma**
    9. Branchlets, flowers rusty pubescent; calyx divided to above the middle; corolla lobes shorter than the tube .... **Ichnocarpus**

4. Corolla lobes overlapping to left:
    10. Corolla salver-shaped:
        11. Plant herbaceous; corolla-lobes broad .... **Catharanthus**
        11. Shrubs or trees; corolla-lobes lanceolate:
            12. Leaves verticillate or ternate .... **Alstonia**
            12. Leaves opposite decussate .... **Tabernaemontana**
    10. Corolla funnel-shaped:
        13. Corolla yellow; fruit a drupe .... **Cascabela**
        13. Corolla other than yellow; fruit follicular mericarp:
            14. Xerophytic trees; corolla thick, variously coloured .... **Plumeria**
            14. Mesophytic shrubs; corolla thin, white:
                15. Twining liana; flowers large, more than 1.5 cm wide .... **Vallaris**
                15. Erect shrubs; flowers small, less than 1 cm wide .... **Rauvolfia**

AGANOSMA (Blume) G. Don

Aganosma heynei (Spreng.) I.M.Turner, Taxon 62(1): 156. 2013. *Echites heynei* Spreng., Syst. Veg. 1: 632. 1834. *Aganosma dichotoma* (Roth) K.Schum., in Engl. & Prantl, Pflanzenfam. 4 (2): 173. 1895; Gamble, Fl. Madras 2: 819. 1923. *Echites dichotoma* Roth, Nov. Pl. Sp. 136. 1821. *Aganosma caryophyllata* G. Don, Gen. Syst. 4: 77. 1837; Hook.f., Fl. Brit. India 3: 664. 1882.

Vern.: *Palamalle, Guda palateega, Nalla teega, Malathi teega.*

Extensive climbing shrub, up to 15 m long, branchlets tomentose. Leaves ovate or elliptic-oblong, 5.5-10 × 3-5.5 cm, coriaceous, glabrous or tomentose beneath, base rounded or obtuse, apex acute, 3-5-nerved; petiole 6.2 – 18 mm long. Flowers white in laxly corymbose cymes; calyx lobes 5, linear elliptic, 1.5 cm; corolla lobes 5. 1.5 cm, tube 1.5 cm; pedicel to 1 cm long. Follicles curved upwards, to 20 cm long.

Occasionally cultivated in gardens for its sweet scented and showy flowers.

Fl. & Fr.: February – September.

ALLAMANDA L.

Allamanda cathartica L., Mant. Pl. 214. 1771; Gamble, Fl. Madras 2 : 821. 1923.

Straggling or climbing evergreen shrub. Leaves obovate or oblong to lanceolate, 8-12 x 2.5-4 cm, attenuate to cuneate at base, wavy-margined, acuminate, subcoriaceous, petioles to 0.5 cm long. Flowers in cymose terminal panicles; calyx-lobes ovate-oblong, 0.8-1.2 cm long; corolla bright yellow or golden; tube 2-3.5 cm long, greenish at base, cylindrical in lower half, campanulate above, lobes ovate-orbicular, obtuse, 2.5-3.5 cm long. Capsule subglobose, 2-4 cm in diam.; spine 0.5-1 cm; seeds winged, 1.2-1.5 cm long.

Grown in gardens, houseyards as an ornamental.

Fl. & Fr.: July-October.

ALSTONIA R.Br. *nom. cons.*

Alstonia scholaris (L.) R.Br., Mem. Wern. Nat. Hist. Soc. 1: 76. 1811; Gamble, Fl. Madras 2: 810. 1923. *Echites scholaris* L., Mant. Pl. 53. 1767; Hook.f., Fl. Brit. India 3: 642. 1882.

Vern.: *Edakulapala, Palagaruda, Edakula ponna*; Eng.: Shitan wood, Devils tree, Dita-bark tree.

Large, evergreen tree, base buttressed, bark dark grey, nearly smooth, lenticellate, yellow inside, exuding milky latex when fresh cut, branches spreading in tiers of whorls, branchlets copiously lenticellate. Leaves 4-7 in a whorl, coraiaceous, elliptic-oblong, obovate or oblanceolate, 10-25 × 3.8-8 cm, glabrous, green above, paler and smooth beneath, base acute or obtuse, margin entire, apex obtusely acute or emarginated, secondary nerves numerous, conspicuous, parallel, connecting near margin and form a intramarginal vein; petiole to 5 mm. Flowers greenish-white, to 1.5 cm across, sub-capitate at the ends of the panicle branches; calyx-tube short, lobes 5, eglandular; corolla 0.8-1.5 cm across, salver-shaped, tube cylindric. Follicle pendulous, cylindric, narrow, 15-50 × 0.3 cm; seeds numerous, linear-oblong, 6 mm long, flattened, hairs longer at both ends.

Planted in gardens, along roadsides, on dividers as an ornamental and avenue tree.

Fl.: December-March; Fr.: May – September.

ANODENDRON A. DC.

Anodendron paniculatum (Roxb.) A.DC., Prodr. 8: 444. 1844; Hook.f., Fl. Brit. India 3: 668. 1882; Gamble, Fl. Madras 2: 820. 1923. *Echites paniculata* Roxb., Fl. Ind. 2: 17. 1832.

Large, woody climbers with stout stems; branches terete, glabrous. Leaves elliptic or oblong, to 13.7 × 4.5 cm, coriaceous, base narrowed or rounded, margin entire, apex acuminate; petiole stout, up to 15 cm long. Flowers cream coloured, in axillary and terminal paniculate cymes; peduncle 9 cm long, pedicel 4 mm long; calyx cupular, lobes 5, equal; corolla salver-form, tube 2 mm long, inflated over stamens, lobes 5, overlapping to right, twisted to left; stamens 5, connivent, attached 1/3 from the base; ovaries 2, free, style 1 mm long, stigma obconic to disciform. Follicles divaricate, almost at right angle to the fruiting peduncle, to 12 × 2 cm, seeds compressed, crowned by white silky coma.

Occasional in Warangal district.

Fl. & Fr.: October – May.

Mallurgutta (WGL), *RKP* 110981 (BSID).

CARISSA L. *nom. cons.*

1. Corolla large, the tube more than 1.5 cm long:
   2. Leaves elliptic or obovate, apex acute-mucronate .... **C. spinarum**
   2. Leaves ovate, apex obtuse-emarginate .... **C. carandas**
1. Corolla small, the tube less than 1 cm long .... **C. hirsuta**

Carissa carandas L., Mant. Pl. 52. 1767; Hook.f., Fl. Brit. India 3: 630. 1882; Gamble, Fl. Madras 2: 804. 1923.

Vern.: *Pedda kalivi, Pedda vakakaya, Kali-kai, Vaaka.*

Large, evergreen thorny shrub, up to 5 m high, with two stout sharp horizontal spines, spines to 3 cm long, branches terete, alternate. Leaves elliptic-oblong, sometimes orbicular, 2-6.5 × 1-3 cm, base rounded or subacute, margin entire, apex obtuse or emarginate, glabrous on both sides. Flowers white, in axillary or terminal corymbose cymes, peduncle up to 1.1 cm long; calyx lobes 5, unequal; corolla –tube up to 1.3 cm long, glabrous without, villous within. Fruit ellipsoid, ovoid, 6.2 mm diam., seed 1 per cell.

Cultivated in kitchen gardens in small scale for its green fruits used as vegetable in Hyderabad (Ramana, 2010).

Fl. & Fr.: March – June.

Carissa hirsuta Roth, Nov. Pl. Sp. 128. 1821; Gamble, Fl. Madras 2: 805. 1923. *C. spinarum* L. var. *hirsuta* (Roth) Hook.f., Fl. Brit. India 3: 631. 1881.

Vern.: *Kalli.*

A thorny shrub, branchlets grey-tomentose, spines long and straight. Leaves elliptic oblong, 4-5 × 2-2.5 cm, softly grey-tomentose, apex obtuse or acute, base slightly cuneate. Flowers white, in terminal, shortly peduncled cymes; calyx-lobes ovate, villous. Berry globose, black-purple

In dry hills in Khammam district.

Fl. & Fr.: March-June.

Yenkur (KMM), *RR* 107959 (BSID).

Carissa spinarum L., Mant. Pl. 559. 1771; Hook.f., Fl. Brit. India 3: 631. 1882; Gamble, Fl. Madras 2: 805. 1923. *Carissa paucinervia* A.DC. in DC., Prodr. 8: 333. 1844; Hook.f., Fl. Brit. India 3: 631. 1882.

Vern.: *Kalivi, Palaregi, Chinna kalivi, Chinna vaaka.*

Small shrub, branches scandent, diffuse or suberect, subglabrous or pubescent, spines to 2 cm long, some times forked. Leaves ovate-elliptic, 2-3.5 × 1-2 cm, base acute, margin entire, apex acute or mucronate. Flowers white or slightly rose, scented, in few-flowered terminal or axillary cymes; corolla tube 1-1.3 cm long, thinly pubescent, lobes oblong or oblong-lanceolate. Berry globose, 0.6-0.8 cm across, pointed, 4-seeded, black-purple when ripe.

Common in scrub jungles and open degraded forests of Ranga Reddi, Nalgonda, Hyderabad and Mahabubnagar districts.

Fl. & Fr.: February – October.

Central University campus (RR), *MSM* 15154; Nagarjunasagar (NLG), *BVRS* 125 (NU).

CASCABELA Rafin.

Cascabela thevetia (L.) Lippold, Feddes Repert. 91: 52. 1980. *Cerbera thevetia* L., Sp. Pl. 209. 1753. *Thevetia nerifolia* Juss. ex Steud., Nom. Bot. (ed.2.) 2: 680. 1841 (*'nereifolia'*); Gamble, Fl. Madras 2: 821. 1923. *T. peruviana* (Pers.) Merr., Philipp. J. Sci. 9: 130. 1914. *Cerbera peruviana* Pers., Syn. P. 1: 267. 1805.

Vern.: *Pacha ganneru, Patchalari.*

Large, evergreen shrub to small tree, bark greyish. Leaves crowded, alternate, coriaceous, narrowly linear, to 12 × 0.8 cm, base acute, margin entire, apex obtuse, glabrous, green above, paler beneath, midrib prominent below, lateral nerves faint; petiole to 1 cm long. Flowers yellow or white, large, axillary, solitary, or in a few-flowered cymes; calyx-lobes 5, unequal, lanceolate, to 1 cm; corolla funnel-shaped, yellow, 3.5 – 5 cm diam., tube 4 cm, lobes 5; stamens 5, included. Drupe yellow, broadly turbinate, four-angled, fleshy, glabrous.

Cultivated in gardens for its showy flowers, rarely as an escape.

Fl. & Fr.: Throughout the year.

Yedapally (NZB), *BR* & *KH* 6465; Medak (MDK), *CP* 11743; Mohammadabad (RR), *MSM* & *PRSR* 15125; Mannanur (MBNR), *TDCK* 13681.

CATHARANTHUS G. Don

1. Leaves lanceolate, apex acute to acuminate; flowers less than 0.8 cm diameter; mericarps 4.5 cm long; erect annual herb .... **C. pusillus**

1. Leaves elliptic-oblong or obovate, apex rounded; flowers more than 2 cm diam.; mericarps 3 cm long; perennial undershrub .... **C. roseus**

Catharanthus pusillus (Murray) G.Don, Gen. Syst. 4: 95. 1837. *Vinca pusilla* Murr., Nov. Comment. Soc. Regiae Sci. Gott. 3: 66. 1733; Hook.f., Fl. Brit. India 3: 640. 1882. *Lochnera pusilla* K. Schumach., in Engl. & Prantl, Pflanzenfam. 4(2): 145. 1895; Gamble, Fl. Madras 2: 809. 1923.

Vern.: *Erri mirapa, Kaatuka kalugaku, Kanupoolaku, Gulvidi, Sukbanda.*

Small, annual, erect herb, 30 cm high, branchlets terete below, quadrangular above. Leaves lanceolate, 2.5-6 × 0.7-1.5 cm, base acute to attenuate, apex acute to acuminate; petiole 4 mm long. Flowers white in terminal 5-7-flowered terminal or axillary corymbs; calyx-lobes filiform, 3.7 mm long; corolla-tube 0.7-1 cm long. Follicles 4-7.5 cm long, very slender.

Frequent in cultivated fields, occasional on road sides.

Fl. & Fr.: July-December.

Vempalli (ADB), *GO* & *PVP* 4739; Kamareddy (NZB), *BR* & *KH* 7145; Venkatapuram (MDK), *TP* & *CP* 14010; Kusumasamudram RF (RR), *MSM* & *NV* 10349; Narsapur (MDK), *KMS* 6745 (CAL & MH); Medak (MDK), *RG* 104149 (BSID); Narlapuram (MBNR), *BR* & *TSS* 29837; Pakhal (WGL), *KMS* 13105 (MH); Kodimial (KRN), *GVS* 20065 (MH); Durachipalli (NLG), *BVRS* 36 (NU).

Cathranthus roseus (L.) G. Don, Gen. Syst. 4: 95. 1837. *Vinca rosea* L., Syst. Nat. ed. 10. 944. 1759; Hook.f., Fl. Brit. India 3: 640. 1882. *Lochnera rosea* (L.) Reich., Consp. Regni Veg. 134. 1828; Gamble, Fl. Madras 2: 809. 1923.

Vern.: *Billa ganneru, Nitya kalyani, Nityasumangali*; Eng.: Periwinkle.

Perennial subshrub, to 75 cm high. Leaves elliptic-oblong or obovate, base acute, margin entire, apex rounded, glabrous; petiole up to 1 cm long. Flowers white or pink in axillary pairs, shortly pedicellate; calyx-lobes 5; corolla tube up to 3 cm long, lobes 5, obovate-triangular; stamens 5, adnate to the corolla-lobes towards apex of the corolla tube, sessile, anthers up to 0.25 cm long; ovary of 2 carpels, style filiform, up to 2 cm long, stigma columnar. Follicles paired, 3-4 x 0.3 cm, seeds black.

Cultivated as ornamental and naturalized.

Fl. & Fr.: Round the year.

Domakonda (NZB), *BR* 9692; Mombojipally (MDK), *BR* & *CP* 11640; Central University campus (RR), *MSM* 19210; Dimmagudem (KMM), *RCS* 99063 (BSID); Peddvura (NLG), *BVRS* 507 (NU).

CERBERA L.

Cerbera odollam Gaertn., Fruct. Sem. Pl. 2: 193. t.124. 1791; Hook. f., Fl. Brit. India 3: 638. 1882.

Small to moderate-sized tree, bark whitish-grey, papery, branchlets whorled, stout, marked with leaf scars. Leaves lanceolate, obovate-oblong or oblanceolate, 10-28 × 3-7 cm, glabrous, base tapered, margin entire, apex abruptly acuminate, secondary nerves numerous, slender, horizontal, connected by an intra-marginal nerve; petiole to 3 cm. Flowers white with yellow throat, turning purple, 3.5-4.5 cm long, odorous, in terminal, peduncled paniculate cymes. Fruit green, subglobose, 5-10 cm long, smooth; seeds ovoid, 2-2.5 cm across, smooth, glabrous.

Cultivated as an ornamental.

Fl.: March-April; Fr.: August.

HOLARRHENA R.Br.

Holarrhena pubescens (Buch.-Ham.) Wall., ex G. Don, Gen. Hist. 4: 78. 1837. *Echites pubescens* Buch.-Ham., Trans. Linn. Soc. London 13: 521. 1821. *Holarrhena antidysenterica* Wall., ex DC., Prodr. 8: 413. 1844; Hook.f., Fl. Brit. India 3: 644. 1882; Gamble, Fl. Madras 2: 811. 1923.

Vern.: *Kolamukki, Palavareni, Pala chettu, Tedlapala, Palankudu, Kodaga, Palakodishe, Istari-pala, Kodisipala.*

Deciduous small tree, up to 8 m high, bark thick, brown, peeling off in irregular flakes. Leaves chartaceous, elliptic-oblong, 8.5-14 × 5.5-9 cm, base rounded, margin entire, apex shortly acuminate, secondary nerves ca. 14 pairs; petiole to 4 mm. Flowers white to yellowish, in terminal and axillary, paniculate or corymbose cymes; calyx 5-lobed, lobes linear, 2 mm long; corolla tube pubescent, 1 cm long, lobes 1 mm wide, 1 cm long; stamens 5, inserted at base of corolla tube, anthers included, free from stigma; stigma bilobed, reaching to middle. Follicles distinct, linear, smooth, terete, slender, curved, to 20 cm long; seeds oblong-linear, coma apical, pale brown, testa thin, some what brittle, closely adhering.

Common, on hill slopes in deciduous forests and open lands in most of the districts.

Fl.: April-July. Fr.: Through out the year.

Chennoor RF (ADB), *TP* & *GO* 5417; Neredukonda (ADB), *KH* 5569; Mamidipally RF (NZB), *BR* & *KH* 9017; Narsapur RF (MDK), *CP* 11576, *KMS* 7649 (MH); Nayanigutta (RR), *MSM* & *NV* 10328; Gourammasari (MBNR), *TDCK* 13698; Mannanur (MBNR), *TDCK* 15359; Eturnagaram (WGL), *SSR* 18549; Rollapadu (KMM), *RR* 108030 (BSID); Aklaspur (KRN), *GVS* 22506 (MH).

ICHNOCARPUS R.Br.

Ichnocarpus frutescens (L.) W.T.Aiton, Hortus Kew. 2: 69. 1811. *Apocynum frutescens* L., Sp. Pl. 213. 1753. *Quirivelia frutescens* (L.) M.R. & S.M.Almeida, J. Bombay Nat. Hist. Soc. 90: 427. (1993) 1994. *Ichnocarpus frutescens* (L.) R.Br., Mem. Wern. Nat. Hist. Soc. 1: 62. 1811; Hook. f., Fl. Brit. India 3: 669. 1882; Gamble, Fl. Madras 2: 820. 1923; Hanumanthappa in Pulllaiah & Moulali, Fl. Andhra Pradesh 2: 574. 1997.

Vern.: *Nalla teega, Illukatte, Karra tivva, Mani tivva, Korampala, Kaaram pala, Pala tivva.*

Straggling shrub or liana, branched profuse, younger parts rusty pubescent. Leaves oblong or elliptic-ovate, 3-8 × 1-2 cm, margin entire, apex acute. Cymes paniculate, pubescent, terminal and axillary; calyx cupular, lobes 5, alternating with glandular scales; corolla cream, 6 mm wide, 5-lobed, salver-shaped, lobes 5,

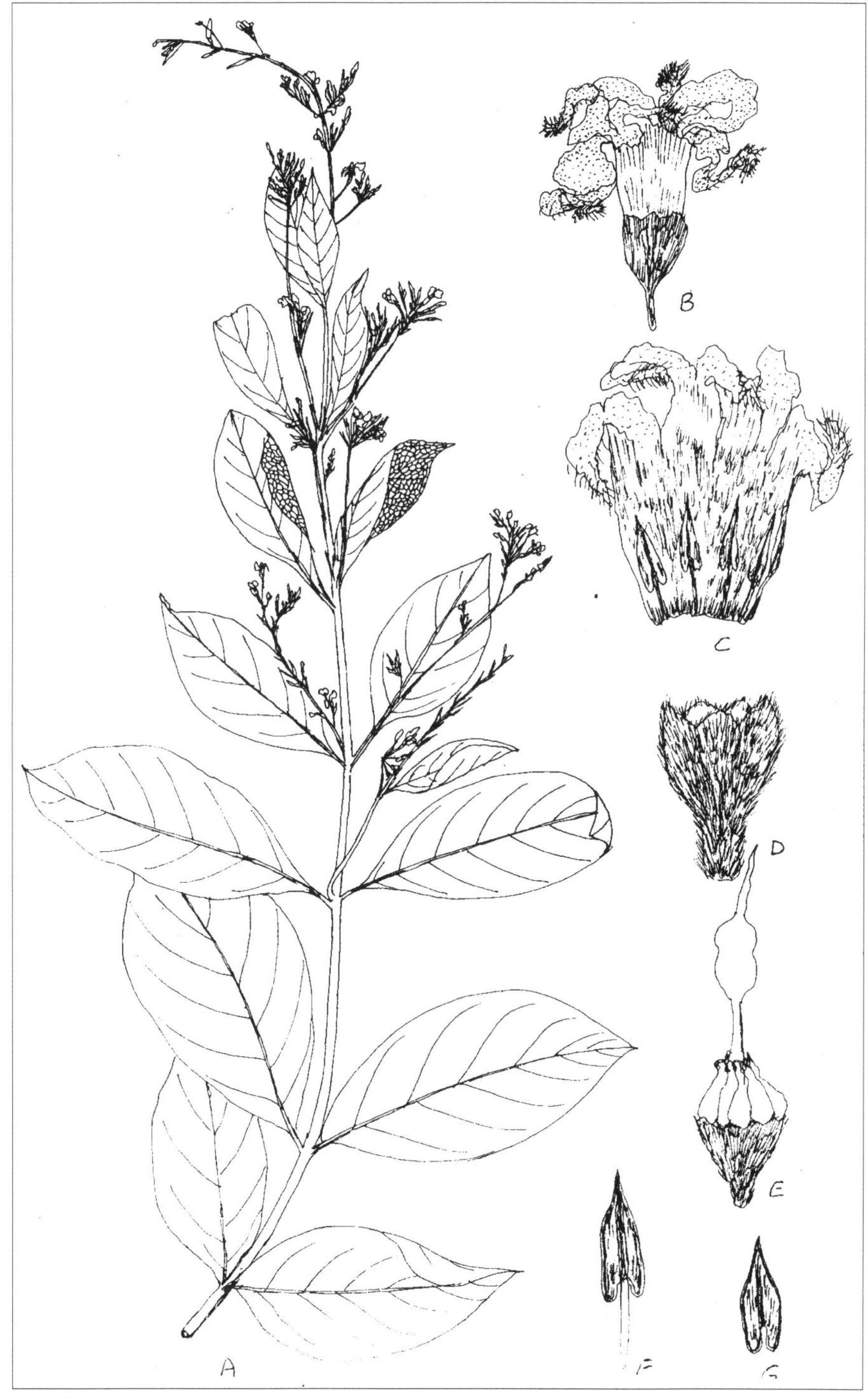

Figure 58: Ichnocarpus frutescens (L.) W.T. Aiton

A. Twig, B. Flower, C. Corolla split open, D. Calyx, E. Pistil, F. & G. Stamens.

overlapping to right, pubescent; stamens 5, included. Fruit follicular mericarps, 8-12 × 0.3 cm, seeds linear, basally furrowed, coma apical, silky.

Common in all districts in thickets in forests.

Fl. & Fr.: September – February.

Rajganjala RF (ADB), *GO* 4383; Janakampalli (NZB), *TP* & *BR* 6325; Gandivet RF (NZB), *BR* 9542; Gangapur RF (MDK), *BR* & *CP* 11436; Anantasagar RF (RR), *MSM* & *KH* 10510; Gunavaram RF (KMM), *RCS* 98782 (BSID); Thupakulagudem (WGL), *PVS* & *NRR* 76989 (BSID); Osmangunta (NLG), *BVRS* 425 (NU).

NERIUM L.

Nerium oleander L., Sp. Pl. 209. 1753. *N. indicum* Mill., Gard. Dict. 2: 1768. *N. odorum* Soland. in Ait., Hort. Kew. 297. 1789; Hook.f., Fl. Brit. India 3: 655. 1882; Gamble, Fl. Madras 821. 1923.

Vern.: *Ganneru.*

Large, erect, profusely branched, evergreen, xerophytic shrub, branched from the base, to 3 m high. Leaves linear-lanceolate, 7.5-15 × 1.2-2.5 cm, sparsely pubescent when young, later glabrous, thickly coriaceous, midrib very stout, base acute, margin entire, apex acute, thick green when aged, arranged in whorls. Flowers, rose, red or white in terminal many-flowered polychasial cymes, fragrant. Follicles 5.5-17 × 0.7-1.0 cm, cylindric, straight, ribbed.

Ornamental, cultivated in gardens and temples, naturalized.

Fl. & Fr.: Most part of the year.

Neredugonda (ADB), *GO* & *DAM* 5099; Tekrial (NZB), *TP* & *BR* 6421.

PLUMERIA L.

Plumeria rubra L., Sp. Pl. 209. 1753. *P. acuminata* R.Br. in Aiton, Hort. Kew 2: 2. 1789. *P. acutifolia* Poiret in Lam., Encycl. Suppl. 2: 667. 1812; Hook. f., Fl. Brit. India 3; 641. 1882; Gamble, Fl. Madras 2: 821. 1923.

Vern.: *Deva ganneru, Devakanchan, Nuruvarahaalu, Vaada ganneru*; Eng.: Temple of Pagoda tree.

Tree, up to 7 m high, bark greyish, smooth. Leaves obovate, obovate-oblong, oblanceolate, 20-25 × 7-8 cm, glabrous to pubescent beneath, base tapering, margin entire, apex acute, obtuse or shortly acuminate, secondary nerves 10-15, 1.5 cm apart, conspicuous with marginal vein; petiole to 7 cm. Flowers red or white with yellow notch, fragrant, large in short, umbellately divided cymes with long peduncle. Follicles paired, to 30 × 3.5 cm, seeds flat, winged.

Planted as an ornamental in gardens and temples in all districts.

Fl.: Most part of the year.

Banala (MBNR), *TDCK* 13700; Vatavarlapally (MBNR), *SRS* 108972 (BSID); Gangapur to Chownapur (MDK), *RG* 106739 (BSID).

RAUVOLFIA L.

1. Leaves 4 at each node; calyx puberulous; corolla tube not inflated .... **R. tetraphylla**
1. Leaves 2-4 per node; calyx glabrous; corolla tube inflated .... **R. serpentina**

Rauvolfia serpentina (L.) Benth. ex Kurz., For. Fl. Burma 2: 171. 1877; Hook.f., Fl. Brit. India 3: 632. 1882; Gamble, Fl. Madras 2: 807. 1923. *Ophioxylon serpentinum* L., Sp. Pl. 1043. 1753.

Vern.: *Sarpagandha, Pathalagaridi, Pathala gain.*

Small erect undershrub, up to 2 m high. Leaves in whorls of 3, lanceolate, 10-15 × 3-5 cm, glabrous, bright green above, pale beneath, base attenuate, margin entire, apex acute to shortly acuminate, main nerves 8-10 pairs, petiole to 2 cm. Flowers white, often tinged with pink, in many-flowered irregular corymbose cymes. Drupes connate except at top, 0.6-1 cm in diam., purplish black when ripe.

Planted in gardens. Rare in Pandem RF in Warangal district (C.S. Reddy, 2001).

Fl. & Fr.: March – August.

Hanamkonda (WGL), *CSR* 690 (KU); Pandem RF (WGL), *CSR* 1343 (KU).

Rauvolfia tetraphylla L., Sp. Pl. 208. 1753. *R. canescens* L., Sp. Pl. (ed. 2.). 303. 1762; Gamble, Fl. Madras 2: 808. 1923.

Vern.: *Papataku, Bara chandrika.*

Large, spreading dichotomously branched shrub, up to 2 high. Leaves 4 at each node, elliptic-ovate, elliptic-oblong, 1.1-5.8 × 1-3 cm, lateral nerves 10 pairs; petiole 3 mm long. Flowers white or pink in terminal 5-7-flowered corymb; calyx 5-lobed, reddish, 4 mm long; corolla white, turning purple-red, tube long, 0.8-1.5 cm. Drupe brownish-purple, globose, connate to the top, 5-6 mm in diam., red when ripe.

Cultivated and found as an escape in Ranga Reddi district.

Fl. & Fr.: September-February.

Central University campus (RR), *MSM* 19203.

TABERNAEMONTANA L.

1. Lobes of corolla overlapping to left; corolla tube long, dilated below the middle; main nerves 8 pairs .... **T. divaricata**
1. Lobes of corolla overlapping to right; corolla tube slender and narrow, dilated at the middle; main nerves 10 pairs .... **T. gamblei**

Tabernaemontana divaricata (L.) R.Br. ex Roem. & Schult., Syst. Veg. 4: 427. 1819. *Nerium divaricatum* L., Sp. Pl. 209. 1753. *Tabernaemontana coronaria* (Jacq.) Willd., Enum. Pl. Hort. Berol. 275. 1809. *Nerium coronaria* Jacq., Coll. Bot. 1: 38. 1786. *Ervatamia coronaria* (Jacq.) Stapf in Dyer, Fl. Trop. Africa 4: 127. 1902. *E. divaricata* (L.) Burkill, Rec. Bot. Surv. India 10: 320. 925; Hook.f., Fl. Brit. India 3: 646. 1882; Gamble, Fl. Madras 2: 813. 1923.

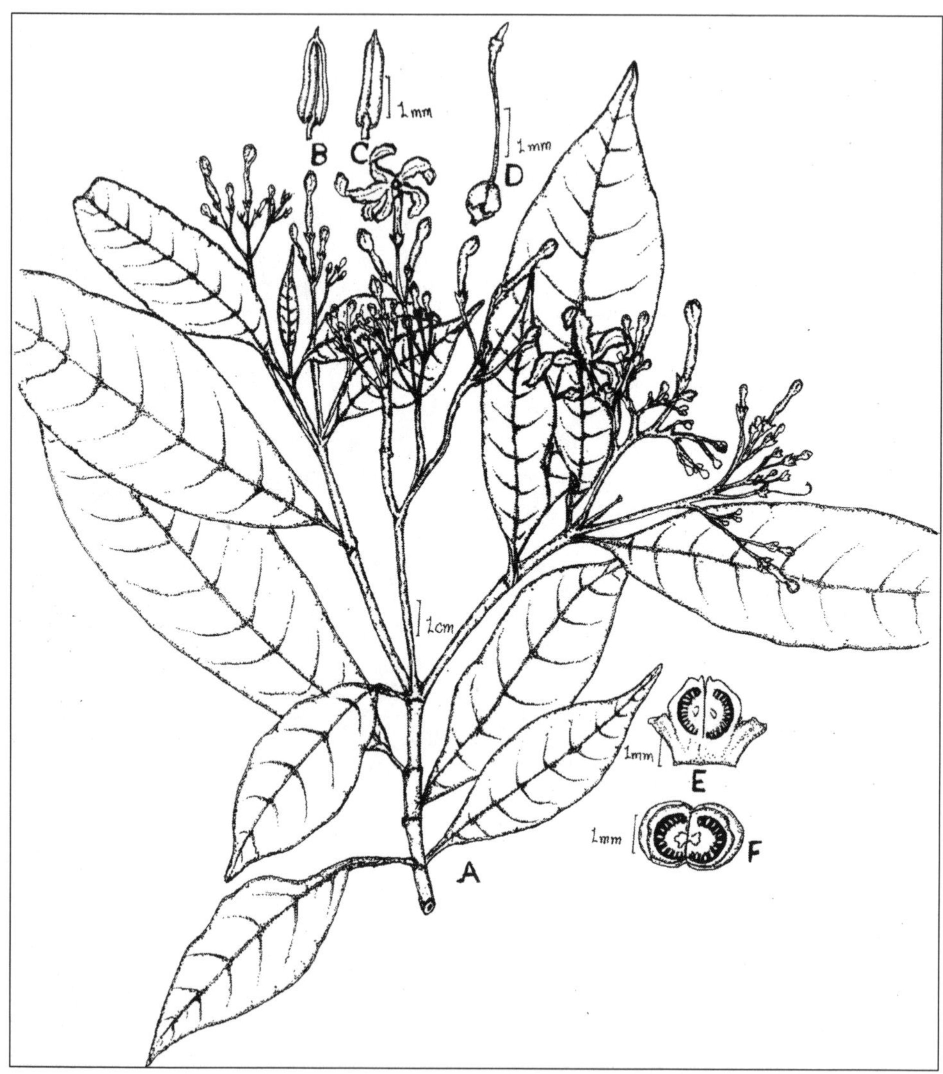

Figure 59: Tabernaemontana divaricata (L.) R.Br. ex Roem. & Schult.
A. Twig, B&C. Stamens, D. Pistil, E. Ovary l.s., F. Ovary c.s.

Vern.: *Nandivardhanam.*

Dichotomously branched, large shrub. Leaves oblanceolate to oblong, 3.5 – 10.5 × 3-4 cm, subcoriaceous, base cuneate to attenuate, margin entire, apex acuminate, lateral nerves 8 pairs; petiole to 1 cm long. Flowers creamish-white with yellow centered spot, in terminal divaricate cymes, fragrant. Mericarps lanceolate, curved and beaked; seeds without coma.

Grown as ornamental throughout.

Fl. & Fr.: May – December.

Inchapalli (ADB), *DAM* & *GO* 5152; Neredukonda, Nirmal (ADB), *KH* 5565; Umamaheswaram (MBNR), *SRS* 108963 (BSID); Bhadrachalam (KMM), *RCS* 99039 (BSID).

Tabernaemontana gamblei Subr. & A.N.Henry, Bull. Bot. Surv. India 12: 5. 1970. *Ervatamia caudata* Gamble, Kew Bull. 310. 1921 non *T. caudata* Merr. 1909; Gamble, Fl. Madras 2: 813. 1923.

Glabrous shrub. Leaves opposite, elliptic-obovate, abruptly caudate, ca 12 × 5cm, apex acuminate, main nerves 11 pairs, arched. Flowers white in terminal cymes; calyx lobes unequal, oblong, ciliate; corolla lobes overlapping to right, tube 1.5 cm long, dilated at the middle. Follicles ovate, long beaked, 1-seeded.

Ornamental plant, cultivated in gardens.

Fl. & Fr.: Round the year.

VALLARIS Burm.

Vallaris solanacea (Roth) Kuntze, Revis. Gen. Pl. 2: 417. 1891; Gamble, Fl. Madras 2: 815. 1923. *Peltanthera solanacea* Roth, Nov. Pl. Sp. 132. 1821. *Vallaris heynei* Spreng. in L., Syst. Veg. 1: 635. 1824; Hook.f., Fl. Brit. India 3: 650. 1882. *V. dichotoma* (Roxb.) Wall., ex G. Don, Gen. Syst. 4: 79. 1837. *Echites dichotoma* Roxb., Fl. Ind. 2: 19. 1832.

Vern.: *Ginne amalati, Pedda malle, Palamalle tivva, Arbimallika.*

Large twining shrub. Leaves opposite, elliptic or lanceolate, 3-7 × 1.5-2.5 cm, gland dotted, membranous, margin entire, apex acute. Cymes dichotomous, many-flowerd, sometimes reduced to fascicles. Flowers large, white, sweet scented; calyx 5-lobed, glandular, lobes narrow, apex acute; corolla salver-shaped, lobes-5, spreading, overlapping to the right; stamens 5, inserted at the top of the corolla, filament short, anthers connivent; ovary bicarpellary, ovules many, style filiform, stigma thick, bilobed. Fruit double follicle, joined along the length, placenta detached when mature; seeds biseriate, rounded, apex attenuate, silky coma attached on the top.

Occasional in dry forests in Ranga Reddi district.

Fl. & Fr.: June – September.

WRIGHTIA R. Br.

1. Flowers greenish or cream coloured; calyx and corolla pubescent, corona scales oblong; follicles connate through out, rough with white tubercles; leaves elliptic-obovate, tomentose, apex caudate-acuminate .... **W. arborea**

1. Flowers white; calyx and corolla glabrous, corona scales fimbriate; follicles jointed at apex only, smooth; leaves elliptic-oblong or oblong-lanceolate, glabrous, apex abruptly acuminate .... **W. tinctoria**

Wrightia arborea (Dennst.) Mabb., Taxon 26: 533. 1977. *Periploca arborea* Dennst., Schluess. Hort. Malab. 13. 23 & 25. 1818 & in Forts. Allg. Tentsch. Gaertn. Mag. 3; 2. 41 & 83. 1818. *Wrightia tomentosa* Roem. & Schult. in Syst. Veg. 4: 414. 1819; Hook.f., Fl. Brit. India 3: 653. 1882; Gamble, Fl. Madras 2: 816. 1923.

Vern.: *Tella pala, Tedla pala, Pala adavi venkudu, Pala bariki.*

Small, deciduous tree, up to 10 m high, bark light grey, corky, branchlets softly white tomentose. Leaves elliptic-obovate, 6.5-15 × 3.8-7.5 cm, slightly pubescent along the nerves above, tomentose beneath, drying dark green, base acute, margin entire, apex caudate-acuminate, main lateral nerves 12-15 pairs; petiole to 5 mm. Flowers greenish or cream coloured, in short, dense, erect, terminal, corymbose cymes; calyx 5-lobed, cup-shaped, tomentose, corolla salver-shaped, tube short, to 4 mm, corona scales laciniate, 5, one-seriate, entire or slightly lobed; stamens 5, exserted, inserted on the corolla tube. Fruit cylindrical, pendent. Follicles connate through out, rough with white tubercles, 11.5-27 cm long, apically beaked, seeds slender, attenuate at the apex.

Occasional in deciduous forest of Karimnagar, Medak, Ranga Reddi and Mahabubnagar districts.

Fl.: April – June. Fr.: December – February.

Rampur (ADB), *SSR* 19308; Narsapur forest (MDK), *KMS* 7953 (MH); Anantagiri (RR), *TP* & *MSM* 11088; Appapur (MBNR), *SSR* 15539; Sarlapally (MBNR), *TDCK* 15326; Bhupatipur (KRN), *GVS* 20176 (MH); Rollapadu (KMM), *RR* 108031 (BSID); 8 km from Pasra (WGL), *RKP* 108235 (BSID).

Wrightia tinctoria (Roxb.) R.Br., Asclepiadaceae 59. 1810; Hook.f., Fl. Brit. India 3: 653. 1882; Gamble, Fl. Madras 2: 815. 1923. *Nerium tinctorium* Roxb., Orient. Repert. 1: 39 cum tab. 1791.

Vern.: *Akupala, Doddipala, Teddapala, Ankudu, Palakordsha, Reppala, Chitti ankudu, Jeddepala.*

Densely foliaceous, deciduous tree, to 12 m high, bark pale, scaly, smooth, branchlets glabrous. Leaves membranous, elliptic-oblong or oblong-lanceolate, 4-12 × 2.4-4.4 cm, glabrous above, sparsely pubescent beneath, base truncate, margin entire, apex bluntly acuminate, secondary nerves 8-9 pairs; petiole to 2 mm long. Flowers white, fragrant, in terminal, lax, trichotomous cymes; calyx up to 0.2 cm long, sometimes puberulous, glandular within, lobes 5, divided more than half way down, oblong, obtuse, margins scarious, often ciliate; corolla white, up to 2.3 cm across, tube up to 0.3 cm long, lobes 5, up to 1.2 cm long, oblong, obtuse, corona scales of two different lengths, shorter ones up to 0.2 cm long, longer ones up to 0.5 cm long; stamens adnate to the throat of the corolla tube, anthers up to 0.7 cm long, pubescent on the margins; ovary up to 0.1 cm long, style up to 0.6 cm long. Follicles 2, distinct, apically joined, cylindric, curved, 30 x 0.5 cm, green, seeds pointed at the apex with a deciduous coma, often more than 2 cm long at the base.

Common in open dry deciduous forests and on dry sandy soils in all districts.

Fl.: Through out the year.

Kadam RF (ADB), *TP* & *PVP* 4074; Manchippa (NZB), *TP* & *BR* 6001; Mudheli RF (NZB), *BR* 9536; Marrikunta (MDK), *CP* 11381; Mohammadabad RF (RR), *MSM* & *KH* 10203; Mannanur (MBNR), *TDCK* 13682, *MVS* & *MB* 38636; Rachepalle (KRN), *GVS* 22252 (MH); Kinnerasani Wild Life Sanctuary (KMM), *RR* 112679 (BSID).

ASCLEPIADACEAE

1. Stamens united into a staminal column:
    2. Pollen masses pellucid on margin or at apex; anthers without membranous appendage:
        3. Leafless herbs with fleshy 4-angled stems and branches:
            4. Inflorescence racemose; leaf rudiments 1 -5 mm long, lanceolate, basally accompanied by pair of stipular glands .... **Caralluma**
            4. Inflorescence umbellate; leaf rudiments broadly cordate, without stipular glands .... **Boucerosia**
        3. Leafy herbs with terete stems:
            5. Corona double, corolline and staminal .... **Leptadenia**
            5. Corona single or double, staminal:
                6. Corolla rotate; with terete stems, twining undershrubs .... **Heterostemma**
                6. Corolla tubular; twining herbs .... **Ceropegia**
    2. Pollen masses without pellucid layer, anthers with membranous appendage:
        7. Erect plants:
            8. Shrubs or small trees; corona processes laterally compressed; corolla whitish to violet, lobes spreading .... **Calotropis**
            8. Herbs; corona processes hooded; corolla orange red, lobes reflexed .... **Asclepias**

7. Plants twining:

9. Corona single:

10. Corolla divided half way down, rotate; flowers large, lateral, in lax racemes:

11. Corolla lobes thin; corona processes dilated at base; leaves linear or linear-lanceolate or oblong .... **Oxystelma**

11. Corolla lobes thick; corona processes annular, fleshy; leaves broadly ovate- cordate .... **Holostemma**

10. Corolla divided near to the base, rotate or campanulate; flowers small, axillary in umbellate cymes:

12. Corona segments laterally compressed, produced into spur at base .... **Pentatropis**

12. Corona segments not compressed, spur absent:

13. Straggling shrubs, usually leafless; follicles smooth .... **Cynanchum**

13. Twining shrubs, leafy; follicles echinate .... **Pergularia**

9. Corona double

14. Corolla rotate:

15. Corolla non green .... **Tylophora**

15. Corolla green .... **Dregea**

14. Corolla urceolate, campanulate or salver-shaped:

16. Corolla base cylindrical; stem and leaves tomentose .... **Marsdenia**

16. Corolla base inflated; stem and leaves glabrous or nearly so; flowers fragrant .... **Telosma**

1. Stamens free, not united into a staminal column:

17. Corona processes connate with the stamens .... **Decalepis**

17. Corona processes connate with the corolla:

18. Corolla tube as long as or longer than the lobes, velvety .... **Cryptostegia**

18. Corolla tube shorter than the lobes:

19. Corolla lobes valvate:

20. Plants glabrous .... **Hemidesmus**

20. Plants pubescent or tomentose .... **Finlaysonia**

19. Corolla lobes imbricate, overlapping to the right .... **Cryptolepis**

ASCLEPIAS L.

Asclepias curassavica L., Sp. Pl. 215. 1753; Hook.f., Fl. Brit. India 4. 18: 1883; Gamble, Fl. Madras 2: 833. 1923.

Vern.: *Jilledu mandaram, Agni jilledu.* Eng.: Indian root, Bastard Ipecacuanha, Blood flower.

Erect herb, 90 cm high, stem terete, latex milky. Leaves decussate, chartaceous, linear-lanceolate, puberulous along nerves, base acute, margin ciliate, apex acute; petiole 1.5 cm, lateral veins 15-18 pairs. Flowers in 8-12-flowered, subterminal umbellate cymes, bright orange red; calyx 4.5 mm long, glandular within, pubescent without; corolla rotate, lobes elliptic, margin thin, corona staminal, uniseriate, 5-lobed, yellow, adnate to stipe of staminal column; stamens 5, pollinia five; gynostegium long, stipitate, cylindrical. Follicles paired, erect, 8 x 1 cm, lanceolate, glabrous; seeds dark brown, coma silky-white, 1.7 – 3 cm long.

Grown as an ornamental, occasionally as an escape.

Fl. & Fr.: Throughout the year.

Mombojipally (MDK), *BR* & *CP* 11645; Katasintha (HYD), *KMS* 5999 (MH).

BOUCEROSIA Wight & Arn.

Note: Karuppusamy & *al.* (2013) brought out monograph on Indian *Carallumas* where they segregated the genus *Boucerosia* and *Caralluma* based on inflorescence.

Boucerosia umbellata (Roxb.) Wight & Arn., Contr. Bot. Ind. 34. 1834. *Stapelia umbellata* Roxb., Pl. Coromandel 3: 36. t. 241. 1795. *Caralluma umbellata* Haw., Syn. Pl. Succ. 47. 1812; Hook.f., Fl. Brit. India 4: 77. 1883. *Desmidorchis umbellata* (Haw.) Decne. Ann. Sci. Nat. Ser. 2(9): 266. 1838. *Boucerosia campanulata* Wight, Icon. Pl. Ind. Orient. 4(2): 1.t. 1287. 1848. *Caralluma campanulata* (Wight) N.E.Br., Gard. Chron. 12: 369. 1892.

Vern.: *Kundeti kommulu, Jinka kommulu.*

Stem fleshy, quadrangular, angle sharp, sides concave, glabrous. Leaves reduced to teeth like, young stem only has small cordate leaves, caducous, leaf scar permanent, protruded. Flowers in terminal umbels, umbel many-flowered, peduncle not many-branched. Flowers large, campanulate or rotate, dark brown to reddish pink coloured, odorous; calyx lobes 5, greenish, imbricate, lanceolate, apex acute; corolla lobes 5, spreading, 1.5 – 2 cm long, triangular, outer side pale pinkish green, inner side dark brown to pinkish, usullay yellow cross bands with concentric rings, glabrous, if hairy only fulvous inner side, glandualar; corona staminal, biseriate, outer 5-lobed, attached to staminal column, 3-4 mm long, glabrous; inner lobes 5, 4-5 mm long, apex acute, incumbent on the anther lobes; stamens 5, pollen masses 5, solitary in each cell, yellow waxy, pellucid transparent, caudicle brown, corpuscle dark brown. Follicles paired, 9-13 × 2-3 cm, lanceolate, tapering towards the apex, tip blunt; seeds glabrous, 8-10 × 3-3.5 mm, oblong, base rounded, margin dark brown, coma silky, 3-4 cm long.

Rare in Nalgonda district along rocky boulder corners and crevices etc. with association of *Prosopis juliflora*.

Fl. & Fr.: March – August.

Kambalpally (NLG), *BVRS* 324 (NU).

CALOTROPIS R.Br.

1. Corona lobes erect, pink and purple spotted, glabrous, apex of corona, bifid and without auricles; smaller shrubs .... **C. procera**
1. Corona lobes spreading, whitish or pale lavender, hairy, apex of corona rounded with 2 obtuse auricles; larger shrubs .... **C. gigantea**

Calotropis gigantea (L.) Dryand, Hortus Kew. 2: 78. 1811. *Asclepias gigantea* L, Sp. Pl. 214. 1753. *Calotropis gigantea* (L) R.Br. ex Schult. Syst. Veg. 6: 91. 1820; Hook.f., Fl. Brit. India 4: 17. 1883; Gamble, Fl. Madras 2: 832. 1923.

Vern.: *Tella jilledu, Racha jilledu.* Eng.: Gigantic swallow-wort.

Large bushy shrub, up to 5 m high; stems woody, terete, pubescent, grooved at the middle, latex milky; bark yellowish white. Leaves decussate, coriaceous, elliptic-oblong, pubescent on both sides, 6-15 × 3.5-9.5 cm, base cordate with auricles, margin entire, apex acute, lateral nerves 4-7 pairs, petiole 3-8 mm long. Flowers in axillary or lateral, many-flowered cymes or umbels, white or pale lavender; calyx 5-lobed, 4-5 mm long, saucer-shaped, glandular within; corolla rotate, 5-lobed, spreading, glabrous; corona staminal, fleshy, completely adnate and shorter than the staminal column, glabrous with a recurved spur at the base, pollinia 5, yellow, waxy; gynostegium 1.2-2.3 cm long. Follicles paired, horizontal, 3.8-10 cm long, oblong, glaucous green, cottony pubescent; seeds flat, brown.

Common in waste lands of plains.

Fl. & Fr.: Through out the year.

Sattenapalle (ADB), *TP* & *PVP* 4155; Gowraram (NZB), *TP* & *BR* 6393; Narsingi (MDK), *BR* & *CP* 11522; Central University campus (RR), *MSM* 15182; Sri Rangapur (MBNR), *TDCK* 13614; Appapur (MBNR), *BR* & *SKB* 34452; Peddavuru (NLG), *BVRS* 506 (NU); Hussainsagar (HYD), *KMS* 5960 (MH); Eturnagaram (WGL), *RKP* 110826 (BSID).

Calotropis procera (Aiton) Dryand, Hort. Kew. 2: 78. 1811; Hook.f., Fl. Brit. India 4: 18. 1883; Gamble, Fl. Madras 2: 832. 1923. *Asclepias procera* Aiton, Hort. Kew. ed. 1.1: 305. 1789.

Vern.: *Jilledu, Erra jilledu.*

Bushy shrub, up to 2.5 m high, stem terete, latex milky. Leaves decussate, coriaceous, 5-15 × 2-10 cm, cottony pubescent on both sides, base cordate with auricled base, margin entire, apex acute; lateral nerves 4-8 pairs; petiole 2-5 mm long. Flowers in many-flowered umbellate cymes; peduncle fleshy, fugaciously pubescent; pedicels in pair; calyx 5-lobed, 4 mm long, margin papery, glabrous; corolla rotate, pink, lobes

5, glabrous; corona staminal, five, fleshy, completely adnate to staminal column and equaling or exceeding gynostegium, 5-6 mm long. Follicles paired, horizontal, turgid, tip blunt, rounded; seeds ovate, brown.

Common in waste lands throughout the plains in drier areas.

Fl. & Fr.: Throughout the year.

Sone (ADB), *TP* & *PVP* 4010; Chevella (RR), *MSM* 19239; Ramanagutta (KMM), *RR* 108004 (BSID); Nagarjunasagar (NLG), *BVRS* 530 (NU); Bothchalama (NLG), *KMS* 9792 (MH).

CARALLUMA R. Br.

Caralluma adscendens (Roxb.) R. Br.

Key to Varieties

1. Corolla lobes glabrous .... var. **adscendens**
1. Corolla lobes hairy:
   2. Distal portion of the stem clearly attenuate; flowers only on the attenuated portion, singly or doubled on the axis .... var. **attenuata**
   2. Distal portion of the stem not attenuate; flowers on cylindrical axis near the tip .... var. **fimbriata**

Caralluma adscendens (Roxb.) R.Br., Mem. Wern. Nat. Hist. Soc. 1: 25. 1809; Hook.f., Fl. Brit. India 4: 76. 1883; Gamble, Fl. Madras 2: 861. 1923; var. **adscendens**. *Stapelia adscendens* Roxb., Pl. Coromandel 1: 28. 1820.

Vern.: *Kundeti kommulu.*

Succulent herb, up to 1.5 m high, stem often with reddish streaks, apex slightly attenuate, the angles rounded, leaf-scar acute, produced tubercle, leaves small, only on young parts, lanceolate, glabrous, purplish, acute at apex. Flowers more or less pendulous, hairless; calyx 5-lobed, glabrous, greenish; corolla 5-lobed, lobes lanceolate, finely spotted with purple on base of inner side, transversely striped with dark purple inside the lower half or sometimes pale coloured with out stripes, dark purple with upper half inside, margin subulate in the lower half; corona staminal, biseriate, outer annular, with two filiform filamentous, 1.5 mm long, alternating with anther lobes, glabrous, recurved tips, inner corona flap-like, overlapping anther-lobes, 1 mm long, oblong, apex truncate or crenate; stamens 5, 1.5 mm long, pollinia five, solitary in each cell, yellow waxy, caudicle brown, corpuscle dark brown. Follicle usually solitary, 6-10 x 0.3-0.6 cm, linear-lanceolate, tapering towards the apex, apex beaked, glabrous; seeds many, oblong or obovate, base rounded, dark-brown, margin light brown, 1 mm wide, coma silky, 2-3 cm long.

Rare among the bushes of Medak and Mahabubnagar districts.

Fl. & Fr.: Throughout the year.

Pocharam tank (MDK), *TP* & *CP* 12048; Medak tank (MDK), *CP* & *BR* 11795; Molachintapally forest (MBNR), *TDCK* 15320.

Caralluma adscendens (Roxb.) R.Br. var. **attenuata** (Wight) Gravely & Mayur., Bull. Madras Govern. Mus. Ser. Nov. 4(1): 13. 1931. *Caralluma attenuata* Wight, Icon. Pl. Ind. Orient. 4: t. 15. 1848; Hook.f. Fl. Brit. India 4: 76. 1883; Gamble, Fl. Madras 2: 861. 1923.

Vern.: *Kundeti kommulu, Sanna kundeti kommulu.*

Succulent herb, up to 45 cm high, stem 4-angled, rapidly grows when young, strongly attenuate and usually much branched distally, streaked with reddish purple on the angles, angle rounded, lower stem stout. Flowers pendulous, hairy, opening widely; calyx glabrous, green, 5-lobed, lobes lanceolate; corolla 5-lobed, rotate or campanulate, corolla-tube short, 2 mm long, lobes streaked with dark purplish transverse lines inside, margins hairy, hairs spindle-shaped.

Common in all dry forests of the foot hills in Medak district.

Fl. & Fr.: Throughout.

Dichpally (NZB), *BR & KH* 6438; Patharajampet (NZB), *BR* 9702; Kodimial (KRN), *GVS* 20148 (MH); Karkalpadu, Amangal RF (MBNR), *SRS* 104548 (BSID); near Domalapenta (MBNR), *VBH* 84946 (BSID).

Caralluma adscendens (Roxb.) R.Br. var. **fimbriata** (Wall.) Gravely & Mayur., Bull. Madras Govern. Mus. Ser. Nov. 4(1): 13. 1931. *Caralluma fimbriata* Wall., Pl. Asiat. Rar. 1: 7. t. 8. 1830; Gamble, Fl. Madras 2: 861. 1923.

Erect, succulent herb, basally branched, green with brownish streaks, quadrangular, angles rounded. Leaves reduced, small, scale-like, green, glabrous, lanceolate, 0.1 mm long, apex acute, caducous, leaf scar protruded slightly on old stems. Flowers solitary in leaf axils on younger stems, pedicels glabrous, cylindrical, 2-4 mm long; calyx 5-lobed, divided up to base, lobes 1-1.5 mm long, lanceolate, apex acute, glandular within, glabrous; corolla rotate, yellowish with purplish striations on lower half and dark purple on upper half on inner side, lobes 7-8 × 1.5-2 mm, lanceolate, spathulate, rounded-apiculate, expanded near the base, long purple hairs along the margin; corona staminal, biseriate, outer lobes 5, deeply cleft, 1 mm long, inner 5 lobes basally adnate to the outer corolla, apex crenate, incumbent on the anther lobes; stamens 5, pollinia masses solitary in each cell, yellow, waxy, erect lobes, attached by light brown caudicle, corpuscle light brown. Follicles paired, 7-10 × 0.5-0.8 cm, linear-lanceolate, tapering from the base, slightly curved at the apex, glabrous; seeds many, obovate, base rounded, glabrous, 7-8 x 3-4 mm, dark brown, margin light brown, coma silky, 2-4 cm long.

Fl.& Fr.: Through the year.

Mallelatheertham (MBNR), *AU & SSR* 31762.

CEROPEGIA L.

1. Stem erect:
   2. Leaves 2.5-5 cm long; follicles fusiform, to 3.5 cm long .... **C. pusilla**
   2. Leaves 7-10 cm long; follicles slender, ca 12.5 cm long .... **C. spiralis**

1. Stem twining:
   3. Leaves fleshy:
      4. Leaves very small or absent; corolla lobes as long as the tube, connate at tip .... **C. juncea**
      4. Leaves more than 2 cm long; corolla lobes shorter than tube .... **C. bulbosa**
   3. Leaves not fleshy:
      5. Leaves and peduncles hispid pubescent .... **C.hirsuta**
      5. Leaves and peduncles glabrous .... **C. candelabrum**

Ceropegia bulbosa Roxb.

Key to Varieties

1. Leaves orbicular, ovate; pollen masses large, 284-296 × 196-200 µm, caudicles 32-40 µm long, corpuscle 92 × 32 µm .... **C. bulbosa** var. **bulbosa**
1. Leaves linear to lanceolate; pollen masses 220 × 148 µm, caudicles ca 220 × 160 µm, corpuscle 40 × 20 µm .... **C. bulbosa** var. **lushii**

Ceropegia bulbosa Roxb., Pl. Coromandel 1: 11. t. 7. 1795 & Fl. Ind. 2: 28. 1824; Hook.f., Fl. Brit. India 4: 67. 1883; Gamble, Fl. Madras 2: 857. 1923; var. **bulbosa.**

Vern.: *Manchi manda.*

Fleshy, twining, slender, perennial, glabrous herb, root tuber rounded. Leaves chartaceous, fleshy, orbicular, ovate-oblong, 2.5-4.5 × 1.5-2.6 cm, glabrous, apex acutely apiculate, base acute, subsessile or petiolate. Flowers small in 3-7-flowered, pedunculate, umbellate cymes, greenish, pedicel short; corolla greyish purple, 2.5 cm long, tube long, inflated at base, hairy inside, corona double, outer corona minute, saucer-shaped, inner slender, linear or sickle-shaped. Follicles slender, 8.5-9 cm long, glabrous, tapering towards an acute apex.

Occasional perennial in dry localities.

Fl. & Fr.: July-December.

Ceropegia bulbosa Roxb. var. **lushii** (Grah.) Hook.f., Fl. Brit. India 4: 68. 1883; Gamble, Fl. Madras 2: 857. 1923. *C. lushii* Grah. in Bot. Mag. t. 3300. 1834.

Fleshy twiners, tubers flattened with several fibrous rootlets from its base; stems slender, jointed, glabrous. Leaves very narrowly linear-lanceolate, falcate, 5-11 × 0.3-0.7 cm, fleshy, glaucous, base narrowed, margin entire, apex acuminate, glabrous on both surfaces, midrib thick; petiole ca 5 mm long. Flowers 6-10, in umbellate cymes, 2.8 cm long; peduncles arising between the petioles, ca 8 mm long; pedicels slender, 5-7 mm long; calyx divided up to base; sepals lanceolate, acute, ca 2 mm long; corolla greenish, tube inflated at base, ca 13 mm long; lobes linear from a deltoid base, ca 8 mm long, connate at tips, ciliate inside, violet purple; outer corona lobes 5, minute, inner corona lobes narrowly linear, ca 2 mm long; pollinia five, pollen masses solitary in each anther cell, yellow, waxy with pellucid layer, attached by light-brown caudicle, dark brown corpuscle; pistil ca 2 mm long. Follicles cylindrical; seeds ovate-oblong, flattened, with broad membranous margins.

Occasional in Hyderabad district (Ramana & *al.*, 2011).

Fls.: July-September; Fr.: September-October.

Rajendranagar (HYD), *MVR* 2052 (BSID).

Ceropegia candelabrum L., Sp. Pl. 211. 1753; Hook.f., Fl. Brit. India 4: 70. 1883. var. **candelabrum;** Gamble, Fl. Madras 2: 857. 1923. *C. ellotti* Hook.f., Fl. Brit. India 4: 70. 1883.

Vern.: *Nimmatai, Langatai, Bachalimanda, Pullamanda, Manchimanda, Tiyyamanda.*

Glabrous, perennial twiners, stems glabrous. Leaves variable, chartaceous, elliptic-ovate or lanceolate, 2-8 × 0.8-4 cm, glabrous, apex acutely apiculate, base rounded, acute or subcordate, lateral veins 4-5 pairs, conspicuous beneath; petiole 1.2 cm long. Flowers in many-flowered, lateral, umbelliform cymes; peduncle 2.8 cm long, pedicel 0.8 cm long; calyx 5-lobed; corolla green below, the tips purplish, the veins prominent, to 4.5 cm long, basal part inflated, then cylindric, the lobes ciliate within, corona biseriate, outer very small, inner processes spathulate. Follicles paired, 10-14 × 0.5-0.7 cm, cylindric, tapering at apex, glabrous; seeds 0.8 × 0.2 cm, ovate-oblong, margined with 2-2.5 cm long coma.

Occasional, along hilly slopes of deciduous forests Nizamabad, Ranga Reddi and Mahabubnagar districts.

Fls.: August – January. Fr.: November – March.

Gandivet RF (NZB), *BR* & *CPR* 7198; Anantagiri RF (RR), *TP* & *MSM* 11059.

Ceropegia hirsuta Wight & Arn. in Wight Contrib. Bot. India 30. 1834; Hook. f., Fl. Brit. India 4: 71. 1883; Gamble, Fl. Madras 2: 859. 1923. *C. vincaefolia* Hook., Curtis Bot. Mag. t. 3740. 1840.

Hispid climber with tuberous globose roots. Leaves membranous, variable, lanceolate to linear or broadly oblong or ovate, 4-6 × 2.5-3.5 cm, thinly pubescent on either sides, base acute to truncate or rounded, apex apiculate; petiole 1-1.5 cm. Flowers in 2-few-flowered umbellate cymes, peduncle hairy, 1.5 cm long; calyx 5-8 mm long, hirsute; corolla yellowish or greenish, 2.8-5.8 cm long, blotched with purple, depressed, inflated at base, funnel-shaped above, outer corona of 5-fid, lobes hairy, inner linear, erect; pollinia 5, pollen masses solitary in each anther cell, yellow, waxy.

Occasional in hills of Warangal and Mahabubnagar districts.

Fl.: July-November; Fr.: September onwards.

Tadwai (WGL), *CSR* 1210 (KU); Mallus (WGL), *CSR* 1654 (KU).

Ceropegia juncea Roxb., Pl. Coromandel t. 10. 1795; Hook. f., Fl. Brit. India 4: 68. 1883; Gamble, Fl. Madras 2: 856. 1923.

Vern.: *Bellagada.*

Fleshy twining herb, glabrous, root tuberous. Leaves scale like, lanceolate, 1 × 0.2 cm, subsessile, fleshy, deciduous. Flowers in 3-5-flowered umbellate cymes, greenish, speckled and veined with pink and purple, pedunculate, corolla 4-5.5 cm long, inflated at base, funnel-shaped above the middle, hairy inside near the mouth, outer corona of 5 bidentate-deltoid, ciliate lobes, inner linear, erect, hooked at apex; pollinia five. Follicle narrow, 12.5 cm long, long-beaked.

Common in dry deciduous forests of Warangal, Hyderabad (Ramana, 2010) and Mahabubnagar districts.

Fl. & Fr.: December – August.

Musthyal (WGL), *CSR* 1077 (KU).

Ceropegia pusilla Wight & Arn. in Wight Contrib. Bot. India 31. 1834 & Wight, Icon. Pl. Ind. Orient. t. 1261. 1848; Hook. f., Fl. Brit. India 4: 66. 1883; Gamble, Fl. Madras 2: 856. 1923.

Very small herb, root tuberous, stems erect, puberulous above. Leaves linear, crowded on a very short stem close to the ground, 2.5-5 cm long, glabrous, apex and base acute. Flowers solitary, 2 cm long, peduncle minute; corolla subcylindric, base inflated, glabrous inside, outer corona 5-lobed, ciliate, inner long, linear, pinkish, erect. Follicles fusiform, to 3.5 cm long.

Very rare, in moist deciduous forests of Adigmet (Sayeduddin, 1935; Sudhakar Reddy & *al.*, 2001).

Fl.: June – August; Fr.: September.

Ceropegia spiralis Wight, Icon. Pl. Ind. Orient. t. 1267. 1848; Hook.f., Fl. Brit. India 4: 66. 1883; Gamble, Fl. Madras 2: 856. 1923.

Vern.: *Teega nimmatai, Teega ganapa.*

Slender, erect herb, tuberous, stem glabrous. Leaves chartaceous, linear, glabrous, 7-10 × 0.2-0.3 cm, apex and base acute. Flowers in 1-flowered cymes; corolla 4-5 cm long, tube subcylindric, hairy within above from ovate or orbicular base, generally spirally twisted, corona biseriate, outer corona of bifid, deltoid lobes. Follicles very slender, about 12.5 cm long.

Rare in hills of Ranga Reddi district (Seshagiri Rao, 2012).

Fl.: August – October; Fr.: November onwards.

CRYPTOLEPIS R. Br.

1. Main lateral nerves 10-15 pairs, parallel and horizontal, uniting into a intramarginal vein; follicles to 7 cm long .... **C. dubia**
1. Main lateral nerves 5-9 pairs, not parallel and horizontal; follicles more than 8 cm long .... **C. sinensis**

Cryptolepis dubia (Burm.f.) M.R.Almeida, Fl. Maharashtra 3A: 221. 2001. *C. buchananii* R.Br. ex Roem & Schult., Syst. Veg. 4: 409. 1819 ('*buchanani*'); Hook. f., Fl. Brit. India 4: 5. 1883; Gamble, Fl. Madras 2: 826. 1923.

Vern.: *Adavi pala tiga, Gunji teega, Gurugupalateega.*

Large, much branched, climbing shrub with plenty of latex, bark smooth, dark brown, papery, peeling off in scales. Leaves coriaceous, oblong to elliptic, 2-15 × 1.2-8 cm, glabrous, shining above, pale glaucous beneath, apex obtuse-apiculate, sometimes emarginated-mucronate, base acute or rounded; lateral veins 10-22 pairs, slender, nearly parallel, united in a intramarginal nerve; petiole to 1 cm. Flowers in many-flowered axillary, paniculate cymes, with divaricate fleshy branches constricted at the nodes, yellow, 1.8 cm across; pedicel 3 mm; calyx 2.5 mm, margin hairy; corolla shortly campanulate, 1.2 cm long, margin revolute; corona corolline, 5-lobed, arising from the base of corolla, 1 mm long. Follicles paired, 7-10 × 1.5 cm, basally inflated, glabrous, ridged, tapering to a point; seeds many, flat, apically narrowed.

Common in open and scrub forest in Adilabad, Nizamabad, Medak, Karimnagar, Khammam and Mahabubnagar districts.

Fl.: April-June; Fr.: December –February.

Neelwai RF (ADB), *TP* & *GO* 5420; Jaipur RF (ADB), *KH* 5559; Neredukonda RF (ADB), *KH* 5566; Gandhari (NZB), *TP* & *BR* 6346; Gowraram RF (NZB), *TP* & *BR* 6384; Pathur RF (MDK), *TP* & *CP* 12114; Aklaspur (KRN), *GVS* 25615 (MH); Amrabad (MBNR), *BR* & *SKB* 32560; Lakshmipuram RF (KMM), RCS 99007 (NSID).

Cryptolepis sinensis (Lour.) Merr., Philipp. J. Sci. 13: 254. 1919. *Pergularia sinensis* Lour., Fl. Cochinch. 169. 1790. *Cryptolepis elegans* Wall., ex Don, Gen. Hist. 4: 82. 1837; Hook. f., Fl. Brit. India 4: 6. 1883; Gamble, Fl. Madras 2: 826. 1923.

Vern.: *Pala teega.*

Slender, wiry climber, stems branched, glabrous, slightly lenticellate, latex milky-white. Leaves oblong-lanceolate, 3.7-6.2 × 1.8 cm, thin, glaucous beneath, base obtuse, apex apiculate or mucronate, lateral veins 7-8 pairs, curved upwards, spreading; petiole slender, 8 mm long. Flowers yellow in few-flowered axillary and terminal cymes; peduncle 1-2.4 cm long, pedicel 1.5 to 2.5 cm long, glabrous; corolla shortly campanulate, 2 cm long, corona corolline, 5-lobed, inserted near the middle of the corolla tube, subulate. Follicles paired, divaricate, 10-20 cm long, linear, tip blunt; seeds oblong, brown.

Occasional near foot hills of deciduous forests of Khammam district.

Fl.: June-September; Fr.: December-February.

Way to Tallagunti (KMM), *PVS* 84072 (BSID).

CRYPTOSTEGIA R. Br.

Cryptostegia grandiflora Roxb. ex R.Br., Bot. Reg. t. 435. 1819; Hook.f., Fl. Brit. India 4: 6. 1883; Gamble, Fl. Madras 2: 862. 1923.

Vern.: *Rubber teega, Gantapala, Garudapala, Vishbudhi, Rabbaru chettu.*

Woody, climbing shrub, glabrous, lenticelled, latex milky-white, plenty. Leaves elliptic-oblong, coriaceous, 7-10 × 3-5 cm, glabrous, apex subacute, acute to tapering, lateral veins 8-10 pairs; petiole 1.3 cm long. Flowers in terminal or axillary di- or trichotomous cymes, light purple, to 4.5 cm diam; calyx lobes lanceolate, corona funnel-shaped, corona corolline, at the base of the throat, 5-lobed, each dividing into two filiform segments, glabrous; stamens 5, 5 mm long. Follicles paired, 8-12 × 2-2.5 cm, trigonous, tip blunt, curved, glabrous; seeds many, winged, glabrous.

Occasional, cultivated but naturalized and growing wild in all districts.

Fl. & Fr.: August-December.

Mancheryal (ADB), *GO* & *PVP* 4941; Nirmal (ADB), *GO* & *BR* 9402; Domakonda (NZB), *BR* 9689; Hyderabad, *E.C.Edwards* 36650 (DD); Lingampalli (RR), *MSM* 19218; Cherukupally (NLG), *BVRS* 344 (NU).

CYNANCHUM L.

Cynanchum viminale (L.) Bassi, Bononiensi Sci. Inst. Acad. Comment. 6 (Opusc.) 17. 1768. subsp. **viminale**. *Euphorbia viminalis* L., Sp. Pl. 1: 452. 1753. *Sarcostemma viminale* (L.) R.Br. in Mem. Wern. Soc. 1: 51. 1809 printed in 1811 & Prodr. Nov. Holl. 464. 1810; Ali, Fl. Pakistan 150: 30. 1983. *S. acidum* (Roxb.) Voigt, Hort. Sub. Cal. 542. 1845; Hanumanthappa in Pullaiah & Moulali, Fl. Andhra Pradesh 2: 595. 1997. *Asclepias acida* Roxb., Fl. Ind.2: 31. 1832. *Sarcostemma brevistigma* Wight & Arn. in Wight Contrib. Bot. Ind. 59. 1834; Hook.f., Fl. Brit. India 4: 26. 1883; Gamble, Fl. Madras 2: 838. 1923.

Vern.: *Konda pala, Somalatha, Aku jemudu, Chittipala, Dudhi, Pandiri jamudu, Akulenijamudu.*

Much branched, leafless, climbing, succulent, jointed shrub with straggling branches, sometimes trailing over bushes, reaching to 2 m, stem 5 mm in diam, green, glabrous. Leaves when present linear-oblong, 6-10 mm long. Flowers creamy-white, fragrant, in terminal umbels; bracts lanceolate, pedicel to 1.2 cm; calyx 5-lobed, lobes free, pubescent; corolla rotate, divided nearly up to base, to 6 mm long, margins revolute, glabrous, corona staminal, biseriate, outer cupular, inner erect, 5-lobed, white, concealing anthers. Follicles paired, slightly divergent, to 7 cm long, lanceolate, narrowed towards both ends, apex blunt; seeds glabrous.

Occasional in dry and often rocky places in Medak, Ranga Reddi, Nalgonda and Mahabubnagar district.

Fl. : June – January; Fr.: October-December.

Pochammaralu (MDK), *CP* 11754; Anantasagar (RR), *MSM* & *NV* 10404; Cherukupally (NLG), *BVRS* 181 (NU).

Note: Since molecular analyses have demonstrated that *Sarcostemma* R. Br. is deeply nested in the predominantly Madagascan stem-succulent clade of *Cynanchum* L., the genus has been treated by Meve and Liede-Schumann (2012) as a synonym of *Cynanchum*. Some of the former *Sarcostemma* species have been transferred to *Cynanchum* in the course of various Flora treatments, and some new species belonging to this radiation have been described under *Cynanchum*.

The protologue of *Cynanchum viminale* refers to an illustration of a plant from Egypt (Felfel Tavil) as the type and states explicitly 'Habitat in Africae maritimis'. However, Indian taxa such as *Sarcostemma acidum* (Roxb.) Voigt and *S. brevistigma* Wight & Arn. were synonymised under *S. viminale* subsp. *viminale* by Ali (1983), based on material from Pakistan. Later, this synonymy has been adopted by Jagtap & Singh (1999) for India. Other regional Floras, such as the Flora of China and the Flora of Nepal (www.eFloras.org) have not followed this usage and recognised *S. acidum* and *S. brevistigma* as separate species. The present author followed the view of Ali (1983) and considered these two species as synoynyms of *C. viminale* subsp. *viminale*, thus extending the area of the latter from West Africa and Egypt to India.

DECALEPIS Wight & Arn.

Decalepis hamiltonii Wight & Arn. in Wight, Contrib. Bot. India 64. 1834; Hook.f., Fl. Brit. India 4: 11. 1883; Gamble, Fl. Madras 2: 828. 1923.

Vern.: *Nannari, Maredu kommulu, Madina kommulu.*

Twining shrubs, deciduous, branchlets terete, nodes swollen, glabrous, pubescent when young, latex milky-white. Leaves orbicular or elliptic-obovate, subcoriaceous, 3-7 × 3.5 -5.4 cm, glabrous, pubescent along the nerves when young, apex rounded or obtuse, margin entire to undulate, base acute or rounded, secondary nerves 6-7 pairs, arched, conspicuous beneath, petiole to 1 cm, nerves and stem brown after drying, Flowers in axillary trichotomous cymes, peduncle to 2 cm long, bracteate; calyx 5-lobed, linear-oblong, tinged with brown, lobes 2 mm long; corolla cream, campanulate, 3 mm long, glabrous without, white villous within; corona staminal, biseriate, adnate to back of the base of filaments, outer scale like, inner flat; stamens 5, connivent, pollinia horizontal, pollen mass granular; ovaries 2 mm, subglobose. Follicles paired, woody, 6 × 3 cm, lanceolate, apex acute, epicarp thick, glabrous; seeds ridged.

Found growing on the slopes among rocks by stream banks in Ranga Reddi and Mahabubnagar districts.

Fl.: May – July; Fr.: January onwards.

Farahabad (MBNR), *BR* & *SKB* 34430; Pullayapally (MBNR), *MVS* & *MB* 38651.

DREGEA E. Mey.

Dregea volubilis (L.f.) Benth. ex Hook. f., Fl. Brit. India 4: 46.1883. *Asclepias volubilis* L.f., Suppl. Pl. 170. 1781. *Marsdenia volubilis* (L.f.) T.Cooke, Fl. Bombay 2: 166. 1904; Gamble, Fl. Madras 2: 846. 1923. *Wattakaka volubilis* (L.f.) Stapf in Curtis Bot. Mag. Sub t. 8876. 1922; Hanumanthappa in Pullaiah & Moulalli, Fl. Andhra Pradesh, 2: 599. 1997.

Vern.: *Bandi-gurga, Pedda-gurga, Kalasaku, Bondiguruginja*; Eng.: Green wax flower, Sneezing silk cotton.

Extensive perennial woody, stout, large, hoary twiner with watery sap, branches lenticellate, bark brown, rough, somewhat corky. Leaves broadly ovate, 8-17 × 1.5 – 12 cm, glabrous, base rounded, oblique to cordate, apex acuminate, lateral veins 4-6 pairs, lowest 1 or 2 pairs arising from base, clearly marked, petiole to 5 cm long.

Flowers in drooping, many-flowered umbellate cymes, arising in between the petioles nearer to one of them, greenish yellow, pedicel terete, slender, woody in fruit; calyx 2.5 mm long, pubescent, corolla rotate, 6.5 mm long, margin ciliolate; corona staminal, uniseriate, adnate at the base of the staminal column. Follicles solitary or paired and divaricate, oblong, 9 × 2.5 cm, gradually narrowed and apically blunt, fugaciously tomentose, seeds many, ovate, yellow with pinkish margin.

Common throughout the state in hedges and open forests in the hills in most districts.

Fl.: April -June; Fr.: December– February.

Neelwai (ADB), *TP* & *GO* 5431; Lingaram (NZB), *BR* & *GO* 9061; Gadirayal (RR), *MSM* & *KH* 10271; Sarlapally (MBNR), *TDCK* 15325; Nagarjunasagar (NLG), *BVRS* 147 (NU).

FINLAYSONIA Wall.

Finlaysonia wallichii (Wight) Venter, Ann. Missouri Bot. Gard. 88: 565. 2001. *Hemidesmus wallichii* Wight, Contr. Bot. India 63: 1834. *Hemidesmus indicus* (L.) R.Br. var. *pubescens* (Wight & Arn.) Hook.f., Fl. Brit. India 4: 5. 1883; Gamble, Fl. Madras 2: 825. 1923. *H. pubescens* Wight & Arn. in Wight Contrib. Bot. Ind. 63.1834.

Vern.: *Sugandipala, Peddakula sugandhapalateega*.

Slender, twining undershrub, stem softly dense pubescent, latex milky-white, internodes to 10 cm long. Leaves elliptic to broadly oblong, 2.5-8.5 × 1.4-6.5 cm, pubescent above, velvety tomentose beneath, base rounded, margin entire, apex with prominent apicule, lateral nerves 5 pairs; petiole to 6 mm. Flowers in axillary, dense cymes, peduncle very short, pubescent, bracteate; calyx 5-lobed, 2.5 mm long, hairy, glandular within; corolla dark purple, campanulate, glabrous, 4 mm long; corona corolline, 5 tuber like processess, adnate to corolla-tube, alternating with lobes; stamens 5, 1.5 mm long, pollinial mass granular, yellow, ovaries 2 mm long. Follicle divaricate, linear, tip incurved, pubescent; seeds many, linear-oblong, dark brown.

On the hills, mainly in dry deciduous forests in all districts.

Fl. : August – December; Fr.: January onwards.

Wankidi RF (ADB), *TP* & *GO* 5487; Bellampalli RF (ADB), *KH* 5561; Manchippa RF (NZB), *TP* & *BR* 6034; Ghanapur RF (MDK), *TP* & *CP* 12101; Narsapur (MDK), *KMS* 6817 (CAL); Aklaspur (KRN), *GVS* 25609; Kusumasamudram RF (RR), *MSM* & *NV* 10333; Vatavarlapally (MBNR), *TDCK* 13663; Poosagundi (KMM), *PVS* 84047 (BSID); Ramavaram RF (KMM), *RR* 112672; Veerapuram (WGL), *PVS* 76913 BSID).

GYMNEMA R.Br.

Gymnema hirsutum Wight & Arn. in Wight, Contrib. Bot. India 44. 1834; Hook. f., Fl. Brit. India 4: 29. 1883; Gamble, Fl. Madras 2: 839. 1923.

Climber, stems branched, spirally twisted, rusty pubescent. Leaves thick, ovate, 1.3-3.3 × 1.2-2.1 cm, rusty tomentose, apex acute, base rounded or cordate, lateral veins 3-4 pairs; petiole terete, thick, to 8 mm long. Flowers yellow, 4 mm long, 2-many-flowered axillary, pubescent cymes, peduncle 2-5 mm long, pedicel 1-3 mm long; calyx divided up to base; corolla campanulate, 4 mm long, glabrous, corona coralline, uniseriate. Follicle single, slender, up to 6 cm long, linear, glabrous, seeds ovate-oblong.

In hills of moist deciduous forests.

Fl.: April - July, Fr.: May-August.

Rollapadu (KMM), *RR* 108014 (BSID).

HEMIDESMUS R.Br.

Hemidesmus indicus (L.) R.Br. ex Schult., Syst. Veg. 6: 126. 1820; Hook.f., Fl. Brit. India 4: 5. 1883; Gamble, Fl. Madras 2: 825. 1923. *Periploca indica* L., Sp. Pl. 211. 1753.

Vern.: *Sugandapala, Pala sugandhi, Adavi ippateega, Chinnasugandha.* Eng.: Indian sarsaparilla.

Slender, glabrous twiner, root-stock aromatic, branchlets terete, lenticelled, latex milky-white. Leaves most variable in shape, linear, oblong, linear-lanceolate, lanceolate, elliptic-oblong, 2-13 × 0.2-4 cm, glabrous, pale beneath, apex acute, apiculate, margin entire, lateral veins 5-15 pairs; petiole 2.5 to 6 mm. Flowers in axillary subsessile cymes, greenish purple, bracts numerous; calyx lobes 2 mm long, ovate-lanceolate; corolla campanulate, 5 mm long; corona corolline, uniseriate, scales small, fleshy; stamens 5, 2.5 mm long. Follicles paired, divaricate, slender, cylindric, gradually narrowed towards tip, 10-12.5 cm long.

Common in forests and wastelands.

Fl.: Chiefly August-October; Fr.: December-February.

Sone (ADB), *TP* & *PVP* 4007; Kadam RF (ADB), *TP* & *PVP* 4093; Gandivet RF (NZB), *BR* & *CPR* 7192; Sivanoor (MDK), *BR* & *CP* 11547; Pochammaralu (MDK), *CP* 11764; Central University campus (RR), *MSM* 15152; Vatavarlapally (MBNR), *TDCK* 13663; Mallelatheertham (MBNR), *BR* & *SKB* 30637; Chandoor (NLG), *BVRS* 354 (NU).

HETEROSTEMMA Wight & Arn.

Heterostemma deccanense (Talb.) Swarup & Manglay, J. Linn. Soc. 101: 255. 1989. *Oianthus deccanensis* Talb., For. Fl. Bombay Sind. 2: 260. 1911.

Vern.: *Peddapala joola, Pedda gana joola.*

A wiry climber with pubescent branches, latex milky. Leaves opposite, thick, obovate, to 20 × 17 cm; petiole 3-8 cm long, base obtuse or rounded, apex acute to acuminate, glabrous, basal nerves 7 pairs, spreading. Flowers small, in subsessile to pedunculate cymes arising between petioles; calyx lobes ovate, obtuse, 2 mm long; corolla ovoid-urceolate, lobes 2.5 mm long, coronal scales rounded. Follicles 2, divergent, 7-10 cm long; seeds 8-12, imbricated into two rows.

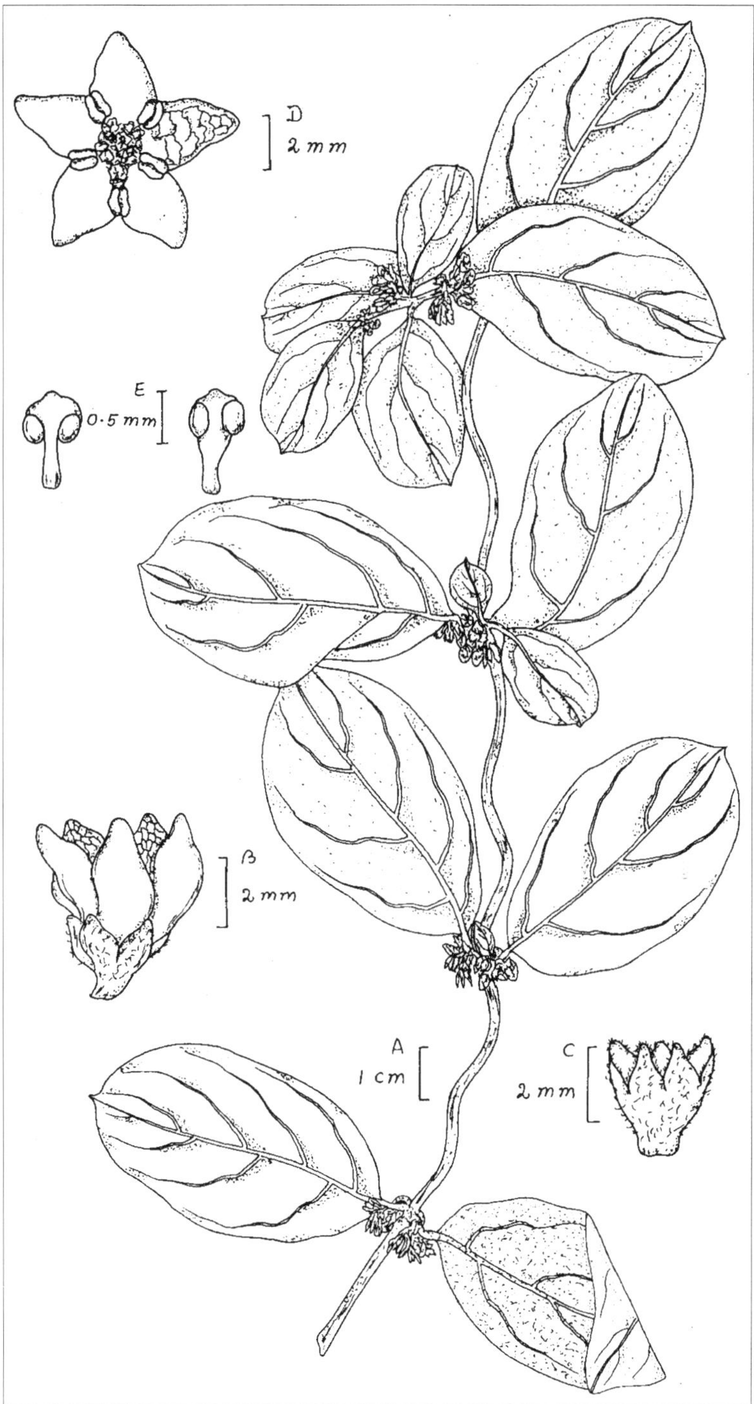

Figure 60: Hemidesmus indicus (L.) R. Br. ex Schult.

A. Twig, B. Flower, C. Calyx, D. Flower top view, E. Pollinia.

Rare in moist deciduous forests in Sukkumamidi in Khammam district (K.N.Reddy & *al.*, 2001).

Fl. & Fr.: July – December.

Sukkumamidi (KMM), *KNR* 77453 (FRLHT, KU).

HOLOSTEMMA R.Br.

Holostemma ada-kodien Schult., Syst. Veg. 6: 95. 1820. *H. annulare* (Roxb.) K. Schum., in Engl. & Prantl, Nat. Pflanzenfam. 4 (2): 250. 1895. *Asclepias annularis* Roxb., Fl. Ind 2: 37. 1832. *Holostemma rheedei* Wall., Pl. Asiat. Rar. 2: 51. 1831; Hook.f., Fl. Brit. India 4: 21. 1883; Gamble, Fl. Madras 2: 834. 1923.

Vern.: *Dudipalatige, Polagurugu, Palatura, Palagaddalu, Palajilledu.*

Extensive perennial climber, to 8 m long, stems shining, faintly ridged, glabrous, latex milky-white. Leaves chartaceous, ovate, 4.5-13 × 2.9-9 cm, puberulous below, base deeply cordate, apex acuminate, lobes rounded, lateral nerves 4-7 pairs, nerves conspicuously raised and looping within the margin, 5-7 glands at base of the midrib above. Flowers in axillary, few-flowered umbellate cymes, pinkish red or white or pale pink out side, crimson inside; calyx 5-7 mm long; corolla rotate, 1.5 cm across, lobes 5, leathery with papery margin; corona staminal, uniseriate, annular ring arising from the base of staminal column; stamens 5. Follicles 1-2, 5-7 × 3.5 cm, turgid, boat-shaped, glabrous; seeds ovate, brown.

In hedges and along open forest paths in bushes in most districts.

Fl.: June-September; Fr.: September-January.

Neelwai RF (ADB), *TP* & *GO* 5443; Bellamapalli (ADB), *KH* 5560; Dharmareddy - Yellareddy road (NZB), *BR* 9515; Muttapally (MDK), *CP* 11781; Aklaspur (KRN), *GVS* 25620 (MH).

LEPTADENIA R.Br.

Leptadenia reticulata (Retz.) Wight & Arn. in Wight, Contrib. Bot. India 47. 1834; Hook.f. Fl. Brit. India 4: 6. 1883; Gamble, Fl. Pres.Madras 2: 850. 1923; Bullock, Kew Bull. 10: 291. 1955. *Cynanchum reticulatum* Retz., Observ. Bot. 2: 15.1781.

Vern.: *Mukkutummudu, Pala theega, Gutti pala teega.*

Twining shrub with watery sap and deeply fissured bark, puberulous. Leaves chartaceous, 3-8 × 1.5-5 cm, pubescent above, densely pubescent beneath, base truncate, apex obtuse or rounded, lateral veins 4-8 pairs, joined at the margin and form intramarginal vein, conspicuous beneath; petiole to 2.5 cm long. Flowers in axillary or terminal umbellate cymes, peduncle to 1 cm long; calyx cupular; corolla yellow, 6 mm across; corona double, coralline and staminal; ovaries half inferior. Follicles cylindric, linear-oblong, bluntly acute at both ends; seeds many, narrowly ovoid, dark brown.

Common in hedges and waste places and in wastelands and forests in Warangal, Hyderabad, Nalgonda and Mahabubnagar districts.

Fl.: April – December. Fr.: October – May.

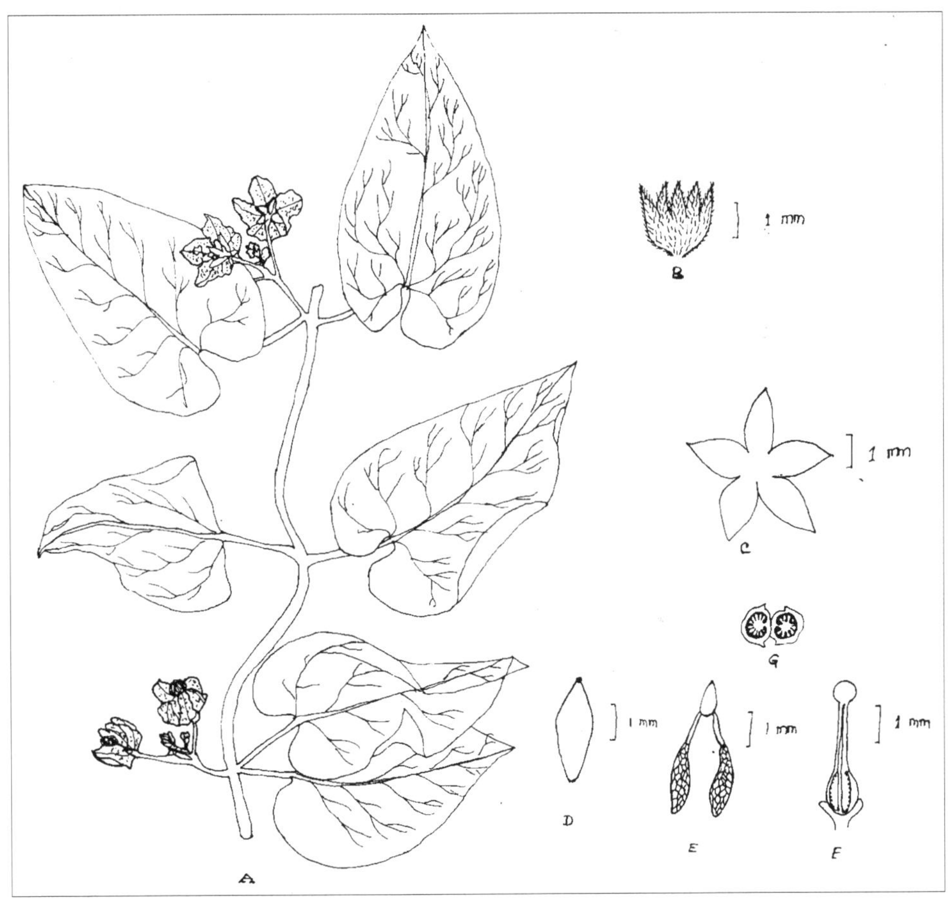

Figure 61: Holostemma ada-kodien Schult.

A. Twig, B. Calyx split open, C. Corolla, D. Corona, E. Pollinia, F. Pistil, G. Ovary c.s.

Hussainsagar (HYD), *KMS* 5958 (MH); Pasra (WGL), *RKP* 105389 (BSID); Mannanur core zone (MBNR), *VBH* 86604 (BSID); Rayagir (NLG), *BVRS* 1 (NU).

MARSDENIA R.Br.

1. Corona coralline .... **M. sylvestris**
1. Corona staminal .... **M. tenacissima**

Marsdenia sylvestris (Retz.) P.I.Forst., Austral. Syst. Bot. 8: 694. 1995. *Periploca sylvestris* Retz., Observ. Bot. 2: 15. 1781; Hook.f., Fl. Brit. India 4: 29. 1883; Gamble, Fl. Madras 2: 839. 1923. *Gymnema sylvestre* (Retz.) R.Br. ex Schult. in Roem. & Schult., Veg. 6: 57. 1819.

Vern.: *Podapatri, Chinna palateega, Putta podara, Puttabhadra*; Eng.: Sugar killer.

Large woody much branched straggler to liane running over from crowns of trees, young branches pubescent, latex milky-white. Leaves ovate or elliptic-oblong or ovate-elliptic, 1.8-5 × 1.1-3.2 cm, pubescent along veins beneath, base obtuse or rounded, apex acute, lateral veins 3-4 pairs; petiole to 1.2 cm long. Flowers in extra-axillary cymes, pedunculate; calyx 5-lobed, 1.5 mm long, corolla greenish-yellow, campanulate, 3 mm long, fleshy, corona coralline, uniseriate, 5-lobed, ridged; stamens 5, 1.5 mm long. Follicles paired, lanceolate, apex beaked, to 7 cm long, seeds with a thin broad marginal wing, brown, glabrous.

Occasional in forests in all districts.

Fl.: April – September. Fr.: October – March.

Manchippa RF (NZB), *TP* & *BR* 6010; Ibrahimpet RF (NZB), *TP* & *BR* 6125; Narsapur RF (MDK), *TP* & *CP* 11989; Sardhana (MDK), *BR* & *CP* 11619; Kusumasamudram RF (RR), *MSM* & *NV* 10367; Vatavarlapalli (MBNR), *TDCK* 13669; Bhadrachalam RF (KMM), *RCS* 99077 (BSID); Lawal beat, Pasra (WGL), *RKP* 108156 (BSID).

Marsdenia tenacissima (Roxb.) Moon., Cat. Pl. Ceylon 21. 1824; Hook.f., Fl. Brit. India 4: 35. 1883; Gamble, Fl. Madras 2: 845. 1923. *Asclepias tenacissima* Roxb., Pl. Coromandel t. 240. 1815.

Vern.: *Penujittu, Murva, Karudu shtupatige.*

Extensive, woody, stout twiner, bark ash or grey coloured, densely soft pubescent or tomentose, latex milky white, bark thick, yellowish green. Leaves broadly ovate, 7-17 × 5.5-12 cm, pubescent above, densely and softly tomentose beneath, base cordate, apex acuminate, lateral veins 5-7 pairs; petiole to 5 cm long, tomentose. Flowers in axillary, branched, corymbose cymes, pubescent, greenish yellow; calyx 5-lobed; corolla campanulate, 6.5 mm long, corona staminal, uniseriate, 5-lobed; stamens 5, yellow. Follicles paired, 12 × 2 cm, ovoid-lanceolate or cylindrical, apex acute with blunt tips, velvety pubescent; seeds many, broadly ovate, brown.

Rare in hills in Adilabad, Karimnagar and Mahabubnagar districts.

Fl.: April – July; Fr.: August- February.

Tiryani (ADB), *KH* 5563; Aklaspur (KRN), *GVS* 20217 (MH).

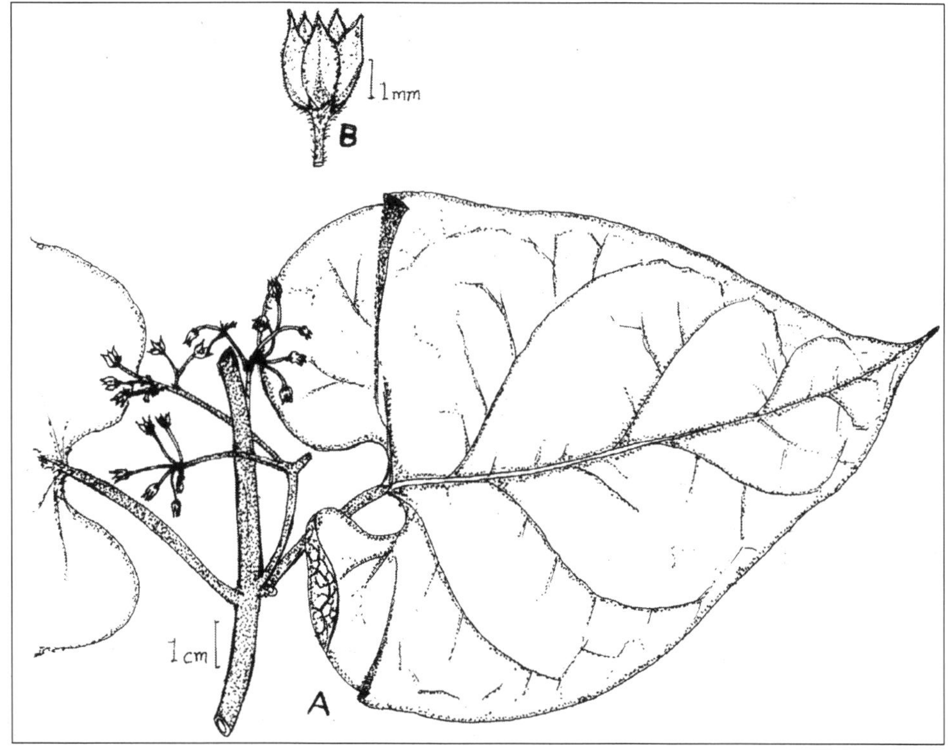

Figure 62: Marsdenia tenacissima (Roxb.) Moon.
A. Twig, B. Flower-entire

OXYSTELMA R.Br.

Oxystelma esculentum (L.f.) Sm., Cycl. 25: 1813. *Periploca esculentum* L.f., Suppl. Pl. 168. 781. *Oxystelma esculentum* R.Br. ex Schult. in L. Syst. Veg. 6: 89. 1820; Hook.f., Fl. Brit. India 4: 17. 1883; Gamble, Fl. Madras 2: 833. 1923.

Vern.: *Dudi pala.*

Very pretty, graceful, twining herb, stem terete, glabrous, latex milky-white. Leaves subcoriaceous, oblong, linear to linear-lanceolate, 3-12 × 0.3-0.8 cm, glabrous, base rounded or subacute, apex finely acuminate, lateral nerves 8-12 pairs. Flowers white or rose in lateral drooping cymes, peduncle terete, to 8 cm long, pedicel filiform, calyx 3.7 mm, glabrous, glandular within, corolla subcampanulate or saucer-shaped, white or pink with purple near the margins, villose, corona double, outer corolline, inner staminal. Follicles glabrous, 3.5 – 7 × 1-1.8 cm, ovate-lanceolate, curved, tapering at the apex.

Less common, in low altitudes in hedges among bushes usually near water courses in Adilabad, Ranga Reddi, Nalgonda, Warangal and Mahabubnagar districts.

Fl. & Fr.: August – December.

Penganga river (ADB), *GO* & *PVP* 4286; Mosra – Malkapur Road (NZB), *TP* & *BR* 6088; Gacchibowli (RR), *MSM* 15198; Amrabad (MBNR), *BSS* & *KP* 32542; Nagarjunakonda (NLG), *BVRS* 109 (NU); Pakhal RF (WGL), *KMS* 13191 (MH); Kompally cheruvu (KMM), *MHR* & *MCK* 17044.

PENTATROPIS Wight & Arn.

Pentatropis capensis (L.f.) Bullock, Kew Bull. 10: 284. 1955. *Cynanchum capense* L.f., Suppl. Pl. 169. 1781. *Pentatropis microphylla* (Roth) Wight, Contrib. Bot. India 52. 1834; Hook.f. Fl. Brit. India 4: 19. 1883; Gamble, Fl. Madras 2: 835. 1923.

Vern.: *Chekurutivva, Eddupullateega, Chitapataku, Kakanasika, Poola pala.*

Slender glabrous twiner, latex watery. Leaves chartaceous when dry, ovate to ovate-oblong or elliptic, 1.2-3 × 0.3-1.5 cm, glabrous, base sub-cordate or rounded, apex obtuse-apiculate, lateral veins 3 pairs, indistinct; petiole 6 mm long. Flowers axillary, in 3-4-flowered umbellate cymes, purplish pink, 1 cm across on slender filiform pedicels; calyx 5-lobed, free, 2 mm long; corolla rotate, 9 mm long, corona staminal, uniseriate, 5-lobed, adnate to the back of the stamens; stamens 5. Follicle usually single, 5 x 1 cm, lanceolate, acuminate, minutely winged, glabrous.

Common on bushes of forests and open forest land in Warangal and Mahabubnagar districts.

Fl.: April-December; Fr.: October –January.

PERGULARIA L.

Pergularia daemia (Forssk.) Chiov., Result. Sci. Miss. Stefan. Paoli Somal. Ital. 1:115. 1916. *Asclepias daemia* Forssk., Fl. Aegypt-Arab. 51. 1775. *Cynanchum extensum* Jacq., Misc. 2: 353. 1782. *Daemia extensa* (Jacq.) R.Br., Mem. Wern. Nat. Hist. Soc.

1: 50. 1809; Hook.f., Fl. Brit. India 4: 20. 1883. *Pergularia extensa* (Jacq.) N.E. Br. in Thistleton-Dyer, Fl. Cap. 4: 758. 1908; Gamble, Fl. Madras 2: 836. 1923.

Vern.: *Dushtaupa tige, Jittupaku.*

Perennial, twining, subhispidly hairy twiner, with foetid smell, latex milky-white. Leaves thin, chartaceous, broadly ovate or suborbicular, 2-8 × 2-5.5 cm, pubescent on both surfaces, mostly along nerves beneath, base cordate, apex acuminate, basal lobes rounded, lateral veins 4-6 pairs, veins dichotomously arched, conspicuous beneath; petioles 3-7 cm long, pubescent. Flowers in axillary, many-flowered racemose cymes, peduncle 5-12 cm long, greenish yellow; calyx 3 mm long; corolla campanulate, 1.2 cm long, tube slightly inflated, fleshy, lobes folded back; corona staminal; stamens 5, 8 mm long. Follicles paired, curved, lanceolate, tapering to a long beak, softly echinate, pubescent; seeds many, brown.

Common on bushes of open forests and waste places in all districts .

Fl.: August – December; Fr.: December – April.

Jannaram (ADB), *TP* & *PVP* 4211; Nasrullababd (NZB), *TP* & *BR* 6176; Mombojipally (MDK), *BR* & *CP* 11643; Anantasagar (RR), *MSM* & *NV* 10404; Molachintapally (MBNR), *TDCK* 13692; Kodimial (KRN), *GVS* 21868 (MH); Subedari (WGL), *RKP* 105248 (BSID); Yenkanapalem (KMM), *RR* 113826 (BSID); Cherukupally (NLG), *BVRS* 665; Nagarjunasagar (NLG), *KMS* 9716 (MH).

TELOSMA Coville

1. Flowers in interpetiolar umbellate racemes; sap milky .... **T. minor**
1. Flowers in axillary cymes; sap watery .... **T. pallida**

Telosma minor (Andrews) Craib, Bull. Misc. Inform. 1911: 418. 1911; Gamble, Fl. Madras 2 : 844. 1923. *Pergularia minor* Andrews, Bot. Repos. 3 : t. 184. 1801; Hook.f., Fl. Brit. India 4 : 38. 1883.

Climber. Leaves cordiform, chartaceous, base cordate, apex acuminate; petioles 5-6 cm long. Flowers white, yellow within, in terminal condensed umbellate racemes; calyx lobes lanceolate; corolla tube campanulate, inflated below; lobes equal, oblong. Follicles not seen.

Cultivated as an ornamental for its fragrant flowers.

Fl. & Fr.: September-March.

Telosma pallida Craib, Bull. Misc. Inform. Kew 1911: 418. 1911; Gamble, Fl. Pres.Madras 2: 844. 1923. *Pergularia pallida* Wight & Arn. in Wight, Contrib. Bot. India 42. 1834; Hook.f., Fl. Brit. India 4: 38. 1883.

Extensive twining undershrub, sap watery, stem suffruticose, cylindric, branched, slightly ridged, sparingly pubescent. Leaves membranous, ovate or ovate-oblong, 3.8-10 × 2.3-7 cm, glabrescent, apex shortly acuminate, base obtuse or subcordate to cordate, lateral veins 4-6 pairs, petiole 1-3.7 cm long. Flowers in many-flowered, simple or branched umbellate cymes, pale green or yellowish white; bracts linear, 3 mm long; pedicel terete, densely pubescent; calyx 5-lobed; corolla salver-shaped, 1.4 cm long, tube 7 mm long, slightly inflated at the base, lobes linear-oblong, corona staminal, uniseriate, 5-lobed, arising from the base of staminal column; stamens 5. Follicle usually single, 8-10 × 1.5-2 cm, lanceolate, deeply grooved on one side, shortly acuminate at apex; seeds brown, ovate-oblong.

Rare in Karimnagar and Mahabubnagar districts.

Fl.: May-September; Fr.: September-December.

Kodimial (KRN), *GVS* 20123 (MH).

TYLOPHORA R. Br.

1. Plants erect, short undershrubs:
   2. Flowers in umbels; corona lobes free from staminal column; pedicels filiform .... **T. fasciculata**
   2. Flowers in spikes; corona lobes adnate to staminal column; pedicels thick .... **T. macrantha**
1. Plants twining, trailing or climbing:
   3. Stems twining; leaves not rotund .... **T. indica**
   3. Stems procumbent, leaves rotund .... **T. rotundifolia**

Tylophora fasciculata Buch.-Ham. ex Wight, Contrib. Bot. India 50. 1834; Hook.f., Fl. Brit. India 4: 40. 1883; Gamble, Fl. Madras 2: 842. 1923.

Slender, erect, suffruticose, undershrub, slightly twining, 40 cm high, latex watery, roots fascicled. Leaves coriaceous, ovate or lanceolate, 1.5-6 × 1-2.7 cm, glabrous above, pubescent on the nerves beneath, base rounded to cordate, apex acute to acuminate, diminishing in size upwards, lateral veins 3-6 pairs, petiole 0.3-2.5 cm. Flowers in extra-axillary or lateral, simple or branched umbels, chocolate brown; pedicels filiform; calyx 5-lobed; corolla rotate, apex minutely bifid, corona staminal, uniseriate, 5-lobed. Follicles usually solitary, 3-5 × 2-2.5 cm, ovate or spindle-shaped, striate, glabrous; seeds many.

Occasional, usually at low levels among grass in Adilabad, Nizamabad and Ranga Reddi districts.

Fl. & Fr.: Apri – November, Fr.: August – December.

Madharam RF (ADB), *TP* & *GO* 5471; Neredukonda (ADB), *KH* 5568; Mudheli RF (NZB), *BR* & *KH* 7123; Rudraram RF (RR), *TP* & *MSM* 12718.

Tylophora indica (Burm.f.) Merr., Philipp. J. Sci. 19: 373. 1921. *Cynanchum indicum* Burm.f., Fl. Ind. 70. 1768. *T. asthmatica* (L.f.) Wight & Arn. in Wight, Contrib. Bot. India 51. 1834; Hook.f., Fl. Brit. India 4: 44, 1883; Gamble, Fl. Madras 2: 843. 1923. *Asclepias asthmatica* L.f., Supl. 171. 1781.

Vern.: *Verri paala, Kaka paala, Kukka pala, Telegapala, Poda pacchali.*

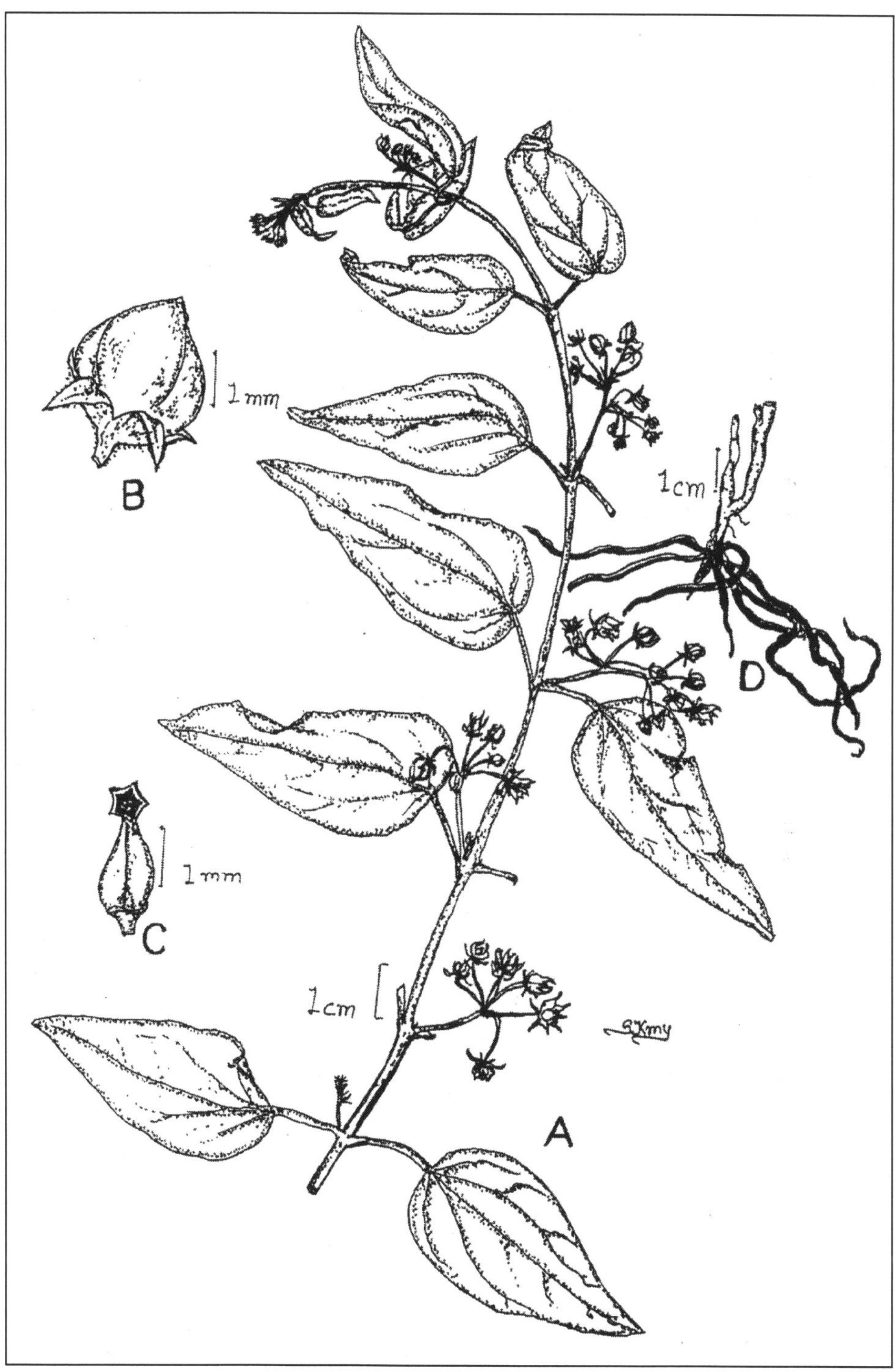

Figure 63: Tylophora fasciculata Buch.-Ham. ex Wight

A. Twig, B. Flower, C. Pistil, D. Root

Much branched climber, stem minutely ridged, subtomentose, with long fleshy roots, latex watery. Leaves subcoriaceous, ovate-oblong or elliptic-oblong, 1.3-8 × 0.5-4.2 cm, pilose, shining above, pubescent along nerves above, pubescent beneath, base rounded to cordate, apex acute to obtusely mucronate, lateral nerves 4-7 pairs; petiole to 1.4 cm long, pubescent. Flowers in lateral umbellate to paniculate cymes, pale white or greenish yellow to purple; calyx 5-lobed, lobes free, to 3.5 mm long; corolla rotate, 9 mm across; corona staminal, uniseriate, 5-lobed, fleshy; stamens 5. Follicles paired, divaricate, cylindric, glabrous, acute at both ends; seeds many, brown.

Common throughout the state amongst bushes, in hedges in open forests in most districts.

Fl. & Fr.: June- December.

Janakampalli (NZB), *TP* & *BR* 6326; Pocharam tank (MDK), *TP* & *CP* 12021; Anantasagar RF (RR), *MSM* & *NV* 10390; Mohammadabad RF (RR), *MSM* & *KH* 10216; Thummalabailu (MBNR), BR, *TSS* & *MVS* 28205; C.T.Padu (MBNR), *MVS* & *DV* 41393; Rusthempet (MDK), *KMS* 8003 (MH).

Tylophora macrantha Hook.f., Fl. Brit. India 4: 40. 1883; Gamble, Fl. Madras 2: 843. 1923.

Slender erect undershrub, roots thick, fascicled, the upper branches twining, more or less finely pubescent. Leaves ovate, ovate-lanceolate or elliptic-lanceolate, 3.5-6.1 × 1-3.2 cm, pubescent, lateral veins 5-6 pairs; petiole 1.2 cm long. Flowers in spikes, 5-6 in each spike; pedicels thick, terete; calyx 5-lobed; corolla rotate, 8 mm across; corona uniseriate, 5-lobed, adnate below to base of staminal column; stamens 5, 1.5 mm long. Follicles solitary, lanceolate, glabrous; seeds many.

Occasional on hills in Medak and Mahabubnagar districts.

Fl.: May – December; Fr.: June – February.

Tylophora ovata (Lindl.) Hook. ex Steud. Nomencl. Bot. (ed.2) 2: 726. 181. *T. mollissima* Wall. ex Wight Contr. Bot. India 49. 1834.

Liana, to 5 m long, rusty hispid, villous or pubescent. Leaves ovate, 2-12 × 1-8 cm, base cordate, apex acute. Flowers yellowish or yellow-green in many-flowered, dense raceme; pedicel threadlike; sepals subulate to ovate; corolla rotate, lobes oblong-ovate; corona lobes ovoid; stigma depressed. Follicles lanceolate to oblong-lanceolate; seeds oval, coma 2-3.5 cm.

Occasional in Medak district.

Fl. & Fr.: June - December

Nagsanpally RF (MDK), *RG* 104193 (BSID).

Tylophora rotundifolia Buch.-Ham. ex Wight, Contrib. Bot. India 50. 1834; Hook.f., Fl. Brit. India 4: 43. 1883; Gamble, Fl. Madras 2: 843. 1923.

Perennial twiner, somewhat fleshy, pubescent or tomentose, ridged, somewhat fleshy. Leaves coriaceous, round to broadly ovate, 2.5–8 × 1.8-7 cm, glabrous above, more or less pubescent beneath, apex rounded to obtuse, occasionally mucronate or emarginated, lateral nerves 4-6 pairs, petiole terete, to 7 mm long. Flowers in terminal many-flowered umbellate cymes, pedicels filiform, 1.2-3.2 cm long; bracts very small, persistent; calyx-lobes linear-lanceolate, to 3.7 mm long, pubescent; corolla rotate, lobes 5, ovate, apex slightly cleft, corona staminal, uniseriate, corona lobes swollen, globose with free tips above. Follicle single, lanceolate, 4-7.5 x 1.5-2.5 cm, ridged on one side, somewhat compressed, glabrous; seeds brown, nearly orbicular, 6.2 mm long.

Occasional in forests in damp places in Medak district.

Fl.: May – September; Fr.: November – December.

Pocharam Wild Life Sanctuary (MDK), *RG* 116416 (BSID); Pillutla RF (MDK), *RG* 106777 (BSID).

LOGANIACEAE

1. Herbs; corolla valvate; capsule didymous:
    2. Flowers in 2-3-chotomous cymes, 5-merous .... **Cynoctonum**
    2. Flowers solitary or clustered, 4-merous .... **Mitrasacme**
1. Shrubs, often scandent or trees .... **Strychnos**

CYNOCTONUM Gmel.

Cynoctonum mitreola (L.) Britton, Mem. Torrey Bot. Club 5: 258. 1894; Heine, Kew Bull. 17: 171-172. 1963. *Ophiorhiza mitreola* L., Sp. Pl. 150. 1753. *Mitreola oldenlandioides* Wall., ex DC., Prodr. 9: 9. 1845; Hook.f., Fl. Brit. India 4: 79. 1883; Gamble, Fl. Madras 2: 863. 1923.

Annual, slender, erect herb, up to 50 cm high. Leaves membranous, ovate-elliptic or elliptic-lanceolate, 4-7 × 2-3 cm, glabrous, base acute running down into the petiole, margin entire, main nerves 8-10 pairs, petiole 0.7 – 1.2 cm long. Flower minute, white in axillary dichotomous scorpiod cymes. Capsule unilateral, 0.4-0.5 cm broad, two-horned, glabrous.

Occasional in moist places in Khammam district.

Fl. & Fr.: August- November.

Edrallapalli RF (KMM), *RCS* 98754, 102463 (BSID).

MITRASACME Labill.

Mitrasacme indica Wight, Icon. Pl. Ind. Orient. t. 1601. 1850 & Ill. t. 156. ff. 1-13. 1850; Leenh. in Steenis Fl. Males. I. 6: 384. 1962. *M. alsionoides auct. non* R.Br.; Hook.f., Fl. Brit. India 4: 80. 1883; Gamble, Fl. Madras 2: 864. 1923.

Very slender, annual, erect herb, up to 15 cm high, stem glabrous. Leaves opposite, in distant pairs, sessile, linear, linear-lanceolate or lanceolate, 2-7 × 0.8-2 mm, acute, glabrous. Flowers white, 2.5-3.7 mm long, ca 2.5 mm across, solitary or fascicled 2-3, axillary, pedicels filiform, 3-7 (in fruit 5-25) mm long; calyx 2 mm long, lobes somewhat cuspidate in fruit. Capsule subglobose, 1.5-2.5 mm diam, seeds with reticulate testa.

Occasional in moist places in hills in Mahabubnagar district.

Fl. & Fr.: August – November.

STRYCHNOS L.

1. Bark grey, smooth; petiole to 1.2 cm; flowers 8-12 mm long; berry orange, 2.5-5 cm diam.; seeds numerous .... **S. nux-vomica**
1. Bark blackish, cracked; petiole to 5 mm; flowers 4-5 mm long; berry black, 1.3 cm diam.; seeds 1 or 2 .... **S. potatorum**

Strychnos nux-vomica L., Sp. Pl. 189. 1753; Hook.f., Fl. Brit. India 4: 90. 1882; Gamble, Fl. Madras 2: 868. 1923.

Vern.: *Mushti, Mushini, Visha mushti, Pedda mushti.* Eng.: Snake wood tree.

Deciduous tree, up to 15 m high; bark thin, grey, smooth, covered with tubercles and containing chlorophyll tissue inside. Leaves coriaceous, elliptic, ovate, orbicular, 3.5-7.8 × 3-6 cm, base rounded, 5-nerved, transverse nerves irregular, inconspicuous; petiole to 1.2 cm. Flowers greenish-white, fragrant in terminal compound cymes. Fruit globose, 2.5-5 cm diam., thick-shelled, slightly rough, but shining, orange-red when ripe; seeds numerous, discoid, 4 × 2.16 cm, much compressed, ashy-grey, with silky hair radiating from the centre, concave on one side and convex on the other.

Common in deciduous forests in most districts.

Fl.: March – April; Fr.: March – August.

Chinnoor (ADB), *GO* 4498; Gangapur RF (MDK), *BR* & *CP* 11425; Vatavarlapally (MBNR), *TDCK* 13656; Mallelatheertham (MBNR), *BR, TSS* & *MO* 28267; Pakhal west (WGL), *ANH* 15963 (MH); Eturunagaram (WGL), *RKP* 110811 (BSID); Kudurpalli (KRN), *SLK* 70717 (NBG); Sathiyampet RF (KMM), *RR* 112652 (BSID).

Strychnos potatorum L.f., Suppl. Pl. 148. 1782; Hook.f., Fl. Brit. India 4: 90. 1882; Gamble, Fl. Madras 2: 868. 1923.

Vern.: *Chilla ginja, Chilla, Induga, Indiba chettu, Chilaka mushti, Katakamu;* Eng.: Cleaning nut tree.

Deciduous tree, to 15 m high; bark dark, black, deeply vertical cracked, branches fluted. Leaves coriaceous, ovate or elliptic, 4-7 × 2.5-4 cm, glabrous, shining, bright green, apex acute-subacuminate, margin entire, base rounded, 3-5-nerved, transverse nerves parallel, conspicuous, petiole to 5 mm. Flowers whitish in axillary cymes; calyx cupular, lobes 5, imbricate; corolla 4 mm wide, salver-form, tube 4.5 mm, villous within, lobes 5, reflexed, 2.5 mm; stamens 5, subexserted; ovary 2-celled, ovules a few. Berry globose, 1.3 cm diam., black when ripe, seed 1, rarely 2, in whitish pulp, bluntly lenticular, shining with short adpressed silky hairs.

Common in dry deciduous forests in all districts.

Fl.: February- May. Fr.: October-March.

Satharmat (ADB), *TP* & *PVP* 4105; Sirnapally RF (NZB), *BR* & *PSPB* 9585; Gangapur RF (MDK), *BR* & *CP* 11425; Anantagiri RF (RR), *TP* & *MSM* 11068; Jalupenta

(MBNR), *TDCK* 15313; Nagarjunakonda (NLG), *BVRS* 147 (NU); Kondagutta (KRN), *GVS* 25683 (MH).

GENTIANACEAE

1. Flowers violet; ovary 2-celled; stamens connivent around style, exserted, dehiscence porous .... **Exacum**
1. Flowers yellow or white; ovary 1-celled; stamens away from style, included, dehiscence longitudinal:
    2. Corolla regular:
        3. Flowers sessile or subsessile, corolla white; leaves lanceolate-oblong .... **Enicostema**
        3. Flowers pedicelled, 1-1.2 cm long, corolla pink; leaves obovate or elliptic .... **Centaurium**
    2. Corolla irregular:
        4. Flowers yellow; stigma subentire .... **Hoppea**
        4. Flowers pink or white; stigma deeply 2-lobed .... **Canscora**

CANSCORA Lam.

1. Bracts at the branches of the inflorescence not perfoliate:
    2. Calyx not winged; leaves subsessile-petiolate:
        3. Stem 4-angled .... **C. diffusa**
        3. Stem 4-winged .... **C. heteroclita**
    2. Calyx prominently winged; leaves sessile .... **C. alata**
1. Bracts at the branches of the inflorescence perfoliate .... **C. roxburghii**

Canscora alata (Roth) Wall., Numer. List. 4363. 1831. *Exacum alatum* Roth ex Roem. & Schult. Syst. Veg. 3: 159. 1818. *Canscora decussata* (Roxb.) Roem. & Schult.f., Mant. 3: 229. 1827; C.B.Clarke in Hook.f., Fl. Brit. India 4: 104. 1883; Gamble, Fl. Madras 2: 879. 1923. *Pladera decussata* Roxb., Fl. Ind. 1: 418. 1820.

Vern.: *Chitti akchinta, Sankhaphuli, Janna jaada*.

Annual erect herb, up to 30 cm high, stem 4-winged. Leaves elliptic-lanceolate, 1.5-2 × 0.5-1 cm, 3-nerved, base rounded to acute, scabrid along margin, apex gradually acute, sessile. Flowers white or pale pink in axillary or terminal dichasial cymes; calyx lobes 4, equal, winged; corolla lobes 4, oblong; stamens 4; ovary bicarpellary, style filiform, stigma bifid. Capsule oblong, cylindric, obscurely 2-valved, seeds angular.

Occasional, both in forests and cultivated fields in Nizamabad, Karimnagar, Khammam and Mahabubnagar districts.

Fl. & Fr.: November – March.

Gandhari RF (NZB), *BR* & *CPR* 7236; Aklaspur (KRN), *GVS* 22484 (MH); Poochavaram RF (KMM), *RCS* 98735 (BSID).

Canscora diffusa (Vahl) R.Br. ex Roem. & Schult. Syst. Veg. 3: 301. 1818. *Gentiana diffusa* Vahl, Symb. Bot. 3: 47. 1794; C.B.Clarke in Hook.f., Fl. Brit. India 4: 103. 1883; Gamble, Fl. Madras 2: 878. 1923. *Canscora decurrens* Dalzell, Hooker's J. Bot. Kew Gard. Misc. 2: 136. 1850; C.B.Clarke in Hook.f., Fl. Brit. India 4: 103. 1993; Gamble, Fl. Madras 2: 878. 1923.

Vern.: *Janjada.*

Very pretty, slender, annual, erect herb, up to 35 cm high, stems obtusely quadrangular, obscurely winged. Leaves membranous, upper ones broadly ovate, 1-3.5 × 0.5-1.5 cm, base rounded, apex acute, apiculate, sessile; the lower ones elliptic-ovate, base much narrowed, apex acute, subsessile-petiolate. Flowers pink, numerous, in lax diffuse paniculate cymes; pedicels filiform, 1.2 cm long, leaf like bracts to the forks of the branches of the panicle. Capsule oblong, compressed, seeds angular, minutely rugose.

Occasional in moist shady places along ravines in most districts.

Fl. & Fr.: October – March.

Satnella (ADB), *GO* & *PVP* 4220; Janakampally (NZB), *TP* & *BR* 6321; Sirnapally river (NZB), *BR* & *GO* 9078; Gangapur RF (MDK), *TP* & *CP* 12084; Laxmapur (MDK), *BR* & *CP* 11448;; Kanmankalva RF (RR), *MSM* & *KH* 10305; Rangammagudem RF (RR), *MSM* & *KH* 10577; Aklaspur (KRN), *GVS* 22530 (MH); Rusthempet (MDK), *KMS* 8006 (MH, CAL); Thiruppumetta RF (KMM), RCS 98735 (BSID); Pakhal (WGL), *RKP* 108262 (BSID).

Canscora heteroclita (L.) Gilg in Engler& Prantl, Pflanzenfam. 4(2): 76. 1895. *Gentiana heteroclita* L., Mant. 2: 560. 1770. *Canscora sessiliflora* Roem. & Schult.f., Mant. Syst. Veg. 3: 230. 1827; C.B.Clarke in Hook.f., Fl. Brit. India 4: 104. 1883; Gamble, Fl. Madras 2: 879. 1923.

Annual, erect herb, up to 30 cm high, stem 4-winged. Leaves broadly ovate, 0.7-1.5 × 0.6-1 cm, base rounded to obtuse, apex acutely apiculate, upper ones linear-lanceolate. Flowers rose coloured in the forks in dichasial cymes, becoming monochasial. Capsule oblong-compressed, brownish.

Occasional in perennially moist areas (C.S. Reddy, 2001).

Fl. & Fr.: December – February.

Madikonda (WGL), *CSR* 1115 (KU).

Canscora roxburghii Arn. ex Miq., Anal. Bot. Ind. 3: 11. 1852. *Canscora wallichii* C.B.Clarke in Hook.f., Fl. Brit. India 4: 105. 1883. *C. perfoliata* Wight, Icon. PL. Ind. Orient. t. 1327. 1848; Gamble, Fl. Madras 2: 879. 1923.

Annual, slender, dichotomously branched, erect herb, up to 25 cm high, stem angular, faintly winged. Radical leaves numerous, spathulate, rosulate, persistent, obovate or oblong, 0.5 – 1.5 × 0.2 cm, base subcordate, margin entire, apex acute, sessile, cauline leaves smaller, linear-oblong. Flowers rose coloured or white in repeatedly dichotomous cymes with a flower in each fork; bract large, perfoliate, orbicular-rhomboid; bracteoles 2, linear-lanceolate; calyx lobes 4, 1 cm; corolla lobes 4, unequal, 1.2 cm; stamens 4, filaments 1.5 mm; ovary unilocular, ovules numerous, style 4 mm, stigma bifid. Capsule narrowly oblong.

Weed in cultivated fields of Nizamabad district.

Fl. & Fr.: December – April.

Manchippa (NZB), *BR* & *KH* 5544.

CENTAURIUM Hill.

Centaurium meyeri (Bunge) Druce, Rep. Bot. Exch. Club. Brit. Isles 1916: 613. 1919; Agarwal, J. Econ. Tax. Bot. 3: 981. 1982. *Erythraea meyeri* Bunge in Ledeb., Fl. Alt. 1: 220. 1829; C.B. Clarke in Hook.f., Fl. Brit. India 4: 102. 1883. *Chironia centauriodies* Roxb., Fl. Ind. 2: 283. 1824. *Centaurium roxburghii* (G.Don) Druce, Bot. Exch. Club. Soc. Brit. Isles 4: 614. 1916. *Erythraea roxburghii* D. Don, Lond. & Edinb. Phil. Mag. & J. Sci. 8: 77. 1836; C.B.Clarke in Hook.f., Fl. Brit. India 4: 102. 1883 incl. var. *saxatilis* C.B.Clarke; Gamble, Fl. Madras 2: 875. 1923. *Centaurium centaurioides* (Roxb.) Rolla Rao & Hemadri, J. Bombay Nat. Hist. Soc. 67: 357. 1970.

An erect herb, 4-20 cm high. Radical leaves obovate or elliptic, 2.5 cm long, usually obtuse, cauline leaves smaller, narrower. Flowers pink, long-pedicelled, 1-1.2 cm long, in repeatedly branched dichasial cymes, bracts small, pedicels in the forks of cymes, 0.6-1.2 cm long; calyx 5 mm long; corolla-tube slender, 6.5-7.5 mm long, lobes spreading, very small.

In moist places in Nizamabad and Medak districts.

Fl. & Fr.: March- May.

Nagaram (NZB), *BR* 9709; Manchippa tank (NZB), *BR* & *KH* 9039; Gundaram RF (NZB), *BR* & *KH* 9680; Medak (MDK), *BR* & *CP* 11419.

ENICOSTEMA Blume *nom. cons.*

Enicostema axillare (Poir. ex Lam.) A. Raynal, Adansonia n.s., 9: 75. 1969. subsp. **axillare**. *Gentiana axillaris* Poir. ex Lam., Tabl. Encycl. 1 (2): 487. 1793 non Raf. 1828. nec Reichb. 1828. *Adenema hyssopifolium* Don, Gen. Hist. 4: 201. 1838. *Enicostema hyssopifolium* (Willd.) Verdc., Bothalia 7: 462. 1961. *Exacum hyssopifolium* Willd., Sp. Pl. 1: 640. 1798. *Enicostema littorale* auct. non Blume; C.B.Clarke in Hook.f., Fl. Brit. India 4: 101. 1883; Gamble, Fl. Madras 2: 875. 1923.

Vern.: *Vellaragu, Gulvidi, Nela-guli, Nelagulimidi.*

Perennial, erect, glabrous herb, up to 40 cm high; stem 4-angular, tender parts glaucous. Leaves decussate, lanceolate-oblong, 5-7 × 0.5-1.5 cm, 3-nerved, base cuneate, apex acute, petiole to 5 mm long. Flowers white, sessile, in dense axillary clusters round the stem; calyx cupular, lobes 5, subequal, recurved with age; corolla 2 mm wide, salver-form, lobes 5, twisted, 7 mm; stamens 5, equal and fertile, included, anthers apiculate; ovary 1-celled, ovules numerous on parietal-intruded, bifurcate placentae. Capsule 3.5 x 3 mm, septicidally 2-valved.

Occasional in moist places in all districts.

Fl. & Fr.: June – January.

Kadthala (ADB), *GO* & *DAM* 4993; Vempalli (ADB), *GO* & *PVP* 4799; Manchippa RF (NZB), *TP* & *BR* 6012; Ghanapur RF (NZB), *BR* & *KH* 7116; Sangareddy tank (MDK), *BR* & *CP* 11292; Lingampalli (RR), *MSM* 12761; Jalupenta (MBNR), *TDCK* 15333; Allinarum (MBNR), *BSS* & *SKB* 30801; Way to Pagarikutta (MDK), *KMS* 6631 (MH); Bhadrachalam RF (KMM), *RCS* 102425 (BSID); Mukunur (KRN), *PVS* 84029 (BSID); Durachipalli (NLG), *BVRS* 34 (NU).

EXACUM L.

Exacum pedunculatum L., Sp. Pl. 111. 1753; C.B.Clarke in Hook.f., Fl. Brit. India 4: 97. 1883; Gamble, Fl. Madras 2: 874. 1923.

Annual, erect, glabrous herb, up to 30 cm high, stem quadrangular. Leaves opposite, elliptic-lanceolate, 1-4.5 × 0.5-1.2 cm, glabrous, 3-nerved, base cuneate, margin entire, apex acute; petiole 3 mm long. Flowers blue in terminal much branched cymes. Capsule globose, surrounded by persistent winged calyx; seeds minute, pale brown.

Occasional in open moist places, along stream banks in all districts.

Fl. & Fr.: September – March.

Sone (ADB), *TP* & *PVP* 4028; Gandivet RF (NZB), *BR* & *CPR* 7200; Manjira barrage (MDK), *BR* & *CP* 14089; Naskal (RR), *MSM* & *PRSR* 15109; Rechapalli (KRN), *GVS* 21897 (MH); Rusthempet (MDK), *KMS* 8004 (CAL, MH); Khammam to Dhaniyaigudem (KMM), *RR* 108517 (BSID); Mudigonda (NLG), *BVRS* 614 (NU).

HOPPEA Willd.

Hoppea dichotoma Willd., Ges. Naturf. Freunde Berlin Neue Schriften 3: 435. 1801; C.B.Clarke in Hook.f., Fl. Brit. India 4: 100. 1883; Gamble, Fl. Madras 2: 877. 1923.

Small, divaricately branched herb, to 15 cm high, stem and branches quadrangular. Leaves ovate, 6-7 × 2-4 mm, glabrous, base cuneate, apex apiculate, sessile. Flowers pale yellow in terminal, dense, dichotomous cymes; bracts 2, foliaceous, oblong; calyx campanulate, lobes linear-oblong or lanceolate; corolla lobed, lobes 1.5-2 mm long; stamen 4, anisomorphic; ovary globose, stigma simple. Capsule ellipsoid, glabrous, enclosed in persistent sepals; seeds numerous, minute, pale to dark green, muricate.

Occasional in grass in fallow fields, by river banks and often moist exposed places in all districts.

Fl. & Fr.: October – March.

Sone (ADB), *TP* & *PVP* 4016; Near Potchera (ADB), *GO* & *DAM* 5070; Gandhari RF (NZB), *BR* & *CPR* 7237; Belal tank (NZB), *BR* 7268; Pochammaralu (MDK), *BR* & *CP* 11480; Anantasagar RF (RR), *MSM* & *KH* 11010; Rechapalli (KRN), *GVS* 22264 (MH); Rusthempet (MDK), *KMS* 8005 (MH); Karkalpadu beat, Amangal RF (MBNR), *SRS* 104550 (BSID).

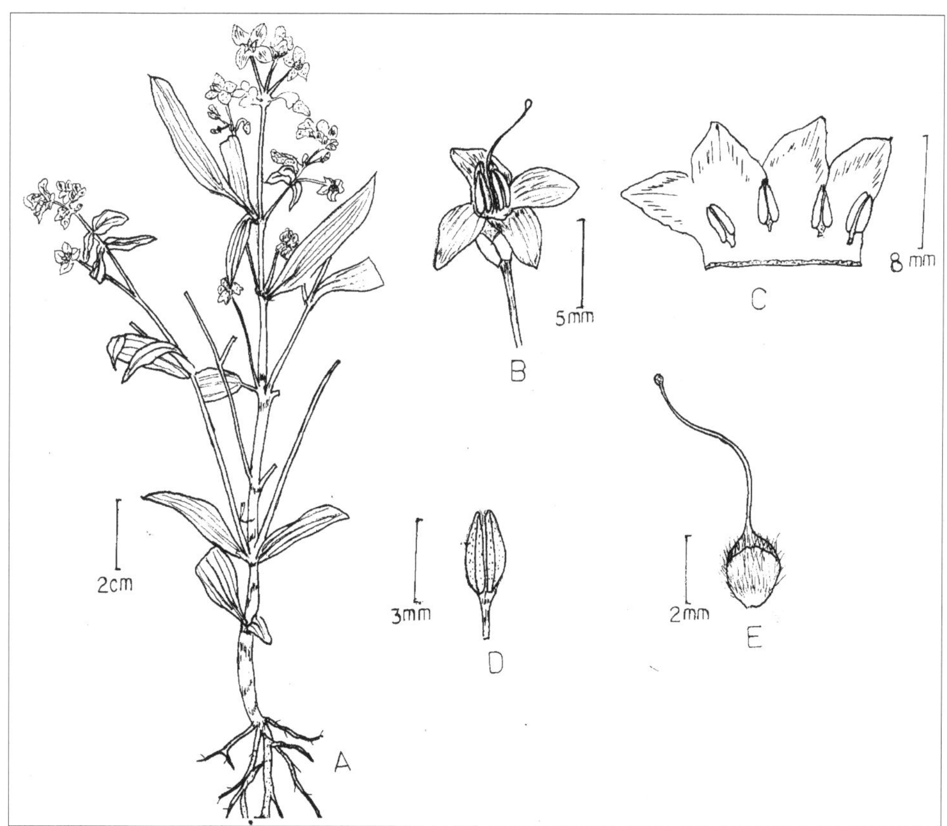

Figure 64: Exacum pedunculatum L.

A. Twig, B. Flower, C. Corolla split open, D. Stamen, E. Pistil.

MENYANTHACEAE

NYMPHOIDES J. Hill

1. Corolla lobes not fimbriate, glandular, crested down the middle with 3 longitudinal folds; seeds 6-10, tuberculate .... **N. hydrophylla**
1. Corolla lobes fimbriate, eglandular, not crested down the middle; seeds more than 10, smooth .... **N. indica**

Nymphoides hydrophylla (Lour.) Kuntze, Revis. Gen. Pl. 2: 429. 1891. *Menyanthes hydrophylla* Lour., Fl. Cochinch. 1: 129. 1790. *Nymphoides cristatum* (Roxb.) Kuntze, Revis. Gen. Pl. 429. 1891. *Menyanthes cristata* Roxb., Pl. Coromandel 2: 3 t. 105. 1798. *Limnanthemum cristatum* (Roxb.) Griseb., Gen. & Spec. Gent. 342. 1839; Hook.f., Fl. Brit. India 4: 131. 1883; Gamble, Fl. Madras 2: 883. 1923.

Vern.: *Pitta kaluva, Allitamara, Kukka tamara.*

Aquatic herb with floating leaves. Leaves orbicular, 6-7 × 6-9 cm, glabrous, minutely scaberulous, purplish and with green veins beneath, base deeply cordate, margin entire, petiole 2.5 cm long. Flowers white, axillary, solitary or in clusters, at the distal node of the stem along with roots; pedicel ca 5 cm; calyx lobes 5; corolla lobes 5; ovary triangular, stigma capitate. Capsule broadly ovoid, few-seeded, 4 × 3 mm; seeds tuberculate.

Occasional in ponds, slow flowing streams, tanks and ditches in most districts.

Fl. & Fr.: April – September.

Nirmal tank (ADB), *TP* & *PVP* 4068; Kammarpally (NZB), *TP* & *BR* 6236; Gundaram (NZB), *BR* & *KH* 9684; Narsapur tank (MDK), *TP* & *CP* 12007; Narsapur (MDK), *RG* 11656 (BSID); Pakhal (WGL), *KMS* 11705 (CAL, MH); Gunavaram (MDK), *RG* 116456 (BSID).

Nymphoides indica (L.) Kuntze, Revis. Gen. Pl. 2: 429. 1891. *Menyanthes indica* L., Sp. Pl. 145. 1753. *Limnanthemum indicum* (L.) Griseb. emend. Thwaites, Enum. Zeyl. 205. 1860; Hook.f., Fl. Brit. India 4: 131. 1883; Gamble, Fl. Madras 2: 883. 1923.

Stoloniferous herb, to 60 cm long. Leaves floating, orbicular, deeply cordate, with obtuse basal lobes and a triangular sinus, 12-16 × 10-20 cm, thick, pubescent, sinuate. Flowers dimorphic, with an yellow centre, appearing between the basal lobes of the leaves in umbellate clusters at both proximal and distal nodes of the stem. Capsule sub globose, 0.4 × 0.2 cm, 15-seeded; seeds shining, lenticular, minutely tuberculate.

Rare in ponds in Nizamabad, Warangal and Karimnagar districts.

Fl. & Fr.: January – November.

Gundaram (NZB), *BR* & *KH* 9685; Pakhal lake (WGL), *KMS* 11613 (MH).

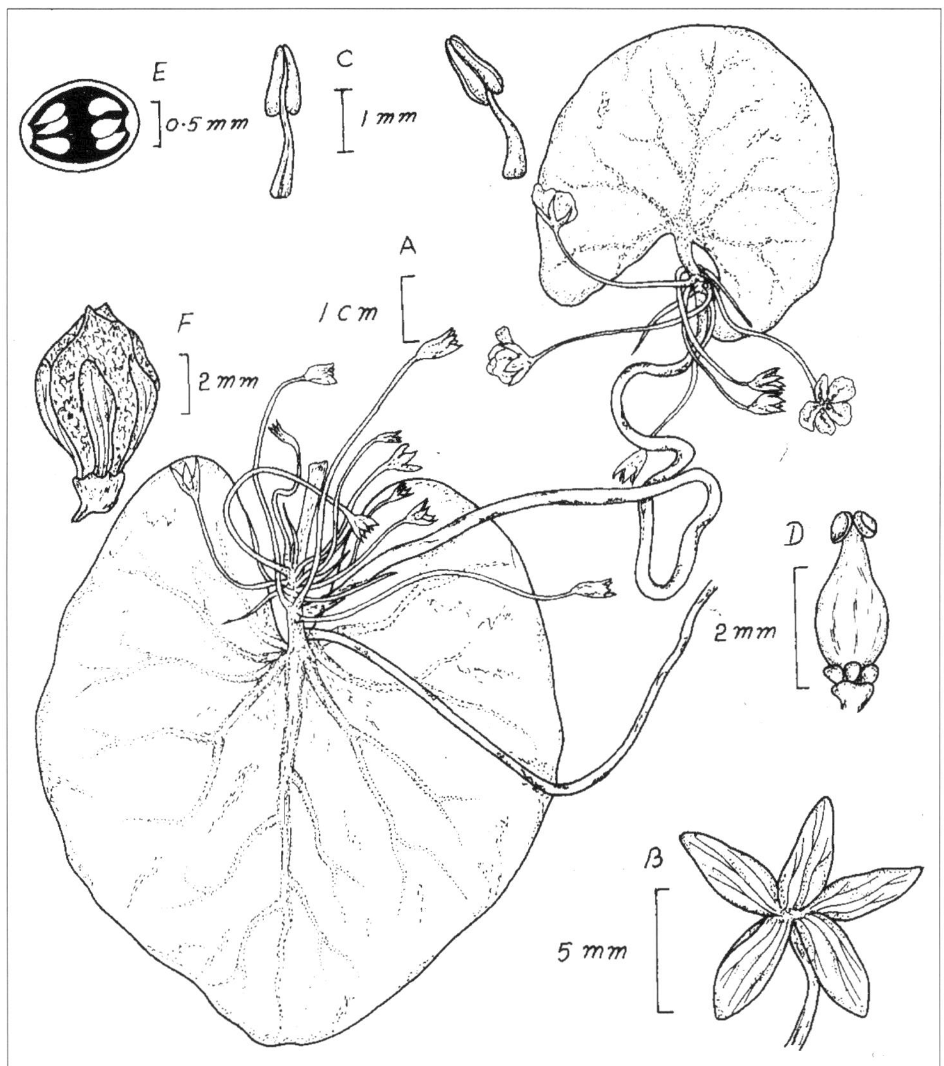

Figure 65: Nymphoides hydrophylla (Lour.) Kuntze

A. Twig, B. Calyx Split open, C. Corolla split open, D. Stamens, E. Pistil, F. T.S. of ovary.

HYDROPHYLLACEAE

HYDROLEA L. *nom. cons.*

Hydrolea zeylanica (L.) Vahl, Symb. Bot. 2: 46. 1791; Hook.f., Fl. Brit. India 4: 133. 1883; Gamble, Fl. Madras 2: 884. 1923. *Nama zeylanica* L., Sp. Pl. 226. 1763.

Vern.: *Neeli nakshatralu, Pedda ponaganti.*

Annual, subsucculent, spreading herb, up to 60 cm long, often rooting at nodes; stems and branchlets pubescent when young. Leaves simple, elliptic lanceolate, 0.6 – 3 × 0.3 – 1 cm, base obtuse or truncate, margin entire, apex acute, glabrous above, glaucescent below, chartaceous; petioles short, up to 4 cm long. Flowers axillary or terminal racemes; peduncle up to 2 cm long; pedicels up to 0.5 cm long; calyx glandular pubescent, lobes 5, elliptic-lanceolate, 5 mm, hirsute, acute; corolla deep blue, 1.5 cm across, lobes 5, ovate, 8 mm; stamens 5, filaments basally dilated, 2 mm long, anthers sagittate, 2 mm long; ovary oblong-glandular, 1.2 mm, styles 1.8 mm, stigma capitellate. Capsule globose, hispid; seeds minute, wrinkled.

Common in moist places, river and canal banks, fallow fields in Warangal, Karimnagar, and Nalgonda districts.

Fl. & Fr.: October – March.

Kudurpalli (KRN), *SLK* 70769 (NBG); Pakhal lake (WGL), *KMS* 11617 (MH); Mudigonda (NLG), *BVRS* 615 (NU).

BORAGINACEAE

KEY 1

1. Leaves entire, symmetric:
    2. Flowers in terminal helicoid spikes or racemes:
        3. Nutlets glochidiate .... **Cynoglossum**
        3. Nutlets glabrous or pubescent .... **Heliotropium**
    2. Flowers usually axillary, solitary or in terminal paniculate cymes .... **Trichodesma**
1. Leaves crenate-serrate, oblique .... **Coldenia**

KEY 2

1. Style divided (bipartite/style arms 2) .... **Coldenia**
1. Style undivided:
    2. Ovary not lobed; fruit a drupe of 4 pyrenes or when of 4 nutlets, free, not attached to the carpophore .... **Heliotropium**
    2. Ovary 4-lobed; fruit of 4 nutlets attached to a carpophore:
        3. Style gynobasic, stigma depressed-capitate; carpophore conical; nutlets glochidiate on the outer face .... **Cynoglossum**
        3. Style (sub)terminal, stigma simple; carpophore pyramidal; nutlets smooth on the outer face .... **Trichodesma**

COLDENIA L.

Coldenia procumbens L., Sp. Pl. 125. 1763; Hook.f., Fl. Brit. India 4: 144. 1883; Gamble, Fl. Madras 2: 892. 1923; Kazmi, J. Arnold Arbor. 51: 148. 1970.

Vern.: *Hamsapadu, Chepputattaku, Bukkinaku.*

Annual prostrate scabrid herbs with spreading branches; branchlets densely pilose, closely appressed to substratum. Leaves simple, obovate-oblong, 1-2 × 0.5-1 cm, base cuneate, margin crenate-serrate, apex obtuse-emarginate, pilose above, densely so below; sometimes white-glandular-pubescent along the nerves; petiole up to 1 cm long. Flowers white or creamy, minute, scarcely 5 mm across, axillary, solitary, subsessile, stalk enlarging in fruit; calyx lobes 4 (5), ovate, 1.5 mm, thick, scabrous without; corolla white, 2 mm across, tubular, tube 1.5 mm, lobes 4 (5), ovate, 0.5 mm; stamens 4 (5), included, attached at the middle of the tube; ovary conical, 4-lobed, style- arms 2, capitate. Fruit drupe of 4 sub(connate) pyrenes, pubescent; seed solitary in each pyrene.

Common weed of fallow fields, waste lands, river banks, dried up tanks, ponds in most districts.

Fl. & Fr.: December - August.

Kadam dam (ADB), *GO* & *DAM* 5197; Lakshmapur (NZB), *TP* & *BR* 6289; Medak (MDK), *RG* 104023 (BSID); Ramayampet (RR), *BR* & *CP* 11518; Kusumasamudram (RR), *MSM* & *NV* 10339; Molachintapally (MBNR), *TDCK* 15317; Hussainsagar (HYD), *KMS* 5955 (CAL, MH); Yetapaka (KMM), *RR* 106074 (BASID); Kodimial (KRN), *GVS* 20081 (MH); Ramanagutta (KMM), *RR* 107968 (BSID).

CYNOGLOSSUM L.

Cynoglossum zeylanicum (Lehm.) Brand, Pflanzenr. IV, 252: 134. 1921. *Echinosperma zeylanicum* Lehm., Pl. Asperif. Nucif. 116. 1818. *Cynoglossum zeylanicum* (Hornem.) Thunb. ex Lehm., Neue Schriften Naturf. Ges. Halle 3(2): 20. 1817 *invalid*; Kazmi, J. Arnold Arbor 52: 344. 1971. *Anchusa zeylanica* Vahl ex Hornem., Enum. Pl. Hort. Hafn. 3. 1807 *invalid*. *Cynoglossum denticulatum* A. DC. var. *zeylanicum* (Hornem.) C.B.Clarke in Hook.f. Fl. Brit. India 4: 157. 1883; Gamble, Fl. Madras 2: 900. 1923.

Vern.: *Kada anthrinta.*

Annual, erect, hispid herb, up to 60 cm high, stems and branchlets coarsely or strigose. Leaves simple, elliptic-lanceolate, 3-6 × 1-3 cm, base rounded to acute, margin entire, apex acute, strigose above (hairs with bulbous bases), shortly hairy beneath without tubercles; petiole up to 0.7 cm long, lower leaves long-petiolate. Flowers white with blue centre, terminal in elongated sparingly branched or loosely paniculate racemes; pedicels up to 0.5 cm long; calyx lobes 5, obovate, sericeous without, spreading in fruit; corolla pale blue, funnel-form, throat glandular, glands 5, lobes 5, suborbicular; stamens 5, inserted on the throat, anthers oblong, (sub)sessile; ovary 4-lobed, ovule one per cell, style gynobasic, stigma capitate. Fruits pyramids of 4-rounded nutlets, nutlets glochidiate, small, hairy, sticky, green; seeds flat.

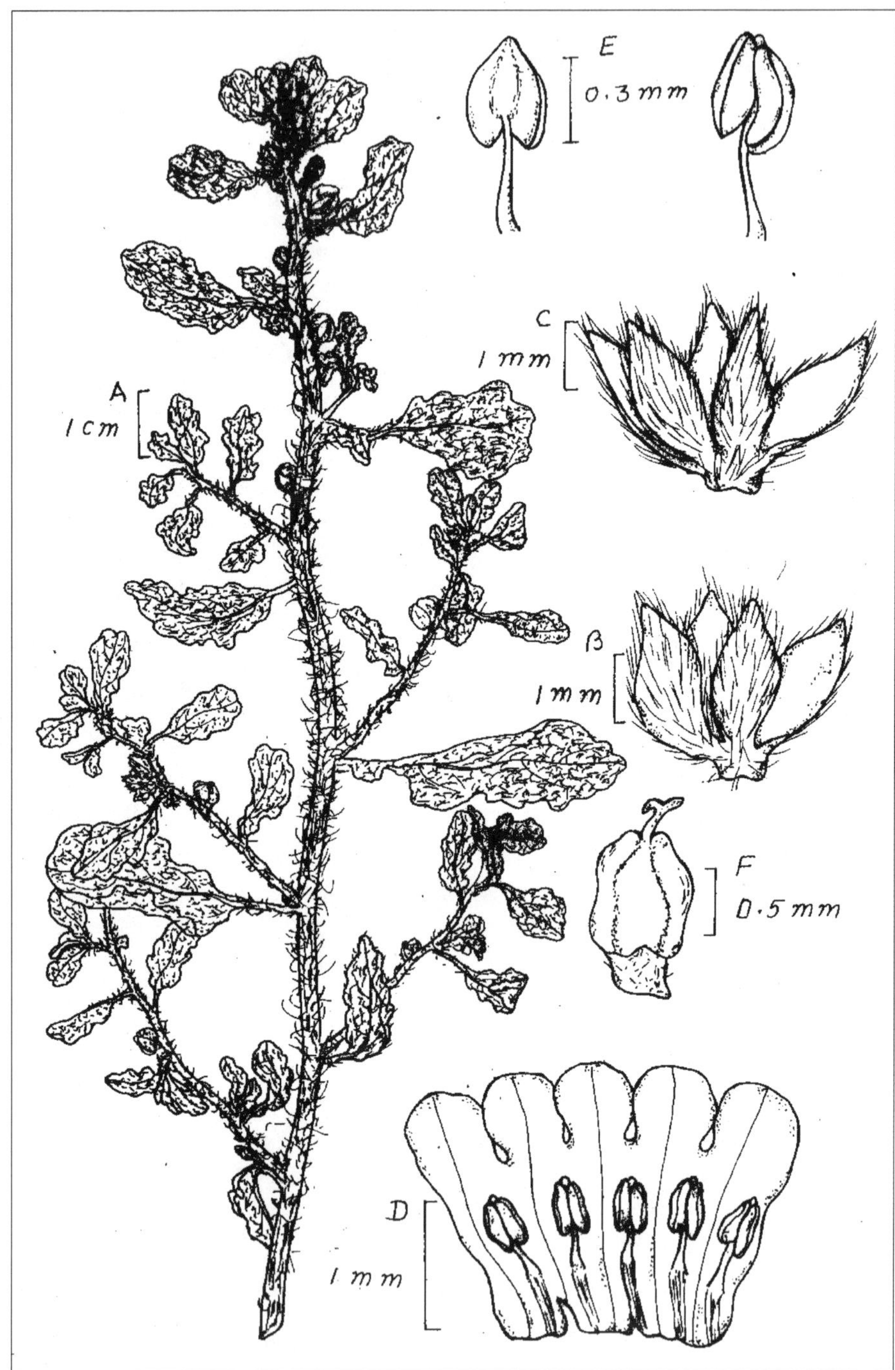

Figure 66: Coldenia procumbens L.

A. Twig, B. Calyx with 4 sepals, C. Calyx with 5 sepals, D. Corolla split open with stamens, E. Stamens, F. Pistil.

Occasional in waste lands, forests and hills in Mahabubnagar district (Raghava Rao, 1989).

Fl. & Fr.: September – April.

HELIOTROPIUM L.

1. Erect herbs:
   2. Inflorescence a raceme; stamens apically connivent:
      3. Raceme profusely branched; corolla >5 mm across; bracts linear .... **H. zeylanicum**
      3. Raceme sparingly branched; corolla <3 mm across; bracts lanceolate .... **H. bracteatum**
   2. Inflorescence a spike; stamens free:
      4. Nutlets connate in pairs, angled, apically beaked .... **H. indicum**
      4. Nutlets free, rounded or tuberculate, obtuse:
         5. Leaves glabrous, fleshy, stigma 2-fid .... **H. curassavicum**
         5. Leaves tomentose or strigose, not fleshy, stigma acute:
            6. Leavs obovate-spathulate, white woolly tomentose; stigma acute, minutely pubescent .... **H. ovalifolium**
            6. Leaves lanceolate, densely strigose; stigma short, oblong, densely hairy .... **H. subulatum**
1. Spreading or procumbent herbs or undershrubs:
   7. Fruiting calyx woolly, enclose the drupe; nutlets included; stamens free; inflorescence spike .... **H. supinum**
   7. Fruiting calyx other than woolly, spreading; nutlets exserted; stamens connivent; inflorescence raceme or spike:
      8. Bracts inconspicuous .... **H. strigosum**
      8. Bracts conspicuous .... **H. scabrum**

Heliotropium bracteatum R. Br., Prodr. Fl. Nov. Holl. 493. 1810; C.B.Clarke in Hook.f., Fl. Brit. India 4: 151. 1883; Gamble, Fl. Madras 2: 897. 1923.

Vern.: *Gutta kandi.*

Annual, erect, hairy herb, up to 50 cm high, stems and branchlets softly sericeous. Leaves simple, linear-lanceolate, 2-4 × 0.3-0.5 cm, base cuneate, margin revolute, apex acute, softly strigose-hirsute on both sides, 1-nerved, lateral nerves obscure; petiole up to 0.3 cm long. Flowers white, terminal, in racemes; peduncle sparingly branched, up to 9 cm long; bracts large, lanceolate; pedicels up to 0.4 cm long; calyx lobes 5, ovate to obovate, 2.5 mm, scabrous without, acute; corolla white, salver-form, 2.5 mm across, tube 2 mm, sparsely strigose or almost glabrous, lobes 5, broadly ovate, acute; stamens 5, anther ovate-cordate, apically connivent; ovary globose, style disciform above, tip pointed. Nutlets rounded, smooth, sparsely hirsute.

Comon in dry rocky hills, hill slopes, plains and fields.

Fl. & Fr.: July – December.

Narsapur RF (MDK), *TP* & *CP* 14019; Kulcharam RF (MDK), *TP* & *CP* 14078; Kusumasamudram (RR), *MSM* & *NV* 10337.

Heliotropium curassavicum L., Sp. Pl. 130. 1753; Gamble, Fl. Madras 2: 896. 1923; Kazmi, J. Arnold Arbor. 51: 178. 1970.

An annual sub-erect or ascending herb, up to 50 cm high, branchlets glabrous. Leaves linear – lanceolate, 2-3 × 0.2-0.8 cm, base cuneate to decurrent, margin entire, apex acute to subacute, glabrous, fleshy, aromatic, veins obscure; petiole to 0.2 cm long. Flowers white in terminal, helicoid spikes; calyx lobes 5, oblong, 2 mm, glabrous, acute; corolla 2.5 mm across, funnel-shaped, campanulate, tube to 2 mm, lobes ovate, 1 mm; stamens 5; ovary sub-globose, 1 mm across, stigma conical, shortly 2-fid. Fruits pyramidal with 4 nutlets, nutlets smooth, 1 mm across, glabrous, margin thickened, covered with calyx lobes; seeds dark brown, 1 mm across.

Common in waste lands, fallow fields and plains in Mahabubnagar district.

Fl. & Fr.: December – May.

Heliotropium indicum L., Sp. Pl. 139. 1753; C.B.Clarke in Hook.f., Fl. Brit. India 4: 152. 1883; Gamble, Fl. Madras 2: 896. 1923.

Vern.: *Nagadanti.*

Annual erect hispid herb, up to 50 cm high. Leaves simple, alternate or sometimes sub-opposite, elliptic-ovate, 5-8 × 2-4 cm, base truncate to attenuate, margin undulate, apex acute, pubescent; petiole up to 8 cm long. Flowers pale violet, terminal or leaf-opposed, often furcated, in helicoid spikes; calyx lobes 5, linear-lanceolate, 2.5 mm, sparsely scabrous without, acute; corolla pale violet, salver-form, 2 mm across, tube 4 mm, lobes 5, ovate, obtuse; stamens 5, anther lanceolate, anther tips protruded; ovary globose, style disciform above, stigma obtuse. Fruit separating into 2-seeded, 2-pointed pyrenes, nutlets connate in pairs, angled, apically beaked.

Occasional in plains, on road sides or in waste lands, less on the hills, up to 800 m in most districts.

Fl. & Fr.: Round the year.

Jaipur RF (ADB), *GO* & *PVP* 4956; Borgaun (NZB), *BR* & *KH* 9040; Narsapur (MDK), *TP* & *CP* 12000; Peddkunta, Mohammadabad RF (RR), *NV* 10418; Malkalcheruvu (RR), *MSM* & *KH* 10275; Vempalli (KRN), *GVS* 20177 (NH); Along Moosi river bank (HYD), *KMS* 5941 (MH); Vempalli (KRN), *GVS* 20177 (MH); Ramanagutta (KMM), *RR* 107986 (BSID); Khammam to Dhaniyaigudem (KMM), *RR* 108532 (BSID).

Heliotropium ovalifolium Forssk., Fl. Aegypt.-Arab. 38. 1775; C.B.Clarke in Hook.f., Fl. Brit. India 4: 150. 1883; Gamble, Fl. Madras 2: 896. 1923; Kazmi, J. Arnold Arbor. 51: 178. 1970.

Vern.: *Nugu danti.*

Annual erect or decumbent softly white villous herb, up to 30 cm high; branchlets softly white sericeous, arises from thick root-stock. Leaves simple, obovate-spathulate or elliptic, 1-2 × 0.5-0.8 cm, base cuneate, margin entire, apex obtusely apiculate, white woolly tomentose; lateral nerves obscure; petiole up to 2 cm. Flowers white, terminal, in helicoid or scorpioid cymes; calyx lobes 5, obovate-lanceolate, unequal, strigose in and out, acute, persistent; corolla white; campanulate, 2 mm across, tube 1.5 mm, strigose without, lobes 5, ovate, acute; stamens 5, anthers linear-lanceolate, free, bluntly apiculate; ovary globose, stigma acute, minutely pubescent. Nutlets 4, single-seeded, densely pubescent, fruiting calyx spreading.

Common in moist waste places, cultivated fields and foot hills in most districts.

Fl. & Fr.: April - December.

Nirmal (ADB), *GO* & *BR* 9410; Kandakurthi (NZB), *BR* 7257; Manchippa (NZB), *BR* & *GO* 9217; Rampur (MDK), *BR* & *CP* 11651; ICRISAT (MDK), *LJGV* 3206 (CAL); Kusumasamudram RF (RR), *NV* 10335; Peddakunta (RR), *NV* 10420; Vatavarlapalli (MBNR), *BSS* & *SKB* 30606; Bhupatipur (KRN), *GVS* 22289 (MH); Pakhal lake (WGL), *KMS* 13166 (MH): Yenkur (KMM), *RR* 107916; Rayagir (NLG), *BVRS* 10 (NU).

Heliotropium scabrum Retz., Observ. Bot. 2: 8. 1781; C.B.Clarke in Hook.f., Fl. Brit. India 4: 152. 1883; Gamble, Fl. Madras 2: 897. 1923. *H. marifolium* Retz., Observ. Bot. 2: 8. 1781; Hook.f., Fl. Brit. India 4: 152. 1883; Kazmi, J. Arnold Arbor. 51: 154. 1970.

Vern.: *Nela danti.*

Annual much branched procumbent or spreading woody herb, growing in tufts; branchlets strigose-hairy, closely appressed to substratum. Leaves simple, ovate-lanceolate or linear-lanceolate, 1-1.5 × 0.2-0.4 cm, base attenuate, margin entire, apex acute, flat, scabrous; lateral nerves obscure; petiole up to 0.2 cm long. Flowers white, (sub)terminal, racemes crowded among the upper leaves; bracts conspicuous, lanceolate; pedicels up to 0.3 cm long; calyx lobes 5, obovate to lanceolate, strigose on either side, acute; corolla white, salver-form, 5 mm across, tube 2 mm, lobes 5, broadly ovate, strigose-hirsute without; stamens 5, anthers ovate-cordate, connivent; ovary globose, style disciform above, stigma conical, pointed. Fruits brown, shortly bristly with 4 nutlets.

Common in sandy rocky plains, waste places and road sides in Medak, Karimnagar, Nalgonda, Hyderabad, Mahabubnagar and Ranga Reddi districts.

Fl. & Fr.: July-December.

Malkalcheruvu (RR), *MSM* & *NV* 10423; Nagarjunasagar (NLG), *BVRS* 529 (NU); Narsapur (MDK), *KMS* 7983 (CAL, MH).

Heliotropium strigosum Willd., Sp. Pl. 1: 743. 1798; C.B.Clarke in Hook.f., Fl. Brit. India 4: 151.1883. *H.brevifolium* Wall., in Roxb., Fl. Ind. 2: 2. 1824; Gamble, Fl. Madras 2: 897. 1923; Kazmi, J. Arnold Arbor. 51. 153. 1970.

Eng.: Indian forget-me-not, Red jasmine.

Perennial, prostrate or often procumbent, small, much branched, strigose-hirsute herb, up to 20 cm long; branches very white-strigose hispid. Leaves simple, linear or linear-lanceolate, 2 × 0.4 cm, base acute, margin revolute, apex acute, white strigose. Flowers white, in axillary spicate racemes; bracts narrow, small, inconspicuous; calyx up to 2 mm long, strigose; lobes 5, one lobe broadly lanceolate, others narrow-lanceolate; corolla white, strigose, tube about as long as calyx, lobes 5, ovate, short, plicate in the sinuses; stamens 5, anthers minute, ovate, connective produced into a long acumen; style minute, swelling large with bluntly conical stigma about as long as the style. Fruit broad, depressed, conical at the apex; nutlets ovoid, with minute grey hairs.

Common weed in open forests in Nizamabad, Medak, Ranga Reddi and Karimnagar districts.

Fl. & Fr.: May – November.

Mamidipally RF (NZB), *BR* & *PSPB* 9598; Pocharam RF (MDK), *TP* & *CP* 12058; Peddakunta – Mohammadabad RF (RR), *NV* 10423; Kodimial (KRN), *GVS* 20076 (MH); Ramanagutta (KMM), *RR* 107999 (BSID); Tallada – Kothagudem (KMM), *RR* 106100 (BSID).

Heliotropium subulatum (Hochst. ex A.DC.) Vatke, Oesterr. Bot. Zeitschr. 25: 166. 1875; Gamble, Fl. Madras 2: 896. 1923. *Tournefortia subulata* Hochst. ex A. DC., Prodr. 9: 528. 1845. *Heliotropium zeylanicum* auct. non Lam.; C.B.Clarke in Hook.f., Fl. Brit. India 4: 148. 1883.

Annual erect scabrous hispid herb, up to 50 cm high. Stems and branches covered with trichomes, arising from tuberculate bases. Leaves simple, lanceolate, 2.5-5 × 0.5 – 1.2 cm, base cuneate, margin entire or slightly undulate, apex acute, densely strigose; sessile. Flowers yellow, sessile in terminal elongate spikes, unilateral or sometimes 2-ranked; peduncle up to 45 cm long; ebracteate; calyx deeply divided, lobes 5, lanceolate, acute, densely hairy on the margins; corolla white or dull yellow, with minute appressed trichomes, lobes incurved when young, later patent, tube cylindrical, lobes 5, caudate-acuminate; stamens 5, inserted in the swollen part of the corolla, surrounding the style, anthers sessile, ovate-oblong, bifid at the apex; style glabrous, stigma short, oblong, densely hairy, included in the corolla tube. Nutlets 4, 2 mm long, brown, rounded or tuberculate on the back, separating first in pairs, later free.

Occasional weed in waste lands and foot hills Nalgonda and Mahabubnagar districts.

Fl. & Fr.: Round the year.

Cherukupally (NLG), *BVRS* 346 (NU); Nagarjunakonda (NLG), *BVRS* 142 (NU).

Heliotropium supinum L., Sp. Pl. 130. 1753; C.B.Clarke in Hook.f., Fl. Brit. India 4: 149. 1883; Gamble, Fl. Madras 2: 896. 1923; Kazmi, J. Arnold Arbor. 51: 179. 1970.

Annual prostrate softly white villous herb, up to 20 cm long; branchlets softly hispid-pubescent. Leaves simple, elliptic-ovate, 1.5-2 × 0.8-1.5 cm, base attenuate, margin entire, apex obtuse, softly white woolly on either side, petiole up to 0.7 cm long. Flowers white, in terminal scorpiod spikes; peduncle up to 2.5 cm long; bracts 0; calyx lobes 5, oblong-lanceolate, 3.5 mm, softly sericeous on either side, acute; corolla white, 2.5 mm across, tube 2.5 mm, lobes 5, broadly ovate, subacute; stamens 5, anthers linear-lanceolate, shortly apiculate; ovary globose, style disciform above. Nutlets ovoid, usually margined, rounded and tuberculate on the back, enclosed by persistent woolly calyx.

Rare in cultivated fields, waste lands, river beds and foot hills in Adilabad, Nizamabad and Mahabubnagar districts.

Fl. & Fr.: December – May.

Kunthala water falls (ADB), *GO* & *DAM* 5134; Mudheli (NZB), *TP* & *BR* 6370; Kandakurthi (NZB), *BR* 7254.

Heliotropium zeylanicum (Burm.f.) Lam., Encycl. 3: 94. 1789; Gamble, Fl. Madras 2: 896. 1923; subsp. **zeylanicum**; Kazmi, J. Arnold Arbor. 51: 156. 1970. *H. paniculatum* R.Br., Prodr. 494. 1810; C.B.Clarke in Hook.f., Fl. Brit. India 4: 151. 1883. *H. curassavicum* L. var. *zeylanicum* Burm.f., Fl. Ind. 41. t. 16. f. 2. 1768.

Annual erect herb with stout rootstock, up to 60 cm high; branchlets glabrescent to sparsely scabrous. Leaves simple, linear-lanceolate, 1-2 × 0.3 - 0.4 cm, base cuneate, margin revolute, apex acute, minutely scabrous above, appressed strigose below; 1-nerved; petiole up to 0.5 cm long. Flowers white-yellow, axillary or terminal, in long slender racemes; peduncle profusely branched, up to 8 cm long; bracts linear; pedicels up to 0.3 cm long; calyx lobes 5, ovate, 1.5 mm, glabrous, acute; corolla whitish yellow, >5 mm across, tube 1.5 mm, lobes 5, broadly ovate, 2.5 mm, acute, stamens 5, anthers ovate-cordate, apically connivent; ovary globose, style disciform above, tip pointed. Nutlets globose, sparsely hairy.

Rare in plains in open forests, road sides and fallow fields in Ranga Reddi, Nalgonda, Karimnagar, Hyderabad and Mahabubnagar districts.

Fl. & Fr.: December – February and August – September.

Kusumasamudram RF (RR), *MSM* & *NV* 10337; Vijayapuri (NLG), *KMS* 9836 (CAL, MH); Nagarjunasagar (NLG), *BVRS* 111 (NU).

TRICHODESMA R. Br. *nom. cons.*

KEY 1

1. (Lower) leaves petiolate; calyx lobes rounded; corolla throat without glandular spots .... **T. zeylanicum**
1. Leaves (sub) sessile throughout; calyx lobes hastate or cordate; corolla throat with glandular spots:

2. Leaf base cordate; corolla light violet; glands of the throat 4 .... **T. sedgewickianum**

2. Leaf base auriculate; corolla pinkish blue; glands of the throat 10 .... **T. indicum**

KEY 2

1. Stamens inserted; anther tips not twisted .... **T. sedgewickianum**

1. Stamens exserted; anther tips twisted:

   2. Leaf base auriculate; pedicels not drooping; calyx lobes hastate, 1-ribbed; corolla throat with glandular spots .... **T. indicum**

   2. Leaf base acute to (sub)cordate; pedicels drooping; calyx lobes rounded, 3-ribbed; corolla throat without glandular spots .... **T. zeylanicum**

Trichodesma indicum (L.) R.Br., Prodr. Fl. Nov. Holl. 149. 1810; C.B.Clarke in Hook.f., Fl. Brit. India 4: 153. 1883; Gamble, Fl. Madras 2: 899. 1923; Kazmi, J. Arnold Arbor. 52: 516. 1971. *Borago indica* L., Sp. Pl. 137. 1753.

Vern.: *Guvva gutti.*

Annual erect, grey-villous hispid herbs, up to 75 cm high. Leaves simple ovate-oblong, 3-5 × 1-1.5 cm, base auriculate, margin entire, apex obtuse or subacute, pubescent, subsessile. Flowers bluish or pinkish-white, axillary solitary, pedicels up to 5 cm long; calyx 1 cm long, accrescent in fruit, lobes 5, lanceolate, hastate at base, margins strongly ribbed, villous; corolla blue, 1 cm across, tube 4 mm, throat with yellow glandular spots, lobes 5, rounded, apiculate and recurved in flower; stamens 5, exserted, filaments 2-3 mm long, anthers 1.2 cm, pilose; ovary 1 mm, style 7.5 mm. Fruits pyramidal, 4-ribbed, nutlets ovoid, smooth.

Weed of waste places and cultivated fields in all districts.

Fl. & Fr.: July – March.

Sattenapalli (ADB), *TP* & *PVP* 4153; Gupalapatnam (ADB), *GO* 4393; Kammarapally (NZB), *TP* & *BR* 6246; Neela (NZB), *BR* & *GO* 9221; Venkatapuram (MDK), *TP* & *CP* 14008; Mohammadabad RF (RR), *NV* 10429; Central University campus (RR), *MSM* 15178; Rangapur RF (MBNR), *AJR* 20477; Bhadrachalam RF (KMM), *RCS* 102426 (BSID); Durachipalli (NLG), *BVRS* 28 (NU); Kodimial (KRN), *GVS* 20096, 20138 (MH).

Trichodesma sedgewickianum S.P.Banerjee, Bull. Bot. Soc. Bengal 16: 10. 1962. *T. amplexicaule* auct. non Roth; C.B.Clarke in Hook.f., Fl. Brit. India 4: 153. 1883; Gamble, Fl. Madras 2: 899. 1923.

Annual, erect, hispid herb, up to 75 cm high. Leaves simple, broadly oblong-lanceolate, 4-9 × 2-4 cm, base cordate, margin entire, apex obtuse to acute, coriaceous, prominently hispid but not villous, subsessile. Flowers light violet, axillary, solitary; pedicels up to 2.5 cm long; calyx lobes 5, hastate or cordate at base, scabrous along nerves without, sparsely hirsute within; corolla light violet, 0.7 cm across, salver-form, tube inflated over stamens, throat with 4 unequal spots towards one side, lobes 5, broadly ovate, obtuse, apiculate not recurved; stamens 5, included, anthers 3 mm, sericeous on the back; ovary 1.5 mm, style 4 mm. Fruits spinous, hairy, nutlets smooth.

Occasional weed of cultivated fields and roadsides in most districts.

Fl & Fr.: August-November.

Nasarullabad RF (NZB), *TP* & *BR* 6148; Medak (MDK), *CP* 11730; Narsapur (MDK), *KMS* 6811 (CAL, MH); Osmangunta (NLG), *BVRS* 283 (NU).

Trichodesma zeylanicum (Burm.f.) R.Br., Prodr. Prodr. Nov. Holl. 496. 1810; C.B.Clarke in Hook.f., Fl. Brit. India 4: 154. 1883; Gamble, Fl. Madras 2: 899. 1923; Kazmi, J. Arnold Arbor. 52: 515. 1971. *Borago zeylanica* Burm.f., Fl. Indica 41. 1768.

Vern.: *Peddanosanugu.*

Annual erect hispid herb, up to 1 m high; branchlets nearly glabrous. Leaves simple, opposite, elliptic-lanceolate, 1.5-10 × 0.5-4 cm, base acute to (sub)cordate, margin entire, apex acute, densely pubescent above, white hirsute beneath, lower leaves petiolate, petiole up to 0.7 cm long, upper leaves sessile. Flowers pale blue with red streak near throat without glandular spots, axillary or terminal, paniculate cymes by reduction of leaves; pedicels up to 3 cm long, drooping; calyx lobes 5, rounded at base, scabrous in and out; corolla pale blue, salver-form, lobes 5, broadly ovate, apiculate, no glandular spots; stamens exserted. Nutlets ovoid, smooth.

Frequent along hill slopes and waste places in most districts.

Fl. & Fr.: November - February.

Keslapur (ADB), *GO* & *PVP* 4549; Sirnapally (NZB), *BR* & *PSPB* 9573; Mohammadabad RF (RR), *MSM* & *KH* 10324; Nagarjunasagar (NLG), *KMS* 9724 (CAL, MH); Rechapalli (KRN), *GVS* 22240 (MH); Umamaheswaram (MBNR), *GS* 18300; Mannanur (MBNR), *SRS* 109733 (BSID).

CORDIACEAE

Recent taxonomists place *Carmona, Cordia, Ehretia* and *Rotula* under a separate family Cordiaceae while in previous Floras they were placed under the family Boraginaceae.

1. Leaves closely alternate or fascicled, less than 3 cm long; flowers solitary or clustered .... **Rotula**
1. Leaves distantly alternate or (sub)opposite, atleast some over 5 cm long; inflorescence a cyme:
    2. Calyx lobes divided to below the middle, deciduous when fruit setting; style once forked, arms 2 .... **Ehretia**
    2. Calyx lobes connate more than half, accrescent in fruit; style twice forked, arms 4. .... **Cordia**

CORDIA L.

KEY 1

1. Leaves fulvous-tomentose or scabrous below:
   2. Leaf margin distantly, sharply dentate .... **C. macleodii**
   2. Leaf margin entire .... **C. domestica**
1. Leaves glabrescent or thinly pubescent below:
   3. Corolla < 0.5 cm across; leaves rough with white disc above, filaments glabrous at base .... **C. monoica**
   3. Corolla > 0.7 cm across; leaves glabrescent above; filaments hairy at base .... **C. dichotoma**

KEY 2

1. Fruiting calyx not ribbed .... **C. dichotoma**
1. Fruiting calyx ribbed:
   2. Leaves 10 – 24 x 10 – 20 cm .... **C. macleodii**
   2. Leaves 2-8 x 1 – 6.5 cm:
      3. Filaments glabrous at base; leaves rough, glabrous .... **C. monoica**
      3. Filaments hairy at base; leaves pubescent or tomentose beneath .... **C. domestica**

Cordia dichotoma G. Forst., Fl. Ins. Austr. 18. 1786. *C. obliqua* Willd., Phytographia 4. t 4. 1794; C.B.Clarke in Hook.f., Fl. Brit. India 4: 137. 1883 *p.p.*: Gamble, Fl. Madras 2: 887. 1923. *C. myxa* auct. plur. (non L. 1753). *Cordia wallichii* G. Don, Gen. Syst. 4: 379. 1838; Gamble, Fl. Madras 2: 887. 1923. *C. tomentosa* Wall., in Roxb., Fl. Ind. 2: 339. 1824; non Roem. & Schult. 1819. *C. obliqua* Willd. var. *wallichii* (G.Don) C.B. Clarke in Hook.f., Fl. Brit. India 4: 137. 1883.

Vern.: *Iriki, Nakkeri, Bankiriki.*

Moderate-sized deciduous tree, up to 10 m high; bark greyish black, with deep vertical lines; branches drooping, hairy when young; bark thick, grey or brown with shallow furrows and wrinkles; branchlets glabrous. Leaves simple, variable in size and shape, but usually elliptic-oblong or orbicular, 4 – 11 × 3 – 8 cm, base obtuse or cuneate, margin repand-crenate, apex (sub) acute, glabrescent on both sides, sometimes hairy in the nerve axil; petiole up to 4 cm long. Flowers white to cream-coloured, fragrant, axillary or terminal, in corymbose glabrous panicles; peduncle up to 5 cm long; calyx 6 mm long, ribbed, lobes 5, dentate, not recurved; corolla white or cream, 8 mm across, lobes 5, recurved; stamens 5, villous at base, filaments 4 mm long, anthers 1.5 mm long; ovary globose, 2 mm long, style 2 mm long. Drupe ovoid, cream-coloured, mucilaginous, supported by accrescent calyx; seeds solitary.

Frequent in mixed forests and hedges of fields Adilabad, Nizamabad, Medak, Hyderabad, Nalgonda and Mahabubnagar districts.

Fl. & Fr.: December – January.

Satnella river (ADB), *GO* & *PVP* 4223; Manchippa (NZB), *BR* & *KH* 9038; Sirnapalli RF (NZB), *BR* & *GO* 9093, 9095; Narsapur RF (MDK), *CP* 11579; Rusthempet (MDK), *KMS* 8001 (MH); Kudichintalabailu (MBNR), *TDCK* 15387; Cogudum (KMM), *RCS* 104223 (BSID); Kunavaram RF (KMM), *RCS* 98784 (BSID); Peddavuru (NLG), *BVRS* 511 (NU).

Note: Merr., Trans. Ann. Phil. Soc. (n.s.) 24: 329-330. 1935 treated *Cordia obliqua* Willd. as synonym of *Cordia dichotoma* Forst.f. But Johnston, J. Arn. Arb. 32: 8-10. 1951 has shown that the two are quite distinct. Kazmi, J. Arn. Arb. 51: 142. 1970 had separated *Cordia dichotoma* and *C. obliqua* on the length-breadth ratio of leaf and leaf apex and base. The characters considered by him, especially in respect of leaf apex and base are found in different leaves on the same twig. Hence *C. dichotoma* and C. *obliqua* are combined here.

Cordia domestica Roth, Nov. Pl. Sp. 123. 1821; Gamble, Fl. Madras 2: 888. 1923. *C. myxa* L. var. *domestica* (Roth) C.B.Clarke in Hook. f., Fl. Brit. India 4: 137. 1883.

Vern.: *Urunakkera.*

Small tree, up to 8 m high. Leaves simple, ovate, 2-5 × 1-2.5 cm, base truncate to rounded, margin entire, apex obtuse to (sub)acute, fulvous pubescent above, densely so below; petiole up to 2 cm long. Flowers yellowish white, terminal in few-flowered small corymbs; peduncle up to 6 cm long; calyx lobes 5, fulvous-pubescent without; corolla yellowish white, lobes 5; stamens 5, filaments 6 mm, anthers oblong, 3.5 mm; ovary globose, style with 4 arms, stigma capitate. Drupes ovoid, seated on the cup-shaped, ribbed calyx.

Rare in thorny scrub jungles in Medak district.

Fl. & Fr.: February – July.

Cordia macleodii Hook.f. & Thomson, J. Proc. Linn. Soc., Bot. 2: 128. 1858; C.B.Clarke in Hook.f., Fl. Brit. India 4: 139. 1883; Gamble, Fl. Madras 2: 888. 1923.

Vern.: *Pedda botava, Iriki.*

Small deciduous tree, up to 10 m high; branchlets white-tomentose; bark thick, grey, corky. Leaves simple, alternate (rarely sub-opposite), ovate, 10-24 × 10-20 cm, base truncate or cordate, margin distantly, sharply dentate, apex obtuse or slightly acute, rugose above, fulvous tomentose below; petiole up to 7 cm long. Flowers white, large, terminal or axillary corymbs; calyx tubular, lobes 5; corolla white, lobes 5, oblong; stamens 5, anthers oblong; ovary globose, style with 4 arms, stigma capitate. Drupes ovoid, up to 2 x 1 cm, glabrous, acute, yellowish, 1-2-seeded, seated on tomentose, ribbed fruiting calyx.

Rare in forests in Nizamabad and Karimnagar (Naqvi, 2001) districts.

Fl. & Fr.: February – August.

Mudheli RF (NZB), *TP* & *BR* 6357, 6358.

Cordia monoica Roxb., Pl. Coromandel 1: 43. t. 58. 1796; C.B. Clarke in Hook. f., Fl. Brit. India 4: 137. 1883; Gamble, Fl. Madras 2: 888. 1923.

Vern.: *Bankeera.*

Small tree or shrub, up to 8 m high; branches white or fulvous-tomentose. Leaves simple, variable in shape and amount of tomentum, but usually elliptic-ovate, 3.8–8 × 2.1-4.1 cm, base rounded, margin crenate, apex acute, rough with white discs above, thinly pubescent below; petiole up to 1.5 cm long. Flowers yellow, fragrant, 5 mm long, dense in branched, lax cymes; calyx tube cylindric, lobes 5, tomentose without; corolla white, 4 mm across, tube 5 mm, lobes 5, (sub)acute; stamens 5, filaments glabrous at base; ovary ovoid. Drupes ovoid, bright orange red, gummy, not supported by accrescent calyx.

Occasional in forests, waste lands, hedges of fields and foot hills up to 600m in Adilabad and Nalgonda districts .

Fl. & Fr.: October –April.

Devapur (ADB), *GO* & *PVP* 4243; Bhongir fort wall (NLG), *BVRS* 3 (NU); Nagarjunasagar (NLG), *BVRS* 134 (NU).

EHRETIA P. Br.

1. Shrubs; leaves obovate, to 2.5 × 0.7 cm .... **E. microphylla**
1. Trees; leaves other than ovate, 8-18 × 5-8 cm:
    2. Leaf margin serrate; drupe with 2 pyrenes, pyrenes without cavity .... **E. acuminata**
    2. Leaf margin entire; drupe with 4 pyrenes, pyrenes with a cavity ...... **E. laevis**

Ehretia acuminata R. Br., Prodr. Fl. Nov. Holl. 497. 1810; C.B.Clarke in Hook.f., Fl. Brit. India 4: 141. 1883; Gamble, Fl. Madras 2: 890. 1923.

Small-medium-sized trees, up to 10 m high; bark pale-grey, longitudinally fissured. Leaves simple, elliptic or ovate, to 18 × 8 cm, base obtuse, margin serrate, apex acuminate, glabrous or minutely hispidulous above; petiole up to 4 cm long. Flowers white, axillary or terminal, in subsessile fascicles; peduncle up to 10 cm long, densely pubescent; calyx lobes 5, obtuse, ciliate; corolla white, lobes 5, broadly oblong, soon reflexed; stamens 5, anthers exserted; ovary globose, style 2-fid. Drupe yellow, ellipsoid with 2 pyrenes without cavity, 4-seeded.

Frequent in University campuses and National parks in Hyderabad district (Prasanna & *al.*, 2012).

Fl. & Fr.: January - May.

Ehretia laevis Roxb., Pl. Coromandel t. 56. 1796 & Fl. Ind. 2: 341. 1824; C.B.Clarke in Hook.f., Fl. Brit. India 4: 141. 1883; Gamble, Fl. Madras 2: 891. 1923. *E. canarensis* (C.B.Clarke) Gamble, Fl. Madras 2: 891. 1923. *E. laevis* Roxb. var. *canarensis* C.B.Clarke in Hook.f. Fl. Brit. India 4: 142. 1883. *Ehretia pubescens* Benth. in Royle, Ill. Bot. Himal. Mts. 306. 1836; Gamble, Fl. Madras 2: 891. 1923 . *E. laevis* Roxb. var. *pubescens* (Benth.) C.B.Clarke in Hook.f., Fl. Brit. India 4: 142. 1883.

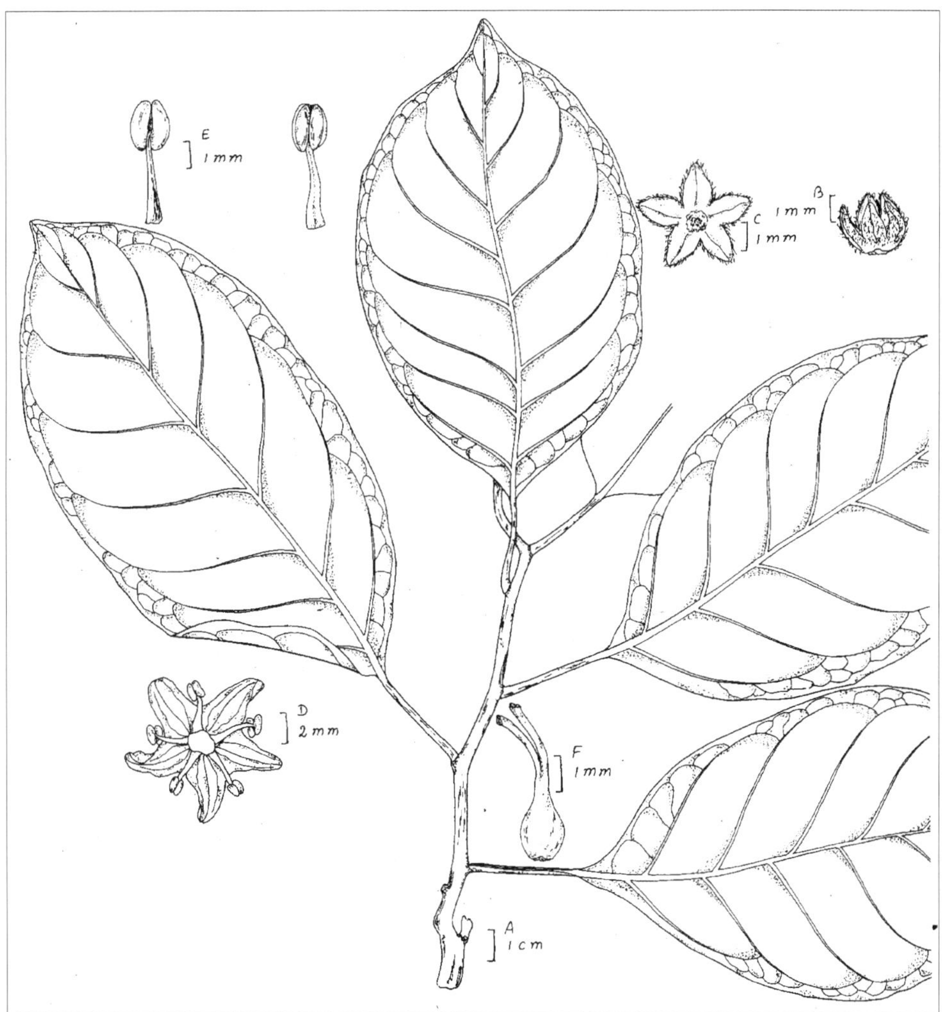

Figure 67: Ehretia laevis Roxb.

A. Twig, B. Calyx (side view), C. Calyx (upper view), D. Corolla with stamens, E. Stamens, F. Pistil.

Vern.: *Paldatam, Peddapulmera, Pogari, Pogadi chettu, Tella pisini.*

Moderate-sized or small deciduous tree, up to 10 m high; tender parts hirsute-pubescent; bark white or pale grey. Leaves unsymmetric, ovate or ovate-oblong, 8-18 × 5-8 cm, base cuneate, margin entire, apex obtuse or obtusely acute, sparsely hirsute above, thinly pubescent below; petiole up to 1.5 cm long. Flowers white, scented, in terminal, unilateral cymes; peduncle up to 4 cm long; calyx lobes 5, lanceolate, 2.5 mm, thick-herbaceous, thinly pubescent without; corolla white, lobes 5, broadly oblong, obtuse; stamens 5, filaments 5 mm, anthers 1.5 mm; ovary globose, style divided from near the middle. Drupe globose with 4 pyrenes, red when ripe.

Common in deciduous forests in Adilabad, Nizamabad, Hyderabad, Warangal, Mahabubnagar and Nalgonda districts.

Fl. & Fr.: September – April.

Pothangal (NZB), *BR* & *KH* 9648; Mallelatheertham (MBNR), *SKB* & *BSS* 30707; Nagarjunasagar (NLG), *BVRS* 526 (NU); Pakhal (WGL), *ANH* 15932 (MH), *RKP* 108281, 108286 (BSID); Perantapalli (KMM), *PVS* 84060 (BSID); Uskavagu road, Mannanur (MBNR), *SRS* 107440, 108915 (BSID).

Ehretia microphylla Lam., Tabl. Encycl. 1: 425. 1792; Gamble, Fl. Madras 2: 891. 1923. *E.buxifolia* Roxb., Pl. Coromandel t. 57. 1796 & Fl. Ind. 2: 343. 1824; C.B.Clarke in Hook.f., Fl. Brit. India 4: 144. 1883. *Carmona retusa* (Vahl) Masam., Trans. Nat. Hist. Soc. Taiwan 30: 61. 1940; Pullaiah & Moulali, Fl. Andhra Pradesh 2: 617. 1997. *Cordia retusa* Vahl, Symb. Bot. 2: 42. 1791.

Vern.: *Bapana buri, Nomu chettu.*

Profusely branching shrubs, up to 1.5 m high; stems and branchlets strigosely pubescent. Leaves simple, sub-sessile, alternate or in whorls at nodes, obovate, 1.5 – 2.5 × 0.5 – 0.7 cm, base cuneate, margin entire, apex obtuse, scabrid above, glabrous beneath. Flowers creamish-white, axillary, solitary or in clusters on slender hairy peduncles; peduncle up to 7 mm long; bracteate; pedicels up to 0.5 cm long; calyx lobes 4 (5), elliptic-oblong, unequal, 5 mm, scabrous below; corolla cream coloured, subrotate, 7 mm across, tube 2 mm, lobes 5, oblong-elliptic, stamens 4 (5), anthers oblong; ovary globose, styles 2, stigma capitellate. Drupe globose, red when ripe of 4 pyrenes.

Common in deciduous forests and dry scrub jungles Nizamabad, Medak, Khammam, Hyderabad, Nalgonda and Ranga Reddi districts.

Fl. & Fr.: October - March.

Ghanapur RF (NZB), *BR* & *KH* 6494; Pocharam (MDK), *BR* & *CP* 11612; Pochammaralu (MDK), *CP* 11756; Rudraram RF (RR), *TP* & *MSM* 12721; Zulurpeda (KMM), *RR* 113819 (BSID); Lankapalli hills (KMM), *RR* 106005 (BSID); Mallepally (NLG), *BVRS* 532 (NU).

ROTULA Lour.

Rotula aquatica Lour., Fl. Cochinch. 121. 1790; Gamble, Fl. Madras 2: 893. 1923. *Rhabdia lyciodes* C. Martius, Nov. Gen. Sp. Pl. 2: 137. t. 195. 1827; C.B.Clarke in Hook.f., Fl. Brit. India 4: 145. 1883.

Vern.: *Jamchi, Jamoch.*

Erect or trailing subshrubs, up to 90 cm long with virgate branches. Leaves simple, fascicled, spathulate, 0.5–1.8 × 0.3–0.8 cm, base attenuate, margin entire, apex obtuse, pubescent; petiole up to 3 mm long. Flowers pink, axillary or terminal, solitary or 2-3 in clusters; shortly pedicellate; bracts 2; calyx lobes 5, equal, almost free, lanceolate, thinly pubescent without, rounded, ciliate, acuminate; corolla pink, sub-rotate, 8 mm across, tube 2 mm, lobes 5, elliptic-oblong, 4 mm, obtuse; stamens 5, inserted, anthers oblong or ovate; ovary ovoid, 4-locular, each 1-ovuled, style 1, stigma capitellate. Drupe red or orange when ripe, with 4 one-seeded pyrenes.

Common on the rocky river beds in Adilabad, Nizamabad, Karimnagar and Mahabubnagar districts .

Fl. & Fr.: February – June.

Kadam dam (ADB), *GO* & *BR* 9418; Pothangal (NZB), *BR* & *KH* 9650; Godavari river bed, Sriramsagar project area (NZB), *BR* 7928; Krishna river bank (NLG), *KMS* 9826 (MH); Godavari (KRN), *GVS* 22300 (MH).

CONVOLVULACEAE

KEY 1

1. Seed 1; fruit indehiscent .... **Poranopsis**
1. Seeds more than 1; fruits dehiscent (rarely indeshiscent):
   2. Stigma filiform to elliptic, linear or oblong:
   3. Flowers nocturnal; corolla salver-form, white, tube up to 9 cm long, stigma linear-oblong; seeds enveloped with a mealy pulp .... **Rivea**
   3. Flowers diurnal; corolla funnel-shaped or campanulate, tube short, less than 3 cm long; stigma filiform to elliptic or oblong:
      4. Stem and leaves with stellate indumentum .... **Jacquemontia**
      4. Stem and leaves without stellate indumentum:
         5. Outer 2 sepals shorter or equal to the inner 3, up to 1 cm long, oblong-elliptic, ovate-elliptic or lanceolate; ovary glabrous .... **Convolvulus**
         5. Outer 2 sepals large, up to 2.5 cm long, much broader than inner ones, broadly ovate; ovary glabrous or hairy .... **Hewittia**
   2. Stigma biglobular:
      6. Capsule circumscissile, upper part of the epicarp separating from the lower part; stem terete or often winged .... **Operculina**

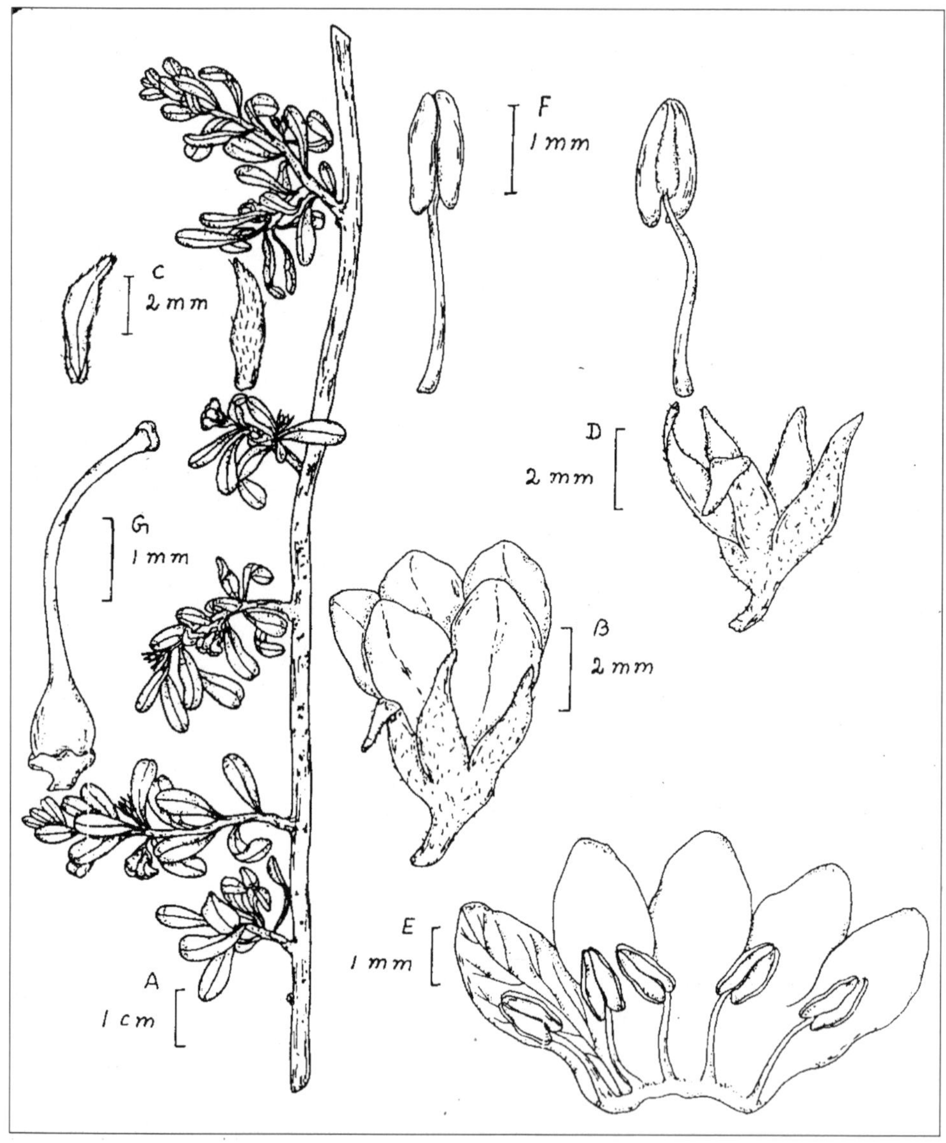

Figure 68: Rotula aquatica Lour.

A. Twig, B. Flower, C. Bracts, D. Calyx, E. Corolla split open with stamens, F. Stamens, G. Pistil.

6. Capsule valvular or irregularly dehiscent; stem terete:
    7. Corolla with hairy mid-petaline bands outside; fruits fleshy or leathery, rarely dehiscent .... **Argyreia**
    7. Corolla mostly without hairy mid-petaline bands; fruits dehiscent:
        8. Pollen grains spinulose; corolla variously coloured, often purple or reddish purple .... **Ipomoea**
        8. Pollen grains smooth:
            9. Corolla nearly always yellow or white, with or without a red or purple centre:
                10. Anthers mostly spiraled when dehisced; seeds rounded to truncate apically; stigma undulate or lobed, the papillae clavate or cylindrical .... **Merremia**
                10. Anthers straight when dehisced; seeds obtuse to rounded apically; stigma smooth, lobed to undulate, papillate .... **Xenostegia**
            9. Corolla blue or white, rotate or funnel-shaped .... **Evolvulus**

KEY 2

1. Pollen grains spinulose:
    2. Fruit dehiscent .... **Ipomoea**
    2. Fruit indehiscent, woody mealy or fleshy:
        3. Corolla salver-shaped; fruiting calyx woody .... **Rivea**
        3. Corolla tubular or funnel-shaped; fruiting calyx not woody .... **Argyreia**
1. Pollen grains smooth:
    4. Style 1:
        5. Outer sepals much longer and broader than the inner ones:
            6. Ovary glabrous:
                7. Capsule indehiscent .... **Poranopsis**
                7. Capsule dehiscent .... **Jacquemontia**
            6. Ovary hairy .... **Hewittia**
        5. Outer sepals not as above:
            8. Stigma filiform .... **Convolvulus**
            8. Sigma globose or biglobose:
                9. Capsules within the enclosed calyx, dehiscing circumsciss or irregularly; stems winged .... **Operculina**
                9. Capsules not enclosed in the calyx, dehiscing regularly (4-valved); stems not usually winged:

10. Anthers mostly spiraled when dehisced; seeds rounded to truncate apically; stigma undulate or lobed, the papillae clavate or cylindrical .... **Merremia**

10. Anthers straight when dehisced; seeds obtuse to rounded apically; stigma smooth, lobed to undulate, papillate .... **Xenostegia**

4. Styles 2 .... **Evolvulus**

ARGYREIA Lour.

1. Leaves whitish or golden tomentose below:

2. Outer bracts large, foliaceous, broadly ovate to ovate-elliptic, more than 2 cm broad, tomentose beneath:

3. Outer bracts persistent; sepals linear-oblong, less than 4 mm long, more than 8 mm broad, velutinous below, margin ciliolate .... **A. sericea**

3. Outer bracts early caducous .... **A. nervosa**

2. Outer bracts small, not broadly ovate, less than 2 cm broad .... **A. kleiniana**

1. Leaves glabrous to hirsute or woolly beneath:

4. Leaves glabrous or remotely puberulent above:

5. Suberect shrubs; leaves obovate, apically obtuse, glabrous above, silky pubescent below; fruits yellowish brown when dry .... **A. cuneata**

5. Climbing shrubs; leaves lanceolate, ovate to broadly ovate or elliptic-ovate .... **A. cymosa**

4. Leaves pubescent to villous or woolly on both sides .... **A. pilosa**

Argyreia cuneata (Willd.) Ker-Gawl., Bot. Reg. 8: t. 661. 1822; Hook.f., Fl. Brit. India 4: 191. 1883; Gamble, Fl. Madras 2: 909. 1923. *Convolvulus cuneatus* Willd., Sp. Pl. 1: 873. 1798.

Vern.: *Nettarulu.*

Perennial, bushy erect shrub, up to 1.5 m high; stem woody, herbaceous towards tip, somewhat twining, glabrous to shortly pubescent in younger parts. Leaves simple, obovate or elliptic, 4-8 x 2-4 cm, base cuneate, margin entire, apex obtuse, mucronulate or emarginated, glabrous above, grey silky-villous beneath; midrib raised beneath, lateral veins inconspicuous on both sides; petiole up to 1 cm long, tomentose. Flowers axillary, 3-5-flowered cymes; peduncle up to 5 cm long, pubescent; bracts small, 3-4, linear-lanceolate, deciduous, 5-7 × 2-2.5 mm, apex acute, glabrous above, pubescent below, pedicels up to 1 cm long, pubescent like peduncle; sepals unequal, ovate to broadly ovate, 3-4 x 3-5 mm, outer 3 apex acute, glabrous inside, shortly hirsute outside, inner 2 apex obtuse, pubescent outside except the margins; corolla bright purple, funnel-shaped, mouth slightly lobed, tube up to 3 cm long, 2.3 cm across, midpetaline bands shortly hairy; stamens inserted, anther up to 4 mm long, white, filaments unequal, hairy at dilated base; ovary conical, glabrous, style single, inserted, stigma biglobose. Berry indehiscent, yellowish-brown when dry, ovoid, crowned with short style base, fruiting calyx enlarged or reflexed; seeds dull black, pubescent.

Occasional in forests, road sides and on hill tops in Mahabubnagar district.

Fl. & Fr.: August – January.

Mannanur RF (MBNR), *VBH* 86613 (BSID).

Argyreia cymosa (Roxb.) Sweet, Hort. Brit. 289. 1826; Hook.f., Fl. Brit. India 4: 190. 1883; Gamble, Fl. Madras 2: 908. 1923. *Lettsomia cymosa* Roxb., Fl. Ind. 2: 82. 1824.

Vern.: *Maddi tige, Gantapala.*

Climbing shrubs, stem woody, herbaceous towards the tip, twining, pubescent. Leaves simple, deltoid to cordiform, 6-11 × 4-9 cm, base truncate to cordate, margin entire, apex acute to shortly acuminate, pubescent on both sides; midrb and lateral veins raised beneath; petiole up to 4.5 m long, pubescent like stem. Flowers axillary, several-flowered cymes; peduncle up to 10 cm long, pubescent, bracts leafy, 2-4 in number, orbicular, 0.5–1.5 × 0.2-1.5 cm, apex acute or obtuse, pubescent on both sides, persistent; pedicels short, up to 1.5 cm long, pubescent; sepals unequal, outer 3 large, ovate to orbicular, apex obtuse, inner 2 small, oblong, apex acute, pubescent outside, glabrous inside; corolla pale purple, darker near the centre, funnel-shaped, corolla-tube up to 4.5 cm, mouth 5-lobed, 3–3.2 cm across, densely silky on the midpetaline bands; stamens inserted, anthers up to 5 mm long; ovary globular, glabrous, style single; stigma biglobose. Berry globose, glabrous; seeds black, 2, 3 or 4, ovate to elliptic, glabrous to pubescent.

Occasional in bushes in forests, but common in open forests in Adilabad, Medak and Mahabubnagar districts.

Fl. & Fr.: September – January.

Jaipur RF (ADB), *GO* & *PVP* 4825; Narsapur RF (MDK), *RG* 113472 (BSID); Mannanur (MBNR), *SRS* 109726 (BSID).

Argyreia kleiniana (Roem. & Schult.) Raiz., Indian Forester 92: 302. 1966. *Ipomoea kleiniana* Roem. & Schult. in L. Syst. Veg. 4: 789. 1819. *Argyreia bracteata* Choisy, Convolv. Or. 99. 1834 & in DC., Prodr. 9: 328. 1845; C.B.Clarke in Hook.f., Fl. Brit. India 4: 188. 1883; Gamble, Fl. Madras 2: 907. 1923

Climbing shrubs, stem robust, hollow, shortly tomentose. Leaves simple, broadly ovate, 8-12 × 7-10 cm, base truncate to sub-cordate, margin entire, apex obtuse or sometimes emarginated, glabrous above, densely silky beneath, midrib and lateral veins raised beneath, petiole up to 8 cm long, tomentose. Flowers axillary, 1-5-flowered, corymbose cymes; peduncle up to 8.5 cm long, tomentose; bracts oblong, 2-2.5 × 0.8 – 1.5 cm, apex obtuse to acute, tawny villous, pedicels up to 5 mm long, sepals subequal,

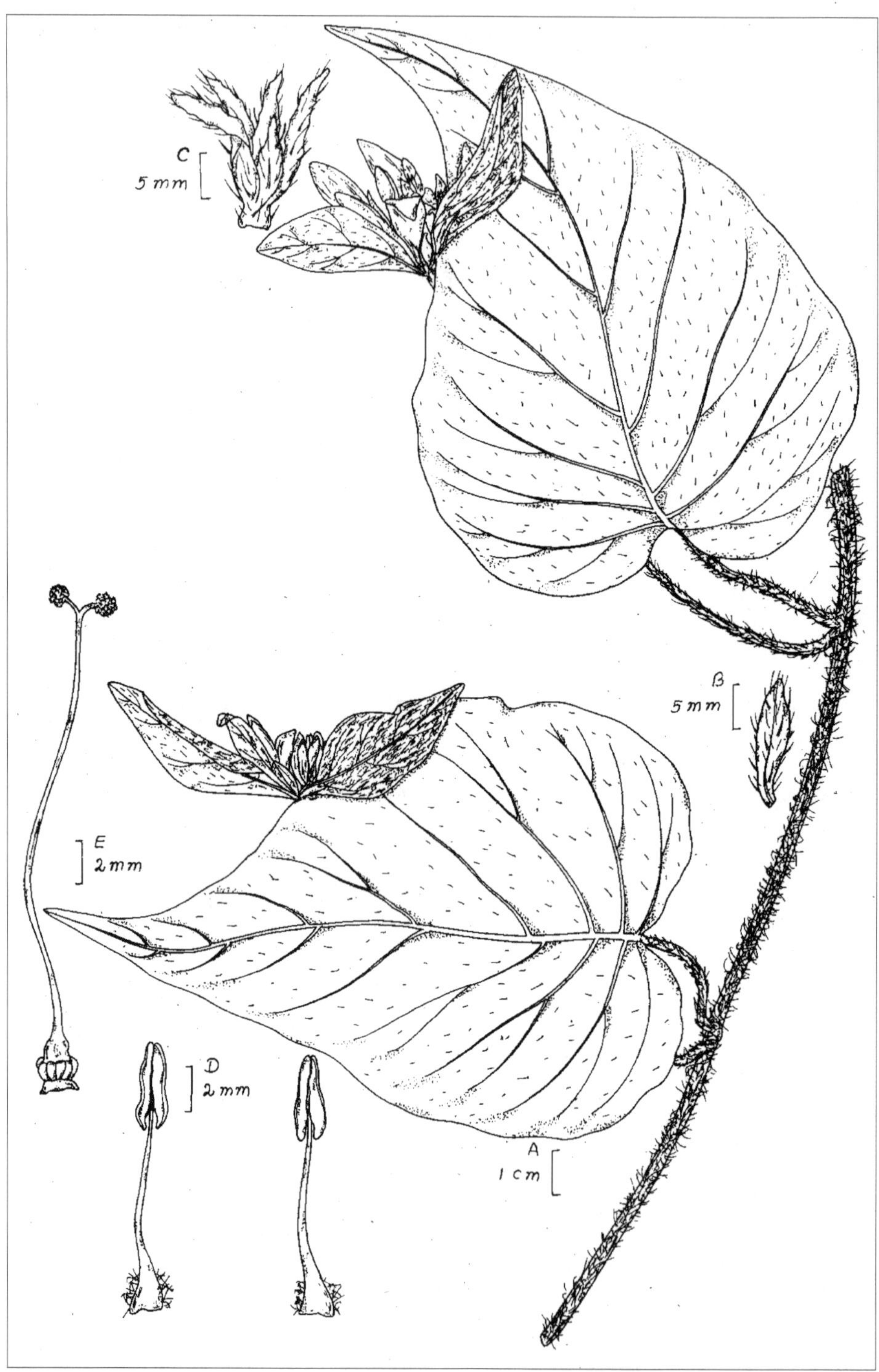

Figure 69: Argyreia cymosa (Roxb.) Sweet
A. Twig, B. Bract, C. Calyx, D. Stamens, E. Pistil.

ovate-lanceolate, 8 x 4 mm, apex acute, scarious, outer 2 slightly ciliate, shortly pilose outside in the middle portion only, glabrous within; corolla red-purple, campanulate, mouth slightly 5-lobed, tube up to 4.5 cm long, 6 cm across, midpetaline bands pilose outside; stamens inserted, anthers up to 4 mm long, filaments subequal, dilated base hairy; ovary 1.5 mm, glabrous, style single, stigma biglobose. Fruit globose, yellowish, 1.5 cm across.

Common in open forests in Mahabubnagar district.

Fl. & Fr. : August – December.

Khammam to Dhaniyaigudem (KMM), *RR* 108514 IBSID); Pillutla RF, Narsapur (MDK), *RG* 106785 (BSID).

Argyreia nervosa (Burm.f.) Boj., Hort. Maurit. 224. 1837. *Convolvulus nervosus* Burm.f., Fl. Ind. 48. 1768. *Argyreia speciosa* (L.f.) Sweet, Hort. Brit. 289. 1827; C.B.Clarke in Hook.f., Fl. Brit. India 4: 185. 1883; Gamble, Fl. Madras 2: 907. 1923. *Convolvulus speciosus* L.f., Suppl. 137. 1781.

Vern.: *Chandrapala, Samudrapala.*

Climbing shrubs, stem woody, herbaceous towards the tip, hollow, young branches densely white-tomentose; milky latex. Leaves simple, large, cordate, 10-32 × 9-28 cm, base cordate, margin entire, apex obtuse or acute, glabrous above, densely fulvous or white tomentose below, shining; midrib and lateral veins raised beneath; petiole up to 30 cm long, tomentose like stem. Flowers axillary, 1-few-flowered subcapitate cymes; peduncle up to 28 cm long, pubescent like stem; bracts 3-5, large, ovate-lanceolate, 5 × 3 cm, apex acuminate, softly pilose outside, glabrous inside; pedicels up to 1.2 cm long, pubescent like peduncle; sepals subequal, outer 2 slightly larger, elliptic-ovate, 1-1.9 × 1-1.4 cm, apex acute, densely white tomentose outside, glabrous inside, inner 3 little shorter, broadly elliptic, 1.4–1.5 × 0.8–1.2 cm, apex obtuse, tomentose outside, glabrous inside; corolla rose-purple, throat darker, funnel-shaped, mouth 5-lobed, tube up to 1.4 cm long, 5 cm across, midpetaline bands pubescent; stamens inserted, anthers up to 7 mm long, pinkish white, filaments subequal, dilated base with papillose; ovary conical, glabrous, style single, stigma biglobose. Fruit indehiscent, dry baccate, green, turns brown when dry, crowned with short style base, fruiting sepals leathery, accrescent and persistent; seeds 2-4, glabrous, brown.

Occasional in the open forests of Warangal and Medak districts.

Fl. & Fr.: September – February.

Narsapur (MDK), *KMS* 6762 (MH).

Argyreia pilosa Wight & Arn., Pug. Pl. Ind. Or. 38. 1836; C.B.Clarke in Hook.f., Fl. Brit. India 4: 189. 1883; Gamble, Fl. Madras 2: 908. 1923.

Vern.: *Paradu teega.*

Climbing shrubs, up to 6 m long, stem herbaceous towards the tip. Leaves simple, broadly ovate, 6-12 × 2-7 cm, base rounded or slightly cordate, margin entire, apex acute to acuminate, hirsute above, strigose below; midrib and lateral veins raised

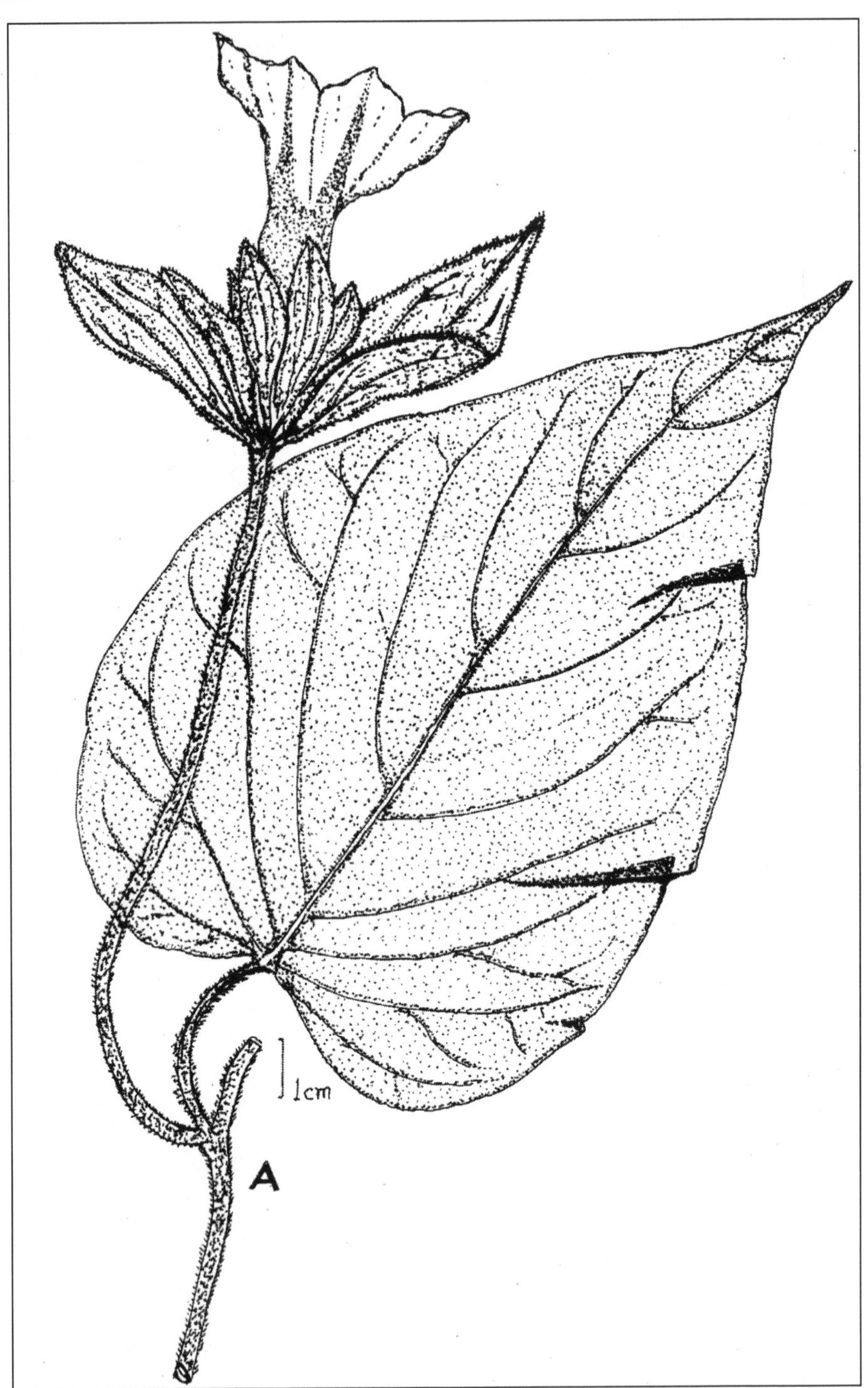

Figure 70: Argyreia pilosa Wight & Arn.
A. Twig.

beneath; petiole up to 5 cm long, pubescent like stem. Flowers axillary, few to many-flowered capitate cymes; peduncle up to 2.8 cm long, pubescent like stem; bracts 2-4, linear-oblong or subspathulate, 0.6–1.5 × 0.2–0.4 cm, apex acute, both sides pilose; pedicels up to 5 mm long, pubescent like stem; sepals subequal, ovate-lanceolate, 0.7–1 × 0.2-0.4 cm, apex acute to acuminate, exposed parts strigose, glabrous inside; corolla dark red or purple, funnel-shaped, tube up to 4.5 cm long; stamens 5, filaments subequal, 2 long, 3 short, dilated at base, pubescent; ovary conical, glabrous, disc small. Fruit not seen.

Occasional in Adilabad district.

Fl. & Fr.: August – November.

Wankidi RF (ADB), *TP* & *GO* 5491;

Argyreia sericea Dalzell in Dalzell & A. Gibson, Bombay Fl. 169. 1861; C.B.Clarke in Hook.f., Fl. Brit. India 4: 188. 1883; Gamble, Fl. Madras 2: 908. 1923

Climbing shrubs, stem robust, herbaceous towards the tip, pilose. Leaves simple, broadly ovate, 5-14 × 4-12 cm, base cordate or truncate, margin entire, apex acute to acuminate, sparsely hispid above, tomentose below, midrib and lateral veins raised beneath; petiole up to 5 cm, velutinous. Flowers axillary, 2-4-flowered subcapitate cymes; peduncle up to 8.5 cm long, velutinous; bracts large, linear or oblong, 4-6 × 1.3 -2.2 cm, inner bracts small, elliptic, 1.5–2.2 × 0.6–1 cm, tawny-silky, persistent; pedicels subsessile, up to 2 mm long, pubescent like peduncle; sepals unequal, outer 2 long, linear-oblong, 1.2–1.6 × 0.2-0.4 cm, inner short, lanceolate, 8-10 × 1-2 mm, glabrous inside, velutinous outside; corolla pale purple, funnel-shaped, mouth slightly 5-lobed, tube up to 2.5 cm long, 3.5 cm across, midpetaline bands pubescent; stamens inserted, anthers up to 4 mm long, white, filaments subequal, dilated at base, hairy; ovary conical, glabrous, disc annular, style single, stigma biglobose. Fruit baccate, globose, crowned with short style base, fruiting calyx slightly enlarged; seeds 2, 3 or 4, orbicular, glabrous, black.

Occasional in deciduous forests in Adilabad, Warangal, Medak and Mahabubnagar districts

Fl. & Fr.: July – November.

Rajganjala (ADB), *GO* 1385; Borapuram (MBNR), *BSS* & *SKB* 32357; Thonbigunla RF, Ramayampet (MDK), *RG* 112493 (BSID).

CONVOLVULUS L.

1. Leaves sessile, linear to oblong, basally attenuate; flowers less than 1 cm long:
    2. Peduncle up to 4 cm long; outer sepals ovate-elliptic; fruits depressed globose .... **C. rottlerianus**
    2. Peduncle absent; outer sepals lanceolate; fruits ovate-elliptic .... **C. prostratus**
1. Leaves petiolate, ovate or ovate-oblong, basally cordate or hastate; flowers more than 1 cm long .... **C. arvensis**

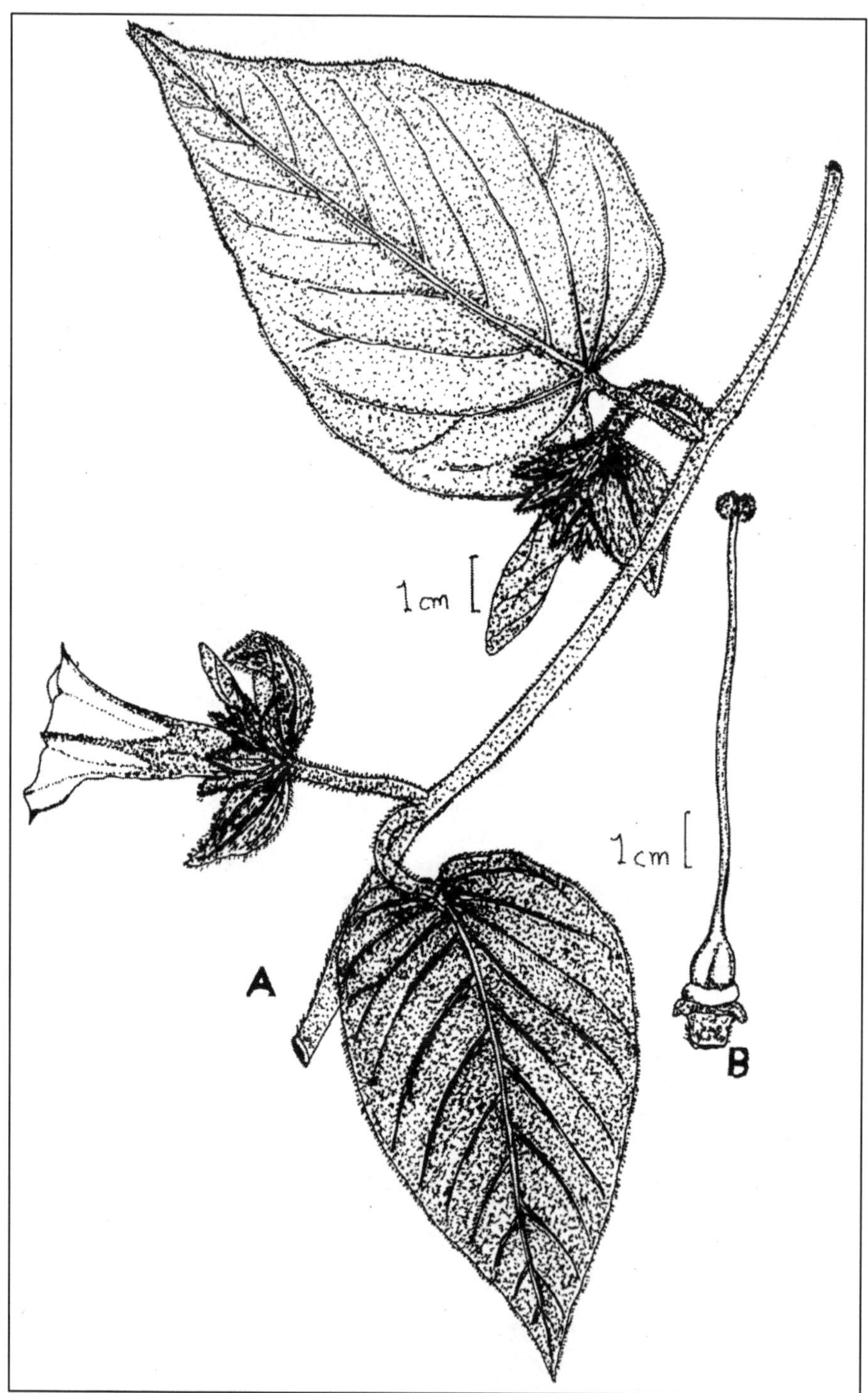

Figure 71: Argyreia sericea Dalzell

A. Twig, B. Pistil

Convolvulus arvensis L., Sp. Pl. 153. 1753; C.B.Clarke in Hook.f., Fl. Brit. India 4: 219. 1883; Gamble, Fl. Madras 2: 925. 1923

Annual herbs, stem trailing or twining, terete or slightly angular, glabrous or sparsely pubescent. Leaves simple, very variable in leaf shape, size and pubescence, linear or triangular-oblong, 1-6 × 0.5–1.5 cm, base hastate or sagittate, margin entire, toothed or lobed, apex acute to obtuse, mucronulate, glabrous or slightly pubescent below; midrib raised beneath, pubescent; petiole shorter than the blade, up to 2.5 cm long, pubescent. Flowers axillary, mostly solitary or 2-3-flowered cymes; peduncle up to 3 cm long, terete or angular; bracts small, 2, linear, about 3 mm long; pedicels up to 1.4 cm long; sepals 5, subequal, outer 2 little shorter, oblong-elliptic, 4-4.2 × 2-2.5 mm, apex obtuse, glabrous on both sides, inner 3 broadly oblong-elliptic, 5-5.2 × 2.8-3.1 mm, apex obtuse, glabrous on both sides; corolla white or pink, broadly infundibuliform, mouth slightly 5-lobed, tube up to 2 cm long, 1.2 cm across, midpetaline bands pubescent outside; stamens inserted, anthers up to 2.2 mm long, white, filaments subequal, 2 long, 3 short, slightly broadened at base; ovary globose, glabrous, disc small, style inserted, glabrous, stigma bifid. Capsule globose, glabrous; seeds 4, dark brown or black.

Occasional in forests, but common in waste lands, marshes, cultivated fields, along hedges in Medak, Nizamabad, Karimnagar and Mahabubnagar districts.

Fl. & Fr.: August – February.

Chinnamallareddy (NZB), *BR* & *KH* 9675.

Convolvulus prostratus Forssk., Fl. Aegypt. Arab. 203. 1775. *C. pluricaulis* Choisy, Mem. Soc. Phys. Geneve 6: 477. 1883; Hook.f., Fl. Brit. India 4: 218. 1883.

Annual herbs, stems terete, thinly fulvous or fulvous-villous. Leaves simple, alternate or whorled, linear, linear-oblong, 0.3-2 × 0.2-0.4 cm, apex acute or slightly obtuse, mucronulate, appressed hairy above, densely long trichomes below. Flowers axillary, mostly solitary or up to 3-flowered cymes, white, rose or rose-yellow, funnel-shaped. Capsule oblique, ovate-elliptic, straw-coloured, seeds 1 or 2, black.

In drier regions of Warangal district (Biju, 1997).

Fl. & Fr.: September – December.

Convolvulus rottlerianus Choisy, Convolv. Orient 95. 1834 & in DC., Prodr. 9: 403. 1845; C.B.Clarke in Hook.f., Fl. Brit. India 4: 219. 1883; Gamble, Fl. Madras 2: 925. 1923.

Perennial undershrubs, stem erect, terete, silky-villous. Leaves simple, linear, 2-3 × 0.1-0.4 cm, base attenuate, margin entire, apex acute to acuminate, silky-villous on both sides; midrib raised beneath; petiole absent. Flowers axillary, 1-3-flowered cymes; peduncle up to 4 cm long, thin, terete, pilose; bracts small, 4-4.5 × 1 – 1.5 mm, hirsute; pedicels shorter than peduncle, up to 4 mm long, silky-villous; sepals 5, subequal, outer 2 slightly narrower, ovate-elliptic, 3.8-4.2 × 2-2.3 mm, silky villous outside, glabrous inside, inner 3 slightly broader, ovate-elliptic, 4-4.3 × 3-3.8 mm, apex apiculate, glabrous except margin and centre region; corolla campanulate, mouth

5-lobed, tube up to 3 mm long, up to 8 mm across, midpetaline bands hairy outside; stamens inserted, anthers ± 1 mm long, filaments subequal, 2 long, 3 short, slightly dilated at base; ovary conical, glabrous, disc small, style glabrous, stigma linear, papillate. Capsule depressed globose, fruiting calyx nearly enclosed; seeds 4, black, minutely pubescent.

Occasional in forests of Mahabubnagar district.

Fl. & Fr.: October – March.

Thumukunta (MBNR), *BR* & *MV* 40730.

EVOLVULUS L.

1. Corolla blue, rotate; leaves elliptic to linear-oblong or lanceolate, longer than broad, apex apiculate; peduncle long; seeds glabrous . **E. alsinoides**
1. Corolla white, funnel-shaped, leaves orbicular or broadly ovate, as broad as long, apex rounded; peduncle absent or very short; seeds muricate .... **E. nummularius**

Evolvulus alsinoides (L.) L., Sp. Pl. (ed. 2) 392. 1762; C.B.Clarke in Hook.f., Fl. Brit. India 4: 220. 1883; Gamble, Fl. Madras 2: 923. 1923. *Convolvulus alsinoides* L., Sp. Pl. 157. 1753.

Vern.: *Vishnukrantham.*

Annual or perennial herbs, woody at base, stem prostrate or ascending or spreading, slender, terete, pale green, thinly or densely silky. Leaves simple, elliptic to linear-oblong or lanceolate, 0.4-1.5 × 0.1-1 cm, base acute, margin entire, apex apiculate, silky pilose on both surfaces, midrib and lateral veins not conspicuous on both sides; petiole very short, ± 2 mm long. Flowers axillary, 1-8-flowered; peduncle up to 3 cm long, slender, longer than leaf; bracts 2, linear-lanceolate, 2 mm long, apex acute to acuminate, hairy on outer surface; pedicels up to 8 mm, with characteristic bending, hairy; sepals subequal, outer 3 large, lanceolate, 3 x 0.5 mm, apex acute to acuminate, greenish brown in colour, villous from middle to apex on the outer surface, ± 1 mm long, glabrous within; corolla blue, up to 1 cm across, rotate or funnel-shaped, tube exserted; stamens 5, adnate to the base of the corolla tube, anthers small, 1 mm long, white, filaments subequal, 1.5 – 2.2 mm long, not dilated at base, glabrous; ovary orbicular, glabrous, disc very small, styles 2, exserted, stigmas 4 (each style bifid). Capsule globose, glabrous; seeds usually 4, rarely less, 2 or 3, glabrous, pale brown to black.

Common weed in grassy areas, open places and along roadsides, occasional in waste lands, scrub jungles and on hills in all districts.

Fl. & Fr.: Round the year.

Bersaipet (ADB), *GO* & *PVP* 4589; Nasarullabad RF (NZB), *TP* & *BR* 6154; Peddagutta (NZB), *BR* & *GO* 9100; Pillutla RF (MDK), *BR* & *CP* 11259; Narsapur (MDK), *KMS* 6623 (MH); Kusumasamudram (RR), *NV* 10342; Kodimial (KRN), *GVS* 20035 (MH); Pakhal RF (WGL), *KMS* 11691 (MH); Vijayapuri (NLG), *KMS* 9833(CAL,MH); Kothagudem (KMM), *RR* 112572 (BSID); Thaskapally (MBNR), *SRS*

109075 (BSID); Jakaram (WGL), *RKP* 105283 (BSID); Mudigonda (NLG), *BVRS* 157 (NU).

Evolvulus nummularius (L.) L., Sp. Pl. (ed. 2) 391. 1762; Oostr. in Steenis, Fl. Males.I. 5: 558. 1958. *Convolvulus nummularius* L., Sp. Pl. 157. 1753.

Vern.: *Chiti elukacheviaku, Sitammavari savaralu.*

Perennial herbs, stem prostrate, rooting at nodes, terete, green, pilose with short, patent curved hairs, glabrescent. Leaves simple, orbicular or broadly ovate, 0.4-2.5 × 0.7 – 1.5 cm, base subcordate or truncate, margin entire, apex rounded, glabrous on both surfaces; midrib and lateral veins raised beneath; petiole short, up to 7 cm long. Flowers axillary, solitary or 2-3 flowers in leaf axils; peduncle none or very short; bracts 2 or 4, linear-lanceolate, 1-2 mm long, apically acute; pedicels 1-2 mm long, pilose to glabrous; sepals subequal, elliptic-ovate to ovate-oblong, 2-4 × 1-1.5 mm, apex acute, sparsely pilose or glabrescent; corolla white, funnel-shaped, mouth deeply 5-lobed, tube up to 2.5 mm long, up to 4 mm across, midpetaline bands well defined with pilose outside; stamens subexserted, anthers ± 1 mm long, filaments unequal, 2 short, 3 long (rarely 3 short and 2 long), glabrous, ovary 0.7 mm, orbicular, disc very small, style 1 or 2, stigmas 4 (each style bifid). Capsule globose, 4-valved, reflexed; seeds 4, young seeds brown with pink spots, muricate, mature ones brown to black.

Common near moist places of Medak, Khammam and Ranga Reddi districts.

Fl. & Fr.: Round the year.

Mothugudem (KMM), *NV* 15635: Narsapur tank (MDK), *TP* & *CP* 12005; Central University campus (RR), *TP* & *MSM* 15635; Musalivera RF (KMM), *RR* 112629 (BSID).

HEWITTIA Wight & Arn.

Hewittia malabarica (L.) Suresh in Nicolson & *al.*, Inter. Van Rheed's Hort. Malab. 88. 1988. *Convolvulus malabaricus* L., Sp. Pl. 155. 1753. *C. scandens* Milne, Descr. Cat. 2: 1773. *C. bicolor* Vahl, Symb. Bot. 3: 25. 1794. *Hewittia bicolor* (Vahl) Wight & Arn., Madras J. Lit. Sci. 5: 22. 1837; C.B.Clarke in Hook.f., Fl. Brit. India 4: 216. 1883; Gamble, Fl. Madras 2: 924. 1923. *H. scandens* (Milne) Mab. in Manilal, Bot. Hist. Hort. Malab. 84. 1980.

Annual or biennial climbing herbs, stem twining or prostrate, terete, densely pubescent. Leaves simple, ovate, 4-13 x 2-10 cm, base cordate, margin entire or broadly angled, apex acute or acuminate, shining on both surfaces, hairy beneath; midrib raised beneath, hairy; petiole up to 7 cm long. Flowers in axillary, 1-3-flowered cymes; peduncle up to 8 cm long, pubescent like stem; bracts lanceolate, up to 1.3 × 0.2 cm, acuminate, pubescent; pedicels up to 0.6 cm long; sepals unequal, outer 2 larger, ovate, 1.5 -2.5 × 0.8 – 1.2 cm, third single medium sized, lanceolate to ovate, 1-1.7 × 0.5-0.9 cm, innermost pair small, narrowly ovate, 1-1.5 × 0.2 – 0.5 cm; corolla pale yellow to white with purple throat, campanulate, mouth slightly 5-lobed, tube up to 2.2 cm long, up to 2.6 cm across, midpetaline bands pilose; stamens inserted, anthers up to 4 mm long, yellowish white, filaments equal, 4 mm long, inserted, hairy at base; ovary ovoid, 1.2 mm, disc annular, style single, 6 mm long, inserted, hairy at base, stigma papillate. Fruit capsule with persistent calyx; seeds 4, often covered with ash coloured wrinkled membrane.

Occasional in forests in Medak, Ranga Reddi and Mahabubnagar districts.

Fl. & Fr.: August – March.

Pargi (RR), *MSM* 12796; ICRISAT (MDK), *LJGV* 3298 (CAL); Amrabad (MBNR), *BR* & *SKB* 32525.

IPOMOEA L.

1. Inflorescence few to several-flowered scorpiod or dichasial cymes, corolla salver-shaped or tubular, stamens and style exserted:
    2. Leaves ovate to orbicular, entire, angular or slightly lobed; many-flowered helicoid cymes; corolla scarlet red, tube inside yellowish orange .... **I. hederifolia**
    2. Leaves deeply pinnatisect to the midrib with 18-21 alternate or opposite pairs of linear to filiform lobes, glabrous; few-flowered cymes; corolla red or white, tube inside red .... **I. quamoclit**
1. Inflorescence solitary, compound, paniculate or capitate cymes; corolla funnel-shaped or tubular; stamens and style inserted or subexserted:
    3. Plants muricate; flowers nocturnal; sepals long aristate; corolla large, tube long:
        4. Corolla white, tube 7-13 cm long, midpetaline bands pale green outside .... **I. alba**
        4. Corolla purplish, tube 3-6.5 cm long, midpetaline bands purplish on both sides .... **I. turbinata**
    3. Plants glabrous, pubescent, pilose or hirsute; flowers diurnal; sepals vary; corolla small or medium sized:
        5. Sepals verrucose or cristate on the back, glabrous:
            6. Leaves conspicuously bilobed at apex .... **I. pes-caprae**
            6. Leaves not conspicuously bilobed at apex:
                7. Seeds white tomentose, long arachinoid hairs along the margin .... **I. marginata**
                7. Seeds pubescent:
                    8. Semi-succulent swamp plants; leaves mostly hastate (rarely cordate) .... **I. aquatica**
                    8. Herbaceous vines; leaves not hastate:
                        9. Leaves coriaceous, rounded-cordate, apically obtuse or acuminate, corolla lavender to purple .... **I. asarifolia**

9. Leaves membranaceous, cordate or palmately compound:
   10. Leaves simple, entire, reniform, apically acute to slightly apiculate, flowers yellowish white with deep purplish throat .... **I. obscura**
   10. Leaves palmately compound, pseudostipulate:
       11. Bracts palmately laciniate like leaves; flowers up to 1 cm long .... **I. coptica**
       11. Bracts simple, entire; flowers more than 5 cm long .... **I. cairica**

5. Sepals smooth, glabrous or hirsute:
   12. Plants glabrous or rarely pubescent; sepals mostly obtuse or orbicular, convex; seeds tomentose to villous:
       13. Shrubs; corolla finely tomentose outside; style minutely pubescent .... **I. carnea**
       13. Herbaceous or woody vines; corolla glabrous; style glabrous .... **I. staphylina**
   12. Plants pubescent to hispid; sepals oblong, ovate, lanceolate or linear:
       14. Sepals oblong, mucronulate, subcoriaceous .... **I. batatas**
       14. Sepals ovate, lanceolate or linear-lanceolate:
           15. Plants pilose; flowers in subsessile capitate cymes (up to 7 mm long); sepals oblong-spathulate, basally not hastate, hispid or pilose except inner side (middle to above) .... **I. eriocarpa**
           15. Plants pubescent or hispid; flowers in pedunculate cymes; sepals lanceolate or linear-lanceolate, hirsute:
               16. Flowers in dense capitate pedunculate heads; outer bracts large, involucrate:
                   17. Leaves pubescent to strigose beneath; outer bracts more than 5 mm wide; flowers white; seeds greyish pubescent .... **I. pestigridis**
                   17. Leaves cottony-tomentose beneath; outer bracts less than 5 mm wide; flowers magenta or mauve; seeds glabrous or pubescent .... **I. wightii**
               16. Flowers not in capitate heads; bracts small, lanceolate or linear .... **I. nil**

Ipomoea alba L., Sp. Pl. 161. 1753. *I. bona-nox* L., Sp. Pl. (ed. 2) 228. 1762; C.B.Clarke in Hook.f., Fl. Brit. India 4: 197. 1883. *Calonyction bona-nox* (L.) Boj., Hort. Maurit. 228. 1837; Gamble, Fl. Madras 2: 920. 1923.

Vern.: *Nagaramukkatte, Panditivankaya.*

Perennial, large twiners, stem twining or prostrate, rooting at nodes, woody at base, herbaceous towards tip, glabrous. Leaves simple, cordate, 5-16 × 4-13 cm, base cordate, margin entire or angled, apex acuminate, glabrous, midrib and lateral veins raised beneath; petiole 5-18 cm long. Flowers axillary, few to several (up to 4)-flowered; peduncle up to 9 cm long; bracts 2, glabrous; pedicels up to 3 cm long; sepals 5, subequal, outer 2 small, inner 3 large; corolla white, fragrant, salver-shaped, mouth 5-lobed, tube up to 13 cm long, up to 8 cm across, midpetaline bands well marked; stamens exserted, anthers up to 8 mm long, white, filaments subequal, 2 long, 3 short, glabrous; ovary conical, disc small, glabrous, style up to 11.5 cm long, glabrous, stigma biglobose. Fruit capsule, persistent style base at apex, fruiting calyx enlarged; seeds 4, black, glabrous.

Cultivated in Hyderabad.

Fl. & Fr.: September - February.

Ipomoea aquatica Forssk., Fl. Aegypt.-Arab. 44. 1175; C.B.Clarke in Hook.f., Fl. Brit. India 4: 210. 1883. *I. reptans* Poiret in Lam., Encycl. (Suppl.3): 460. 1814, non L.: Gamble, Fl. Madras 2: 916. 1923.

Vern.: *Thutikura.*

Annual herbs, stems thick, fistular or spongy, rooting at nodes, trailing on moist soil or mud, or floating on water. Leaves simple, variable in shape and size, mostly hastate, some times triangular, ovate-oblong, lanceolate or linear, 1-9 × 0.2-7 cm, base truncate, cordate to sagittate or hastate, margin entire, apex acute or acuminate, glabrous; midrib and lateral veins raised beneath; petiole up to 9.3 cm long, thin, glabrous. Flowers axillary, solitary or in few-flowered (up to 3) cymes; peduncle up to 8 cm long, glabrous; bracts small; pedicels up to 6 cm long, glabrous; sepals 5, subequal, outer 2 short, inner 3 slightly bigger; corolla purple, funnel-shaped, mouth slightly 5-lobed, tube up to 3 cm long, up to 3.5 cm across; stamens inserted, anthers up to 4 mm long, white, filaments unequal, 2 long, 3 short; ovary 1.5 mm, conical, disc small, glabrous, style inserted, up to 1.5 cm long, glabrous, stigma biglobose. Capsule globose, apiculate, persistent calyx embracing the fruit; seeds 1 or 2, orbicular, black.

Occasional in stagnant ponds and river banks in clayey soil in most districts.

Fl. & Fr.: Round the year.

Sone (ADB), *TP* & *PVP* 4050; Sarvaipet (ADB), *GO* 4509; Annaram (NZB), *BR* & *CPR* 7168; Medak (MDK), *BR* & *CP* 11464; Narsapur (MDK), *KMS* 6805(CAL, MH); Kusumasamudram (RR), *MSM* & *KH* 11027; Pakhal lake (WGL), *KMS* 11619 (MH); Kodimial (KRN), *GVS* 21864 (MH); Ravannapadu (KMM), *RR* 113856 (BSID); Near Dindi project (NLG), *BVRS* 672 (NU).

Ipomoea asarifolia (Desr.) Roem. & Schult., Syst. Veg. 4: 251. 1818. *Convolvulus asarifolius* Desr. in Lam., Encycl. 3: 562. 1792. *Ipomoea repens* Lam., Tabl. Encycl. 1: 467. 1791. (non Roth 1821); Gamble, Fl. Madras 2: 916.1923; Venkatappa in Pullaiah & Moulali, Fl. Andhra Pradesh 2: 638. 1997. *I. beladamboe* Roem. & Schult. in L., Syst. Veg. 4: 233. 1819; C.B.Clarke in Hook.f., Fl. Brit. India 4: 209. 1883.

Perennial herbs, stem prostrate or twining, rooting at nodes, terete, hollow, glabrous, latex milky white. Leaves simple, cordate or reniform, 3-7 × 3-8 cm, base truncate or cordate, margin entire, apex obtuse, subcoriaceous; midrib and lateral veins raised beneath; petiole up to 9.5 cm long. Flowers axillary, few to many (up to 20)-flowered simple or compound cymes; peduncle up to 3.5 cm long; bracts small, ovate; pedicels up to 1 cm long; sepals 5, subequal, outer 2 small, elliptic-oblong, inner 3 large, elliptic-oblong; corolla purple or white, funnel-shaped, tube up to 4 cm long, up to 5 cm across; stamens inserted, anthers up to 4 mm long, filaments unequal, 2 long, 3 short, ciliolate at base; ovary conical, disc small, style inserted, stigma biglobose. Capsule globose, seeds 4, brown to dark brown, glabrous.

Occasional by moist banks of ponds, streams in Ranga Reddi district.

Fl. & Fr.: August – March.

Mohammadabad RF (RR), *MSM* & *KH* 10597.

Ipomoea batatas (L.) Lam., Tab. Encycl. 1: 465. 1793; C.B.Clarke in Hook.f., Fl. Brit. India 4: 202. 1883; Gamble, Fl. Madras 2: 919. 1923. *Convolvulus batatas* L., Sp. Pl. 154. 1753.

Vern.: *Chilagada dumpa, Chelagada, Genusu gadda.*

Annual herbs, stem prostrate, ascending or twining, terete or angular, rooting at nodes, glabrous or pubescent, with subterranean tubers. Leaves variable in shape and size, triangular or ovate, 4-14 × 4-11 cm, base cordate to truncate, margin entire, apex acute or acuminate, angular or deeply palmately 3-5-lobed, glabrous; midrib and lateral veins raised beneath; petiole up to 17 cm long. Flowers axillary, solitary to few-flowered cymes, peduncle up to 16 cm long; bracts very small, triangular; pedicels up to 1 cm long; sepals 5, unequal, outer 2 short, oblong, elliptic-oblong, inner 3 long, elliptic-oblong; corolla purple, campanulate to funnel-shaped, mouth slightly 5-lobed; tube up to 5.8 cm long, up to 4.8 cm across; stamens inserted to subexserted, anthers 4 × 2 mm, white or purple, filaments white or purple, equal to unequal; ovary globose, style inserted to exserted, glabrous, stigma biglobose. Capsule ovoid, with persistent style base; seeds 1 to 4, brown to dark brown, glabrous or pubescent.

Commonly cultivated for its edible tubers.

Fl. & Fr.: October – March.

Kadthala (ADB), *GO* & *DAM* 5001.

Ipomoea cairica (L.) Sweet, Hort. Brit. 287. 1827; Gamble, Fl. Madras 2: 918. 1923. *Convolvulus cairicus* L., Syst. Nat. (ed. 10). 922. 1759. *Ipomoea palmata* Forssk., Fl. Aegypt.-Arab. 43. 1775; C.B.Clarke in Hook.f., Fl. Brit. India 4: 214. 1883.

Large climbers, stem twining, woody at base, herbaceous towards tip, smooth or muricate, glabrous, terete, latex milky white. Leaves palmately 5-7-foliolate (lobed) almost to the base, lobes elliptic to lanceolate, 3-5 × 1-3 cm, base decurrent or cuneate, margin entire, apex acute, glabrous; midrib raised beneath; petiole up to 7 cm long. Flowers axillary, few to several (up to 14)-flowered, cymose panicle; peduncle up to 2.5 cm long; bracts small, glabrous; pedicels up to 1.8 cm long; sepals 5, subequal, outer 2 slightly smaller, inner 3 larger; corolla purple, funnel-shaped, mouth slightly 5-lobed, up to 4.5 cm long, up to 6.2 cm across; stamens inserted, anthers up to 6 mm long, filaments unequal, 2 long, 3 short; ovary small, conical, glabrous, disc small, style inserted, stigma bilobed. Capsule globose, glabrous; seeds 3-4, ovoid, white, pubescent.

Often cultivated in gardens, sometimes as an escape and occasional on hills in Nizamabad, Medak, Karimnagar, Ranga Reddi and Hyderabad districts. Probably run wild.

Fl. & Fr.: August – April.

Chandoor (NZB), *BR* & *GO* 9203; Akkannapet (MDK), *BR* & *CP* 11554; Patancheru (MDK), *CP* 11684; Anantasagar RF (RR), *MSM* & *KH* 11004.

Ipomoea carnea Jacq., Enum. Syst. Pl. 13.1760; Gamble, Fl. Madras 2: 919. 1923. subsp. **fistulosa** (Choisy) D. Austin, Taxon 26: 237. 1977. *I. fistulosa* Mart. ex Choisy in DC., Prodr. 9: 349. 1845.

Vern.: *Thutu kada.*

Perennial, straggling shrubs, up to 2.5 m long, stem woody at base, herbaceous towards tip, hollow, terete, glabrous, with milky latex. Leaves simple, ovate, ovate-cordate, 6-15 × 4-9 cm, base cordate to truncate, margin entire, apex acuminate, glabrous above, pubescent below; midrib and lateral veins raised beneath; petiole up to 10 cm long, pubescent. Flowers axillary or terminal, few to several-flowered (up to 20) cymose-paniculate clusters; peduncle up to 7 cm long, pubescent; bracts elliptic, oblong, deciduous; pedicels up to 1.5 cm long, pubescent; sepals 5, subequal, outer 2 narrow, sub-orbicular, inner 3 broader, suborbicular; corolla pink to rose, funnel-shaped, mouth slightly 5-lobed, up to 7 cm long, up to 7 cm across; stamens inserted, anthers 8 mm long, filaments unequal, 2 long, 3 short; ovary conical, glabrous, disc small, style inserted, stigma biglobose. Capsule 4-valved, glabrous; seeds 4, sericeous.

Common in waste lands, roadsides, along the streams and hedges throughout the state. Planted for fencing.

Fl. & Fr.: August – March.

Kagaznagar road (ADB), *GO* 4361; Sone (ADB), *TP* & *PVP* 4031; Mosra (NZB), *TP* & *BR* 6086; Gnanapur (MDK), *CP* 11790; Mohammadabad canal bank (RR), *NV* 10432; Amangal to Shadnaar (MBNR), *SRS* 104556 (BSID); Maraigudem RF (KMM), *RCS* 104260 (BSID); Gunavaram (KMM), *RCS* 98938 (BSID).

Ipomoea coptica (L.) Roth ex Roem. & Schult., Syst. Veg. 4: 208. 1819. *Convolvulus copticus* L., Mant. Pl. 559. 1771. *Ipomoea dissecta* Willd., Phytographia 5. t.2.f.3.1794; C.B.Clarke in Hook.f., Fl. Brit. India 4: 200. 1883; Gamble, Fl. Madras 2: 918. 1923.

Annual herbs, stem twining or prostrate, slender, glabrous, terete, hollow. Leaves digitately 5-lobed, midlobe long, side lobes small, lobes ovate, oblong-obovate or oblanceolate, base cuneate, margin serrate, apex acute, glabrous on both sides; midrib and lateral veins raised beneath; petiole up to 1 cm long. Flowers axillary, 1-3-flowered cymes; peduncle up to 3.5 cm long; bracts palmately laciniate like leaves; pedicels up to 4 mm long; calyx up to 6 mm long, sepals 5, subequal, outer 3 large, inner 2 small; corolla white, funnel-shaped, mouth slightly 5-lobed, tube up to 9 mm long, up to 1 cm across; stamens inserted, anthers up to 1.2 mm ong, filaments subequal, 2 long, 3 short; ovary 0.5 mm, conical, glabrous, disc small, style 3 mm, inserted, stigma biglobose. Capsule ovoid, glabrous, fruiting sepals slightly enlarged, seeds 4, densely pubescent.

Rare in open forests in Nizamabad, Warangal, Ranga Reddi and Mahabubnagar districts.

Fl. & Fr.: September-April.

Gandivet (NZB), *BR* & *CPR* 7194; Lingampalle (RR), *MSM* 12740.

Ipomoea eriocarpa R.Br., Prodr. 484. 1810; C.B.Clarke in Hook.f., Fl. Brit. India 4: 204. 1883. *Convolvulus hispidus* Vahl, Symb. Bot. 3. 29. 1794. *Ipomoea hispida* (Vahl) Roem. & Schult., Syst. Veg. 4: 238. 1819; Gamble, Fl. Madras 2: 915. 1923.

Vern.: *Puriti tiga.*

Herbaceous twiner, stem pubescent or hispid, trichomes up to 1.5 mm long. Leaves simple, oblong, linear-oblong, hastate, 4-8 × 1.5-3 cm, base cordate, margin entire, apex acute, apiculate, glabrescent above, pubescent below; midrib and lateral veins raised on both sides, petiole up to 5 cm long, pubescent like stem. Flowers axillary, few to many-flowered (up to 7) cymes; peduncle very short, up to 7 mm long, hirsute like stem; bracts 2, lanceolate; pedicels absent or very short, up to 7 mm long; sepals 5, subequal, outer 2 long, inner 1 medium sized, inner most 2 small; corolla pink, funnel-shaped, mouth very shallowly lobed, tube up to 5 mm long, up to 1 cm across, midpetaline bands hirsute; stamens subexserted, anthers up to 1 mm long, filaments subequal, 3 long, 2 short; ovary 2 mm long, oblong, sericeous, style 2.5 mm long, subexserted, stigma biglobose. Capsule globose, hairy, seeds 4, grey, minutely pitted.

Occasional in waste lands among bushes in most of the districts.

Fl. & Fr.: August – February.

Itkial (ADB), *TP* & *PVP* 4131; Gupalapatnam (ADB), *GO* 4408; Mosra forest nursery (NZB), *TP* & *BR* 6077; Narsapur (MDK), *KMS* 6737 (CAL, MH); Naskal (RR), *MSM* & *PRSR* 15119; Rangammagudem RF (RR), *MSM* & *KH* 10575; Kodimial (KRN), *GVS* 21871 (MH); Sathiampet to Pennugadapa (KMM), *RR* 112649 (BSID); Yenkur (KMM), *RR* 108551 (BSID).

Ipomoea hederifolia L., Syst. Nat. (ed. 10) 925. 1759. *Quamoclit phoenicea* (Roxb.) Choisy, Mem. Soc. Phys. Geneve 6: 433. 1834; Gamble, Fl. Madras 2: 919. 1923. *Ipomoea coccinea* auct.; C.B.Clarke in Hook.f., Fl. Brit. India 4: 199. 1883.

Vern.: *Kasiratnamu.*

Annual herbs, stem twining, woody at base, herbaceous at tip, strongly twisted, hollow, glabrous or sparsely pubescent. Leaves simple, ovate-cordate, 3-12 × 2.5 – 11, base cordate, margin entire or obscurely lobed, apex acuminate, glabrous; midrib and lateral veins raised on both sides; petiole up to 10 cm long. Flowers scarlet red in axillary, few to many (4-20)-flowered cymes; peduncle up to 10 cm long; bracts small, triangular; pedicels angular, glabrous; sepals 5, subequal, outer 2 small, inner 3 large, to 2.5 cm across; corolla 5-lobed, ca 3 cm long, tube ca 2 cm, yellowish orange inside, lobes to 1 cm; stamens exserted, anthers up to 2 mm long, filaments subequal; ovary conical, glabrous, disc small, style exserted, stigma biglobose. Capsule globose, glabrous, 4-valved; seeds 4, pubescent.

Common in open forest and open places over hedges, along road sides.

Fl. & Fr.: September – March.

Kammarpally (NZB), *TP* & *BR* 6247; Navipet (NZB), *BR* & *PSPB* 9591; Narsapur forest nursery (MDK), *BR* & *CP* 11224; Narsingi (MDK), *BR* & *CP* 11525; Aklaspur (KRN), *GVS* 22518 (MH); on the way to Billyguttalu (KMM), *RR* 105983 (BSID); Bhongir (NLG), *BVRS* 555 (NU).

Ipomoea marginata (Desr.) Manitz., Feddes Repert. 83: 638. 1974. *Convolvulus marginatus* Desr. in Lam., Encycl. 3: 558. 1792. *Ipomoea sepiaria* Koen ex Roxb. Fl. Ind. 2: 90. 1824; C.B.Clarke in Hook.f., Fl. Brit. India 4: 209. 1883; Gamble, Fl. Madras 2: 916. 1923; Venkaptappa in Pullaiah & Moulali, Fl. Andhra Pradesh 2: 638. 1997.

Vern.: *Mettatuti, Purititiga, Tella thuti.*

Annual, twining, villous shrub, stem prostrate or twining, terete, hirsute, prostrate ones rooting at nodes, latex milky white. Leaves simple, highly variable, ovate, triangular, triangular-cordiform, lanceolate, 2-10 × 2-12 cm, base truncate or cordate, margin entire, apex acute to acuminate, glabrous on both sides; petiole up to 6 cm long. Flowers axillary, few to many-flowered (3-10) subumbellate cymes; peduncle up to 20 cm long; bracts 2, small, early caducous; pedicels up to 1.2 cm long; sepals 5, equal in length, elliptic-oblong to ovate; corolla pink with purple tube, salver-shaped, mouth slightly 5-lobed, tube up to 4 cm long, up to 2.8 cm across; stamens subexserted, anthers purple, up to 3 mm long, filaments subequal, 2 long, 3 short, ciliolate at base; ovary conical, glabrous, style subexserted, glabrous, stigma biglobose. Capsule globose, seeds 4, black, tomentose.

Occasional climber on bushes along road sides, especially on banks of streams and rivers, growing over hedges and thickets in Medak and Warangal districts .

Fl. & Fr.: Round the year.

Patancheru (MDK), *CP* 11689; Pakhal RF (WGL), *KMS* 11687 (MH).

Ipomoea nil (L.) Roth, Catal. Bot. 1: 36. 1797. *Convolvulus nil* L., Sp. Pl. ed. 2. 1: 219. 1762. *Ipomoea hederacea* auct. non (L.) Jacq.; C.B.Clarke in Hook.f., Fl. Brit. India 4: 199. 1883; Gamble, Fl. Madras 2: 917. 1923.

Vern.: *Jirika kolli, Kallivittulu.*

Annual herbs, stem twining, woody at base, herbaceous towards tip, retrorsely villous, latex milky white. Leaves simple, broadly ovate, palmately 3-lobed, 5-9 × 6-12 cm, base cordate, margin entire, apex acuminate, sparsely hairy; midrib and lateral veins raised beneath, hirsute; petiole up to 5 cm long, hirsute. Flowers axillary, few to many (up to 5)-flowered cymose clusters; peduncle ± 2.3 cm long, hirsute; bracts small, lanceolate; pedicels up to 1 cm long, hirsute; sepals 5, subequal, outer 2 long, broadly lanceolate, inner 3 small, lanceolate, sparsely pilose; corolla bluish-purple, funnel-shaped, mouth slightly 5-lobed, tube 3-4 cm long, up to 3 cm across; stamens subexserted, anthers up to 2.8 mm long, filaments unequal, ciliolate at base; ovary conical, disc small, style subexserted, stigma biglobose. Capsule subglobose, glabrous; seeds 5-6, black, puberulent.

Most common in jungles, hedges and along streams in all districts.

Fl. & Fr.: July – February.

Nasarullabad RF (NZB), *TP* & *BR* 6163; Gangapur RF (MDK), *TP* & *CP* 12102; Narsapur town (MDK), *KMS* 6713 (MH); Gadirayal RF (RR), *MSM* 10517; Bhupatipur (KRN), *GVS* 25709 (MH); Palempet (WGL), *RKP* 105338 (BSID); Karepally (KMM), *RR* 105960 (BSID); Mallepally (NLG), *BVRS* 533 (NU).

Ipomoea obscura (L.) Ker.-Gawl., Bot. Reg. 3: t. 239. 1817; C.B.Clarke in Hook.f., Fl. Brit. India 4: 207; Gamble, Fl. Madras 2: 916. 1923. *Convolvulus obscurus* L., Sp. Pl. ed. 2. 220. 1762.

Vern.: *Nallakokkita.*

Annual twining or trailing herb, stem twining or prostrate, herbaceous, terete, hairy, latex milky white. Leaves simple, ovate to broadly ovate or cordate, 2-6 × 2-5 cm, base cordate, margin entire, apex acute, acuminate or apiculate, glabrous or sparsely pilose on both sides; midrib and lateral veins raised beneath; petiole up to 7 cm long. Flowers axillary, 1 to few-flowered (up to 3) cymes; peduncle up to 5 cm long; bracts narrow, apically acute, glabrous; pedicels up to 1.8 cm long; sepals 5, subequal, outer 2 slightly smaller, ovate, inner 3 slightly bigger, broadly ovate; corolla yellow or white with purple throat, funnel-shaped, mouth slightly 5-lobed, tube up to 1.5 cm long, up to 1.8 cm across; stamens subexserted, anthers up to 2 mm long, filaments unequal, 2 long, 3 short, ciliolate at base, ovary 0.5 mm long, conical, glabrous, disc small, style 7 mm long, subexserted, stigma biglobose. Capsule globose, apically beaked; seeds 4, ovoid, black, pubescent.

Common on hedges and roadsides, occasional in waste lands in all districts.

Fl. & Fr.: Round the year.

Jannaram (ADB), *TP* & *PVP* 4185; Laxmapur (NZB), *TP* & *BR* 6292; Choutkoor (MDK), *TP* & *CP* 11911; Akkannapet (MDK), *BR* & *CP* 11552; Giddakunta,

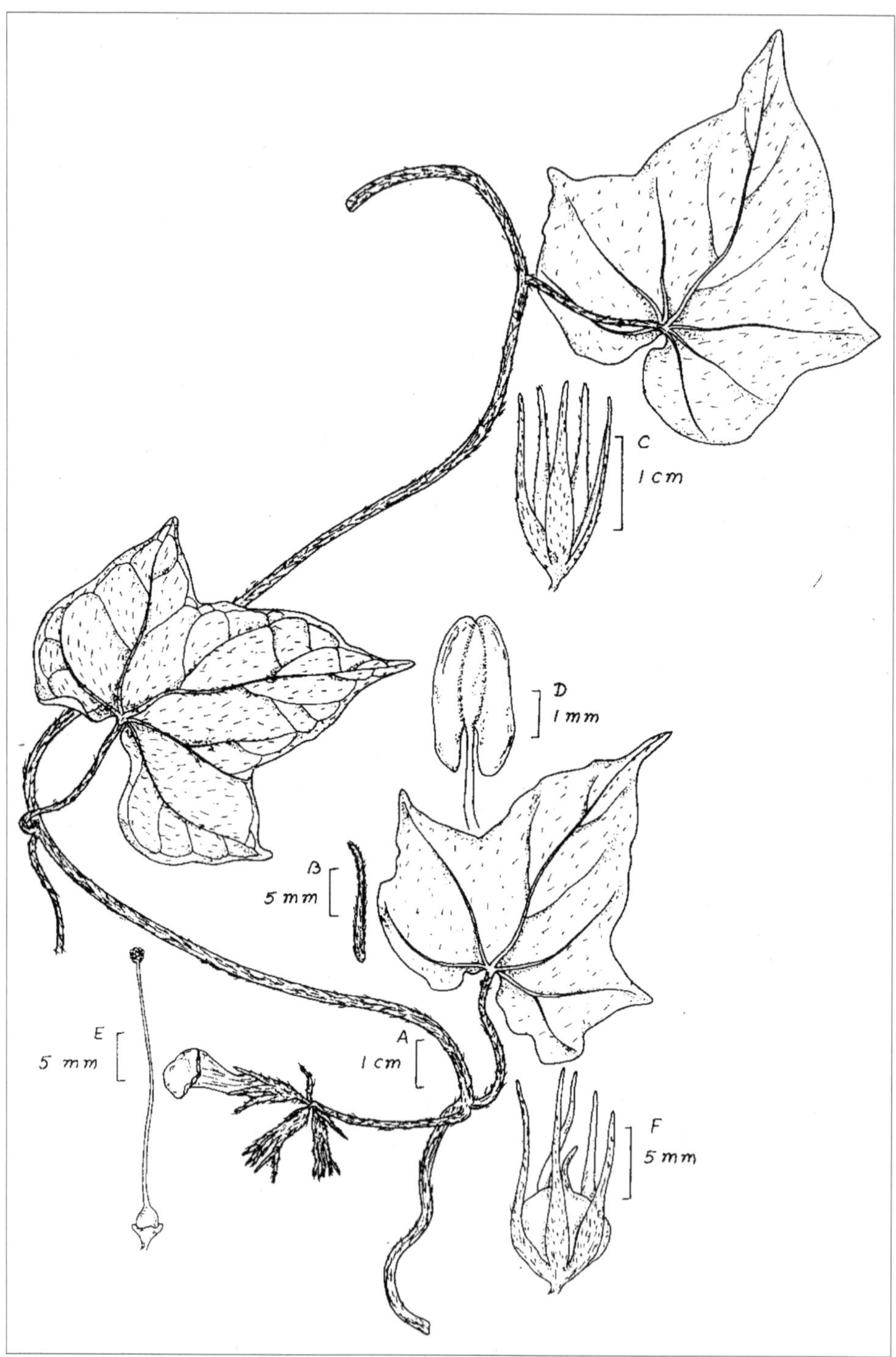

Figure 72: Ipomoea nil (L.) Roth

A. Twig, B. Bract, C. Calyx, D. Stamen, E. Pistil, F. Fruit.

Mohammadabad RF (RR), *NV* 10424; Gacchibowli (RR), *MSM* 12771; ICRISAT (MDK), *LJGV* 3860; Rechapalli (KRN), *GVS* 22282 (MH); Zulurpetta (KMM), *RR* 113823 (BSID); Bhongir (NLG), *BVRS* 552 (NU).

Ipomoea pes-caprae (L.) R.Br. in Tuckey Narr. Exped. Zaire 477. Mar. 1818; Gamble, Fl. Madras 2: 917. 1923. *Convolvulus pes-caprae* L., Sp. Pl. 159. 1753. *Ipomoea biloba* Forssk., Fl. Aegypt.- Arab. 44. 1775; C.B.Clarke in Hook.f., Fl. Brit. India 4: 212. 1883 *p.p.*

Vern.: *Chevulapilli theega, Bedathiga, Balabanditiga.*

Perennial, prostrate, creeping herbs in sandy soils, stem long, trailing, often rooting at nodes, terete, glabrous, latex milky white. Leaves simple, bilobed, 2-3 × 5-6 cm, base truncate, margin entire, apex deeply emarginated, fleshy, glabrous; midrib raised beneath, lateral veins conspicuous but not raised; petiole up to 5 cm long. Flowers axillary, mostly solitary, rarely 2 or even 3-flowered cymes; peduncle up to 6 cm long; bracts small, ovate; pedicels up to 4.5 cm long; calyx up to 1.6 cm long; sepals 5, subequal, outer 2 slightly small, ovate, inner 3 slightly larger, broadly elliptic; corolla pink, funnel-shaped, mouth slightly 5-lobed, tube up to 4 cm long, up to 4.5 cm across; stamens inserted, anthers up to 7 mm long, filaments unequal, 2 long, 3 short, ovary 2 mm long, conical, disc small, style 8 mm, inserted, stigma biglobose. Capsule globose, glabrous, seeds 4, large, densely tomentose, black.

Occasional Karimnagar (Naqvi, 2001) and Mahabubnagar districts.

Fl. & Fr.: October – March.

Note: Sachet (Smithsonian Contrib. Bot. 23: 30. 1975) points out that R.Br. 's contribution is earlier than Sweet's.

Ipomoea pes-tigridis L., Sp. Pl. 162. 1753; C.B.Clarke in Hook.f., Fl. Brit. India 4: 204. 1883; Gamble, Fl. Madras 2: 918. 1923.

Vern.: *Mekamadugu, Chikunuvvu.*

Annual, twining, hispid herbs, stem terete, hollow, covered with long spreading golden bristly trichomes. Leaves palmately 5-7-foliolate, lobes obovate, lanceolate or elliptic, 3-5 × 1.5-2.5 cm, base obtuse, margin entire, apex acuminate, sericeous on both surfaces; midrib and lateral veins raised beneath; petiole up to 6 cm long, pubescent like stem. Flowers axillary, few (1-4)-flowered heads, enclosed with foliaceous bracts, outer 2 bracts large, oblong, inner 4 small, oblong-lanceolate; peduncle up to 14 cm long; pedicels absent or up to 2 mm long, pubescent like stem; calyx 2.5 cm long, sepals 5, subequal, outer 2 large, lanceolate, inner 3 small, linear-lanceolate; corolla white or pink, funnel-shaped, mouth slightly 5-lobed, tube up to 3 cm long, up to 1.6 cm across, midpetaline bands pilose outside; stamens inserted, anthers up to 1 mm long, filaments subequal, ciliolate at base; ovary 1 mm long, conical, glabrous, disc small, style up to 1 cm long, glabrous, stigma biglobose. Capsule ovoid, enclosed in calyx, seeds black, pubescent.

Common climber along the hedges and in forests in all districts.

Fl. & Fr.: August – March.

Ankusapuram (ADB), *GO* 4302; Kammarpally (NZB), *TP* & *BR* 6242; Pathur (MDK), *TP* & *CP* 12106; Narsapur (MDK), *KMS* 6777 (CAL, MH); Lingampalle (RR), *MSM* 12748; Kodimial (KRN), *GVS* 21872 (MH); Billyguttalu (KMM), *RR* 105995; Rampur (MBNR), *BSS* & *SKB* 32637; Osmangunta (NLG), *BVRS* 280 (NU).

Ipomoea quamoclit L., Sp. Pl. 159. 1753; C.B.Clarke in Hook.f., Fl. Brit. India 4: 199. 1883. *Quamoclit pinnata* Boj., Hort. Maurit. 224. 1837; Gamble, Fl. Madras 2: 919. 1923.

Vern.: *Kasiratnamu.*

Annual herbaceous twiners, stem twining, slender, terete, glabrous. Leaves ovate to oblong in outline, 1-8 × 1-6 cm, deeply pinnatisect to the midrib, with 18-21 alternate or opposite pairs of linear to filiform lobes, 2-3.5 × ± 1 mm, glabrous on both sides; petiole up to 4 cm long, glabrous, pseudostipules 1 or 2 at the base. Flowers axillary, 2-5-flowered cymes; peduncle up to 7 cm long; bracts small, deltoid; pedicels up to 3 cm long; sepals 5, slightly subequal, outer pair short, inner 3 slightly long, all are elliptic; corolla red or rarely white, salver-shaped, mouth 5-lobed, tube up to 4 cm long, up to 2.6 cm across; stamens well exserted, anthers up to 3 mm long, filaments subequal, reddish, ciliolate at base; ovary conical, glabrous, disc small, style exserted, stigma biglobose. Capsule ovoid, 4-valved; seeds 4, ovoid-oblong, blackish-brown, with tufts of minute hairs.

Cultivated and escape, widely naturalized.

Fl. & Fr.: July-January.

Darigav (ADB), *GO* 4459.

Ipomoea staphylina Roem. & Schult., Syst. Veg. 4: 249. 1819; C.B.Clarke in Hook.f., Fl. Brit. India 4: 210. 1883; Gamble, Fl. Madras 2: 917. 1923.

Vern.: *Theendratheega.*

Perennial, large straggling or climbing shrub, stem twining or straggling, woody at base, herbaceous towards tip, terete, glabrous, latex milky white. Leaves simple, ovate to ovate- elliptic, 3-15 × 2.5–11.5 cm, base cordate or truncate, margin entire, apex acute to acuminate, glabrous; petiole up to 9 cm long. Flowers axillary, many (up to 80)-flowered paniculate cymes; peduncle up to 15 cm long, bracts small, caducous; pedicels up to 1.3 cm long; sepals 5, subequal, outer 2 small, broadly ovate, inner 3 large, circular; corolla white, purple in the throat, campanulate, mouth slightly 5-lobed, tube up to 1.8 cm across; stamens inserted, anthers 2-3 mm long, filaments more or less equal, ciliolate at base; ovary conical, glabrous, style inserted, stigma biglobose. Capsule ovoid, seeds 4, light brown, clothed with silky hairs.

Occasional on hedges and bushes in waste lands in Mahabubnagar district.

Fl. & Fr.: December – April.

Jalupenta (MBNR), *TDCK* 15310; Billakal (MBNR), *TDCK* 13680, 15395; Rampur (MBNR), *BSS* & *SKB* 32638.

Ipomoea turbinata Lag., Gen. Sp. Pl. 10. 1816; Gunn., Brittonia 24: 163. 1972. *Convolvulus muricatus* L., Mant. 44. 1767. *Ipomoea muricata* (L.) Jacq., Hort. Schoenb. 3(2): 40. t. 323.1803; non Cav. 1794; C.B.Clarke in Hook.f., Fl. Brit. India 4: 197. 1883. *Calonyction muricatum* (L.) G.Don, Gen. Hist. 4: 264. 1838; Gamble, Fl. Madras 2: 920. 1923.

Vern.: *Mookathikaya.*

Annual climbers, stem twining, trailing or prostrate, rooting at nodes in prostrate plants, woody at base, herbaceous towards tip, muricate with soft prickles, latex milky white. Leaves simple, cordate, 5-18 × 4-15 cm, base cordate, margin entire, apex acute to acuminate, glabrous; midrib and lateral veins raised beneath; petiole up to 12 cm long. Flowers axillary, 1-3-flowered; peduncle up to 5 cm long; bracts 2, oblong, up to 8 mm long; pedicels up to 1.5 cm long; sepals 5, subequal, bearing a long thick recurved or patent awn, outer 2 small, inner 3 long, ovate to oblong, glabrous; corolla purple, fragrant, salver-shaped, mouth 5-lobed, tube up to 6.5 cm long, up to 4.2 cm across, midpetaline bands well marked; stamens subexserted, anthers up to 2 mm long, filaments subequal, 2 long, 3 short, pubescent at base; ovary conical, glabrous, disc small, style glabrous, stigma biglobose. Capsule globose, apiculate, seeds 4, black, glabrous.

Occasional in deciduous forests in Warangal and Karimnagar districts (Naqvi, 2001). Mostly cultivated, often as escape and naturalized on damp and waste lands.

Fl. & Fr.: July – December.

Musthyal (WGL), *CSR* 782 (KU).

Ipomoea wightii (Wall.) Choisy, Mem. Soc. Phys. Geneve 6: 470. 1834; C.B.Clarke in Hook.f., Fl. Brit. India 4: 203. 1883; Gamble, Fl. Madras 2: 917. 1923. *Convolvulus wightii* Wall., Pl. As. Rar. 2: 55. t. 171. 1831.

Climbing undershrubs, stem twining or prostrate, terete, yellowish, strigose, latex milky white. Leaves simple, ovate, 5-8 × 6-7 cm, base cordate to truncate, margin entire to obscurely lobed, apex acute to acuminate, pilose above, white cottony tomentose below, coriaceous; midrib and lateral veins not conspicuous on both sides; petiole up to 5 cm long; pubescent like stem. Flowers axillary, few to many (up to 4)-flowered, dense to somewhat lax head-like cymose clusters; peduncle up to 7 cm long, strigose; bracts lanceolate, pilose; pedicels short or absent, up to 3 mm long; sepals 5, nearly equal, linear-lanceolate, 10-12 × 2.5–4 mm, fulvous-hirsute; corolla pink-purple, funnel-shaped, mouth slightly 5-lobed, tube up to 2 mm across, midpetaline bands pilose; stamens inserted, anthers up to 2 mm long, filaments subequal, 2 long, 3 short, ciliate at base; ovary conical, glabrous, disc small, style single, glabrous, stigma biglobose. Capsule globose, slightly pubescent; seeds ovoid, black, minutely puberulous.

Occasional in cultivated fields among grasses in Nizamabad district.

Fl. & Fr.: October – March.

Mingaram (NZB), *BR* 8284.

JACQUEMONTIA Choisy

Jacquemontia paniculata (Burm. f.) Hallier f., Bot. Jahrb. Syst. 16: 541. 1893; Gamble, Fl. Madras 2: 926. 1923. *Ipomoea paniculata* Burm. f., Fl. Ind. 50. t. 21.f. 3. 1768. *Convolvulus parviflorus* Vahl, Symb. Bot. 3: 29. 1794; C.B.Clarke in Hook. f., Fl. Brit. India 4: 220. 1883.

Annual climbing herbs, stem herbaceous, twining, young plant decumbent or erect, terete, pubescent with white tomentum, latex milky white. Leaves simple, ovate or cordate, 3-9 × 2-6 cm, base cordate or truncate, margin entire, apex acuminate with mucronulate tip, glabrescent above, sparsely pubescent below; midrib and lateral veins raised beneath; petiole up to 4 cm long, finely pilose. Flowers axillary, few to many-flowered umbelliform cymes; peduncle up to 7.5 cm long, pubescent like stem; bracts small, subulate; pedicels up to 5 mm long, pubescent like peduncle; calyx 6 mm long; sepals 5, unequal, outer 2 large, ovate-lanceolate, third medium sized, lanceolate, inner most 2 small, broad-ovate to orbicular; corolla white or pink, campanulate, mouth slightly 5-lobed, tube up to 6 mm long, up to 6 mm across; stamens sub-exserted, anthers up to 2.2 mm long, filaments subequal, hairy at base; ovary 1 mm long, ovoid, glabrous, style 6 mm long, single, stigma elliptic-oblong. Capsule globose, 8-valved; seeds 4, slightly winged, brownish or purplish black.

Occasional in scrub jungles among bushes and hill slopes in Medak, Ranga Reddi, Warangal, Khammam, Nalgonda and Mahabubnagar districts.

Fl. & Fr.: September –April.

Narsapur (MDK), *TP* & *CP* 11999; Anantagiri RF (RR), *MSM* & *PRSR* 12780; Pakhal (WGL), *KMS* 11627 (CAL, MH); Mothugudem (KMM), *NV* 15633; Ondikudisai (KMM), *RR* 113837 (BSID); Yadagirigutta (NLG), *BVRS* 550 (NU).

MERREMIA Dennst. ex Endl. *nom. cons.*

1. Leaves simple, entire or lobed:
   2. Leaves entire or trilobed; branchlets turbinate; capsule coarsely wrinkled .... **M. hederacea**
   2. Leaves entire; branchlets smooth, glabrous to sparsely hirsute; capsule smooth .... **M. emarginata**
1. Leaves palmately compound or lobed:
   3. Leaflets appressed-long hairy, outer sepals hairy, peduncle eglandular .... **M. aegyptia**
   3. Leaflets and sepals glabrous, peduncle glandular .... **M. quinquefolia**

Merremia aegyptia (L.) Urban., Symb. Antill. 4: 505. 1910; Gamble, Fl. Madras 2: 928. 1923. *Ipomoea aegyptia* L., Sp. Pl. 162. 1753. *I. pentaphylla* Jacq., Collectanea 2: 297. 1788; C.B.Clarke in Hook.f., Fl. Brit. India 4: 202. 1883.

Vern.: *Elika jemudu.*

Annual twining herbs, stem twining, herbaceous, terete, green or slightly wine red, densely hirsute, strigose, up to 4 mm long, latex gummy, milky white. Leaves palmately compound, leaflets 5, lanceolate, 3-10 × 1-5 cm, base acute, margin entire, apex acuminate, hirsute on both sides; midrib and lateral veins raised beneath; petiole up to 7 cm long, hirsute. Flowers axillary, 1-6-flowered cymes; peduncle up to 3.5 cm long, hirsute like stem, bracts lanceolate, caducous; pedicels up to 2.5 cm long, hirsute; sepals 5, unequal, outermost 2 larger, ovate-lanceolate, 1.5-2.5 cm long, apically acute or acuminate, densely hirsute, third single medium sized, inner most pair small; corolla white, campanulate, mouth slightly 5-lobed, tube up to 2 cm long, up to 2 cm across; stamens inserted, anthers up to 4 mm long, filaments subequal, 3 large, 2 small, ciliolate at base; ovary ovoid, glabrous, nectary disc annular, style inserted, stigma biglobose. Capsule ovoid, glabrous, 4-celled, 4-valved; seeds 4, glabrous, light brown.

Occasional on hedges in Adilabad, Nizamabad, Karimnagar, Khammam, Hyderabad and Nalgonda districts.

Fl. & Fr.: August – February.

Kadam (ADB), *TP* & *PVP* 4125; Laxmapur (NZB), *TP* & *BR* 6293; Kinnerasani Wild Life Sanctuary (KMM), *RR* 112686 (BSID); Sathiyampet RF, Penugadappa (KMM), *RR* 112655 (BSID); Peddavuru (NLG), *BVRS* 512 (NU).

Note: In most of the Indian Floras, this plant has been given under *Ipomoea pentaphylla* of Jacq. 1788 (*loc.cit.*). Hallier f. in Engl., Bot. Jahrb. 1893 made a new combination under *Merremia* as *M. pentaphylla*. But Hallier should have adopted the oldest specific epithet *aegyptia* of L. 1753. Urban in 1910 (*loc.cit.*) has given a new combination based on this which is followed here.

Merremia emarginata (Burm. f.) Hallier.f., Bot. Jahrb. Syst. 16: 552. 1893; Gamble, Fl. Madras 2: 928. 1923. *Evolvulus emarginatus* Burm. f., Fl. Ind. 77.t.30.f. 1768. *Convolvulus reniformis* Roxb., Fl. Ind. 2: 67. 1824. *Ipomoea reniformis* (Roxb.) Choisy, Mem. Soc. Phys. Geneve 6: 446. 1834; C.B.Clarke in Hook.f., Fl. Brit. India 4: 206. 1883. *Merremia gangetica* (L.) Cuf., Bull. Jard. Bot. Etat. 31 (Suppl.): 743. 1961; Venkatappa in Pullaiah & Moulali, Fl. Andhra Pradesh 2: 641. 1997. *Convolvulus gangeticus* L., Cent. Pl. 2: 9. 1756.

Vern.: *Elika jemudu, Yelakajeevaku.*

Annual or perennial creeping herbs, stem prostrate, rooting at nodes, terete, younger parts violet red, sparsely hispid, latex milky white. Leaves simple, reniform to broadly ovate, 1-2.5 × 1.5-3 cm, base cordate, margin crenate, apex obtuse or emarginate, glabrous, midrib and lateral veins slightly raised beneath; petiole up to 3.8 cm long, violet green or green. Flowers axillary, solitary or 2-3-flowered cymes; peduncle very short or absent; bracts small, linear; pedicels absent; sepals subequal, outer 2 small, obovate to orbicular, inner 3 much larger, ovate-orbicular, long ciliate at margin; corolla yellow, campanulate, mouth 5-lobed, tube up to 8 mm, up to 2 cm across; stamens subexserted, anthers ± 1 mm long, filaments ciliolate at base; ovary conical, disc very small, style subexserted, stigma biglobose. Capsule globose, glabrous; seeds 2, glabrous, brown.

Common in marshy places or moist waste lands in all districts.

Fl. & Fr.: October-March.

Sone (ADB), *GO* 4603; Yancha (NZB), *BR* & *PSPB* 9588; Pochammaralu (MDK), *BR* & *CP* 11493; Anantagiri RF (RR), *MSM* & *TP* 11084; Kothagudem (KMM), RR 112571 (BSID); Krishna river bank (NLG), *KMS* 9816 (CAL); Nagarjunasagar (NLG), *BVRS* 527 (NU); Bhupatipur (KRN), *GVS* 22290 (MH).

Merremia hederacea (Burm.f.) Hallier f., Bot. Jahrb. Syst. 18: 118. 1894. *Evolvulus hederaceus* Burm.f., Fl. Indica 77. t.30.f.2.1768. *Ipomoea chryseides* Ker.-Gawl., Bot. Reg. 4: t. 270. 1818; C.B.Clarke in Hook.f., Fl. Brit. India 4: 206. 1883. *Merremia chryseides* Hallier f., Bot. Jahrb. Syst. 16: 552. 1893; Gamble, Fl. Madras 2: 929. 1923.

Annual herbs, stem twining or prostrate, prostrate ones rooting at the nodes, terete, minutely tuberculate, glabrous or hirsute at nodes, latex milky white. Leaves simple, ovate or cordate or shallowly 3-lobed, 3-5 × 3-4 cm, base cordate or truncate, margin entire or crenate, apex acuminate, glabrous on both surfaces; midrib and lateral veins raised beneath; petiole up to 5 cm long. Flowers axillary, 1 to many-flowered (up to 20) dichasial or monochasial cymes; peduncle up to 8 cm long; bracts minute, ovate, caducous; pedicels up to 3 mm long; calyx up to 5 mm long; sepals 5, subequal, outer 2 small, broadly obovate, inner 3 large, broadly obovate to spathulate; corolla yellow, campanulate, mouth slightly 5-lobed, tube up to 9 mm long, 1-3 cm across; stamens subexserted, anthers up to 2 mm long, filaments 5 mm long; ovary 1.2 mm long, conical, glabrous, disc small, style 5 mm, single, exserted, stigma biglobose. Capsule globose, some what 4-angled; seeds 4, reddish brown, pubescent.

Frequent in moist places, on the hills, occasional in bushes in Adilabad, Warangal, Khammam, Mahabubnagar and Hyderabad districts.

Fl. & Fr.: September – February.

Surjapur (ADB), *TP* & *PVP* 4118; Birsaipet (ADB), *GO* & *PVP* 4615; Pakhal (WGL), *KMS* 11623 (CAL, MH); Veerapuram (WGL), *PVS* & *NRR* 76909 (BSID); Sthiyampet RF, Ramavaram Range (KMM), *RR* 112657 (BSID); Farahabad to Vatavarlapally (MBNR), *SRS* 104594 (BSID).

Merremia quinquefolia (L.) Hallier f., Bot. Jahrb. Syst. 16: 552. 1893. *Ipomoea quinquefolia* L., Sp. Pl. 162. 1753.

An annual, slender, glabrous, twining herb. Leaves compound, leaflets elliptic-lanceolate, 1.5-7 × 1-1.8 cm, glabrous. Flowers creamy-white, solitary, axillary, peduncle glandular,. Capsule ovoid, glabrous; seeds oblong, light black, hairy.

Common on hedges in Warangal and Hyderabad (Ramana, 2010) districts.

Fl. & Fr.: September – January.

Narasampet (WGL), *KMS* 11594 (MH).

OPERCULINA Silva Manso

Operculina turpethum (L.) Silva Manso, Enum. Subst. Braz. 16. 1836; Gamble, Fl. Madras 2: 929. 1923. *Convolvulus turpethum* L., Sp. Pl. 155. 1753. *Ipomoea turpethum* (L.) R.Br., Prodr. 485. 1810; C.B.Clarke in Hook.f., Fl. Brit. India 4: 212. 1883. *Merremia turpethum* (L.) Shah & Bhat, J. Bombay Nat. Hist. Soc. 74: 567. 1978; Venkatappa in Pullaiah & Moulali, Fl. Andhra Pradesh 2: 643. 1997.

Vern.: *Tellategada, Nallategada, Errategada.*

Perennial stout climbers, stem robust, herbaceous towards tip, twining, usually winged, 3-5-angled, glabrous or sparsely pilose, latex gummy. Leaves simple, variable in shape, ovate-cordate or oblong, 3-10 × 2.5-12 cm, base cordate or truncate, margin entire, dentate or slightly lobed, apex acute to acuminate, glabrous or pilose above, pubescent beneth; midrib and lateral veins prominent beneath; petiole up to 10 cm long, pubescent. Flowers axillary, 1-few-flowered cymes; peduncle up to 9 cm long; bracts large, oblong; pedicels up to 2 cm long, 4-angled, pubescent; calyx 8 mm long, sepals 5, unequal, outer 3 large, broadly ovate, apically acute to acuminate, pubescent, inner 2 small, ovate; corolla white, funnel-shaped, mouth slightly 5-lobed, tube up to 3.2 cm long, up to 4.3 cm across; stamens inserted, anthers up to 6 mm long, filaments up to 7 mm long, equal; ovary globose, glabrous, style single, inserted, stigma biglobose. Capsule circumscissile, enclosed in enlarged woody calyx; seed usually solitary, glabrous, black.

Occasional twiner on shrubs in waste places in Adilabad, Hyderabad, Khammam, Warangal and Karimnagar districts.

Fl. & Fr.: October – February.

Jannaram (ADB), *TP* & *PVP* 4189; Kadam dam (ADB), *GO* & *DAM* 5405; Pakhal (WGL), *ANH* 15931 (CAL, MH); Rechapalli (KRN), *GVS* 21895, 22239 (MH); Dhaniyaigudem (KMM), *RR* 108504 (BSID); Perantapalli (KMM), *PVS* 84057 (BSID); Eturunagaram (WGL), *RKP* 110821.

Taxonomic Notes: Oostroom (1939) notices that the petiole is much shorter than the blade. From the present study it is found that the petiole and leaf blade are variable and petiole is same as or sometimes larger than the blade.

Gamble (1923), Oostroom (1939, 1953) and others took the winged nature of the stem and petiole as the key character. Meantime, Verdcourt (1963), Austin (1980) and others considered that the winged nature of the stem is the character of this taxon. The studies made by Biju (1997) on different population of *Operculina* in Southern Peninsular India, revealed that this character is highly variable and terete stem is found in some population.

Nomenclatural Notes: Rendle (1905) made a new combination for *Convolvulus turpethum* and established *Merremia tuberosa* (L.) Rendle, without knowing the combination of Rendle, Shah and Bhatt (1977) made the same combination and Rani and Mathew (1983) followed this.

The specific epithet of *Operculina turpethum* was proposed by Linnaeus (1753) from Asiatic sources. Turpethum was a Latinised version of the Greek tourpeth which came from the Arabic word Turbad, which, in turn was taken from Western Indian languages that evolved from Sanskrit.

PORANOPSIS Roberty

Poranopsis paniculata (Roxb.) Roberty, Candollea 14: 26. 1953. *Porana paniculata* Roxb., Pl. Coromandel t. 235. 1815; C.B.Clarke in Hook.f., Fl. Brit. India 4: 222. 1883; Gamble, Fl. Madras 2: 1923.

Perennial large woody climbing shrubs, stem twining, herbaceous towards tip, woody at base, greyish-tomentose throughout. Leaves simple, ovate, 5-10 × 1-8 cm, base cordate, margin entire, apex acuminate or shortly cuspidate, pubescent on both surfaces especially beneath; midrib raised beneath and prominently hairy; petiole 5 – 6 cm long, shorter than blade; greyish tomentose. Flowers axillary or terminal, many-flowered panicle; peduncle up to 17 cm long, pubescent like stem; bracts linear-lancolate, less than 1 mm long, deciduous, pubescent; pedicels short, up to 0.5 mm long; sepals 5, subequal, oblong-obovate, 2-3 × 1 mm, apically acute to acuminate, densely tomentose outside, glabrous within; corolla white, funnel-shaped, mouth slightly 5-lobed, tube up to 4 mm long, up to 6 mm across, midpetaline bands pilose outside; stamens inserted, anthers up to ± 1 mm long, filaments white, subequal; ovary globose, style single, stigma biglobose. Capsule pubescent, fruiting calyx persistent, enlarged; seed one, glabrous, black.

Mostly cultivated, but even some of them escaped from cultivation and are found roadsides and forest edges.

Fl. & Fr.: September – January.

Note: In Southern Peninsular India, the collection is only from cultivation.

Horticultural Note: Commonly cultivated in gardens for its dense mass of fragrant cream flowers and accrescent calyx.

RIVEA Choisy

Rivea hypocrateriformis (Desr.) Choisy, Mem. Soc. Phys. Geneve 6: 408. 1834; C.B.Clarke in Hook.f., Fl. Brit. India 4: 184; Gamble, Fl. Madras 2; 903. 1923. *Convolvulus hypocrateriformis* Desr. in Lam., Encycl. 3: 561. 1792.

Vern.: *Boddikura, Niru boddi.*

Perennial, large, climbing shrubs, stem twining, woody at base, herbaceous towards tip, terete, white tomentose, latex milky white. Leaves simple, broadly ovate to cordate or cordate-orbicular, 3-4 × 3-6 cm, base cordate to truncate, margin entire, apex apiculate, glabrous above, white tomentose beneath; midrib and lateral veins raised beneath; petiole up to 5 cm long, pubescent like stem. Flowers axillary, usually 1-flowered cymes; peduncle up to 1.8 cm long, pubescent like stem; bracts 2, linear-lanceolate, pubescent outside; pedicels up to 4 mm long, sepals 5, more or less same size, ovate, outer 3 apically acute, inner 2 apically obtuse, white tomentose outside, glabrous within; corolla white, fragrant, salver-shaped, glabrous outside, mouth 5-lobed, tube up to 9 cm long, up to 5 cm across, midpetaline bands sparsely pubescent outside; stamens inserted to subexserted, anthers up to 6 mm long, filaments subequal,

3 long, 2 short; ovary conical, glabrous, style subexserted, stigma bifid. Capsule conical, brown; seeds 1-2, enclosed by enlarged woody calyx lobes.

Common on hedges and bushes in dry deciduous forests.

Fl. & Fr.: August – March.

Jaipur RF (ADB), *GO* 4488, *PVP* 9438; Mamidipally RF (NZB), *BR* & *KH* 6436; Mudheli (NZB), *BR* 9533; Pochammaralu (MDK), *CP* 11761; Gacchibowli (RR), *MSM* 12773; Kudichintalabailu (MBNR), *TDCK* 15382; Mannanur (MBNR), *TDCK* 16807; Aklaspur (KRN), *GVS* 25638 (MH); Kodimial (KRN), *GVS* 20060 (MH); Lawal beat, Pasra range (WGL), *RKP* 108179 (BSID); Nagarjunakonda (NLG), *BVRS* 114 (NU).

XENOSTEGIA Austin & Staples

Xenostegia tridentata (L.) Austin & Staples, Brittonia 32(4): 533. 1980. subsp. **tridentata**. *Convolvulus tridentatus* L., Sp. Pl. 157. 1753. *Ipomoea tridentata* (L.) Roth in Roem., Arch. Bot. (Leipzig) 1,2: 38. 1798; C.B.Clarke in Hook.f., Fl. Brit. India 4: 205. 1883. *Merremia tridentata* (L.) Hallier f., Bot. Jahrb. Syst. 16: 552. 1893; Gamble, Fl. Madras 2: 928. 1923; Pullaiah & Moulali, Fl. Andhra Pradeh 2: 643. 1997.

Perennial spreading herbs with thick root stock giving off many elongate prostrate slender branches, stem twining, angular, green, glabrous, latex milky white. Leaves simple, variable, linear, spathulate, linear-hastate or oblong-obovate, 1-3 × 0.2-1 cm, glabrous; base truncate to toothed auriculate, margin entire, apex acute or obtuse; lateral veins slightly raised beneath; petiole short or nearly absent. Flowers axillary, solitary or 2-3-flowered cymes; peduncle up to 3.5 mm long, terete or slightly angled, glabrous; bracts minute, subulate; pedicels up to 1 cm long, tetra or penta-angled; calyx 5 mm long, sepals 5, subequal, outer 2 small, ovate, inner 3 large, lanceolate; corolla pale yellow or white with a purple throat, funnel-shaped, mouth slightly 5-lobed, tube up to 7 mm long, up to 1.1 cm across; stamens inserted, anthers up to 1 mm long, filaments equal, sparsely hairy at the base; ovary 1.5 mm long, 4-celled, glabrous, disc small, style 5.5 mm long, inserted, stigma biglobose. Capsule ovoid, glabrous, 4-valved; seeds 4, globose, dull black.

Most common weed of waste places and occasional in open forests in all districts.

Fl. & Fr.: June- January.

Darigav (ADB), *GO* 4569; Banswada (NZB), *TP* & *BR* 6140; Gandhari (NZB), *TP* & *BR* 6349; Thornal (MDK), *CP* 11398; Sangareddy tank (MDK), *BR* 11300; Kusumasamudram (RR), *NV* 10346; Appapur (MBNR), *BSS* & *SKB* 34459; Nagarjunakonda (NLG), *KT* 9776 (CAL); Nagarjunasagar (NLG), *BVRS* 57 (NU); Kodimial (KRN), *GVS* 20145 (MH); Bhupatipur (KRN), *GVS* 20199 (MH); Musalivera beat, Ramavaram range (KMM), *RR* 112612 (BSID); 10 km towards north of Pasra (WGL), *RKP* 108206 (BSID).

Note: Austin and Staples (1980) based on stigma anatomy, pollen morphology, cotyledon structure and other traits suggest that two species formerly considered members of *Merremia* fall outside an acceptable range of variation for that genus and a new genus *Xenostigma* was proposed by them for these two species.

CUSCUTACEAE

Recent taxonomists (Takhtajan, 1997; Cronquist, 1981 etc.) place *Cuscuta* under a separate family Cuscutaceae while in previous Floras, it was placed under family Convolvulaceae.

CUSCUTA L.

1. Styles 2, stigma mostly globose or depressed-globose; stamens with distinct filaments:
   2. Corolla tube without scales .... **C. hyalina**
   2. Corolla tube with fimbriate scales .... **C. chinensis**
1. Style 1, stigma bilobed, lobes linear-papillate; stamens with almost sessile anthers .... **C. reflexa**

Cuscuta chinensis Lam., Encycl. 2: 229. 1786; C.B.Clarke in Hook.f., Fl. Brit. India 4: 226. 1883; Gamble, Fl. Madras 2: 931. 1923; Yunck., Mem. Torrey Bot. Club 18: 209. 1932. *C. capillaris* Wall., ex G. Don, Syst. 4: 305. 1837. *C. chinensis* Lamk. var. *minor* Choisy, Mem. Soc. Phys. Hist. Nat. Geneve 9: 280. 1841.

Leafless twining parasitic herb, stem twining, slender, less than 0.5 mm diameter, yellow, terete, glabrous. Flowers in fascicles, white, somewhat globular, sessile or shorly pedicellate, up to 6 mm long, dilated at apex; bracts small, oblong, fleshy, glabrous; sepals 5, lobes triangular-ovate, 1.5-2 × $\pm$ 1 mm (outer 2 slightly smaller), apically acute or obtuse, glabrous, fleshy, slightly overlapping; corolla yellowish, campanulate, tube up to 3 mm long, up to 2 mm across, lobes 5, triangular-ovate, up to 2 mm long, apically acute to obtuse, mostly inflexed; stamens 5, subexserted, anthers less than 1 mm long, filaments up to 1 mm long; corolla scales 5, oblong, fimbriate, whitish yellow, fleshy; ovary 1 mm long, globose, glabrous, styles 2, 0.3 mm long, subequal, stigma rounded. Capsule depressed globose, corolla along with stamens remains on the developing capsule, circumscissile, glabrous; seeds 4, glabrous, brown, very minutely muricate.

Commonly parasiting on shrubs and trees in the outskirts of villages and on many weeds in plains and indiscriminately destroying cultivated crops and ornamentals in Nalgonda and in Hyderabad districts.

Fl. & Fr.: October – April.

Kodad (NLG), *BVRS* 575 (NU).

Cuscuta hyalina Roth, Nov. Pl. Sp. 100. 1821; C.B.Clarke in Hook.f., Fl. Brit. India 4: 226. 1883; Gamble, Fl. Madras 2: 931. 1923.

Leafless twining parasitic herb, stem twining, very slender, less than 0.5 mm in diameter, yellow to yellowish brown, much branched, forming often a tangled mass, terete, glabrous. Flowers white, 2.5-3 ×× 2.5 mm, loose umbellate clusters, shortly pedicellate, up to 3 mm long, dilated at the apex; bracts small, lanceolate, glabrous; sepals 5, lobes triangular; up to 2 mm long, apically acute, glabrous, fleshy, often exceeding the corolla; corolla whitish, tube up to 2 mm long, deeply 5-lobed, narrow and acute, erect or reflexed; stamens 5, subexserted, anthers ± 1 mm long, filaments short, inserted just below the sinus, corolla scales absent; ovary depressed globose, glabrous, styles 2, equal or subequal, less than 1 mm long, stigma capitate. Capsule depressed globose, circumscissile, interstylar opening small with a longitudinal groove, seeds 2-4, with few longitudinal grooves.

Occasional in the scrub jungles in Karimnagar and Mahabubnagar districts. Parasite on *Amaranthus*, *Polygonum*, *Tribulus* and *Trianthema*.

Fl & Fr.: September - March.

Ramapur (MBNR), *SRR* 2155 (HH).

Cuscuta reflexa Roxb., Pl. Coromandel 2: 3. t.104. 1798; C.B.Clarke in Hook.f., Fl. Brit. India 4: 225. 1883; Gamble, Fl. Madras 2: 931.1923; Yunck., Mem. Torrey Bot. Club 18: 259. 1932; Oostr., Steenis Fl. Males. Ser. 1. 4: 393. 1953.

Vern.: *Sitamma pounalu, Passitiga, Sitamma savaram.*

Leafless twining parasitic herb, stem twining, coarse, 2-2.5 mm in dimeter, yellow, profusely branched, glabrous. Flowers in lax racemes, white or cream, subsessile, up to 4 mm long; bracts small, broadly ovate to deltoid, fleshy, glabrous, corolla white or cream, tubular, tube up to 1.3 cm long, up to 5 mm across, deeply 5-lobed, lobes oblong-lanceolate, subacute, spreading or reflexed; stamens 5, exserted, anthers up to 1.5 mm long, subsessile, up to 1 mm long; corolla scales 5, attached at the base of corolla, fimbriate, ovary 1.5 mm long, conical, style 1, short, stout, stigma bilobed, lobes linear, papillate. Capsule globose, circumscissile near base, fleshy, seeds 4 or less, glabrous, black.

Rare in open scrub jungles covering over hedges, trees and shrubs.

Fl. & Fr.: October – April.

Boath (ADB), *GO* & *PVP* 4639; Bellampalli (ADB), *PVP* 9447; Nizamabad town (NZB), *TP* & *BR* 6115; Patancheru (MDK), *CP* 11694; Anantasagar (RR), *MSM* & *NV* 10408; Jakaram (WGL), RKP 105324 (BSID);

SOLANACEAE

KEY 1

1. Fruit capsule; corolla trumpet or funnel-shaped, lobes shorter than tube:
   2. Corolla <2 cm across, rose coloured; capsule thin walled <2 cm across, smooth; stamens didynamous; calyx campanulate .... **Nicotiana**
   2. Corolla >4 cm across, white; capsule thick walled, >3 cm across; stamens equal; calyx tubular .... **Datura**
1. Fruit a berry; corolla sub-rotate or campanulate, lobes longer than tube (except *Cestrum*):
   3. Fruiting calyx inflated, overtopping berry:
      4. Flowers solitary .... **Physalis**

4. Flowers fascicled .... **Withania**

3. Fruting calyx neither inflated nor overtopping berry:

5. Stamens connivent, anthers longer than filaments:

6. Anther dehiscence longitudinal; corolla yellow; berry depressed-globose, lobed; leaves pinnatisect; herb .... **Lycopersicon**

6. Anther dehiscence porous; corolla other than yellow; berry globose or ellipsoid, smooth; leaves simple or lobed or pinnatisect; herbs to woody shrubs .... **Solanum**

5. Stamens not connivent, anthers as long as or shorter than filaments:

7. Corolla salver-form, stamens inserted; flowers sub-sessile in paniculate cymes, fragrant .... **Cestrum**

7. Corolla sub-rotate, stamens exserted; flowers pedicellate, 1 or 3, not fragrant .... **Capsicum**

KEY 2

1. Anther dehiscence porous .... **Solanum**

1. Anther dehiscence longitudinal:

2. Stamens exserted; corolla sub-rotate or campanulate; lobes longer than tube:

3. Fruiting calyx inflated, larger than berry; corolla campanulate:

4. Flowers fascicled; corolla lobes valvate in bud; leaves glabrous or pubescent .... **Withania**

4. Flowers solitary; corolla lobes plicate in bud; leaves glabrous or pubescent .... **Physalis**

3. Fruiting calyx not inflated, smaller than berry; corolla sub-rotate; cultivated for fruits:

5. Leaf blade pinnatisect; corolla yellow; stamens connivent, anther tips sterile; berry depressed-globose.................... **Lycopersicon**

5. Leaf blade entire; corolla white; stamens not connivent, anthers fertile throughout; berry elongate .... **Capsicum**

2. Stamens inserted; corolla trumpet or funnel-shaped, lobes shorter than tube:

6. Flowers solitary; corolla > 10 x 4 cm; filaments > 7 cm long, anthers > 1 cm long .... **Datura**

6. Flowers aggregated in panicles; corolla < 6 x 2 cm; filaments < 4.5 cm long, anthers <5mm long:

7. Corolla trumpet-shaped, rose coloured; cultivated for leaves .... **Nicotiana**

7. Corolla salver-form, yellowish green; grown for flowers .... **Cestrum**

CAPSICUM L.

1. Flowers solitary; corolla white (or purplish); fruit pendent, often large, > 3 cm long .... **C. annuum**
1. Flowers 2-5; corolla greenish-white; fruit erect, rather small, <2.5 cm long .... **C. frutescens**

Capsicum annuum L., Sp. Pl. 188. 1753; Gamble, Fl. Madras 2: 941. 1923.

Vern.: *Mirapa, Mirapakaya.*

Annual erect herbs, up to 1.5 m high, stem angular. Leaves simple, ovate to lanceolate, 2-6 × 1-3 cm, base obtuse, margin entire, apex acuminate, glabrous or minutely pubescent; petiole up to 3 cm long. Flowers axillary, solitary, pendent atleast in fruit; pedicels up to 2 cm long, pendent; calyx truncate, undulate or 5-toothed, persistent; corolla white, sub-rotate, mouth 5-lobed, tube up to 2.5 mm long; stamens 5, epipetlous, filaments up to 2.5 mm long; ovary 1.5 mm, style 3.5 mm, stigma capitate. Berry oblong-ovoid or long tapeting, up to 9 cm long and 2 cm across, much longer than broad, pendent, green turning red when ripe; seeds smooth, yellow, compressed, circular, 2.5 mm across.

Commonly cultivated and often as an escape.

Fl. & Fr.: Round the year.

Kadthala (ADB), *GO & DAM* 5002.

Capsicum frutescens L., Sp. Pl. 189. 1753; C.B.Clarke in Hook.f., Fl. Brit. India 4: 239. 1883; Gamble, Fl. Madras 2: 941. 1923. *C.minimum* sensu C.B.Clarke in Hook.f., Fl. Brit. India 4: 239. 1883 *p.p.*, non Roxb.

Vern.: *Golkonda mirapa, Konda mirapa, Aggi mirapa.*

Perennial shrubs, up to 2 m high, stem angular and grooved. Leaves simple, elliptic-ovate, 4-8 × 2-4 cm, base obliquely rounded to acute, margin entire, apex acuminate, glabrous, petiole up to 3 cm long. Flowers axillary or terminal, 2-5 in bunches, pedicels up to 3.5 cm long, erect; calyx cupular, 2 mm, lobes 5, triangular, acute, persistent; corolla greenish-white, sub-rotate, mouth 5-lobed, tube up to 2.5 mm long, up to 1.2 cm across, lobes triangular-ovate, acute; stamens 5, exserted, anthers oblong, up to 2.5 mm long; filaments up to 2.5 mm long; ovary 2 mm, style 4 mm, stigma capitate. Berry ovoid or oblong, dwarf, up to 2.5 cm long, green, turning red to purplish when ripe; seeds smooth, yellow, compressed, circular.

Mostly cultivated, occasionally found as an escape.

Fl. & Fr.: Round the year.

Laxmapur (MBNR), *TDCK* 15339.

CESTRUM L.

1. Corolla white, day flowering; berry black or white .... **C. diurnum**
1. Corolla yellowish green, night flowering, very fragrant; berry white .... **C. nocturnum**

Cestrum diurnum L., Sp. Pl. 191. 1753; Bor & Raiz., Beautif. Ind. Climbers and shrubs 118. 1954.

English: Day queen.

Perennial erect or straggling shrub or small tree, up to 3.5 m high, stem woody at base, herbaceous towards tips, terete, glabrous. Leaves simple, oblanceolate-oblong, 8-15 × 3-5 cm, base rounded to acute, margin entire, apex acute, glabrous, petiole up to 1 cm long. Flowers fragrant, axillary, in cymose panicles, open after sunrise; bracts and bracteoles short; pedicels short; calyx up to 4 mm long, puberulous without, lobes 5, triangular, subacute; corolla white, salver-shaped, mouth 5-lobed, tube up to 7 mm across, lobes rounded, obtuse, reflexed; stamens 5, inserted, anthers up to 1 mm long, filaments up to 2 mm long; ovary globose, style up to 1.5 cm long, stigma capitate. Berry ovoid, black or white, subtended by calyx; seeds smooth.

Cultivated in gardens.

Fl. & Fr.: September – March.

Cestrum nocturnum L., Sp. Pl. 191. 1753; Bor & Raiz., Beautif. Ind. Climbers and shrubs 118. 1954.

English: Night queen.

Perennial erect or straggling shrub or small tree, up to 4 m high, stem woody at base, herbaceous towards tip, terete, glabrous. Leaves simple, ovate-lanceolate, 8-15 × 3-5 cm, base acute, margin entire, apex acuminate, glabrous on both sides; petiole up to 2 cm long. Flowers fragrant, very sweet-scented at night, axillary in cymose panicles, open after sunset; bracts and bracteoles short; pedicels short; calyx up to 4 mm long, tubular; corolla yellowish-green, salver-shaped, mouth 5-lobed, tube up to 1.8 cm long, up to 5 mm across, lobes acute, spreading; stamens 5, inserted, anthers up to 1 mm long, filaments up to 2 mm long with linear short appendage at base; ovary 1 mm, globose, style up to 1.7 cm long, stigma capitate. Berry ovoid, white; seeds smooth.

Often planted and also as an escape.

Fl. & Fr.: September – March.

DATURA L.

1. Branchlets whitish pubescent .... **D. innoxia**
1. Branchlets glabrous or sparsely pubescent:
    2. Branches tinged purple; flowers white or purple; capsule breaking irregularly near the apex, globose, deflexed .... **D. metel**
    2. Branches green; flowers white; capsule splitting regularly at the top into 4 valves, deeply ovoid, erect .... **D. stramonium**

Datura innoxia Mill., Gard. Dict. ed. 8. n. 5. 1768; Sant., J. Bomaby Nat. Hist. Soc. 47: 659. 1948. *D. metel* auct. non L.1759: C.B.Clarke in Hook.f., Fl. Brit. India 4: 243. 1883; Gamble, Fl. Madras 2: 941. 1923.

Vern.: *Ummetha.*

Stout-stemmed, erect shrub, up to 1 m high, stem green, branchlets whitish pubescent. Leaves simple, ovate, 3.5-13 × 2-9 cm, base oblique, margin undulate, apex acute, softly pubescent; petiole up to 6 cm long. Flowers axillary, solitary; calyx tubular, up to 7 cm long, lobes 5, lanceolate, acute; corolla white, funnel-shaped, 5 cm across, tube up to 12 cm long, lobes 5; stamens 5, filaments up to 7 cm long, anthers up to 1 cm long; ovary 0.5 cm long, style up to 12 cm long. Fruit capsule, with long, weak, sharp spines.

Occasional in waste lands, road sides and foot hills in most districts. Less common than *D. metel.*

Fl. & Fr.: Round the year.

Mancheriyal (ADB), *GO* & *PVP* 4935; Mingaram (NZB), *TP* & *BR* 6305; Narsingi (MDK), *BR* & *CP* 11524; Anantasagar (RR), *NV* 10396; Rechapalli (KRN), *GVS* 20168 (CAL, MH); Hussainsagar (HYD), *KMS* 8023 (CAL, MH); Amangal to Shadnagar (MBNR), *SRS* 104555 (BSID); Krishna river (NLG), *KMS* 9824 (CAL).

Datura metel L., Sp. Pl. 179. 1753; Sant., J. Bombay Nat. Hist. Soc. 47: 657. 1948. *D. fastuosa* auct. non L. 1759; C.B.Clarke in Hook. f., Fl. Brit. India 4: 242. 1883; Gamble, Fl. Madras 2: 941. 1923. *Datura fastuosa* auct. non L. var. *alba* (Nees) C.B.Clarke in Hook. f., Fl. Brit. India 4: 243. 1883; Gamble, Fl. Madras 2: 941. 1923.

Vern.: *Ummetha.*

Annual, erect, pubescent shrub clothed with glandular tomentum, up to 1 m high, stem and plant parts tinged purple, branchlets glabrous or sparsely pubescent. Leaves simple, variable, ovate, 10-15 × 7-10 cm, base unequal, truncate, margin frequently lobed, rarely entire, apex acute, tomentose; petiole up to 7 cm long. Flowers axillary, solitary; peduncle up to 2 cm long; calyx tubular, up to 7 cm long, lobes 5, lanceolate, acute; corolla white or purple, funnel-shaped, 7 cm across, tube up to 15 cm long, lobes 5; stamens 5, filaments up to 8 cm, anthers up to 1 cm long; ovary 0.7 cm, style up to 10 cm long. Fruit capsule, with short, stout, blunt spines, breaking irregularly near the apex, deflexed.

Common weed in waste places and road sides near villages and cultivated fields in most districts.

Fl & Fr.: June – December.

Neredugunda (ADB), *GO* & *DAM* 5096; Lingampalle (RR), *MSM* 19217; Rechapalli (KRN), *GVS* 20168 (MH); ICRISAT (MDK), *LJGV* 2745 (CAL); Yenkannapalem (KMM), *RR* 113829 (BSID).

Note: The identity and nomenclature of the taxon has considerably been confused by several earlier authors and as a result most of them consider *Datura metel* L. as an American plant and the Indian plant as *D. fastuosa* L. or *D. alba* Nees or *D. fastuosa* var. *alba* (Nees) C.B.Clarke. However, it is now certain that the centre of origin of *D. metel* L. is Asia and the Indian taxon is *D. metel* L. Santapau (*loc. cit.*) and Satina & Avery (in Blackslee: The genus *Datura* 1959) have discussed the subject at length.

Datura stramonium L., Sp. Pl. 179. 1753; C.B.Clarke in Hook.f., Fl. Brit. India 4: 242. 1883; Gamble, Fl. Madras 2: 941. 1923.

Vern.: *Ummetha.*

Annual erect herb, up to 1 m high, stem and branchlets green, glabrous or sparsely pubescent. Leaves simple, ovate or elliptic, 4.5-12 × 3-6 cm, base oblique, margin many toothed or sinuate with triangular teeth, apex acute, glabrous; petiole up to 6 cm long. Flowers axillary, solitary; pedicels up to 1 cm long; calyx narrowly tubular, up to 5.2 cm long, 5-angular, lobes 5, persistent, puberulous; corolla white, funnel-shaped, 1.5 cm across, tube up to 15 cm long, lobes 5, cuspidate; stamens 5, filaments up to 5 cm long, hairy at base, anthers 1.4 cm long; ovary 6 mm long, style up to 13 cm long. Capsule ovoid, erect, spinescent with very sharp spines, splitting regularly at the top into 4 valves deeply; seeds black, suborbicular, subcompressed, minutely dotted.

Occasional in waste lands and roadsides.

Fl. & Fr.: Round the year.

Sri Rangapur (MBNR), *TDCK* 13615; Cherukupally (NLG), *BVRS* 666 (NU).

LYCOPERSICON Mill.

Lycopersicon esculentum Mill., Gard. Dict. ed. 8. no. 2. 1768 *nom. cons.*; C.B.Clarke in Hook.f., Fl. Brit. India 4: 237. 1883; Gamble, Fl. Madras 2: 941. 1923. *Lycopersicon lycopersicum* (L.) Karsten, Deut. Fl. 966. 1882. *Solanum lycopersicum* L., Sp. Pl. 185. 1753.

Vern.: *Tomato, Rama mulaga.*

Sticky-pubescent herb, up to 75 cm high. Leaves pinnatisect, lobes obliquely triangular-ovate, 2-7 x 1-2.5 cm, lower ones separating from leaf, shortly petiolulate, terminal lobes acute; petiole up to 2 cm long. Flower(s) axillary, solitary, pedicellate; calyx lobes 5, triangular-lanceolate, up to 6 mm long, strigose without, shortly connate, acute, spreading in fruit; corolla yellow, subrotate, up to 2 mm across, lobes lanceolate, acute, finely ciliate; stamens 5, exserted, connivent around style, anthers tapering into a sterile apex, filaments up to 2 mm long; ovary 2-celled, placenta multifid, style up to 7 mm long, stigma obtuse. Berry depressed-globose, pulp juicy, red or yellow when ripe; seeds discoid.

Commonly cultivated, often as an escape.

Fl. & Fr.: Round the year.

Domakonda (NZB), *BR* 9693; Medak (MDK), *BR* & *CP* 11458.

NICOTIANA L.

1. Flowers in terminal racemes; corolla tube 1 cm across .... **N. plumbaginifolia**
1. Flowers in terminal or axillary corymbose panicles; corolla tube 1.5 cm across .... **N. tabacum**

Nicotiana plumbaginifolia Viv., Elench. Pl. 26. t. 1802; Hook.f., Fl. Brit. India 4: 246. 1883.

An erect viscid-pubescent, shallow-rooted, annual herb, up to 80 cm high. Leaves radical and cauline; basal ones form rosette, obovate-spathulate, lyrate, to 20 cm long, base cuneate to decurrent; cauline leaves elliptic, ovate-elliptic or lanceolate, clasping the stem. Flowers in terminal racemes; pedicel to 1 cm; calyx tubular, 10-lined with linear-lanceolate, acuminate unequal lobes; corolla salver-shaped, pinkish, 1 cm across; tube linear, 4-5 cm, lobes 8 cm, spreading; stamens 5, attached near the throat, at different levels; filaments to 2.5 mm, anthers oblong-cordate, 1.6 mm; ovary 3 mm, ellipsoid-globose, style 3 cm. Capsule 1 × 0.6 cm, ovate, glabrous, as long as the calyx; seeds minute, rugose.

Common weed of open habitats and in gardens in Hyderabad (Ramana, 2010) and in Warangal (C.S.Reddy & *al.*, 1999) districts.

Fl. & Fr.: February – May.

Native of Neotropics, naturalized elsewhere.

Nicotiana tabacum L., Sp. Pl. 180. 1753; C.B.Clarke in Hook.f., Fl. Brit. India 4: 245. 1883; Gamble, Fl. Madras 2: 941. 1923.

Vern.: *Pogaku.*

Erect glandular-pubescent subshrubs, up to 1.5 m high. Leaves simple, cauline, sessile, broadly elliptic-ovate, 30-50 × 20-40 cm, base cuneate, margin entire, apex acuminate or acute, lower leaves semiamplexicaul and decurrent, pubescent. Flowers axillary and terminal, corymbose panicles; peduncle up to 15 cm long; pedicels up to 1 cm long; calyx campanulate, lobes 5, lanceolate, 1 cm long; corolla rose, trumpet-shaped, 1.5 cm across, tube 5 cm, narrow below, ventricose above, lobes 5, shorter than tube, 0.7 cm long, acuminate; stamens 5, didynamous, the odd one shorter, inserted, attached to the throat at the same level, filaments up to 4 cm long, anthers oblong, 4 mm long, longitudinal dehiscence; ovary 2-celled. Capsule ellipsoid-globose, 2.5 × 1.5 cm, smooth; seeds minute, angular.

Cultivated occasionally.

Fl. & Fr.: Round the year.

Laxmapur (MBNR), *TDCK* 15338.

PHYSALIS L.

1. Calyx segments glandular hairy .... **P. priminosa**
1. Calyx segments otherwise:
    2. Prostrate herbs; internode cylindrical; corolla yellow, no colouration .... **P. minima**
    2. Erect herbs; internode ribbed; corolla yellow with colouration:
        3. Herbs glabrous .... **P. angulata**
        3. Herbs villous, viscid-pubescent .... **P. maxima**

Physalis angulata L., Sp. Pl. 183. 1753; Ganapathi, Curr. Sci. 57: 98. 1988.

Common weed of cultivated fields (C.S.Reddy & *al*., 1999) throughout Telangana.

Note: Sivarajan and Mathew (1996) consider *P. angulata* L. and *P. minima* L. to be conspecific. Ganapathi (1988) distinguished the two species based on morphology and breeding experiments.

C.S. Reddy & *al*. (1999) distinguished the two species of *Physalis* as follows.

Character	*P. angulata*	*P. minima*
Plant body	Erect, glabrous; stems at times, glabrescent	Prostrate, pubescent
Stem	Internode obviously ribbed	Internode cylindrical
Corolla colour	Yellow with pale purple maculations inside the throat	Yellow, no colouration
Corolla tube	> 6 mm in diam	<6mm in diam.
Anther	Bluish purple	Yellow
Fruiting calyx	Green, purple colouration present on vein network, glabrous	Green, no colouration on vein network, pubescent to glabrescent
Berry	Suborbicular, 0.6-0.8 cm in diam.	Oblong-ovoid, 0.8-1 cm in diam.
Seeds	D-shaped, 1.7 × 1.4 mm	Suborbicular-oblong, 2 × 1.8 mm

Physalis maxima Mill., Gard. Dict. ed. 8. Physalis no. 15. 1768; Gentry & Standley in Flora Guatemala 10: 90. 1974; Ganapathi & *al*., Cytologia 56: 285. 1991; Singh & Pandey, Indian J. Forest. 25: 187. 2002. *Physalis pruinosa* L., Sp. Pl. 184. 1753; Martinez, Taxon 42: 104. 1993.

Erect, stout annual herbs, villous-viscid pubescent all over; stems up to 1 m high, angular-ribbed, fistular. Leaves simple, spirally arranged, 4-9 × 3-6 cm, ovate, base oblique, rounded, margin sinuate-dentate above middle, apex acuminate, light green; petioles 4-10 cm. Flowers solitary, axillary, 1-1.5 cm across, pedicels erect, 1.5-7.0 cm; flowering calyx double the length of the tube, up to 6 mm, divided halfway down, subulate; corolla campanulate, pale yellow, 1 cm across, throat with 5, inconspicuous, pale green blotches; stamens 5, filaments 0.6 cm long, anthers up to 4 mm long, yellow; style terminal, stigma green. Fruiting pedicels drooping, extending up to 6 cm; fruiting calyx 2-2.5 cm long, downy, inflated, deeply sunken at base, strongly 5-angled, reticulately veined, glabrescent, acuminate. Fruits persistent, not falling, alternately borne from the first dichotomy. Berries grey-green, 1.5 cm across, sour; seeds many, obovoid, brownish and pitted.

Occasional, occurs in wastelands and road sides.

Fl. & Fr.: August – January.

Choutuppal (NLG), *CSR* 1771 (KU); Bhongir (NLG), *VSR* & *K.G.Rajarao* 4000(KU); Komuravelle (WGL), *CSR* 1972 (KU); Old Bowenpally, Hyderabad (HYD), *CSR* 2895 (KU).

Physalis minima L., Sp. Pl. 183. 1753; C.B.Clarke in Hook.f., Fl. Brit. India 4: 238. 1883; Gamble, Fl. Madras 2: 939. 1923.

Vern.: *Budama, Budda budama, Budda basara.*

Annual erect herb, up to 60 cm high, stem pubescent, puberulous or glabrous. Leaves simple, ovate, 3-9 × 1.5-5 cm, base obtuse, margin wavy or shallowly toothed, apex acute, pubescent or nearly glabrous; petiole up to 6 cm long. Flowers small, up to 6 cm long, axillary, solitary, pedicellate; calyx lobes 5, triangular, persistent, accrescent in fruit, fruiting calyx less than 2 cm long; corolla pale yellow with dark purple centre, campanulate-tubular, less than 0.6 cm across, plicate, corolla lobes 5, triangular-ovate; stamens 5, exserted, attached near the base of corolla, anthers oblong, 1.5 mm, shorter than the filaments, 2 mm long; ovary 1.2 mm, 2-celled, style 2 mm, linear, stigma obscurely 2-lobed. Berry globose, yellow, less than 0.8 cm across, overtopped by inflated calyx; seeds many, muricate.

Occasional weed in waste lands, road sides and in outskirts of forests in most districts.

Fl. & Fr.: July- January.

Surjapur (ADB), *TP* & *PVP* 4117; Mosra – Malkapur road (NZB), *TP* & *BR* 6096; Lingareddypalli (MDK), *CP* 11718; Kusumasamudram (RR), *NV* 10354; Mothugudem (KMM), *NV* 15634; Bhupatipur (KRN), *GVS* 20197 (CAL, MH); Kazipet (WGL), *RKP* 105219; Rathamhutta hills (KMM), *RCS* 104237; Amangal to Shadnagar (MBNR), *SRS* 104578; Rekularam (NLG), *BVRS* 627 (NU).

Physalis pruinosa L., Sp. Pl. 1: 184. 1753; Waterfall, Rhodoraa 60: 140. 1958; Rydberg, Torrey Bot. Cl. 4: 297-374. 1896.

Vern.: *Peddabuddagosi.*

Erect, viscid glandular-villous annual-perennial herbs; stems angular-ribbed, patent glandular-pubescent. Leaves large, 4-9 × 3-6 cm, ovate – elliptic, base cuneate, rounded or subcordate or unequal, margin variable, subentire to sinuate-dentate with prolongated acuminations, apex acute or acuminate, appressed hairy; petiole 3-7 cm long, patently glandular-pubescent. Pedicels 0.5-6 cm long, patently hairy; calyx segments 0.6 cm long, divided less than halfway down, glandular hairy, lobes ovate-triangular, acuminate; corolla campanulate-rotate, white, 1 cm long; anthers yellow, filaments 0.6 cm long. Fruiting calyx 5-angular, 2-2.5 cm long.

Occasional weed of road sides and fields of Komaravalli in Warangal district (C.S. Reddy, 2001).

Fl. & Fr.: June – February.

Komuravelli (WGL), *CSR* 1972 (KU).

SOLANUM L.

1. Plants unarmed:
    2. Climbers; leaf blade pinnatifid .... **S. seaforthianum**
    2. Erect herbs or shrubs; leaf blade entire:
        3. Annual herbs or shrubs; leaf margin bluntly toothed, leaves glabrous; flowers white; ripe berries black or red .... **S. americanum**
        3. Perennial shrubs; leaf margin entire; ripe berries red or yellow:
            4. Flowers purple; corolla rotate .... **S. pubescens**
            4. Flowers white; corolla campanulate .... **S. erianthum**
1. Plants armed:
    5. Lamina without prickles .... **S. rudepannum**
    5. Lamina with prickles:
        6. Climbers .... **S. trilobatum**
        6. Erect or diffuse (sub) shrubs:
            7. Diffuse subshrub; leaves lacerate; berry yellow or white with green blotches .... **S. virginianum**
            7. Erect shrubs; leaves sinuate; berry uniformly yellow, green, purple or orange:
                8. Berry yellow, green or purple:
                    9. Berry globose, <1.5 cm across; calyx lobes not enlarging in fruit; racemes many-flowered .... **S. anguivi**
                    9. Berry oblong, > 2cm across; calyx lobes enlarging in fruit; racemes 1 or 3-4-flowered .... **S. melongena**
                8. Berry orange .... **S. violaceum**

Solanum americanum Mill., Gard. Dict. (ed. 8) n. 5. 1768; Edmonds, Kew Bull. 27:103. 1972; Hepper in Dassan. & Fosb., Rev. Handb. Fl. Ceylon 6 : 368. 1987. *S. nigrum* auct. non. L. 1753; C.B.Clarke in Hook.f., Fl. Brit. India 4: 229. 1883; Gamble, Fl. Madras 2: 936. 1923.

Vern.: *Kasi, Kamanchi, Gajju chettu.*

Annual erect much branched unarmed herb, up to 1 m high. Leaves simple, ovate, 2-8 × 1-4 cm, base decurrent, margin dentate, apex acute, glabrous or sparsely pubescent; petiole up to 5 cm long. Flowers extra-axillary, umbellate cymes; peduncle up to 2 cm long; pedicels up to 7 mm long; sepals 5, oblong, pubescent or glabrous; corolla white, rotate, 6 mm across, 5-lobed; stamens 5, anthers up to 2 mm long, filaments 0.5 mm long; ovary 1.5 mm, style 2-3 mm long, pubescent below, stigma capitellate. Berry globose, black or red, glabrous; seeds many, ovate-reniform, minutely pitted.

Weed of waste lands, road sides and in cultivated fields, especially in moist places in all districts.

Fl. & Fr.: Round the year.

Bijjur RF (ADB), *GO* & *PVP* 4871,4879; Kammarpally (NZB), *TP* & *BR* 6197; Sirikonda RF (NZB), *BR* 9698; Mombojipally (MDK), *BR* & *CP* 11634; Laxmapur (MBNR), *TDCK* 13648; Narsapur (MDK), *KMS* 6782 (CAL); Kusumasamudram (RR), *NV* 10351; Venkat Reddipalli (RR), *MSM* & *KH* 10308; Mannanur (MBNR), *AJR* 20488; Rollapadu (KMM), *RR* 108073 (BSID); Aklaspur (KRN), *GVS* 22540 (MH).

Solanum anguivi Lam., Tabl. Encycl. 2: 23. 1794; Hepper, Bot. J. Linn. Soc. 76: 290. 1978. *S. indicum* auct. non L.; C.B.Clarke in Hook.f., Fl. Brit. India 4: 234. 1883; Gamble, Fl. Madras 2: 938. 1923.

Vern.: *Tella mulaka, Jiddu munaga chettu, Kakamunchi.*

Armed, diffuse, much branched undershrub, up to 1.5 m high, stems and branches with yellowish, compressed, often curved, short, sharp prickles. Leaves simple, variable in size, shape and pubescence, ovate, shallowly lobed, 4-8 × 3-6 cm, base oblique-truncate, margin wavy, apex subacute, stellate-tomentose beneath, usually less so above; midrib prickly; petiole up to 3 cm long, prickly. Flowers axillary, many-flowered racemes; peduncle up to 5 cm long; pedicels up to 1,5 cm long; calyx prickly, cupular, lobes 5, triangular; corolla blue, 1.5 cm, lobes 5; stamens 5, filaments pubescent, 1.4 cm long; stigma acute. Berry globose, orange or yellow when ripe; seeds minutely pitted.

Occasional in waste lands, open forests and hills above 800 m on degraded slopes.

Fl. & Fr.: August – December.

Madhavaram RF (ADB), *TP* & *GO* 5468; Puttapahad, Ananthasagar (RR), *NV* 10388; Mannanur (MBNR), *TDCK* 13612; Pakhal (WGL), *KMS* 13192 (CAL).

Solanum erianthum D. Don, Prodr. Fl. Nepal 96. 1825; Roe, Taxon 17: 176. 1968. *S. verbascifolium* auct. non L.; C.B.Clarke in Hook.f., Fl. Brit. India 4: 230. 1883; Gamble, Fl. Madras 2: 936. 1923.

Vern.: *Buddama, Vasaganti, Rasagadi maanu.*

Unarmed, large shrub, up to 5 m high, stems and branchlets woolly-tomentose. Leaves simple, elliptic-oblong, or ovate-lanceolate, 7.5-15 × 3.5-7.5 cm, base obliquely acute, margin entire, apex acute-acuminate, densely stellate-tomentose beneath and less so above; petiole up to 6 cm long. Flowers terminal and axillary, dichotomous corymbose panicles; peduncle up to 10 cm long; pedicels up to 0.5 cm long; calyx cupular, stellate-tomentose, 6 mm long, lobes 5, unequal, acute; corolla white, up to 1.5 cm across, lobes 5, stellate-pubescent, acute; stamens 5, filaments 2 mm, anthers oblong, 3 mm long; ovary conical, style 8 mm, stigma acute. Berry globose, stellate pubescent when young, yellow when ripe; seeds triangular or circular, minutely papillose, scaly.

Occasional in forests in Medak, Hyderabad and Mahabubnagar districts.

Fl. & Fr.: June-January.

Narsapur RF (MDK), *TP* & *CP* 12013.

Solanum melongena L., Sp. Pl. 186. 1753. var. **incanum** (L.) Kuntze, Revis. Gen. Pl. 454. 1891. *S. incanum* L., Sp. Pl. 188. 1753. *S. coagulans* Forssk., Fl. Aegypt.-Arab. 47. 1775; C.B. Clarke in Hook.f., Fl. Brit. India 4: 236. 1883.

Vern.: *Konda vankaya chettu.*

Armed erect shrub, up to 1 m high, branches stellate-pubescent, prickles stout, very sharp, recurved, yellow, shining above, usually tomentose at the broad base. Leaves simple, ovate-elliptic, ovate-oblong, angulate, 6-10 × 3-5 cm, base obliquely truncate, margin wavy or shallowly lobed, apex cuneate, prickly along the nerves, stellate hairy with velvety above, woolly beneath; petiole up to 3.5 cm long, prickly. Flowers axillary, in racemes; peduncle extra-axillary, paired, one bearing several male flowers in racemes, other bearing solitary, bisexual flower; calyx cup-shaped, stellate hairy, teeth triangular, acute; corolla blue, stellate hairy outside, up to 2 cm long, lobes ovate, acute, stellate hairy on both sides; stamens 5, anthers narrowly oblong, porous dehiscent, filaments very short, glabrous; ovary globose, hairy at the apex, style hairy below. Berry oblong, ovoid or subglobose, glabrous, yellow when ripe; seeds pale yellow, minutely pitted.

Occasional in outskirts of forests in Warangal and Karimnagar districts

Fl. & Fr.: January- May.

Irlapenta (MBNR), *BR* & *BSS* 32367; Appapur (MBNR), *BR* & *BSS* 30591; Pakhal RF (WGL), *KMS* 11666 (CAL, MH); Rechapalli (KRN), *GVS* 20166, 22279 (MH).

Solanum melongena L., Sp. Pl. 186. 1753. var. **insanum** (L.) Prain, Bengal Pl. 746. 1903; Gamble, Fl. Madras 2: 937. 1923.

Vern.: *Verrivankaya, Mulla vankaya, Siru vanga, Revali mulaga, Niruvanga, Nidupuvanga.* Eng.: Bitter apple.

Armed undershrub, up to 1 m high, stem prickly, prickles yellow. Leaves simple, angulate, 7-12 × 6-8 cm, base obliquely truncate, margin wavy or shallowly lobed, apex (sub) acute, stellate-pubescent, prickly along the nerves; petiole up to 4 cm, prickly. Flowers extra-axillary in racemes; pedicels up to 8 mm long; calyx lobes 5, lanceolate, stellate-pubescent, persistent; corolla purple to violet, lobes 5, triangular; stamens 5, filaments shorter than anthers, yellow; ovary globose, style villous. Berry oblong-globose, yellow, fruiting calyx enlarged; seeds minutely pitted.

Rare in waste lands, outskirts of forests, foot hills and by road sides.

Fl. & Fr.: January - May.

Rechapalli (KRN), *GVS* 20166, 22279 (MH); Mannanur (MBNR), *SRS* 109116 (BSID); Vatavarlapally to Mallelatheertham (MBNR), *SS* 108978 (BSID).

Solanum melongena L., Sp. Pl. 186. 1753; Hook.f. Fl. Brit. India 4 : 235. 1883; Gamble, Fl. Madras 2 : 937. 1923 var. **melongena.**

Vern.: *Vankaya.* English: Brinjal, Egg Plant.

Erect, branched, stellately tomentose, prickly undershrub, 60-100 cm high; prickles straight or slightly curved, 6 mm long. Leaves ovate-oblong, to 8 x 5 cm, base subequal, margins irregularly and shortly lobed, apex acute; petiole 2-10 cm long, often prickly. Flowers violet, 1-2 cm across; extra-axillary, solitary or few-flowered fascicles; calyx cup-shaped, irregularly split into 4-6 ovate-acute prickled, teeth; corolla lobes triangular to acute. Berry globose, yellow, 1.5-2 cm lng.

Cultivated for its fruit, which is used as vegetable.

Fl. & Fr.: October–March.

Solanum pubescens Willd., Phytographia 5. t. 3. 1794; C.B.Clarke in Hook.f., Fl. Brit. India 4: 230. 1883; Gamble, Fl. Madras 2: 936. 1923.

Vern.: *Ushti chettu, Kasivuste.*

Annual erect, unarmed shrub, up to 1.5 m high, younger parts pubescent, older parts glabrous. Leaves simple, ovate-cordate, 4-8 × 2-5 cm, base obliquely truncate to (sub)cordate, margin entire, apex acute, pubescent with yellowish hairs; petiole up to 5 cm long. Flowers axillary, in loose racemose cymes; peduncle up to 4 cm long; pedicel up to 2 cm long; sepals 5, 6-8 mm long, lanceolate, acuminate; corolla purple, up to 2.5 cm across, lobes 5, 1.2 cm long, oblanceolate, acuminate; stamens 5, filaments 2 mm long, anthers linear-oblong, 7-9 mm long; ovary 1.5 × 1 mm, style 1.2 cm long, stigma capitate. Berry globose, glabrous, red; seeds scaly.

Abundant weed of forests of Mahabubnagar and Nalgonda districts.

Fl. & Fr.: July – February.

Domalapenta (MBNR), *TDCK* 15400, 16827; Rangapur RF (MBNR), *AJR* 20480; Nagarjunasagar (NLG), *KMS* 9729 (MH); Kambalpally (NLG), *BVRS* 223 (NU).

Solanum rudepannum Dunal, Prodr. 13(1): 264. 1852. *S. torvum* Sw., Prodr. 47. 1788; C.B.Clarke in Hook.f., Fl. Brit. India 4: 234. 1883; Gamble, Fl. Madras 2: 937. 1923.

Vern.: *Bongeredu, Chitra kaya chettu, Konda vusti.*

Armed, erect undershrub, up to 2.5 m high, stem stellate-pubescent, prickles very few, short with stout base. Leaves simple, variable, ovate, shallowly lobed, 8-15 × 6-12 cm, base obliquely subcordate or truncate, margin sinuate, apex acute, sparsely stellate-pubescent above, densely so below, without prickles or with only one prickle beneath near the base of the midrib; petiole up to 6 cm long, tomentose, occasionally prickly. Flowers extra-axillary, many-flowered corymbose cymes; peduncle up to 5 cm long, pubescent like stem; pedicels up to 1 cm long; calyx unarmed, glabrous or pubescent, cupular, sepals 5, oblong, up to 5 mm, basally connate, acuminate; corolla white, 1 cm across, lobes 5, oblong, 1 cm long, acute; stamens 5, filaments 2 mm long, anthers 5.25 mm long; ovary 1 mm long, style 2.5 mm long, stigma obtuse. Berry globose, glabrous, yellowish when ripe, seated in persistent calyx; seeds sub-reniform, smooth, brown.

Frequent shrub in outskirts of forests and on hills in Hyderabad, Nalgonda and Mahabubnagar districts. Leaves variable in size and shape, probably depending on the season and soil.

Fl. & Fr.: August – February.

Katasintha (HYD), *KMS* 6001(CAL, MH); Mallepally (NLG), *BVRS* 531 (NU).

Solanum seaforthianum Andrews, Bot. Repos. 8: t. 504. 1808.

Eng.: Potato creeper.

Glabrous unarmed climber. Leaves pinnately lobed (divided almost to the base so that the lobes appear as leaflets), lobes ovate-obovate, 2-5 × 1-2 cm, base oblique, margin decurrent, apex acute to shortly acuminate, glabrous; petiole up to 4 cm long. Flowers in leaf-opposed or terminal pendulous panicles; peduncle up to 12 cm long; pedicels up to 1 cm long; calyx reduced to a circular rind, up to 3 × 2.5 mm, 5-toothed, acute; corolla violet, up to 1.5 cm across, lobes 5, 1 cm long, ovate, acute-apiculate; stamens 5, filaments 1-2 mm long, anthers oblong, 2.5 mm long; ovary conical, 0.5 mm long, style curved, 8 mm long, stigma capitate. Berry globose, bright red when ripe; seeds pubescent, pitted.

Occasional on hedges and hills above 800 m in Hyderabad district (Ramana, 2010). Native of South America. Introduced and often runs wild.

Fl. & Fr.: October –April.

Solanum virginianum L., Sp. Pl. 187. 1753. *S. xanthocarpum* Schrad & Wendl., Sert. Hanov. 1: 8. t. 2. 1795; C.B.Clarke in Hook.f., Fl. Brit. India 4: 236. 1883; Gamble, Fl. Madras 2: 938. 1923. *S. surattense* Burm. f., Fl. Ind.57. 1768; Venkatappa in Pullaiah & Moulali, Fl. Andhra Pradesh 2: 654. 1997.

Vern.: *Vankuda, Nelavakuda, Nelamunaka.*

Perennial, armed, diffuse herb or shrub. Leaves pinnatifid, lobes deep, many, ovate-elliptic, 4-9 × 2-4 cm, base attenuate, margin lacerate, apex acute, stellately pubescent; prickles yellow, sharp, straight, up to 2 cm long, numerous along the nerves and on stem and calyx; petiole up to 2.5 cm long. Flowers extra-axillary, 4-5-flowered cymes; peduncle up to 3 cm long; pedicels up to 1 cm long; sepals 5, almost free, obovate, 5.5 mm long, herbaceous, prickly, acuminate; corolla bluish-purple, 2.5 cm across, lobes 5, broadly ovate-triangular, 1.2 cm, acute; stamens 5, filaments 1-2 mm, anthers linear-oblong, 8 mm; ovary globose, 2 mm, style 0.9-1.3 cm, stigma acute, incurved. Berry globose, yellow when ripe; seeds smooth, circular.

Most striking weed in plains and hills throughout the year in waste places, road sides and after harvesting season in fields.

Fl. & Fr.: Round the year.

Sone (ADB), *TP* & *PVP* 4001; Kammarapally (NZB), *TP* & *BR* 6198; Mombojipally (MDK), *BR* & *CP* 11450; Puttapahad, Ananthasagar RF (RR), *NV* 10392; Bhupatipur (KRN), *GVS* 22296 (CAL, MH); Kunavaram RF (KMM), *RCS* 98779 (BSID); Rangapuram (WGL), *PVS* & *NRR* 76901 (BSID); Gaarlapaadu RF (MBNR), *SRS* 109100 (BSID); Nagarjunasagar (NLG), *KMS* 9748(MH); Durachipalli (NLG), *BVRS* 43 (NU).

WITHANIA Pauquy *nom. cons.*

Withania somnifera (L.) Dunal in DC., Prodr. 13: 453. 1852; C.B.Clarke in Hook.f., Fl. Brit. India 4: 239. 1883; Gamble, Fl. Madras 2: 940. 1923. *Physalis somnifera* L., Sp. Pl. 182. 1753.

Vern.: *Aswagandha, Pennerugadda.*

Erect branching undershrub, up to 1.5 m high, all plant parts with stellate tomentose hairs. Leaves simple, ovate, 3-9.5 × 2-6 cm, base truncate, margin entire, apex acute, pubescent; petiole up to 2 cm long. Flowers in axillary fascicles; pedicels up to 0.5 cm long; calyx campanulate, up to 3 mm long, lobes 5, lanceolate, acute, accrescent and subglobose in fruit; corolla greenish or lurid-yellow, campanulate, lobes 5, triangular-oblong, valvate, up to 0.6 cm across; stamens 5, exserted, anthers oblong, up to 1.5 mm long, filaments up to 2.5 mm long; ovary up to 2 mm; style up to 2 mm long, stigma capitate. Berry globose, enclosed in the inflated, accrescent calyx, 5-angled, red when ripe.

Occasional in cultivated fields, waste lands in Nalgonda district.

Fl. & Fr.: June – January.

Kambalpally (NLG), *BVRS* 262 (NU).

SCROPHULARIACEAE

1. Cultigens as garden ornamental .... **Russelia**
1. Wild plants:
    2. Stamens 2:
        3. Staminodes 2:
            4. Fruit septicidal capsule; anthers disjunct .... **Lindernia** *p.p.*
            4. Fruit loculicidal capsule; anthers parallel .... **Dopatrium**
        3. Staminodes 0 .... **Glossostigma**
    2. Stamens 4:
        5. Leaves all alternate .... **Verbascum**
        5. Leaves opposite, subopposite, alternate below, decussate above or decussate below, alternate above:
            6. Corolla distinctly bilabiate:
                7. Corolla with long slender abruptly incurved tube .... **Striga**
                7. Corolla without long slender abruptly incurved tube:
                    8. Calyx 5-partite, the segments very unequal outer 3 ovate-lanceolate, foliaceous, the inner 2 linear ..... **Mecardonia**
                        9. Stamens included within the corolla tube:
                            10. Anther cells congruous, not stipitate .... **Bacopa**
                            10. Anther cells stipitate:

11. Anthers unequal; stigma 2-lobed .... **Stemodia**

11. Anthers equal; stigma 2-lamellate ...... **Limnophila**

9. Anterior pair of stamens attached to the corolla-throat, posterior pair inserted within the corolla tube:

12. Calyx winged or keeled .... **Torenia**

12. Calyx not winged or keeled .... **Lindernia** ***p.p.***

6. Corolla not distinctly bilabiate:

13. Corolla 4-merous .... **Scoparia**

13. Corolla 5-merous:

14. Corolla-tube long slender:

15. Leaves pinnatifid .... **Sutera**

15. Leaves not pinnatifid .... **Buchnera**

14. Corolla-tube short, broad or ventricose above .... **Sopubia**

BACOPA Aubl. *nom. cons.*

1. Prostrate herbs, rooting at nodes; leaves obovate-oblong; capsule ovoid .... **B. monnieri**

1. Erect herbs; leaves linear-lanceolate; capsule globose .... **B. floribunda**

Bacopa floribunda (R.Br.) Wettst. in Engl. & Prantl, Nat. Pflanzenfam. 4(3b). 77. 1891. *Herpestis floribunda* R.Br., Prodr. 442. 1810; Hook.f., Fl. Brit. India 4: 273. 1884. *Monniera floribunda* (R.Br.) T.Cooke, Fl. Pres. Bombay 2: 286. 1905; Gamble, Fl. Madras 2: 953. 1924.

Small, erect, annual herb of 10-30 cm high, rooting at nodes, stem subquadrangular, purplish. Leaves subsessile, decussate, linear-lanceolate, 2-4 × 0.2-0.4 cm, narrowed at the base, margin entire, apex acute, glabrous, glandular punctuate beneath. Flowers solitary, axillary, 5-6 mm across, pedicellate; calyx lobes membranous; corolla white or pale violet blue, 12-15 mm long; stamens 4, didynamous. Capsule globose, 0.5-3 mm across, grooved, enclosed by persistent calyx lobes; seeds many.

Rare in wet places near the water sources in Hyderabad district.

Fl. & Fr.: August – December.

Bacopa monnieri (L.) Wettst., Nat. Pflanzenfam. 4(3b): 77. 1891. *Lysimachia monnieri* L., Cent. Pl. 2: 9. 1756. *Bacopa monnieri* (L.) Pennll, Proc. Acad. Nat. Sci. Philadelphia 98: 94. 1946. (*monniera*). *Monniera cuneifolia* Michaux, Fl. Bor. Am. 2: 22. 1803; Gamble, Fl. Madras 2: 953. 1924. *Herpestis monnieri* (L.) Kunth, Nov. Gen. Sp. 2: 366. 1818, *nom. illegit.*; Hook.f., Fl. Brit. India 4: 272. 1884.

Vern.: *Brahmi, Neeti sambrani, Sambrani chettu.*

Prostrate, annual herb of 5-30 cm long, rooting at nodes, glabrous; branches cylindrical, light pink tinged, succulent. Leaves decussate, sessile, semisucculent, obovate-oblong, 1-2 × 0.3-0.6 cm, base truncate, margin entire, apex obtuse, glabrous. Flowers solitary, axillary, pedicellate; calyx lobes 5, unequal, 0.6 × 0.2 cm, outer 2 ovate, obtuse, glabrous, the third one somewhat falcate, inner two lanceolate, acute, ciliate; corolla blue or white with purple veins, sub-bilabiate, 7-11 mm long, lobes 5, subequal, glabrous; stamens 4, didynamous; ovary oblong-globose, 2 mm long, 5.5 mm, stigma flat. Capsule ovoid, 2 or 4-valved; seeds many, oblong, testa striate.

Common in wet places near water sources in all districts.

Fl. & Fr.: Throughout the year.

Sone (ADB), *TP* & *PVP* 4047; Maddigunta (NZB), *BR* & *PSPB* 9563; Ramayampet (MDK), *BR* & *CP* 11503; Patancheru (MDK), *CP* 11686; Rusthempet (MDK), *KMS* 8002 (CAL); Mohammadabad RF (RR), *MSM* & *KH* 10433; Hyderabad (HYD), *KI* & *NVN* 18742; Yadagirigutta (NLG), *KI* & *NVN* 18745; Jakaram (WGL), *RKP* 105286 (BSID); Bollarum (NLG), *BVRS* 165 (NU); Kodimial (KRN), *GVS* 20109 (MH); Ravannapadu (KMM), *RR* 113857 (BSID).

BUCHNERA L.

Buchnera hispida Buch.-Ham. ex D. Don, Prodr. Fl. Nepal 91. 1825; Hook.f., Fl. Brit. India 4: 298. 1884; Gamble, Fl. Madras 2: 966. 1924.

Erect annual herb, 30-50 cm high. Leaves usually rosette of broad, shorter leaves, cauline leaves in distant pairs, obovate or oblanceolate, 2-6.2 × 0.4-1.5 cm, base narrowed, margin dentate or entire, apex obtuse, hispid. Flowers in spikes, axillary or terminal, sessile; calyx almost 10-ribbed, 5 mm long, hispid without, lobes 5; corolla blue or purple, with long slender tube, 4-6 mm long, lobes 5, obovate; stamens 4, adnate more or less to the middle of corolla-tube, didynamous, anther linear-apiculate. Capsule oblong, 4.5 x 2 mm, longitudinally striate, loculicidal; seeds minute, glabrous.

Rare in grassy places in moist forests of Medak and Mahabubnagar districts.

Fl. & Fr.: September – April.

Narsapur RF (MDK), *RG* 113469 (BSID); Nilgiri viewpoint (MBNR), *SRS* 107429 (BSID).

DOPATRIUM Buch.-Ham. ex Benth.

1. Capsule linear-oblong .... **D. nudicaule**
1. Capsule globose .... **D. junceum**

Dopatrium junceum (Roxb.) Buch.-Ham. ex Benth., Scroph. Ind. 31. 1835; Hook.f., Fl. Brit. India 4: 274.1884; Gamble, Fl. Madras 2: 954. 1924. *Gratiola juncea* Roxb., Pl. Coromandel t. 129. 1800 & Fl. Ind. 1: 142. 1832.

Erect, annual herb of 20-30 cm high, fleshy. Leaves decussate, upper leaves fewer and distant than the lower ones, sessile, oblong to ovate- oblong, 5-14 × 2-4 mm, larger below, margin entire, apex acute, glabrous. Flowers solitary, axillary, pedicellate, 4-5 mm across; calyx lobes 5, linear-lanceolate; corolla pink or bluish, bilabiate, caducous, 3.6 mm long; stamens 2, staminodes 2. Capsule globose, seeds obovoid.

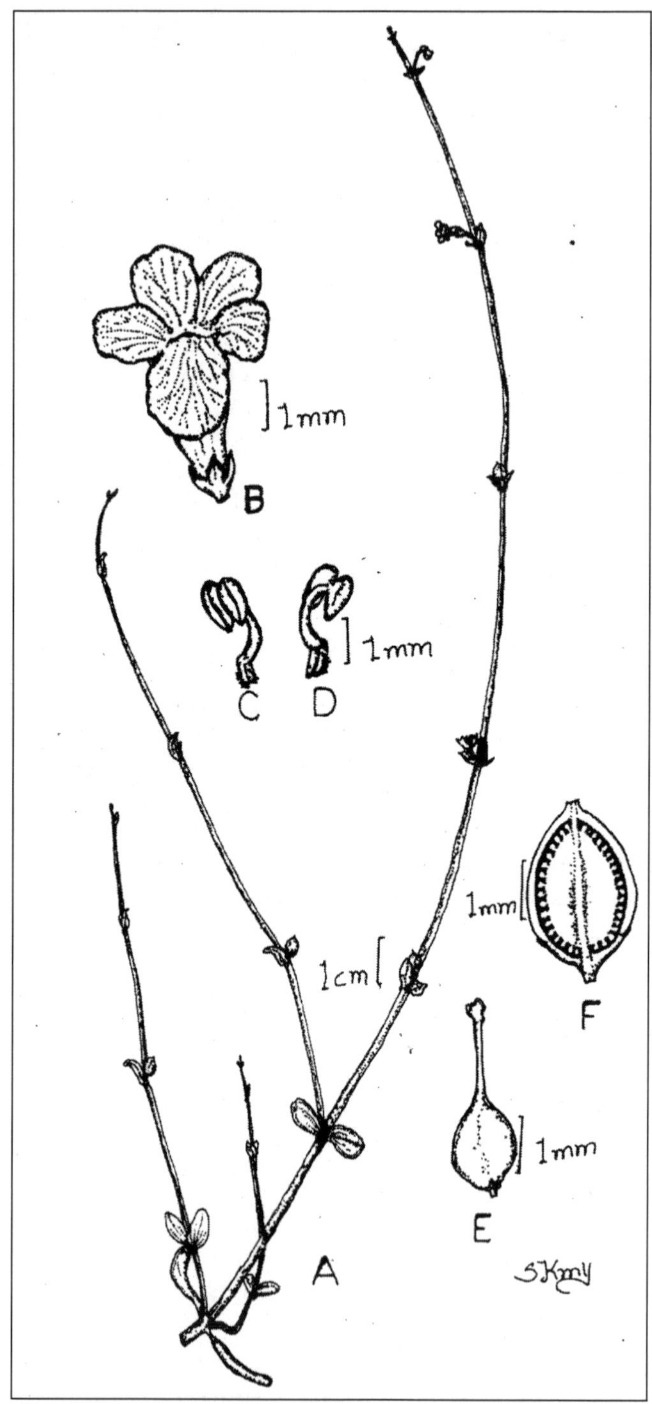

Figure 73: Dopatrium junceum (Roxb.) Buch.-Ham. ex Benth.
A. Twig, B. Flower-entire, C&D. Stamens, E. Pistil, F. Ovary l.s.

Occasional in moist and shady places in Nizamabad, Medak, Ranga Reddi, Warangal and Mahabubnagar districts.

Fl. Fr.: July- September.

Kunthala (ADB), *MHR* 14597; Nirmal – Adilabad (ADB), *MHR* 14450; Maddigunta (NZB), *BR* & *PSPB* 9563; Dosalagunda lake (MDK), *KMS* 6795 (CAL); Kothapalli RF (RR), *MSM* & *KH* 10564; Sagar left canal (KMM), *MHR* & *MCK* 16696; Pakhal forest (WGL), *KMS* 11697 (CAL).

Dopatrium nudicaule (Willd.) Benth., Scroph. Ind. 31. 1835; Hook.f., Fl. Brit. India 4: 274. 1884; Gamble, Fl. Madras 2: 954. 1924. *Gratiola nudicaulis* Willd., Ges. Naturf. Freunde Berlin Neue Schiften 4: 193. 1803.

Erect, annual herb of 15-25 cm high. Leaves chiefly crowded at base of stem with 2-3 distant pairs above, opposite, sessile, broadly oblong to ovate-oblong, 2-7 × 1-3.4 mm, base subamplexicaul, margin entire, apex obtuse or subacute, glabrous. Flowers solitary, axillary, pedicellate; calyx tube with lobes; corolla dull blue, throat minutely hirsute, 6-7 mm long; stamens 2, anther lobes unequal. Capsule linear-oblong; seeds obovoid, minute.

Occasional in moist places in Mahabubnagar district.

Fl. & Fr.: July – October.

GLOSSOSTIGMA Wight & Arn. *nom. cons.*

Glossostigma diandrum (L.) Kuntze, Revis. Gen. Pl. 2: 461. 1891. *Limosella diandra* L., Mant. 2: 252. 1771. *Glossostigma spathulatum* (Hook.) Arn., Nov. Act. Acad. Caeas. Leop.-Carol. German Nat. Cur. 18: 355. 1836; Hook.f., Fl. Brit. India 4: 288. 1884; Gamble, Fl. Madras 2: 964. 1924. *Microcarpaea spathulata* Hook., Bot. Misc. 2: 101. 1837.

Very small, creeping, annual herb of 2 cm long, rooting at nodes. Leaves opposite, petiolate, linear-spathulate, 0.2-0.7 × 0.1-0.24 cm, base narrow, margin crenate, apex obtuse, glabrous. Flowers solitary, axillary, pedicellate; calyx campanulate; corolla pinkish white, subequal, 1.7 mm long; stamens 2. Capsule minute, globose; seeds ellipsoid.

Occasional in moist places in Warangal and Hyderabad districts.

Fl. & Fr.: September – January

Pakhal (WGL), *ANH* 15908 (MH), *KMS* 11698 (MH).

LIMNOPHILA R. Br. *nom. cons.*

1. Bracteoles present; flowers pedicelled or subsessile .... **L. indica**
1. Bracteoles absent; flowers sessile .... **L. heterophylla**

Limnophila heterophylla (Roxb.) Benth., Scroph. Ind. 25. 1835; Hook.f., Fl. Brit. India 4: 270. 1884; Gamble, Fl. Madras 2: 952. 1923. *Columnea heterophylla* Roxb., Fl. Ind. 3: 97. 1832. *Limnophila roxburghii* G. Don, Gen. Syst. 4: 543. 1837. excl. descr. *p.p. L. roxburghii* G. Don var. *reflexa* (Benth.) Hook.f., Fl. Brit. India 4: 270. 1884.

Erect or decumbent annual herb, to 20 cm long, stems glabrous. Submerged leaves capillary multifid, intermediate leaves 4-6 in a whorl, pinnatifid, upper leaves opposite, sessile, semi-amplexicaul or sometimes also alternate, oblong, margin serrulate, apex acute. Flowers sessile, solitary, axillary; calyx lobes ovate-lanceolate, 2 mm long; corolla purplish, bilabiate, 2 mm across, lobes 1 mm long; stamens 4, didynamous. Capsule broadly ellipsoid; seeds minute, brown, shiny, rugose.

Common in shallow stagnant water near streams in Nizamabad and Hyderabad districts (Ramana, 2010).

Fl. & Fr.: August-November.

Annaram RF (NZB), *BR* & *CPR* 7186.

Limnophila indica (L.) Druce, Rep. Bot. Exch. Club. Soc. Brit. Isles 3: 420. 1913; Philcox, Kew Bull. 24: 115. 1970. *Hottonia indica* L., Syst. Nat. ed. 10. 919. 1759. *Cyrilla aquatica* Roxb., Pl. Coromandel t. 189. 1798. *Limnophila gratioloides* R. Br., Prodr. Prodr. Nov. Holl. 442. 1810; Hook.f., Fl. Brit. India 4: 271. 1884; Gamble, Fl. Madras 2: 952. 1924. *L. racemosa* Benth., Scroph. Ind. 26. 1835. *L. aquatica* (Roxb.) Alston, Ann. Roy. Bot. Gard. Peradenia 11: 205. 1929.

Vern.: *Burada bokkena, Kuttra.*

Erect, annual herb, 10-20 cm high, stems glabrous, rooting at lower nodes. Lower leaves pinnately dissected, root like, 1-2 cm long, upper ones 3-fid, upper most often undivided, sessile, linear-spathulate or linear-lanceolate, 7-22 × 1.5-4 mm, base tapering, margin serrate, apex subacute, glabrous. Flowers solitary or sometimes in racemes, 2-3 mm across, axillary, pedicellate, calyx lobes linear-lanceolate; corolla pale yellow or pale purple, bilabiate, 3 mm long; stamens 4, didynamous; ovary 1.5 mm, externally 2-lobed, style 3 mm, stigma pilose. Capsule subovoid, 2-3 mm across, enclosed in persistent calyx; seeds oblong-cuneiform, minute rugose, brown.

Frequent along borders of tanks and ponds in the dry low lands.

Fl. & Fr.: November – February.

Jannaram (ADB), *TP* & *PVP* 4195; Adluryellareddy (NZB), *TP* & *BR* 9058; Mamidipally RF (NZB), *BR* & *PSPB* 9599; Malkapur – Nizamabad Road (NZB), *TP* & *BR* 6110; Narsapur tank (MDK), *TP* & *CP* 11997; Kondapur (RR), *MSM* & *KH* 10586; Pakhal (WGL), *KMS* 11609 (MH); Kodimial (KRN), *GVS* 21852 (MH).

LINDERNIA Allioni

Note: The generic alignment within the group *Lindernia* has long been a matter of discussion and controversy. Linnaeus treated *Lindernia* and *Vandellia*, both with perfect stamens as two distinct genera, though Bentham and many others preferred to unite them. *Ilysanthes* and *Bonnaya* are characterized by two perfect stamens and 2 staminodes but differ in their relative position. In *Ilysanthes*, the posterior pair is fertile and the anterior staminodal, whereas in *Bonnaya* they are in reverse order. Bentham (1876), Hooker (1884) and many others have preferred to keep them under *Ilysanthes*. Haines (1922) reduced the four genera into two but in a different manner. He included *Bonnaya* in *Vandelia* and

Ilysanthes in *Lindernia* thus mixed up taxa with four and two perfect stamens. He recognized these groups mainly based on major vein configurations in their leaves. *Vandelia* with penninerved leaves and *Lindernia* with basally veined leaves.

Neither staminal nor venation characters have provided adequate and reliable taxonomic criteria for circumscribing the different genera within this group. Pennell (1935, 1943) combined all these four taxa under the single generic name *Lindernia*, a natural assemblage characterized by the remarkably uniform corolla, curiously recurving anterior filaments and by similar septicidal dehiscence of capsule. Indian species of this genera have been revised by Sivarajan and Mathew (1983) who recognized 22 species.

1. Stamens 4:
    2. Flowering calyx cleft to less than half of its length:
        3. Leaf blade ovate to almost rhomboid or suborbicular; flowers solitary, axillary; capsules oblong -ovoid .... **L. crustacea**
        3. Leaf blade lanceolate to oblong-ovate; flowers 2-10 in umbel-like clusters or rarely solitary-axillary; capsules ellipsoid .... **L. molluginoides**
    2. Flowering calyx cleft almost to base .... **L. pusilla**
1. Stamens 2:
    4. Leaf margin conspicuously mucronate toothed:
        5. Leaves petiolate, margin acute-serrate; corolla 10-14 mm .... **L. ruellioides**
        5. Leaves sessile or short petiolate, margin aristate-serrate; corolla 7 mm .... **L. ciliata**
    4. Leaf margin without conspicuous mucronate teeth:
        6. Leaves pinnately nerved .... **L. antipoda**
        6. Leaves palmately 3-5-nerved:
            7. Corolla <5 mm long .... **L. parviflora**
            7. Corolla >1 cm long .... **L. hyssopioides**

Lindernia antipoda (L.) Alston in Trimen, Fl. Ceylon 6 (Suppl.): 214. 1931; Sivarajan & Mathew, J. Bombay Nat. Hist. Soc. 80: 133. 1983. *Ruellia antipoda* L., Sp. Pl. 635. 1753. *Gratiola grandiflora* Roxb., Pl. Coromandel t. 179. 1805. *G. veronicifolia* Retz., Observ. Bot. 4: 8. 1786. *Ilysanthes veronicifolia* (Retz.) Urban., Ber. Deutsch. Bot. Ges. 2: 436. 1884; Gamble, Fl. Pres.Madras 2: 959. 1924. *Bonnaya veronicifolia* (Retz.) Spreng., Syst. Veg. (ed. 16) 1: 42. 1824; Hook.f., Fl. Brit. India 4: 285. 1884.

Vern.: *Neella pundu.*

Prostrate or decumbent, annual herb, 10-25 cm long, rooting at lower nodes, stems obtusely quadrangular. Leaves opposite, subsessile or lower shortly petiolate, obovate, linear-oblanceolate, elliptic-linear, base narrowed, margin subentire to

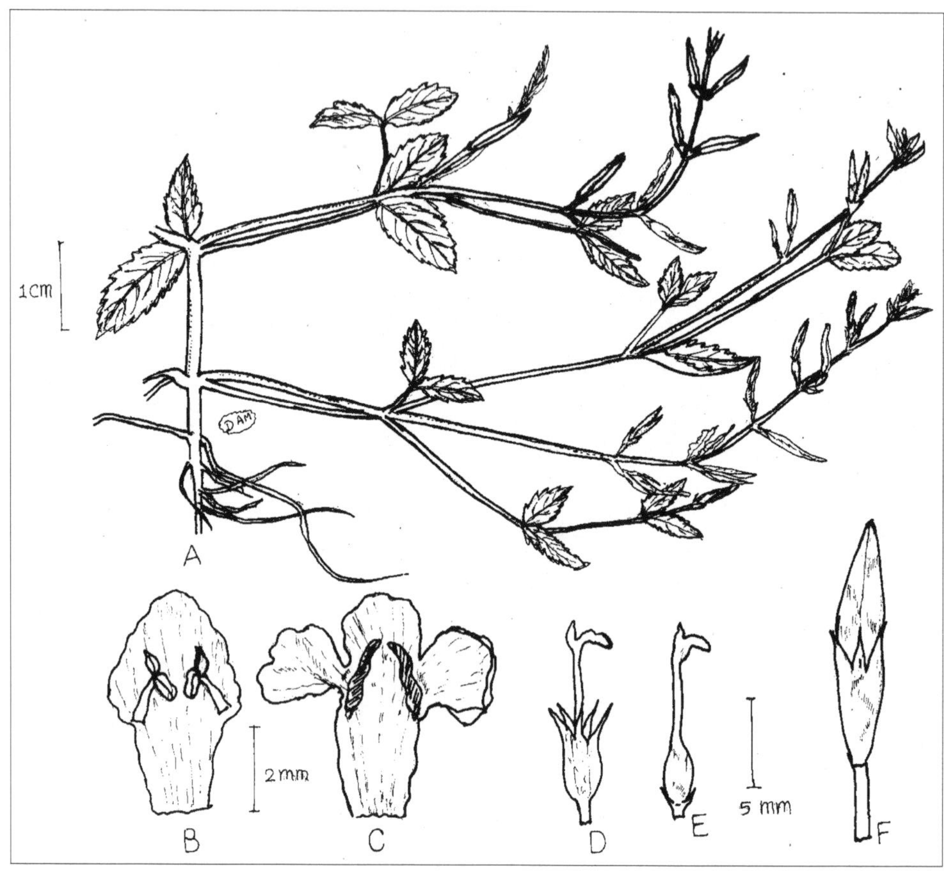

Figure 74: Lindernia antipoda (L.) Alston

A. Habit, B. Upper lip with stamens, C. Lower lip with staminodes, D. Calyx and pistil, E. Pistil, F. Capsule.

serrate, apex obtuse or rounded, glabrous. Flowers solitary, axillary, pedicellate; calyx lobes free, lanceolate, serrate; corolla bluish or purple, 2-fid, 0.8-1.2 cm long; stamens 2, staminodes 2, hooked, glabrescent, hooks convergent. Capsule cylindric, 1.5 cm long; seeds numerous.

Common in marshy places near water sources and paddy fields in Adilabad and Nizamabad districts.

Fl. & Fr.: November – February.

Ankusapuram (ADB), *GO* 4305; Bijjur (ADB), *GO* & *PVP* 4860; Alisagar (NZB), *BR* 7279; Saleswaram (MBNR), *BSS* & *SKB* 33340; Sagar left canal (KMM), *MHR* & *MCK* 17025; Eturnagaram (WGL), *RKP* 110835 (BSID).

Lindernia ciliata (Colsm.) Pennell, J. Arnold Arbor. 24(3): 253. 1943; Philcox, Kew Bull. 22: 51. 1968; Sivarajan & Mathew, J. Bombay Nat. Hist. Soc. 80: 133. 1983. *Gratiola ciliata* Colsm., Prodr. Descr. Gratiol. 14. 1793. *G. serrata* Roxb., Fl. Ind. 1: 140. 1820. *Ilysanthes serrata* (Roxb.) Urb. Ber. Deutsch. Bot. Ges. 2: 436. 1884; Gamble, Fl. Madras 2: 962. 1924. *Bonnaya brachiata* Link & Otto, Icon. Pl. Select. 25. t. 11. 1820; Hook.f., Fl. Brit. India 4: 284. 1884.

Erect, annual herb, 10-25 cm high, usually branched from base, stems quadrangular. Leaves opposite, sessile, oblong or oblong-lanceolate, 1-3 × 0.2-0.6 cm, base obtuse, margin recurved, serrate, serratures acute and bristle tipped, veins arched, apex obtuse, glabrous. Flowers in terminal racemes, pedicellate; calyx lanceolate, 5 mm long; corolla white with pink throat, 6-8 mm long; stamens 2, hooks of the staminodes divergent, filaments short, filiform, anthers divaricate. Capsule obovoid, not exceeding calyx lobes, 1 cm long; seeds numerous, minute, warted.

Common in moist places in Nizamabad, Medak, Karimnagar, Warangal and Mahabubngar districts.

Fl. & Fr.: August – November.

Bijjur (ADB), *GO* & *PVP* 4983; Ibrahimpet RF (NZB), *TP* & *BR* 6137; Narsapur RF (MDK), *TP* & *CP* 14024; Pillutla RF (MDK), *BR* & *CP* 11261; way to Pagirikutta (MDK), *KMS* 6660 (CAL); Aklaspur (KRN), *GVS* 25642 (MH); Jakaram (WGL), *RKP* 105272 (BSID).

Lindernia crustacea (L.) F. Muell., Syst. Census Australia Pl. 97. 1822; Philcox, Kew Bull. 22: 51. 1968; Sivarajan & Mathew, J. Bombay Nat. Hist. Soc. 80: 133. 1983. *Capraria crustacea* L., Mant. Pl. 87. 1767. *Vandellia crustacea* (L.) Benth., Scroph. Ind. 35. 1835; Hook.f., Fl. Brit. India 4: 279. 1884; Gamble, Fl. Madras 2: 959. 1924.

Vern.: *Verla kammi.*

Erect, annual herb, 10-35 cm high, rooting at basal nodes, stems sparsely hairy, ribbed. Leaves opposite, petiole 6 mm long, ovate, 0.8-1 × 0.3–0.7 cm, base truncate, margin crenate-serrate, apex acute, glabrous. Flowers solitary, axillary, pedicellate; calyx 4 mm long, tube cleft to middle, lobes ovate or linear; corolla purplish-blue, bifid, 4-7 mm long; stamens 4, filaments flattened. Capsule obovoid, 4 mm long, covered by calyx, seeds numerous, minute.

Fairly common in moist places in Adilabad, Nizamabad, Medak, Karimnagar and Mahabubnagar districts.

Fl. & Fr.: July- November.

Bijjur (ADB), *GO* & *PVP* 4982; Jaipur RF (ADB), *GO* & *PVP* 4952; Mudheli RF (NZB), *BR* & *CPR* 7211; Gangapur (MDK), *TP* & *CP* 12078; Aklaspur (KRN), *GVS* 25635 (MH).

Lindernia hyssopioides (L.) Haines, Bot. Bihar & Orissa 2: 635. 1922; Sivarajan & Mathew, J. Bombay Nat. Hist. Soc. 80: 136. 1983. *Gratiola hyssopioides* L., Mant. Pl. 174. 1771. *Ilysanthes hyssopioides* (L.) Benth. in DC., Prodr. 10: 419. 1846; Hook.f., Fl. Brit. India 4: 283. 1884; Gamble, Fl. Madras 2: 961. 1924.

Vern.: *Papata janumu.*

Straggling, slender, annual herb, 20-47 cm high, stem angled, striate. Leaves opposite, sessile, oblong-lanceolate, 4-8 × 1.5-2.5 mm, base obtuse, or narrowed, margin entire, sometimes one tooth on either side below the apex, apex obtuse, glabrous. Flowers solitary, axillary, pedicellate, to 1.8 cm long; calyx lobes divided to base, linear; corolla pale blue, bifid, 14 mm long; stamens 2, staminodes 2, disc annular. Capsule obliquely ovoid, almost twice as long as calyx; seeds oblong.

Common along the margin of small ponds in Adilabad and Nizamabad districts.

Fl & Fr.: September - February.

Kadam dam (ADB), *GO* & *DAM* 5404; Jamcheruvu (ADB), *MHR* 14589; Mingaram (NZB), *TP* & *BR* 6308; Armoor (NZB), *BR* 7292; Mohammadabad (RR), *MSM* & *KH* 10558; Kusumthum cheru (MDK), *RG* 104122 (BSID).

Lindernia molluginoides (Benth.) Wettst. in Natur. Pflanzenfam. 4 (3b): 80. 1891. *Vandellia molluginoides* Benth., Scroph. Ind. 35. 1835.

Prostrate or erect herb with ascending branches, rooting at basal nodes; stems 4-angloed. Leaves lanceolate to oblong-ovate, petiolate becoming sessile towards the tips; lanceolate to oblong-ovate, margins coarsely toothed with rounded or pointed teeth, ciliate. Flowers white or pink, 2-10 in axillary umbel like clusters, rarely solitary; bracts leaf-like, becoming smaller above; sepals tubular, united from above the middle, 3-4 mm long; petals white or pink, 4.5 × 4 mm; stamens 4, all with fertile anthers, filament appendages clavate. Capsule ellipsoid, 2.5 × 2 mm, shorter than the sepals, deflexed when ripe; seeds ovate, 0.3-0.5 × 0.25 mm, pitted.

Occasional along the streams and in moist places.

Fl. & Fr.: November – February.

Nirmal (ADB), *MHR* & *KI* 14481.

Lindernia parviflora (Roxb.) Haines, Bot. Bihar & Orissa 4: 635. 1922; Sivarajan & Mathew, J. Bombay Nat. Hist. Soc. 80: 135. 1983. *Gratiola parviflora* Roxb., Pl. Coromandel t. 203. 1811. *Ilysanthes parviflora* (Roxb.) Benth. in DC., Prodr. 10: 419. 1846; Hook.f., Fl. Brit. India 4: 283. 1884; Gamble, Fl. Madras 2: 961. 1924.

Aquatic, annual, erect herb, 15-35 cm high. Leaves opposite, sessile, oblong-elliptic, to 1.4 × 0.4 cm, base rounded, margin entire, apex acute, glabrous. Flowers solitary, axillary, pedicellate; calyx cleft to the base, lobes linear; corolla white or blue, 2-fid, 1 mm long; stamens 2, staminodes 2. Capsule ovoid-oblong; seeds many.

Common in wet places in Adilabad, Nizamabad, Ranga Reddi, Hyderabad, Nalgonda and Mahabubnagar districts.

Fl. & Fr.: September – March.

Bijjur (ADB), *GO* & *PVP* 4854; Narsapur (MDK), *BR* & *CP* 14114; Mohammadabad (RR), *MSM* & *KH* 10559, 10585; Rangammagudem (RR), *MSM* & *KH* 10573; Nagarjunasagar (NLG), *KMS* 9767 (CAL); Near Hussainsagar (HYD), *SNB* 8045(CAL).

Lindernia pusilla (Willd.) Bold., Zakfl. Java 165. 1916; Philcox, Kew Bull. 22: 41. 1968; Sivarajan & Mathew, J. Bombay Nat. Hist. Soc. 80: 139. 1983. *Gratiola pusilla* Willd., Sp. Pl. 1: 105. 1797. *Vandellia scabra* Benth., Scroph. Ind. 36. 1835; Hook.f., Fl. Brit. India 4: 281. 1884; Gamble, Fl. Madras 2: 959. 1924.

Prostrate, annual herb, up to 60 cm long, producing roots from the nodes. Leaves opposite, petiolate, ovate to orbicular-ovate, 0.5-1.7 × 0.5-1.5 cm, base truncate, margin crenate, apex obtuse, glabrous. Flowers solitary or in pairs, axillary, pedicellate; calyx long tube, lobes lanceolate, glandular; corolla white or pale pink, 8 mm long; stamens 4. Capsule ovoid, equaling the calyx; seeds many, oblong-cuneiform.

Common along borders of ponds or tanks in wet and dry low lands and hills in Mahabubnagar district.

Fl. & Fr.: July – November.

Lindernia ruellioides (Colsm.) Pennel, Brittonia 2: 182. 1936. *Gratiola ruellioides* Colsm., Prodr. Descr. Gratiola 12. 1793.

Verm.: *Suleema.*

Prostrate herbs, creeping and rooting at nodes. Leaves petioled, elliptic-oblong, up to 5 × 1.8 cm, base subamplexicaul or broadly cuneate, margin finely serrate, apex rounded. Flowers in axillary and terminal racemes; bracts linear or narrow – lanceolate; corolla light purple, 1.8 cm long, twice as long as the calyx or longer; stamens 2, anterior; stigma flattened. Capsule conical, twice as long as persistent calyx; seeds brown, ellipsoid.

Occasional along streams of forests in Sukkumamidi in Khammam district (C.S. Reddy & *al.*, 2001).

Fl. & Fr.: May – November.

MECARDONIA Ruiz & Pavon

Mecardonia procumbens (Mill.) Small, Fl. S. E. U.S. 1065. 1913. *Erinus procumbens* Mill., Gard. Dict. ed. 8. no. 6. 1768.

Perennial, procumbent herb, 20-30 cm long, rooting at lower nodes. Leaves opposite, petiolate, ovate or elliptic, 7.5-15 × 5-10 mm, base obtuse, margin crenate or serrate, apex obtuse, glabrous. Flowers solitary, axillary, pedicellate; calyx partite, lobes very uneven, foliaceous; corolla yellow, subequally 4-lobed, 7-8 mm long; stamens 4, didynamous. Capsule oblong or ellipsoid; seeds minute, numerous.

Occasional in river beds and sandy moist places (C.S.Reddy, 2001).

Fl. & Fr.: October – March.

Eturnagaram (WGL), *CSR* 1080 (KU).

RUSSELIA Jacq.

Russelia equisetiformis Schlecht & Cham., Linnaea 6 : 377. 1831; Santapau, J. Bombay Nat. Hist. Soc. 49 : 47. 1950. *R. juncea* Zucc., Flora 15(2) : 99. 1832.

Bushy, glabrous shrub, 1-2 m high; stem striate, much-branched, shining, drooping. Leaves linear – lanceolate but often reduced to minute scales. Flowers in lax, axillary and terminal, long peduncled cymes; peduncles reaching 10 cm long; pedicels slender, 2-3 cm long, often pendent; bracts minute; calyx campanulate, 2-3 mm long, deeply divided into lanceolate acute teeth, glabrous; corolla scarlet, tubular, 2-2.5 cm long, glabrous; lobes 5, subequal, rounded. Capsules ovoid-globose, 2-3 mm long, glabrous; seeds many, minute, black.

Grown in gardens, house yards as an ornamental for its showy flowers.

Fl. & Fr.: September-January.

SCOPARIA L.

Scoparia dulcis L., Sp. Pl. 116. 1753; Hook.f., Fl. Brit. India 4: 289. 1884; Gamble, Fl. Madras 2: 964. 1924.

Vern.: *Potti boli, Dakshani, Gurraputhulasi*. Eng.: Sweet broom-weed.

Annual, erect herb, 20-60 cm high. Leaves whorled or decussate, pedicellate, obovate-oblong to oblanceolate, 1.4-3.5 × 0.7-1.5 cm, base tapering, margin serrate, apex subacute, glabrous. Flowers solitary, axillary, pedicellate; calyx 4-lobed, base divided; corolla white, lobes spathulate, 3-4.5 cm long; stamens 4, didynamous. Capsule ellipsoid, globose, seeds angled.

In waste lands and fallow fields and along the sides of railway lines in Nizamabad, Medak, Khammam and Mahabubnagar districts.

Fl. & Fr.: July – December.

Nizamabad (NZB), *BR* & *PSPB* 9623; Pillutla RF (MDK), *BR* & *CP* 14114; Musalivera beat, Ramavaram range (KMM), *RR* 112622 (BSID); Kinnerasani dam area (KMM), *RR* 112682 (BSID).

SOPUBIA Buch.-Ham. ex D.Don

Sopubia delphinifolia (L.) G.Don, Gen. Hist. 4: 560. 1837; Hook.f., Fl. Brit. India 4: 302. 1884; Gamble, Fl. Madras 2: 970. 1924. *Gerardia delphinifolia* L., Cent. Pl. 2: 21. 1756.

Vern.: *Edintajada.*

Perennial, erect herb, 40-60 cm high. Leaves pinnatifid below, upper ones opposite, linear, filiform, 1.5-2.2 cm long, margin entire, apex acute. Flowers axillary, solitary, become racemose above, pedicellate; calyx tube 10-ribbed, glabrous; corolla purple with violet streaks, 8-16 mm long; stamens 4. Capsule oblong-ellipsoid; seeds oblong, cuneate.

Ocasional in open places along the borders of fields, road sides in all districts.

Fl. & r.: September–December.

Bhimnagar (NZB), *TP* & *BR* 6221; Mingaram (NZB), *BR* 9285; Poosanpalle (MDK), *TP* & *CP* 11957; Madikonakutta (MDK), *KMS* 6724 (CAL); Kanmankalva (RR), *MSM* 10456; Aklaspur (KRN), *GVS* 25654; Pakhal (WGL), *KMS* 11677 (MH); Khammam to Dhaniyaigudem (KMM), *RR* 108510 (BSID); Kambalpally (NLG), *BVRS* 611(NU).

STEMODIA L. *nom. cons.*

Stemodia viscosa Roxb., Pl. Coromandel t. 163. 1802; Hook.f., Fl. Brit. India 4: 265. 1884; Gamble, Fl. Madras 2: 949. 1924.

Vern.: *Gunta kamini, Gunta-kakinamu, Bodasaramu.*

Perennial, aromatic, viscid-pubescent, erect herb, 30-55 cm high. Leaves opposite, sessile, oblong to auriculate, 2.5 × 1-1.5 cm, base auricular, margin serrate, apex acute, scabrid. Flower solitary, axillary, pedicellate; calyx lobes subequal, linear, lanceolate; corolla dull violet blue, 1.4 cm long; stamens 4, didynamous. Capsule oblong-globose; seeds numerous, oblong.

Occasional in rice fields and moist places in most districts.

Fl. & Fr.: October – January.

Sone (ADB), *TP* & *PVP* 4024; Sarvapur (NZB), *TP* & *BR* 6371; Narsapur tank (MDK), *TP* & *CP* 12009; Medak (MDK), *BR* & *CP* 11422; Rusthempet (MDK), *KMS* 8007 (CAL); Aklaspur (KRN), *GVS* 22520 (MH); Eturnagaram (WGL), *RKP* 110836 (BSID); Parnasala RF (KMM), *RCS* 98769 (BSID); Krishna river bank (NLG), *KMS* 9817 (CAL).

STRIGA Lour.

1. Stem reddish; flowers pink; leaves scale like .... **S. gesnerioides**
1. Stem not reddish; flowers white or cream; leaves developed:
    2. Calyx 5-ribbed .... **S. densiflora**
    2. Calyx 10-15-ribbed:
        3. Calyx lobes linear lanceolate:
            4. Flowers white .... **S. angustifolia**
            4. Flowers yellow .... **S. sulphurea**
        3. Calyx lobes lanceolate .... **S. asiatica**

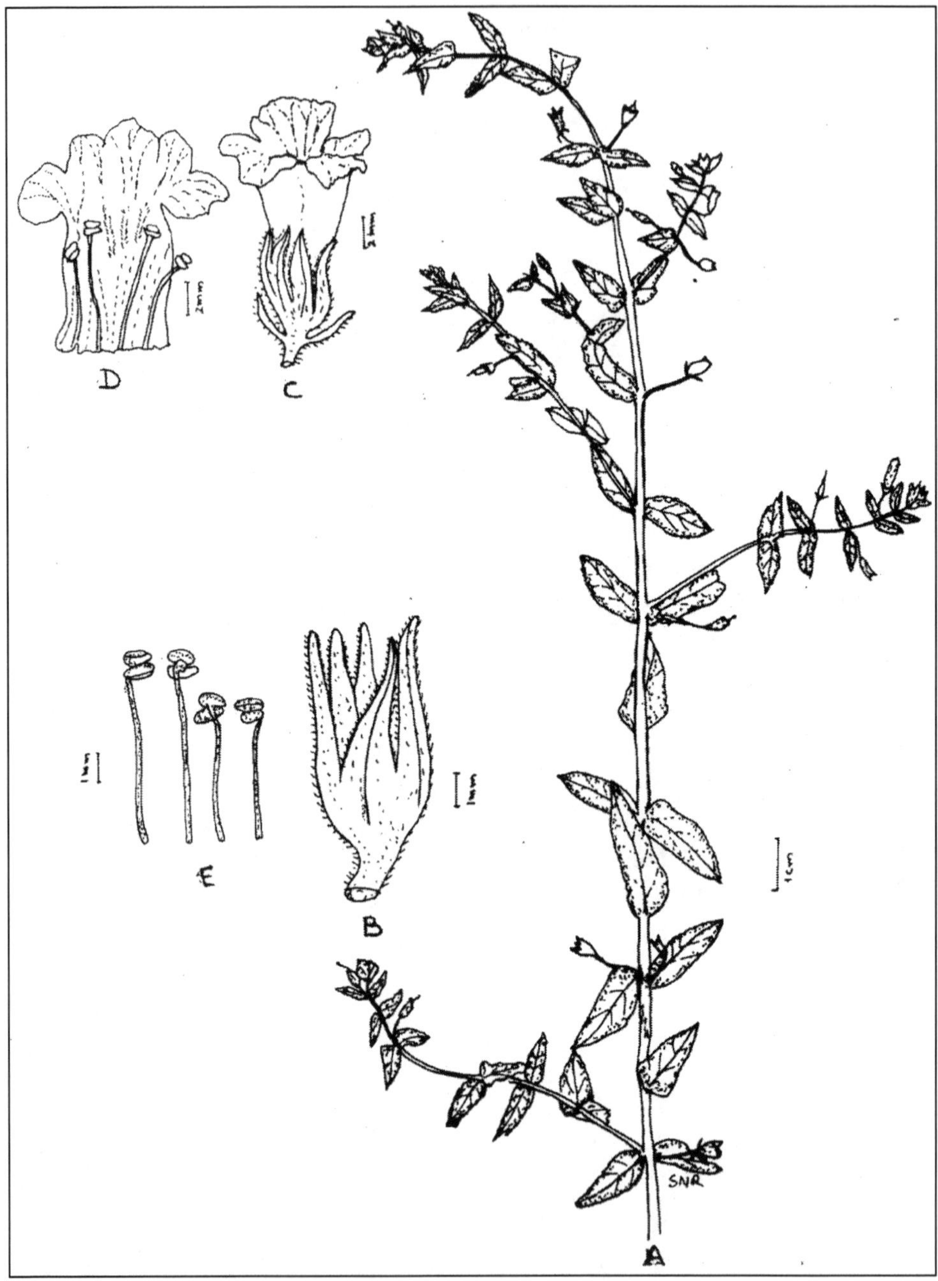

Figure 75: Stemodia viscosa Roxb.

A. Twig, B. Calyx, C. Flower, D. Corolla split open with stamens, E. Stamens.

Striga angustifolia (D.Don) C.J. Saldanha, Bull. Bot. Surv. India 5: 70. 1963. *Buchnera angustifolia* D. Don, Prodr. Fl. Nepal 91. 1825. *Striga euphrasioides* Benth. in Comp. Bot. Mag. 1: 364. 1835; Hook.f., Fl. Brit. India 4: 299. 1884; Gamble, Fl. Madras 2: 967. 1924.

Vern.: *Rati badanika.*

Annual, much branched, erect herb, 30-40 cm high, stem ribbed, strigose hairy. Leaves alternate, sessile, linear, 3-5 × 0.1-0.3 cm, base narrow, margin entire, apex acute, pubescent. Flowers solitary, axillary, pedicellate; calyx 15-ribbed, tube to 1 cm long, lobes linear-lanceolate; corolla white, tube pubescent, 1.5 cm long, lobes 4 (5); stamens 4, didynamous. Capsule obovoid or ovoid; seeds minute, cuneiform.

Common in open grassy places. A root parasite on grasses in all districts.

Fl. & Fr.: July – November.

Sone (ADB), *TP* & *PVP* 4021; Manchippa RF (NZB), *TP* & *BR* 6037; Mingaram RF (NZB), *BR* 9286; Pandivagu (MDK), *BR* & *CP* 11205; Mohammadabad (RR), *MSM* 10441; Jakaram (WGL), *RKP* 105275 (BSID); Mukunur (KRN), *PVS* 84030 (BSID); Rathamhutta hills (KMM), *RCS* 98793 (BSID).

Striga asiatica (L.) Kuntze, Revis. Gen. Pl. 466. 1891. *Buchnera asiatica* L., Sp. Pl. 630. 1753. *Striga lutea* Lour., Fl. Cochin. 22. 1790; Hook.f., Fl. Brit. India 4: 299. 1884; Gamble, Fl. Madras 2: 968. 1924.

Vern.: *Rati badanika.*

Annual, erect herb, 10-40 cm high. Leaves alternate, pedicellate, linear, 1-3 × 0.1-0.3 cm, scabrid, upper bractiform, base acute, margin entire, apex acute. Flowers solitary, axillary, sessile; calyx narrow, tubular, 10-15-ribbed; corolla white or yellow, 7 mm long; stamens 4, didynamous; ovary globose,2.5 mm long, style 7 mm long, stigma 2-lamellate. Capsule oblong, enclosed by calyx tube; seeds minute, oblong, rugose, brown.

Common in open grassy places in all districts. Parasite on Sorghum, bajra and other grasses

Fl. & Fr.: August – December.

Keslapur (ADB), *GO* & *PVP* 4559; Gundaram (NZB), *BR* & *KH* 9686; Nagulabanda (MDK), *CP* 11353; Pocharam tank (MDK), *TP* & *CP* 12032; Mohammadabad (RR), *MSM* & *KH* 10219; Pakhal (WGL), *KMS* 13144 (CAL); Kodimial (KRN), *GVS* 21865 (MH); Jalapenta (MBNR), *BSS* & *SKB* 32610; Koriguttalu (KMM), *RR* 106038 (BSID); Nagarjunakonda (NLG), *BVRS* 91 (NU).

Striga densiflora (Benth.) Benth., Comp. Bot. Mag. 1: 363. 1836; Hook.f., Fl. Brit. India 4: 299. 1884; Gamble, Fl. Madras 2: 967. 1924. *Buchnera densiflora* Benth., Scroph. Ind. 41. 1835.

Annual, erect herb, 12-45 cm high. Leaves opposite, sessile, linear, 1-3.2 × 0.2-0.4 cm, base narrow, margin entire, apex acute. Flowers solitary, axillary, sessile; calyx lobes ribbed, corolla white, tube slender, abruptly incurved, 1.4 cm long; stamens 4, didynamous. Capsule globose or ovoid; seeds oblong.

Occasional in most districts in grass lands and open places, in moist places and along road sides.

Fl. & Fr.: July-December.

Nagarjunakonda (NLG), *BVRS* 86 (NU).

Striga gesnerioides (Willd.) Vatke, Oest. Bot. Z. 25: 11. 1875. *Buchnera gesnerioides* Willd., Sp. Pl. 3: 338. 1800. *Striga orobanchoides* Benth., Comp. Bot. Mag. 1: 361. 1836; Hook.f., Fl. Brit. India 4: 299. 1884; Gamble, Fl. Madras 2: 967. 1924.

Annual, erect herb of 19-42 cm high, total root parasite on *Lepidagathis cristata;* branches reddish-pink. Leaves reduced to scales, to 0.5 × 0.3 mm, apex acute. Flowers in terminal spikes, pedicellate; calyx tube long, scabrid without, 10-ribbed, up to 4 mm long, lobes 5; corolla white or pale pink, 11-15 mm long, glabrous, lobes 5, unequal; stamens 4, didynamous, filaments 5 mm long; ovary oblong, 2.5 mm long, stigma capitate. Capsule oblong, ellipsoid; seeds minute, ovoid.

Occasional in waste lands in Nizamabad, Ranga Reddi, Hyderabad, Nalgonda and Mahabubnagar districts.

Fl. & Fr.: August – January.

Ibrahimpet (NZB), *TP* & *BR* 6126; Anantagiri RF (RR), *TP* & *MSM* 11085; Bottichalami (NLG), *KMS* 19168 (MH); Kambalpally (NLG), *BVRS* 218 (NU).

Striga sulphurea Dalzell ex Dalzell & A. Gibson, Bombay Fl. 182.1861; V.S.Raju, Indian Bot. Reprt. 5: 113. 1986.

An erect, robust, scaberulous herb, stems up to 30 cm high, quadrangular and sparingly branched. Leaves sessile, 2-3.5 × 0.5-1 cm, linear, acute, slender. Flowers axillary, few, subsessile, yellow, the corolla tube not exserted. Capsules less than 1 cm, oblong.

Occasional in paddy fields of Warangal district as a root parasite.

Fl. & Fr.: August – December.

Dharma Sagar (WGL), *VSR* 937 (MH).

SUTERA Roth

Sutera dissecta (Del.) Walp., Repert. Bot. Syst. 3: 271. 1844; Druce, Kew Bull. 1940. 1963-64. *Capraria dissecta* Del., Fl. Egypt. 95. t. 32. f.2. 1812. *Sutera glandulosa* Roth, Nov. Pl. Sp. 291. 1821; Hook.f., Fl. Brit. India 4: 258. 1884; Gamble, Fl. Madras 2: 946. 1924.

Annual, erect or spreading, branched herb, 10-30 cm long. Leaves opposite, ovate, 0.8-2 cm long, pinnatifid, segments toothed. Flowers axillary, solitary; calyx lobes equal, with glandular dots; corolla white or yellow-purple, tube exceeding the calyx, with small spreading limb; stamens 4, didynamous. Capsule ovoid-oblong; seeds minute, spherical or oblong.

Occasional in moist places in Adilabad, Nizamabad, Nalgonda, Karimnagar and Mahabubnagar districts.

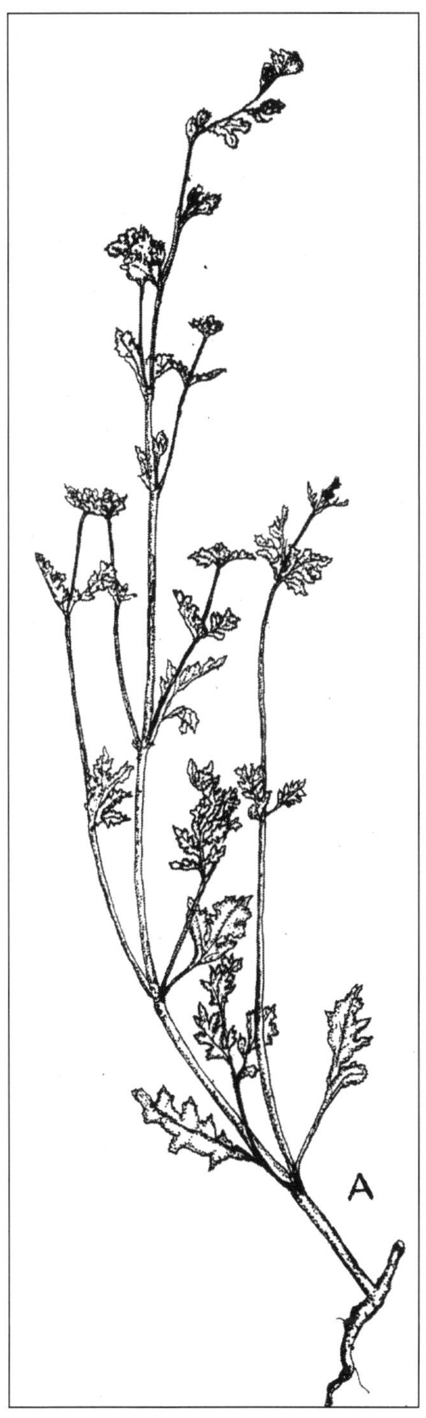

Figure 76: Sutera dissecta (Del.) Walp.

A. Habit

Fl. & Fr.: August – December.

Potchera (ADB), *GO* & *DAM* 5077; Kunthala fields (ADB), *GO* & *DAM* 5140; Adluryellareddy (NZB), *TP* & *BR* 6413; Kandakurthi (NZB), *BR* 7255; Krishna river bank (NLG), *KMS* 9800(CAL); Bhupatipur (KRN), *GVS* 22480(MH);

TORENIA L.

Torenia fournieri Linden ex E. Fourn., Ill. Hort. 23: 129, t. 249. 1876; Gamble, Fl. Madras 2: 957. 1921.

Annual herbs, up to 20 cm high; stems erect, quadrangular, simple or branched above middle, subglabrous; petiole 1-2 cm. Leaves oblong-ovate to ovate, 3-5 × 1.5-2.5 cm, subglabrous, margin coarsely serrate. Racemes often terminal; calyx ellipsoid, 1.3 -1.9 × 0.8 cm, 5-winged; corolla 2.5-4 cm, exceeding calyx by 1-2.3 cm; tube pale violet, upper side yellow; lower lip lobes purple-blue, middle lobe with a yellow patch near base, suborbicular, ca 10 × 8 mm, subequal; upper lip pale-blue, erect, broadly obovate, 1-1.2 × 1.2 -1.5 cm. Capsule narrowly ellipsoid, ca 12 × 0.5 mm; seeds yellow.

Often planted in gardens under shade of trees; runs wild in moist forests.

Fl. & Fr.: June – December.

Public Gardens, Hyderabad (HYD), *CSR* & *C.Pattnaik* 2487 (KU).

Note: It is an invasive species, native of Australia (C.S.Reddy & *al.*, 2008).

VERBASCUM L.

Verbascum chinense (L.) Santapau, Fl. Purandhar 90. 1958. *Scrophularia chinensis* L., Mant. 250. 1771. *Celsia coromandeliana* Vahl, Symb. Bot. 3: 79. 1794; Hook.f., Fl. Brit. India 4: 251. 1884; Gamble, Fl. Madras 2: 945. 1924.

Annual erect herb, 40-60 cm high. Leaves opposite, petiolate, ovate, 1.3-7 × 0.8-3 cm, base pinnatifid and become bract like, margin serrate, apex acute. Flowers in terminal racemes, pedicellate; calyx deeply lobed; corolla yellow, rotate; stamens 4, didynamous. Capsule septicidal, globose; seeds many, rugose.

Common in marshy places in all districts.

Fl.& Fr.: June – December.

Satnella (ADB), *GO* & *PVP* 4219; Sarvapur (NZB), *TP* & *BR* 6368; Pochammaralu (MDK), *BR* & *CP* 11489; Dharmapur (RR), *MSM* & *KH* 10287; Kodimial (KRN), *GVS* 20124 (MH); Way to Gallagant (KRN), *PVS* 84063 (BSID); Laxmipuram (WGL), *PVS* & *NRR* 76935 (BSID); Bhadrachalam (KMM), *RCS* 104319 (BSID); Bollarum (NLG), *BVRS* 168 (NU).

OROBANCHACEAE

OROBANCHE L.

Orobanche cernua Loefl., Iter. Hispan. 152. 1758; Hook.f., Fl. Brit. India 4: 325. 1884; Gamble, Fl. Madras 2: 978. 1924.

Vern.: *Pvgaku malle.*

Annual, erect herb, up to 40 cm high, total root parasite, stem or scape simple or branched. Leaves reduced to scales. Flowers in spikes or raceme, bracteate, sessile; calyx unequally 2-fid; corolla blue, lobes very short, orbicular, 1.2-2 cm long; stamens 4, didynamous. Capsule ellipsoid, 2-valved; seeds many, globose.

Abundant in tobacco fields, parasite on tobacco and brinjal in Adilabad and Nalgonda districts.

Fl. & Fr.: October – January.

Kadthala (ADB), *DAM & GO* 5102; Narayanpur (NLG), *BVRS* 448 (NU).

LENTIBULARIACEAE

UTRICULARIA L.

1. Aquatic, submerged floating herbs, foliar organs dissected into narrow segments; seeds prismatic or lenticular, winged:
    2. Primary foliar segments one or two .... **U. gibba**
    2. Primary foliar segments three or more:
        3. Raceme with a whorl of ellipsoid floats atleast at a short distance above base; spur not constricted; hairs on corolla with globose glands .... **U. stellaris**
        3. Racemes without floats or with fusiform floats at base; spur constricted at middle; hairs on corolla with cylindrical glands .... **U. aurea**
1. Terrestrial or epiphytic herbs; foliar organs entire; seeds not as above:
    4. Bracts medifixed .... **U. caerulea**
    4. Bracts basifixed:
        5. Racemes twining .... **U. scandens**
        5. Racemes erect:
            6. Flowers yellow .... **U. bifida**
            6. Flowers bluish purple .... **U. graminifolia**

Utricularia aurea Lour., Fl. Cochinch. 1: 26. 1790. *U. flexuosa* Vahl, Enum. Pl. 1: 198. 1804; Hook.f., Fl. Brit. India 4: 329. 1884; Gamble, Fl. Madras 2: 980. 1924.

Annual, aquatic, submerged, floating herb, 50 cm long, stolons often floating just below the water, filiform. Leaves numerous, conspicuous, 2-6 cm long, usually in whorls of 4-5, pinnately divided into numerous segments, traps near base of each pinnae. Bract present, bracteole absent. Flowers in raceme, pedicellate; calyx lobes 2, subequal, ovate, corolla yellow, bilabiate, 3-10 mm long, stamens 2, anther one-celled, spur 4 mm, slightly curved, facing downwards. Capsule globose, seeds numerous.

Rare, floating in ponds and lakes in Hyderabad (Rajagopal, 1973; Ramana, 2010).

Fl. & Fr.: November – February.

Osmania University campus (HYD), *TR* 448 (HH).

Utricularia bifida L., Sp. Pl. 18. 1753; Hook.f., Fl. Brit. India 4: 332. 1884; Gamble, Fl. Madras 2: 982. 1924.

Annual, erect, terrestrial herb, 7-15 cm high. Leaves scattered, narrowly linear, 1-1.5 x 0.1 cm, apex rounded, traps on vegetative organs. Bract and bracteoles present. Flowers in racemes, pedicellate; calyx lobes subequal, ovate; corolla yellow, bilabiate, 4-6 mm long; stamens 2, anthers 2-celled. Capsule broadly elliptic; seeds numerous.

Frequent in wet places along streams and near ponds in Medak amd Karimnagar districts.

Fl. & Fr.: July – December.

Rusthempet (MDK), *RG* 96299 (BSID).

Utricularia caerulea L., Sp. Pl. 18. 1753; Gamble, Fl. Madras 2: 983. 1924. *U. racemosa* Wall., ex Walp. in Meyen Observ. Bot. 19: 401. 1843; Hook.f., Fl. Brit. India 4: 333. 1884. *U. nivea* Vahl, Enum. Pl. 1: 203. 1804.

Annual, terrestrial, erect, delicate herb, 6-10 cm high. Leaves few, linear, up to 1.9 cm long, spathulate, rounded at apex, traps few. Bracts and bracteoles present. Flowers in racemes, pedicellate; calyx lobes 2, unequal, upper lobe 2.5 cm long, broadly ovate-acute, lower lobe slightly smaller than upper one; corolla deep purple, 5-8 mm long; stamens 2, spur narrowly conical, straight or curved. Capsule globose; seeds numerous, obliquely oblong.

Frequent in marshy places often among grasses along the ponds, rice fields and ditches in Hyderabad district (Rajagopal, 1973; Ramana, 2010).

Fl. & Fr.: July – December.

Gandipet (HYD), *TR* 512 (HH).

Utricularia gibba L., Sp. Pl. 18. 1753. *U. exoleta* R. Br., Prodr. Fl. Nov. Holl. 430. 1810; Hook.f., Fl. Brit. India 4: 329. 1884; Gamble, Fl. Madras 2: 980. 1924. *U. gibba* L., Sp. Pl. 18. 1753. subsp. *exoleta* (R.Br.) P. Taylor, Mitt. Bot. Stassamml. Munchen 4: 101. 1961.

Annual, small, aquatic, free floating herb, 20 cm long. Leaves numerous on the stolons, up to 1.5 cm long, dichotomously branched, capillary, terete, glabrous or very sparsely setose, traps numerous, lateral on the foliar segments. Bracts and bracteoles present. Flowers in raceme, pedicellate; calyx lobes subequal; corolla yellow, 4-8 mm long; stamens 2, spur narrowly cylindrical from base, straight. Capsule globose, seeds few, lenticular.

Common in ponds and still waters in Warangal and Hyderabad districts.

Fl. & Fr.: July – December.

Gandipet, Hyderabad (HYD), *TR* 834 (HH).

Utricularia graminifolia Vahl, Enum. Pl. 1: 195. 1804; Gamble, Fl. Madras 2: 981. 1924. *U.caerulea* auct. non L.; Hook.f., Fl. Brit. India 4: 331. 1884 *p.p.*

Annual, prostrate, terrestrial herb, 10-15 cm long. Leaves at the base of scape and stolons, linear, 1 × 1.5 cm, somewhat transparent, 3-nerved, base cuneate, apex obtuse, traps either at base or along the rhizoid at random. Bracts and bracteoles present. Flowers in racemes, terminal, pedicellate; calyx lobes 2, unequal; corolla bluish purple, 7-10 mm long, upper lip larger than the lower; stamens 2, filaments adnate to the base of the corolla tube, anther 2-celled, spur 3-5 mm, acute, curved upwards. Capsule globose; seeds few, globose.

Rare in Karimnagar and Mahabubnagar districts in post monsoon ponds and rice fields.

Fl. & Fr.: Throughout the year.

Mahabubnagar, Jatprole (MBNR), *SRR* 1297 (HH).

Utricularia scandens Benj., Linnaea 20: 309. 1847; Hook.f., Fl. Brit. India 4: 332. 1884; Gamble, Fl. Madras 2: 982. 1924. *U.wallichiana* Wight, Icon. Pl. Ind. Orient. t. 1572. 1850. *U. wallichii* Wight, Hook. J. Bot. Kew Gard. Misc. 1: 372. 1850.

Terrestrial, annual herb, 10-15 cm long. Leaves linear, to 6 x 2 mm, cuneate, obtuse, traps few on along the rhizoids. Bract and bracteoles present. Flowers in raceme, terminal, pedicellate; calyx lobes 2, equal, obovate; corolla yellow, 5-10 mm long; stamens 2, spur 4 mm long, subulate. Capsule globose, 2.5 mm across, enclosed by persistent calyx; seeds minute, light brown, obovate, longitudinally reticulate.

Common in marshy places along streams and ponds in Hyderabad district.

Fl. & Fr.: July – December.

Utricularia stellaris L.f., Suppl. Pl. 86. 1782; Hook.f., Fl. Brit. India 4: 328. 1884; Gamble, Fl. Madras 2: 980. 1924. *Utricularia inflexa* Forssk. var. *stellaris* (L.f.) P. Taylor, Mitt. Bot. Stassamml Munchen 4: 96. 1961.

Aquatic, free floating, annual herb, 10-15 cm long. Leaves verticellate or whorled, each pinnately divided and dichotomously forked, traps at the base of filiform segments. Scales and bracteoles absent. Flowers in terminal racemes, pedicellate; calyx lobes subequal; corolla yellow, marked purple; stamens 2, anther cells indistinct, spur 4 mm, straight. Capsule globose; seeds few, winged.

Occasionally floating in ponds and still water in Warangal, Nizamabad, Medak, Karimnagar, Ranga Reddi and Hyderabad districts.

Fl. & Fr.: October – January.

Mosra (NZB), *TP* & *BR* 6102; Laxmapur (NZB), *TP* & *BR* 6282; Pocharam tank (MDK), *TP* & *CP* 12027; Sardhana (MDK), *CP* 11770; Malkalcheruvu (RR), *MSM* & *KH* 10566; Pakhal lake (WGL), *ANH* 15955 (MH).

BIGNONIACEAE

1. Plants climbing by means of tendris .... **Pyrostegia**
1. Plants erects shrubs or trees:
   2. Fruit indehiscent, fleshy berry .... **Kigelia**
   2. Fruit dehiscent, loculicidal capsule:
      3. Leaves pinnately compound:
         4. Fruit usually cylindrical, often subterete, 4-angled .. **Stereospermum**
         4. Fruit compressed, rarely subcylindrical:
            5. Dehiscence septicidal, dehiscing through both the sutures:
               6. Capsule very broad and large, valves smooth, woody; seeds thinly discoid .... **Oroxylum**
               6. Capsule linear, valves thick, not so woody; seeds not discoid .... **Millingtonia**
            5. Dehiscence loculicidal (or) apparently septicidal, rarely through the sutures only:
               7. Capsule 4-locular due to the presence of pseudo-septum:
                  8. Valves glabrous .... **Dolichandrone**
                  8. Valves pubescent .... **Heterophragma**
               7. Capsule 2-locular, pseudo-septum absent:
                  9. Fruit long, linear, valves not woody .... **Tecoma**
                  9. Fruit broad, valves woody .... **Jacaranda**
      3. Leaves palmately compound.. **Tabebuia**

DOLICHANDRONE (Fenzl.) Seem. *nom. cons.*

1. Leaflets <4 cm long; capsule compressed, without speckles .... **D. falcata**
1. Leaflets >5 cm long; capsule subterete with white speckles:
   2. Calyx spathe glabrous .... **D. atrovirens**
   2. Calyx spathe pubescent .... **D. arcuata**

Dolichandrone arcuata (Wight) C.B. Clarke in Hook.f., Fl. Brit. India 4: 380. 1884; Gamble, Fl. Madras 2: 996. 1924. *Spathodea arcuata* Wight, Icon. Pl. Ind. Oient. t. 1340. 1848.

Small, deciduous tree, 10-12 m high. Leaves opposite, pinnate, 15-30 cm long, leaflets 5 or 7, ovate, 5-10 × 3-5 cm, base obtuse, margin entire, apex acute, glabrous. Flowers in terminal racemes, pedicellate; calyx spathaceous; corolla white, 2-lipped, 8 mm long; stamens 4, didynamous. Capsule long, subterete or compressed; seeds minute, many.

Common in dry deciduous forests of Nallamalais.

Fl.: November – February. Fr.: Throughout the year.

Appapur (MBNR), *SSR* 15538; Borapur to Appapur (MBNR), *BR, SKB* & *BSS* 32393.

Dolichandrone atrovirens (Roth) K.Schum., Nat. Pflanzenfam. 4 (3a): 240. 1894; Gamble, Fl. Madras 2: 996.1924. *Bignonia atrovirens* Heyne ex Roth, Nov. Pl. Sp. 284. 1821. *Dolichandrone crispa* Seem., J. Bot. 8: 381. 1870; C.B. Clarke in Hook.f., Fl. Brit. India 4: 379. 1884.

Vern.: *Vadi, Oddi, Niruoddi, Vanka nirodi, Niruddi.*

Small, deciduous tree, 5-6 m high. Leaves opposite, imparipinnate, 15-20 cm long, leaflets 5-7, elliptic, 6-10 × 2.5-6 cm, base obliquely truncate, margin entire, apex acute, glabrous. Flowers in axillary corymbose racemes, pedicellate, calyx spathe acute with a recurved point, stamens 4, didynamous. Capsule subterete, seeds winged at both sides.

Common in dry deciduous forests of Adilabad, Medak and Ranga Reddi districts.

Fl. & Fr.: May – June.

Narsapur RF (MDK), *TP* & *CP* 11985; Rangammagudem RF (RR), *MSM* & *KH* 10316.

Dolichandrone falcata (Wall., ex DC.) Seem., J. Bot. 8: 381. 1870 var. **falcata**; C.B. Clarke in Hook.f., Fl. Brit. India 4: 380. 1884; Gamble, Fl. Madras 2: 996. 1924. *Spathodea falcata* Wall. ex DC., Prodr. 9: 206. 1845.

Vern.: *Chitti-niruddi, Chinna oddi, Oddi.*

Deciduous tree, 5-10 m high, bark smooth, light grey-brownish, exfoliating in irregular scales, young parts and leaves covered with grey stellate hairs. Leaves opposite, pinnate, 5 to 6 cm long, leaflets 7, obovate or obcordate, 2-3.5 × 1.8-2.5 cm, base rounded or cuneate, margin entire, apex acute, glabrous, end leaflets larger than the side leaflets, petiolules very short. Flowers in corymb, axillary, pedicellate; calyx spathulate; corolla white, 2-lipped, 3 cm long; stamens 4, didynamous, anthers oblong. Capsule flat, falcately curved, seeds with membranous wings.

Common in dry localities, often on rocky slopes in all districts.

Fl. & Fr.: May – June.

Potchera (ADB), *GO* & *DAM* 5093; Mothe RF (NZB), *BR* & *GO* 9066; Gangapur (MDK), *BR* & *CP* 11428; Akkannapet (MDK), *BR* & *CP* 11565; Central University campus (RR), *MSM* 15166; Eturnagaram (WGL), *SSR* & *KSM* 18545; Woddipatla (NLG), *BVRS* 340; Rechapalli (KRN), *GVS* 22251 (MH); Ramanagutta (KMM), *RR* 107988 (BSID).

HETEROPHRAGMA DC.

Heterophragma quadriloculare (Roxb.) K. Schum., in Engl. & Prantl, Nat. Pflanzenfam. 4 (3b): 243. 1895. *Bignonia quadriloculare* Roxb., Pl. Coromandel 2: 24. 145. 1798. *Heterophragma roxburghii* DC., Prodr. 9: 210. 1845; C.B. Clarke in Hook.f., Fl. Brit. India 4: 381. 1884; Gamble, Fl. Madras 2: 997. 1924.

Vern.: *Kapu-gargu, Kala-goru, Baray-kalikod, Bondugu.*

Large tree. Leaves large, imparipinnate, 30-60 cm long, leaflets 7-11, elliptic-oblong, 5-12 × 3-5 cm, glabrous, base unequal, margin entire, apex obtuse. Flowers in panicle, terminal, pedicellate; calyx densely tomentose; corolla pale rose or white, tubular, ventricose, 5 cm long; stamens 4, didynamous. Capsule narrowly oblong, pointed, 4-locular due to the presence of pseudo-septum.

Occasional in deciduous forests in Adilabad and Karimnagar districts.

Fl. & Fr.: February-April.

Racherla (ADB), *GO* & *DAM* 5158; Rampur (ADB), *SSR* 19302; Bhadrachalam (KMM), *RCS* 98979 (BSID); Mahadevpur to Kataram (KRN), *SLK* 70810 (MH).

JACARANDA Juss.

Jacaranda mimosifolia D. Don, Bot. Reg. 8: t. 631. 1822.

Vern.: *Swarna sundari.*

Deciduous tree, up to 12 m high; bark brownish, peeling off in small thin flakes. Leaves odd pinnate, 18-20 cm long, petiolate, pinnae 5-10, leaflets 10-15 pairs, narrowly elliptic, coriaceous, 0.7 × 0.3 cm, terminal leaflet to 1.5 × 0.4 cm, base acute, margin entire, apex narrowly acuminate. Flowers in terminal panicles, pedicellate; calyx lobes united; corolla blue, 4 cm long; stamens 4, didynamous. Capsule square-shaped; seeds numerous, winged.

Planted in gardens and along road sides as an ornamental tree.

Fl.: March-May. Fr.: Throughout the year.

KIGELIA DC. *nom. cons.*

Kigelia africana (Lam.) Benth. in Hook. Niger Fl. 463. 1849. *Bignonia africana* Lam., Encycl. 1: 424. 1785. *Kigelia pinnata* DC., Prodr. 9: 247. 1845; Gamble, Fl. Madras 2: 1000. 1924.

Vern.: *Enugu thondamu, Vatavrukshamu, Kijili, Nagamalli.*

Large, spreading, evergreen tree, to 12 m high, bark greyish, smooth, brownish and rough in aged trees. Leaves 1-pinnate, 20-25 cm long, leaflets 5 to 9, oblong, 8-12.5 × 2.5-5 cm, base obtuse, margin entire, apex mucronate, glabrous. Flowers in axillary raceme, pedicellate; calyx bell-shaped; corolla dark red, unequal, 6 cm long; stamens 4. Fruit gourd like, cylindrical, indehiscent, woody, hanging with long stalks.

Planted along road sides.

Fl. & Fr.: April – July.

Jannaram (ADB), *GO* & *DAM* 5413; Kotgir (NZB), *BR* & *KH* 9643; Yedapally (NZB), *BR* & *KH* 6470.

MILLINGTONIA L. f.

Millingtonia hortensis L.f., Suppl. Pl. 291. 1782; C.B. Clarke in Hook.f., Fl. Brit. India 4: 377. 1884; Gamble, Fl. Madras 2: 994. 1924. *Bignonia suberosa* Roxb., Pl. Coromandel 3. 11. t. 214. 1811.

Vern.: *Manumalli, Katumalli, Akasa malli, Punnaga, Manu sampangi, Karaku, Toka malli.*

Evergreen tree, up to 20 m high, bark yellowish, cork brittle. Leaves bipinnate, 20 cm long, leaflets 10-15 pairs, elliptic-ovate, coriaceous, 0.7-1.5 × 0.3-0.4 cm, base acute, margin entire, apex narrowly acuminate, glabrous. Flowers in terminal panicles, pedicellate, fragrant, open at night; calyx small, campanulate; corolla white, tube long, slender, limb obscurely bilabiate; stamens 4, didynamous. Capsule elongate, linear, compressed, septicidally 2-valved; seeds many, flattened, winged.

Planted in villages and towns as an ornamental.

Fl. & Fr.: September – February.

Neelwai (ADB), *GO & DAM* 5148; Patharajampet (NZB), *BR* 7152.

OROXYLUM Vent.

Oroxylum indicum (L.) Kurz, For. Fl. Burma 2: 237. 1877; C.B. Clarke in Hook.f., Fl. Brit. India 4: 378. 1884; Gamble, Fl. Madras 2: 994. 1924. *Bignonia indica* L., Sp. Pl. 625. 1753.

Vern.: *Godugu chettu, Akumanu, Nemali chettu, Pampini, Peddakatti, Sakananasamu, Pampini chekka.*

Deciduous tree, 4-8 m high, bark silvery white, soft, lenticellate. Leaves very large, 2-3 pinnate, 1.2-1.6 m long, leaflets 2-4, ovate to elliptic, 6-13 ×× 3.7-7.5 cm, base rounded or sub-cordate, margin entire, apex acuminate, glabrous. Flowers large, in short terminal racemes, pedicellate; calyx leathery, oblong-campanulate; corolla reddish-purple outside, yellowish inside, fleshy, 10 cm long; stamens 4. Capsule flat, straight, valves woody; seeds compressed, broad with papery transparent wings.

Occasional along the streams, rivers on the cool sites of the hills in the forests in Adilabad and Karimnagar (Naqvi, 2001) districts.

Fl. & Fr.: July – March.

Jaipur RF (ADB), *GO & PVP* 4947.

PYROSTEGIA C. Presl.

Pyrostegia venusta (Ker Gawl.) Miers, Proc. Roy. Hort. Soc. 3 : 188.1863; Chatt., Bull. Bot. Soc. Bengal 2 : 78. 1948. *P. ignea* (Vell.) C. Presl., Bot. Bemerk. 93. 1844; Bailey, Man. Cult. Pl. 903.1949.

Extensive, woody climber; stem terete, glabrous; branches angular; young parts pubescent. Leaves 1-jugate; leaflets ovate-lanceolate, 4-6 x 3-4 cm, rounded or subcordate at base, entire, acuminate, shining green and glabrous above, paler and pubescent beneath; petiolules 1-2 cm long. Flowers in dense, axillary, thyrsoid panicles; pedicels 1-1.5 cm long; calyx cupular, 4-5 mm long, with truncate or obscurely toothed mouth; corolla bright orange, 6-7 cm long; tube narrow in the lower half, widened upwards; limb 2-lipped; lobes oblong-lanceolate, 1.5-1.8 cm long, pubescent outside, with white, ciliate margin. Fruits not seen.

Grown on arches, trellises etc.

Fl. & Fr.: October-January.

SPATHODEA P. Beauv.

Spathodea campanulata P. Beauv., Fl. Oware Benin 1 : 47. t. 27. 1805; Gamble, Fl. Madras 2 : 1000. 1924.

Evergreen tree, branching widely at top, to 22 m high; bark smooth, light brownish grey. Leaves 30-45 cm long; leaflets 13-19, elliptic-oblong to ovate-lanceolate, 5-9 x 3-4.5 cm, rounded at base, abruptly acuminate at apex, entire, glabrous above, paler beneath, shortly petiolate. Flowers large, 9 x 6 cm, in dense, terminal, racemes; calyx golden brown, velutinous outside, 5-6 x 2-2.5 cm; corolla orange-crimson with red spots and stripes, thick, 7-9 x 5-6 cm; tube glandular, cylindrical portion 1.5-2 x 0.8-0.9 cm, inflated portion 6-7 x 5-6 cm; lobes ovate-oblong, 3-5 x 3-4 cm. Capsule erect, linear-lanceolate, flattened parallel to septum, acuminate at both ends, brownish–black, 15-23 x 3-5 cm; seeds white– papery, winged all round, to 2.5 x 2 cm.

Common, planted in gardens, along road sides as an ornamental and avenue tree.

Fl. & Fr.: September-February.

STEREOSPERMUM Cham.

1. Capsule stout, nearly terete, ribbed, to 20 mm in diam., straight; flowers dull crimson in dense panicles .... **S. chelonoides**
1. Capsule slender, angled, not ribbed, to 13 mm in diam; flowers purple or with yellow streaks in lax panicles .... **S. tetragonum**

Stereospermum chelonoides (L.f.) DC. in Bibl. Universelle Geneve 17: 124. 1838 & Prodr. 9: 211. 1845. *Bignonia chelonoides* L.f., Suppl. Pl. 282. 1782. *Stereospermum suaveolens* (Roxb.) DC. in Bibl. Univ. Geneve 17: 124. 1838; C.B. Clarke in Hook. f., Fl. Brit. India 4: 382. 1884; Gamble, Fl. Madras 2: 998. 1924. *Bignonia suaveolens* Roxb., Fl. Ind. 3: 104. 1832.

Vern.: *Kokkesa, Kalagu, Kuberaakshi, Tedla-pala, Patala, Ambuvasini, Kaligotu, Patali.*

Large deciduous tree, 12-15 m high, bark grey, exfoliating in flat scales. Leaves imparipinnate, 30-60 cm long, leaflets 5-9 in number, elliptic, 7-17.5 × 3-12 cm, base obtuse, margin entire, apex acute, glabrescent. Flowers dull purple, fragrant, in large axillary panicles, pedicellate; calyx very glandular; corolla dull crimson with yellow streaks, villous within on lower side; stamens 4. Capsule terete, usually purple spotted; seeds pale yellowish brown, winged, furrowed across the middle.

Occasional in semi evergreen and dry deciduous forests in Adilabad, Khammam and Mahabubnagar districts.

Fl. & Fr.: September –February.

Rampur (ADB), *SSR* 19302; Mothugudem (KMM), *SSR* 15247.

Stereospermum tetragonum DC., Prodr. 9: 210. 1845; Gamble, Fl. Madras 2: 998. 1924. *S. chelonoides* auct. non (L.f.) DC.; C.B. Clarke in Hook.f., Fl. Brit. India 4: 382. 1884. *S. colais* (Buch.-Ham. ex Dillwyn) Mabb., Taxon 27: 553. 1978. *S. personatum* (Hassk.) Chatterjee, Bull. Bot. Soc Bengal 2: 70. 1948; Moulali in Pullaiah & Moulali, Fl. Andhra Pradesh 2: 685. 1997. *Dipterosperma personatum* Hassk., Flora 25. pt.2, Biebl. 28. 1842. *Stereospermum colais* (Dillw.) Mabb. var. *angustifolium* (Haines) Bennet & Raizada, Indian J. Forest. 4: 67. 1981. *S. chelonoides* DC. var. *angustifolium* Haines, Deser. List. Trees etc. South Circle of C.P. 169. 1916. *S. angustifolium* Haines, Kew Bull. 1922: 121. 1922; Gamble, Fl. Madras 2: 999. 1924; Moulali in Pullaiah & Moulali, Fl. Andhra Pradesh 2: 685. 1997.

Vern.: *Girikaya, Magavepa, Tagada, Isikarasi, Pulakevalamu, Pisul, Kaligottu, Godalippulusu Maga vepa, Kala-kod.*

Large deciduous tree, up to 20 m high, bark brown to grey, fairly smooth. Leaves pinnately compound, 15-20 cm long, leaflets 9-11, coriaceous, elliptic, 5-12 × 3-5 cm, base acute, margin entire or minutely crenulate, apex acuminate, glabrous. Flowers yellow, veined with red, in large, corymbose, axillary panicles, pedicellate; calyx campanulate, bilabiate; corolla yellow, bell-shaped, 2.5 cm long; stamens 4. Capsule terete, elongate, twisted.

Common in dry deciduous forests.

Fl. & Fr.: April – October.

Kadam (ADB), *GO & DAM* 5191; Racherla (ADB), *GO & DAM* 5158; Mothgudem (KMM), *SSR* 15950; Manchippa (NZB), *BR & KH* 9026; Eturnagaram (WGL), *SSR* 18539; Narsapur (MDK), *KMS* 8016 (MH).

TABEBUIA Gomes ex DC.

Tabebuia rosea (Bertol.) DC., Prodr. 9 : 215. 1845; A. Gentry, Brittonia 22 : 253. 1970; Steenis, Fl. Males. Ser. I. 8 : 185. 1977. *Tecoma rosea* Bertol., Fl. Guatimal. 25. 1840.

Tree, up to 10 m high. Leaves digitate; leaflets 5, oblong-obovate, terminal, 8-8.5 x 4- 4.3 cm; laterals 4.5-6.5 x 2-3 cm, thin coriaceous, lepidote, base rounded to acute, margins entire, apex obtuse; lateral leaflets smaller than others; petioles 12 cm long. Flowers in terminal racemes; pedicels 2.5 cm long; calyx campanulate, bilabiate, 1. 5 mm long; corolla pinkish white, funnel-shaped; lobes orbicular, obtuse, crisped, 2 x 2.5 cm. Fruits not seen.

Planted in gardens and along road sides as an ornamental and avenue tree.

Fl. & Fr.: July-September.

TECOMA Juss.

Tecoma stans (L.) Juss. ex Kunth, Nov. Gen. Sp. Pl. 3: 144. 1819. *Bignonia stans* L., Sp. Pl. (ed. 2) 871. 1763. *Stenolobium stans* (L.) Seem., Ann. Mag. Nat. Hist. (Ser. 3) 10: 30. 1862; Gamble, Fl. Madras 2: 1000. 1924.

Vern.: *Pasupu ganneru, Ganta ganneru, Swarna puvvu, Suvarna ganneru, Pachagotla.*

A large shrub to small tree. Leaves imparipinnate, leaflets dentate, lanceolate, odd one longest, to 11 × 2.5 cm, others 49 × 1.5-2.5 cm. Flowers golden yellow in terminal racemes.

Cultivated in many gardens and temples in all districts.

Fl. & Fr.: May – December.

Rollapadu (KMM), *RR* 108041 (BSID); Gacchibowli (RR), *MSM* 15190.

PEDALIACEAE

1. Fruit a capsule .... **Sesamum**
1. Fruit a drupe:
    2. Flowers solitary; stamens 4 .... **Pedalium**
    2. Flowers in racemes; stamens 2 .... **Martynia**

MARTYNIA L.

Martynia annua L., Sp. Pl. 618. 1753; Gamble, Fl. Madras 2: 1003. 1924; Hevly, Taxon 18: 527. 1969. *M. diandra* Gloxin, Obs. Bot. 14. t. 1. 1785; Hook.f., Fl. Brit. Indi 4: 386. 1884.

Vern.: *Thelu kondi kaya, Garuda mukku.*

Annual, erect, glandular undershrub, up to 1 m high. Leaves opposite, petiolate, cordate, 5-18 × 6-23 cm, base cordate, margin serrate, apex acute or triangular, pubescent. Flowers in axillary racemes, pedicellate; calyx free, imbricate; corolla purple with yellow throat, ventricose, tube 5 cm long; stamens 2, staminodes 3. Fruit drupe with two sharp hooks at the apex, pericarp pubescent, endocarp woody, black.

Common along road sides and waste lands in almost all districts.

Fl. & Fr.: July – December.

Sarvaipet (ADB), *GO* 4511; Ibrahimpet – Banswada Road (NZB), *TP* & *BR* 6138; Banjapally (NZB), *BR* 9550; Nagulabanda (MDK), *CP* 11342; Narsapur (MDK), *KMS* 6747 (CAL); Kommireddipalli (RR), *MSM* & *NV* 10325; Kannaigudem (KMM), *AJR* & *PAN Reddy* 20470; Rechapalli (KRN), *GVS* 22238 (MH); Pakhal (WGL), *KMS* 13197; Rangapur (WGL), *PVS* & *NRR* 76959 (BSID); Kambalpally (NLG), *BVRS* 233 (NU).

PEDALIUM Royen ex L.

Pedalium murex L., Syst. Nat. (ed. 10) 1123. 1759; C.B. Clarke in Hook.f., Fl. Brit. India 4: 386. 1884; Gamble, Fl. Madras 2: 1001. 1924.

Vern.: *Yenugu palleru, Pedda palleru, Mullagadakaya cherru, Yenugumullu.*

Erect, annual herb, 15-30 cm high. Leaves opposite, petiolate, unequal, 10-15 × 9-18 cm, base truncate, margin crenate, apex retuse or emarginated. Flowers solitary, axillary, pedicellate; calyx free, valvate, lobes equal, linear; corolla yellow, 3 cm long, hispid; stamens 4, staminodes 2.5 mm long. Fruit hard, indehiscent, spiny along the angles, epicarp fleshy.

Occasional in waste lands in Adilabad, Hyderabad (Ramana, 2010), Karimanagr (Naqvi, 2001) and Nalgonda districts.

Fl. & Fr.: April – September.

Mancheriyal (ADB), *GO* & *PVP* 4934; Cherukupally (NLG), *BVRS* 663 (NU).

SESAMUM L.

1. Erect herbs .... **S. indicum**
1. Prostrate herbs .... **S. laciniatum**

Sesamum indicum L., Sp. Pl. 634. 1753; C.B. Clarke in Hook.f., Fl. Brit. India 4: 387. 1884; Gamble, Fl. Madras 2: 1002. 1924. *S. orientale* L., Sp. Pl. 634. 1753; Graham, Cat. Pl. Bombay 126. 1839.

Vern.: *Nuvvulu*; Eng.: Gingelly.

Annual, erect, pubescent herb, 30-90 cm high, stem terete below, tetragonous above. Lower leaves opposite, lobed or deeply divided, upper ones undivided, linear-oblong, 8-10 × 2-4 cm, base attenuate, apex acute. Flowers pink, solitary, axillary; calyx lobes 5, persistent; corolla tubular, bilipped, lobes 5; stamens 4. Capsule1.2-1.8 cm long, oblong, grooved with short beak; seeds obovate, smooth, brown, black or white.

Commonly cultivated, some times as escape.

Fl. & Fr.: Throughout the year.

Vempalli (ADB), *GO* 4439; Banswada (NZB), *TP* & *BR* 6141; Medak (MDK), *CP* 11719; Kanmankalva (RR), *MSM* 10458; Mannanur (MBNR), *TDCK* 13630; Aklaspur (KRN), *GVS* 25601 (MH); on the way to Billyguttalu (KMM), *RR* 105985m(BSID); Palempet (WGL), *RKP* 105349 (BSID).

Sesamum laciniatum Klein ex Willd., Sp. Pl. 3: 359. 1800; C.B. Clarke in Hook. f., Fl. Brit. India 4: 387. 1884; Gamble, Fl. Madras 2: 1002. 1924.

Scabrid, prostrate, annual herb, up to 50 cm long. Leaves opposite, petiolate, elliptic, older leaves deeply pinnatifid, to 1 × 0.7 cm, base truncate, margin dissected, apex subacute, hispid. Flowers solitary, axillary, pedicellate; calyx valvate, unequal, lanceolate; corolla purple, pubescent without, villous at the throat within, 3.5 cm long; stamens 4, didynamous. Capsule ovoid, with 2 apical short beaks; seeds brownish.

Occasional in dry areas mostly near rocky ground in Mahabubnagar district.

Fl. & Fr.: August – December.

ACANTHACEAE

1. Stamens 2:
   2. Plant acaulescent .... **Elytraria**
   2. Plant caulescent:
      3. Corolla lobes 5, subequal:
         4. Calyx 4-partite .... **Barleria**
         4. Calyx 5-partite .... **Eranthemum**
      3. Corolla lobes 2-lipped:
         5. Ovules two or rarely one in each locule:
            6. Calyx 5-partite:
               7. Placenta separating elastically from the base:
                  8. Bracts in one-sided spikes .... **Rungia**
                  8. Bracts in cymose clusters .... **Dicliptera (cuneata)**
               7. Placenta not separating from the valves:
                  9. Anther cells parallel, subequal .... **Ecbolium**
                  9. Anther cells one higher up than the other:
                     10. Anther cells with white basal spur like appendages:
                        11. Bracts elliptic; calyx lobes 4 .... **Justicia**
                        11. Bracts linear; calyx lobes 5 .... **Gendarussa**
                     10. Anther cells without spur like appendages:
                        12. Shrubs .... **Rhinacanthus**
                        12. Herbs .... **Dicliptera (paniculata)**
            6. Calyx 4-partite .... **Rostellularia**
         5. Ovules more than two in each locule:
            13. Calyx 4-partite .... **Nelsonia**
            13. Calyx 5-partite:
               14. Capsules 4-seeded .... **Indoneesiella**
               14. Capsules more than 4-seeded .... **Andrographis**
1. Stamens 4:
   15. Twiners or stragglers:
      16. Calyx minute, annular, 10-15-toothed .... **Thunbergia**
      16. Calyx 5-lobed .... **Meyenia**
   15. Erect or prostrate herbs, shrubs:
      17. Corolla without an upper lip, lower lip large expanded, 3-lobed .... **Blepharis**

17. Corolla two-lipped or with 5 subequal lobes:
 18. Anthers 1-celled .... **Crossandra**
 18. Anthers 2-celled:
  19. Corolla lobes 5, subequal:
   20. Corolla lobes twisted to left in bud:
    21. Ovules more than two in each locule:
     22. Leaf margin dentate or crenate:
      23. Herbs with stiff hairs .... **Gantelbua**
      23. Herbs with soft villous hairs .... **Hemigraphis**
     22. Leaf margin entire or crenulate:
      24. Bracts absent; capsules basally beaked; leaf base truncate or rounded to acute .... **Dipteracanthus**
      24. Bracts present; capsules not basally beaked; leaf base cuneate to attenuate .... **Ruellia**
    21. Ovules 2 in each cell .... **Petalidium**
   20. Corolla lobes imbricate in bud .... **Asystasia**
  19. Corolla 2-lipped:
   25. Corlla lobes twisted to left in the bud:
    26. Ovules more than 2 in each locule .... **Hygrophila**
    26. Ovules two in each locule .... **Dyschoriste**
   25. Corolla lobes imbricate in bud .... **Lepidagathis**

ANDROGRAPHIS Wall.

1. Erect herbs or sub shrubs .... **A. paniculata**
1. Prostrate herbs, with stout root stocks .... **A. serpyllifolia**

Andrographis paniculata (Burm.f.) Nees in Wall., Pl. Asiat. Rar. 3: 116. 1832; C.B.Clarke in Hook.f. Fl. Brit. India 4: 501. 1884; Gamble, Fl. Madras 2: 1048.1924. *Justicia paniculata* Burm. f., Fl. Indica 9. 1768. *Andrographis subspathulata* C.B.Clarke in Hook.f., Fl. Brit. India 4: 502. 1884.

Vern.: *Nela vemu.*

Erect branched herb, 30 to 90 cm high; braches quadrangular, grooved, slightly winged. Leaves linear-obovate, 5-8 × 1-2 cm, glabrous or minutely puberulous beneath, base cuneate or narrowed, margin entire, apex acuminate; petiole to 1.5 cm long. Panicle branches zigzag, to 15 cm; flowers unilateral, distant, bracts lanceolate, acute; bracteoles 2; calyx lobes 5, sparsely glandular–hairy without, acute; corolla white with pink tinge, 5 × 2.5 mm; stamens 2, adnate to the throat of corolla, filaments hairy,

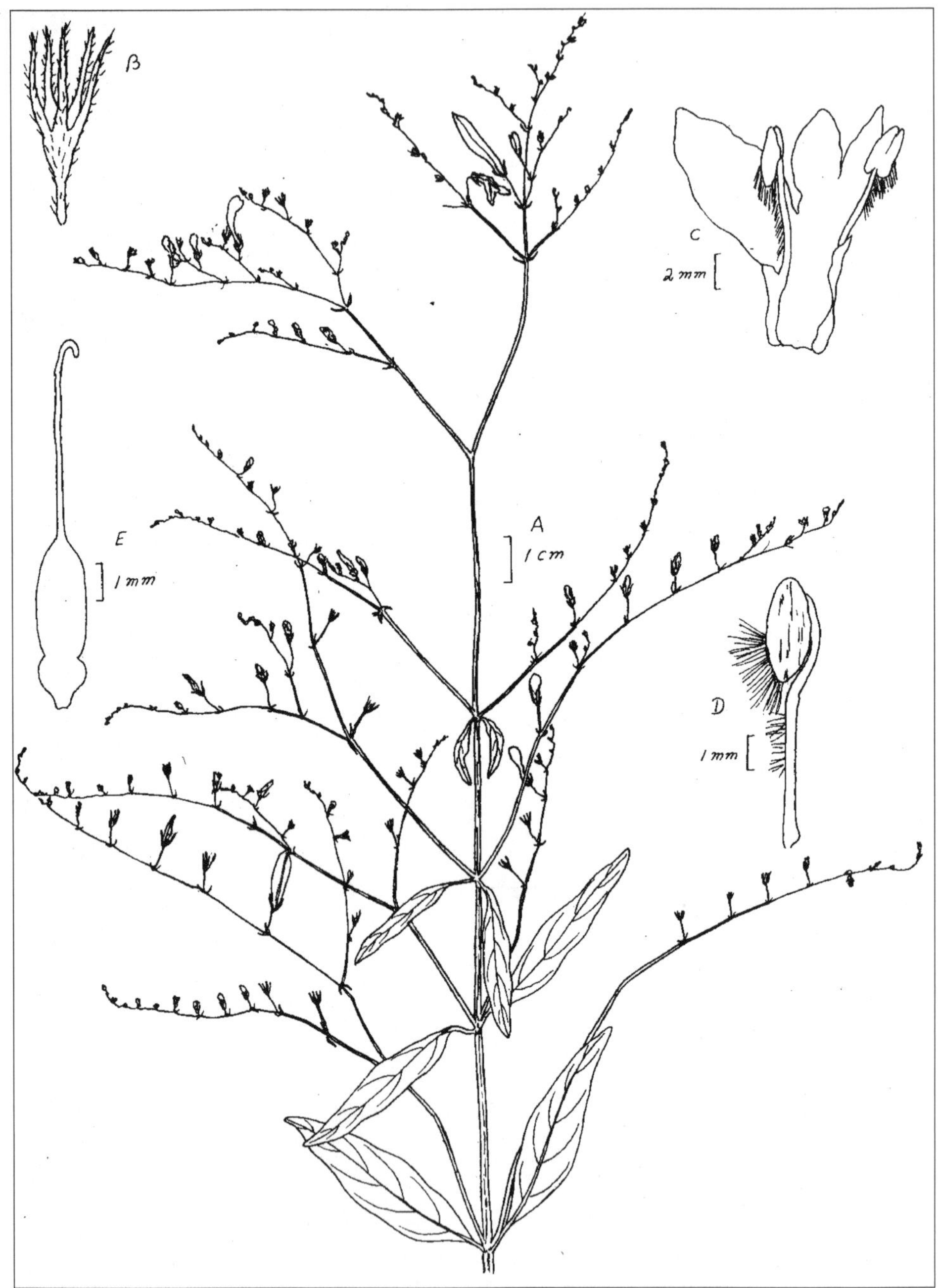

Figure 77: Andrographis paniculata (Burm. f.) Nees
A. Twig, B. Calyx, C. Corolla open, D. Stamen, E. Pistil.

bearded, anthers obovoid, 2 mm, long. Capsule oblong, 1.7 × 0.3 cm, compressed, minutely hairy; seeds subquadrate, rugose, 1 x 0.6 mm

Common in deciduous forests and also open waste lands on rocky soil in Adilabad, Medak, Karimnagar, Ranga Reddi, Nalgonda and Mahabubnagar districts.

Fl. & Fr.: September - April.

Anantagiri RF (RR), *MSM* & *PRSR* 12789; Mohammadabad RF (RR), *MSM* 10445; Panapally (KMM), *RR* 112533 (BSID); Medak (MDK), *RG* 104008 (BSID); Kondagutta (KRN), *GVS* 25684 (MH); Mannanur (MBNR), *TDCK* 13603; Umamaheswaram (MBNR), *STR* 381 (HH); Nagarjunakonda (NLG), *DAM* 5903; Kambalpally (NLG), *BVRS* 612 (NU).

Andrographis serpyllifolia (Vahl) Wight, Icon. Pl. Ind. Orient. t. 517. 1842; C.B.Clarke in Hook.f., Fl. Brit. India 4: 506. 1884; Gamble, Fl. Madras 2: 1051 .1924. *Justicia serpylifolia* Rottl. ex Vahl, Enum. Pl. 1: 169. 1804.

Vern.: *Karumaku, Aku chandrika, Pamu nelavemu.*

Prostrate, trailing herb, branches rising from a woody root stock; branchlets densely hairy. Leaves rotund-orbicular, sub reniform, 2-3 × 2-2.5 cm, villous when young, glabrous at maturity except at margins, base obtuse or subcordate, margin entire, apex obtuse, lateral nerves 3-4 pairs; petiole 0-1 mm. Flowers axillary, solitary or in few-flowered, 2.3 cm long racemes; pedicel subsessile, 3 mm in fruit; bracts orbicular, 3 × 2 mm; calyx glandular pubescent, sepals 5, linear; corolla whitish-pink, tube 7mm, upper lip 6 mm, minutely bifid, lower lip purple, 7 mm; stamens 2, 6 mm, anthers unequal, bearded. Capsule 1.5 cm, elliptic - obovoid, glabrous, with short, blunt acuminate tip; seeds 8, pale, rugose, retinacula spathulate.

Occasional in hilly dry deciduous forests up to 600 m altitude in Mahabubnagar district.

Fl.: June –September and January-March. Fr.: Through out the year.

Mannanur range (MBNR), *STR* 252 (MH).

ASYSTASIA Blume

1. Bracts 3 mm long; corolla white, blue, purple and yellow .... **A. gangetica**
1. Bracts 6 mm long; corolla blue .... **A. dalzelliana**

Asystasia dalzelliana Sant., Kew Bull. 3: 276. 1948. *A. violacea* Dalzell ex C.B.Clarke in Hook.f., Fl. Brit. India 4: 494. 1884 non Dalzell 1850; Gamble, Fl. Madras 2: 1063. 1924.

An erect herb, branchlets glabrous. Leaves ovate elliptic, 1.5-3 × 1-2 cm, base truncate, margin entire, apex acute, thinly pubescent above and below. Flowers blue, axillary, solitary. Capsule pubescent; seeds orbicular.

Occasional in forest in Karimnagar and Ranga Reddi districts.

Fl. & Fr.: September – December.

Anantagiri RF (RR), *MSM* 12784.

Asystacia gangetica (L.) T. Anderson, Enum. Pl. Zeyl. 235. 1860; Gamble, Fl. Madras 2: 1063. 1924. *Justicia gangetica* L., Cent. Pl. 2: 3. 1756. *Asystasia coromandeliana* Wight ex Nees in Wall., Pl. Asiat. Rar. 3: 89. 1832; C.B.Clarke in Hook.f., Fl. Brit. India 4: 493. 1884.

Vern.: *Podabeera.*

Erect or procumbent herb, often some what climbing, to 75 cm long; stems obtusely quadrangular, glabrous. Leaves ovate or elliptic-ovate, 2-4 × 1.5-4 cm, glabrous above, paler and slightly pubescent especially on the nerves beneath, base rounded to truncate, rarely acute in basal leaves, margin entire, apex acute to shortly acuminate; lateral nerves 4-6 pairs, clearly impressed beneath, petiole to 2.5 cm. Racemes sub terminal, secund, peduncle 5 cm, rachis 11 cm, flowers on rachis 1.5-2 cm apart; pedicel 3mm long; bracts lanceolate, 3 mm long; calyx divided near the base, lobes 8-9 mm long, sepals linear-lanceolate, pubescent; corolla white blue, purple or yellow, to 4.2 cm, tube inflated above, limb ventricose, lobes rounded; stamens 4, didynamous, connate at the base in pairs, filament pairs 8 and 12 mm long, anthers oblong, cells parallel, minutely spurred at the base; ovary 3.5 mm, style 2.2 cm, linear, lower part hairy. Capsule up to 3 cm long with a solid base (seedless portion) to 1.6 cm, upper part clavate, constricted above the middle; seeds 4, angular, 4 mm across, corrugated at the edges and tubercled on the faces.

Occasional in Hyderabad (Rajagopal, 1973), Ranga Reddi and Mahabubnagar districts.

Fl. & Fr.: Through the year.

Thaskapally (MBNR), *SRS* 109073 (BSID).

BARLERIA L.

1. Plants without spines; leaf apex not spinous:
   2. Leaves small, hairy:
      3. Outer calyx lobes ovate .... **B. longiflora**
      3. Outer calyx lobes obovate – spathulate .... **B. tomentosa**
   2. Leaves large and glabrescent:
      4. Flowers in axillary and terminal spikes (or) racemes .... **B. cristata**
      4. Flowers solitary in upper leaf axils and glandular, passing into terminal spikes:
         5. Leaves strigose .... **B. strigosa**
         5. Leaves glabrous (or) pubescent .... **B. prattensis**
1. Plants with short intrapetiolar spines:
   6. Bracts linear, spine-tipped .... **B. cuspidata**
   6. Bracts foliaceous .... **B. prionitis**

Barleria cristata L., Sp. Pl. 636. 1753; C.B.Clarke in Hook.f., Fl. Brit. India 4: 488. 1884; Gamble, Fl. Madras 2: 1060.1924.

Vern.: *December puvvulu, Neelambaramu chettu, Gobbi, Kodikannu, Niru goranta.*

Herbaceous hairy herb to undershrub, to 1 m high; stem quadrangular, hairy. Leaves elliptic-oblong to lanceolate or linear-oblong, 4-11 × 1-3 cm, thinly puberulous beneath, some what dense on nerves, base attenuate, margin entire, apex acute, rarey shortly acuminate; lateral nerves 6 pairs, petiole 2-15 mm. Flowers 1-4 in axillary spikes, spike dense strigose hairy going upwards; bracts linear, spinous tipped, thick; calyx scarious, outer ones elliptic, 2-3 × 0.8 cm, prominently nerved, spinous dentate, inner ones linear, 1.5 × 0.2 cm; corolla purplish pink, violet, 5.5 cm long, fertile stamens 2, rudimentary stamens 2, anthers 3 mm long, staminodes 5 mm long, with sterile anther cells. Capsule oblong or ellipsoid, 1.5 × 0.4 cm; seeds 4, 0.4 cm across, beaked, appressed hairy.

Common, in grass lands and hill sides in most districts. Often grown in gardens.

Fl. & Fr.: December- March.

Sirpur RF (ADB), *GO* 4390; Nasarullabad RF (NZB), *TP* & *BR* 6151; Sirnapally RF (NZB), *BR* & *PSPB* 9575; Narsapur (MDK), *BR* & *CP* 14085; Rathamhutta hills (KMM), *RCS* 98790 (BSID); Nilgiri view point (MBNR), *SRS* 109705 (BSID).

Barleria cuspidata F. Heyne ex Nees in Wall., Pl. Asiat. Rar. 3: 93. 1832; C.B.Clarke in Hook.f., Fl. Brit. India 4: 483. 1884; Gamble, Fl. Madras 2: 1059. 1924.

Small shrub with sharp intra petiolar spines, up to 50 cm high; stem and branches terete, 4-angled, nodes with small bristle-like hairs, spines 4, white, 2.5 cm. Leaves linear-lanceolate, 3-5 × 0.3-0.7 cm, thinly scabrous, base cuneate, margin entire, spiny-mucronate, subsessile, raphides minute, irregular. Flowers solitary, axillary, sessile; bract linear, rigid, spine-tipped; calyx 4-lobed, outer lanceolate, 2 × 0.3 cm, spine-tipped, inner lanceolate, 1 × 0.3 cm, acute; corolla yellowish, 2.5 cm across, tube 2.9 cm long, lobes 1.5 cm long, acute; stamens 5, anthers 3 cm long, staminodes 2, with sterile anther cells. Capsule ovoid; seeds 2, ovoid, appressed hairy.

Occasional in waste lands in Adilabad district.

Fl. & Fr.: December - February.

Nirmal (ADB), *GO* & *DAM* 5029.

Barleria longiflora L.f., Suppl. Pl. 290. 1782; C.B.Clarke in Hook.f., Fl. Brit. India 4: 485. 1884; Gamble, Fl. Madras 2: 1059.1924.

Vern.: *Pinna gorinta.* Eng.: White long-flowered Nail dye.

Pale yellowish grey pubescent shrub, to 1.5 m high, much branched. Leaves chartaceous, ovate-elliptic or ovate-deltoid, 2-5.3 × 1.5 – 4 cm, closely densely greyish tomentose, some what less dense on upper surface in mature ones, base rounded or acute, margin entire, apex acute, rarely shortly acuminate; lateral nerves 4-6 pairs, obscure above, prominent beneath; petiole to 2 cm. Flowers solitary, axillary and also in short dense few-flowered 1-sided spikes; bracts linear-lanceolate, to 2 × 0.2 cm;

outer calyx lobes ovate, to 2.5 × 1.7 cm, densely pubescent, scarious with age, prominently reticulately nerved, apex obtuse, inner lobes elliptic-ovate, 6 × 4 mm; corolla white, tube 9 cm long, slender, lobes 1.5 cm long; stamens 2, anthers 3 mm long. Capsule obovoid with short acuminate tip, brown, 1.8 × 0.5 cm; seeds 4, 3.5 mm hairy.

Occasional in Hyderabad, Mahabubnagar and Nalgonda districts.

Fl. & Fr. October - February.

Hyderabad (HYD), *MRS* 81284 (DD); Umamaheswaram way (MBNR), *GS* 18292; Appapur (MBNR), *BSS* & *SKB* 30593 (BSID); Nagarjunasagar (NLG), *KMS* 9751 (MH), *BVRS* 75 (NU).

Barleria prattensis Santapau, Kew Bull. 3: 487. 1949. *B. montana* sensu Bedd., Ic. t. 257. 1868-1874, non Nees 1832.

Small shrub; branches glabrous. Leaves ovate or lanceolate, to 17.5 × 7.5 cm, base cuneate, apex acute–acuminate. Flowers solitary, bracteoles linear, outer calyx-lobes broadly elliptic, acute, glabrous, inner narrow, linear; corolla pink, funnel-shaped. Capsule oblong-acuminate; seeds black, flat, glabrous.

Occasional in forests of Mahabubnagar and Warangal districts.

Fl. & Fr.: September – January.

Mannanur range (MBNR), *STR* 100 (HH).

Barleria prionitis L., Sp. Pl. 636. 1753; C.B.Clarke in Hook.f., Fl. Brit. India 4: 482. 1884; Gamble, Fl. Madras 2: 1058. 1924.

Vern.: *Mullu gorinta, Gobbi, Mundlagobbi, Mullakanakambaralu.*

Armed undershrub, to 1.5 m high, stem glabrous, greyish, channeled. Leaves sub coriaceous, ovate-elliptic or obovate, 1.2-13 × 0.4-5 cm, glabrous above, sparsely hirsutulous beneath, base attenuate and forms a decurrent petiole, margin entire, apex acute, mucronate, rarely forms a spinose tip; lateral nerves 3-5 pairs, more conspicous at lower surface than upper; petiole to 2 cm. Flowers axillary, solitary, or in cymose clusters or spicate towards upper most axils; bracts 2, spine-tipped; calyx lobes 4, unequal, spinose at apex; corolla yellowish-orange, tube 2 cm long, lobes 1.8 cm long; fertile stamens 2, to 3.4 cm long, staminodes 2, very small, villose; style 4 cm long. Capsule ovoid, beaked; seeds 2, hairy.

Occasional in waste lands, found in dry areas; also grown as ornamental.

Fl. & Fr.: September-January.

Jannaram (ADB), *TP* & *PVP* 4178; Biknoor (NZB), *BR* & *CPR* 7247; Narsapur – Hyderabad Road (MDK), *BR* & *CP* 11210; Kanmankalva RF (RR), *MSM* 10464; Achampeta (MBNR), *STR* 318 (HH); Pakhal RF (WGL), *KMS* 11624 (MH); Kaleswaram (KRN), *SLK* 70881 (NBG); Bhadrachalam RF (KMM), *RCS* 98972 (BSID); Nagarjunasagar (NLG), *BVRS* 77 (NU).

Barleria strigosa Willd., Sp. Pl. 3: 379. 1800; C.B.Clarke in Hook.f., Fl. Brit. India 4: 498. 1884; Gamble, Fl. Madras 2: 1061. 1924.

Vern.: *Nilambaram.*

Erect herb to under shrub, 0.5-1(2) m high; stem jointed, with stiff hairs on the whole surface. Leaves ovate or elliptic – oblong, 5-23 × 2-12 cm, base decurrent and forms a petiole to 9 cm, glabrous, strigose on nerves beneath, base long decurrent, margin entire, apex acuminate; lateral nerves 6 pairs, ascending. Flowers axillary and terminal, unilateral, to 5 cm long spikes; bracteoles ovate – lanceolate or elliptic, 1.2-2 cm long; outer calyx lobes subequal, 2.5-3 cm long, straight, ciliate along margins, inner linear-lanceolate, white-hirsute, 1.3 cm long; corolla blue to violet, funnel-shaped, 4-7cm long, tube with glandular hairs; stamens 4, 2 fertile, filaments to 2.5 cm long, anther lobes 4 mm, staminodes 3; ovary 2 mm, style 5 mm long. Capsule oblong, ellipsoid, 2.1 × 0.6 cm, beaked glabrous, 4-seeded; seeds orbicular, hairy,

Fairly common in moist and shady localities, near streams and also grows as under growth in forests in Khammam and Mahabubnagar districts.

Fl.: August- March, Fr.: October- April.

Thiruppumetta RF (KMM), *RCS* 99015 (BSID).

Barleria tomentosa Roth, Nov. Pl. Sp. 314. 1821; C.B.Clarke in Hook.f., Fl. Brit. India 4: 485. 1884; Gamble, Fl. Madras 2: 1059. 1924.

Shrub, to 1.5 m high, branches (also petioles, rachis, bracts) glandular pubescent, less dense with age. Leaves elliptic-ovate or rarely elliptic-oblong, 2.2-9.1 × 0.8-3.9 cm, hispid hairy beneath especially on the nerves, base acute to shortly acuminate, margin entire, hispidulous, apex acuminate with mucro; lateral nerves ca 5 pairs, conspicuous beneath; petiole to 3 cm. Raceme one–sided, 1-3-flowered; bracts oblong-spathulate, to 20 × 1.5 mm; calyx lobes 4, unequal, obovate, pale and scarious at fruiting stage; corolla violet coloured, tube 4.5 cm long, cylindric, lobes 2.3 cm long; stamens 2; style 6.8 cm long, stigma 1 mm. Capsule obovoid, 4-seeded, appressed hairy.

Occasional, in deciduous forests in Mahabubnagar and Nalgonda districts.

Fl. & Fr.: September- March.

Lingal range (MBNR), *STR* 485 (MH); Nagarjunakonda (NLG), *KMS* 9793 (CAL), *KT* 9775 (CAL).

BLEPHARIS Juss.

1. Leaves thin, membranous; corolla white with pink nerves . **B. maderaspatensis**
1. Leaves thick; corolla bluish white .... **B. integrifolia**

Blepharis integrifolia (L.f.) E. Mey & Drege ex Schinz, Vierteljahrsschr. Naturf. Ges. Zurich 60: 416. 1915. *B. repens* (Vahl) Roth, Nov. Pl. Sp. 321. 1821. *Acanthus repens* Vahl, Symb. Bot. 2: 76. 1791. *Blepharis molluginifolia* Pers., Syn. Pl. 2: 180. 1806; C.B.Clarke in Hook.f., Fl. Brit. India 4: 479. 1884. Gamble, Fl. Madras 2: 1013. 1924.

Prostrate, diffusely branched spreading herb, rooting at nodes, to 50 cm high, patently hairy. Leaves 4 at each node, thick, obovate-oblong to oblanceolate, thick, 1-3.4 × 0.4 – 0.8 cm, strigose above, glaucous beneath, base acute or cuneate, margin entire, apex obtuse; lateral nerves ca 3 pairs, obscure, to 1 mm. Inflorescence axillary, 1-few-flowered; bracts and bracteoles spinose-ciliate; bracts in 4 opposite pairs, 2.5 × 1.5, 4 x 2, 4 × 1.75, 7 × 3 mm each; bracts obovate, oblong or oblanceolate, spines stellate hairy at apex; bracteoles oblong-lanceolate, 7 × 1.5 mm; calyx lobes 4, villous at margins, one outer calyx lobe forked or 2-toothed at apex, inner pair linear-lanceolate, 7 × 0.5 mm; corolla violet to purple; stamens 4. Capsule 5 mm long; seeds 4.

Common in dry places, crevices of rocks and sandy places in most districts.

Fl. & Fr.: September- February.

Manchippa (NZB), *TP* & *BR* 6021; Toopran (MDK), *BR* & *CP* 11277; Eklaspuram (MBNR), *BR* & *DV* 41302; Nagarjunakonda (NLG), *DAM* 5906; Narayanpur (NLG), *BVRS* 447 (NU).

Blepharis maderaspatensis (L.) B. Heyne ex Roth, Nov. Pl. Sp. 320. 1821. *Acanthus maderaspatensis* L., Sp. Pl. 639. 1753. *Blepharis boerhaviifolia* Pers., Syn. Pl. 2: 180. 1806 (*'boerhaviaefolia'*); C.B.Clarke in Hook.f., Fl. Brit. India 4: 478. 1884; Gamble, Fl. Madras 2: 1013.1924.

Prostrate herb to undershrub, branches brownish, wiry, divaricately branched, obtuse, quadrangular, scattered with puberulous hairs, nodes distant, Leaves 4 in a whorl, unequal, elliptic-ovate or elliptic, 2-10 × 1.1-4.2 cm, glaucous beneath, glabrous, base obtuse or acute, margin coarsely, distantly toothed, apex acute with apicule; lateral nerves ca 3 pairs, conspicuous beneath; petiole to 4mm. Flowers 4-merous, axillary, few-many in clusters among the leaves; bracts 3-4 pairs, 4.3, 6.5, 10 mm each with glochidiate spinous teeth; bracteoles spathulate, 12 mm long, spinous; calyx lobes 4, oblong, outer one unequal, 12 and 17mm, inner ones 7 mm long, reticulate; corolla white, tube 4.5 mm long, cartilaginous, upper lip absent, lower lip 3- lobed at apex, 1.5 cm long, sericeous; stamens 4, filaments 3.5 mm, anther 3 mm; ovary 2 mm, style 8 mm, stigma bifid, 0.5 mm. Capsule oblong or obovoid, to 10 x 5 mm, glossy; seeds 2, orbicular, 4 × 3.5 mm, echinate, flattened.

Common in plains to hills in all districts among the rocks in open forests.

Fl. & Fr.: October - March.

Gandivet RF (NZB), *BR* & *CPR* 7196; Medak (MDK), *BR* & *CP* 11417; Anantasagar RF (RR), *MSM* 10508; Kambalpally (NLG), *BVRS* 206; Aklaspur (KRN), *GVS* 22510 (MH); Pakhal RF (WGL), *KMS* 11625 (MH, CAL); Ramavaram (KMM), *RR* 112555 (BSID); Farahabad road to Pullaechelma (MBNR), *SRS* 109194 (BSID).

CROSSANDRA Salisb.

Crossandra infundibuliformis (L.) Nees in Wall., Asiat. Rar. 3: 98. 1832. *Justicia infundibuliformis* L., Sp. Pl. 21. 1753. *Crossandra undulaefolia* Salisb. in W. Hooker, Parad. Lond. t. 12. 1805; C.B.Clarke in Hook.f., Fl. Brit. India 4: 492. 1884. Gamble, Fl. Madras 2: 1055. 1924.

Vern.: *Kanakambaralu.*

An unarmed undershrub, up to 1.5 m high. Leaves elliptic- oblong, 8-10 × 2-3 cm, glabrous, margin entire, apex acute. Flowers in axillary and terminal dense spikes, sessile, bracteoles linear-subulate; calyx-lobes 5, unequal, glandular hairy; corolla orange-scarlet with red or yellow throat; stamens 4. Capsule oblong; seeds 4.

Commonly cultivated in gardens for its flowers.

Fl. & Fr.: Round the year.

DICLIPTERA Juss.

1. Placenta separating elastically from the base .... **D. cuneata**
1. Placenta not separating from the valves .... **D. paniculata**

Dicliptera cuneata Nees in Wall., Pl. Asiat. Rar. 3: 111. 1832; C.B.Clarke in Hook. f., Fl. Brit. India 4: 552. 1885; Gamble, Fl. Madras 2: 1073.1924.

Vern.: *Pichi-kusuma.*

Slender, erect, hairy herb, up to 90 cm high; stems branched, glabrous, terete or very slightly quadrangular. Leaves elliptic or ovate, 3-11 × 1.2-5 cm, glabrous, raphides spreading, base acute, margin entire to sub entire, apex shortly acuminate, lateral nerves 5-6 pairs, prominent. Cymes 3-5, axillary, in sub capitate clusters, peduncle to 9 cm long, bracts of the inflorescence linear; floral bracts 2, spathulate, 1.2 × 0.3 cm, bracteoles linear, 5 mm, apiculate; calyx lobes subequal, shortly basally connate, scarious, hairy, lanceolate, 4 mm, acuminate, corolla pink, tube 6 mm, cylindrical, bilipped, 2+3, oblong-obovate, hairy without; stamens 2, attached at the throat, filaments 3mm, puberulous, anther lobes superposed, 1.25 mm; ovary 1 × 0.7 mm, ovules 4, style linear, filiform, 11.5 mm, stigma shortly bifid, 0.5 mm. Capsule ellipsoid; seeds 4.

Occasional, in moist localities of dry deciduous forests at above 600 m altitudes in Nallamalais in Mahabubnagar district.

Fl. & Fr.: November- April.

Mallelatheertham (MBNR), *SRS* 108984 (BSID).

Dicliptera paniculata (Forssk.) I. Darbysh, Kew Bull. 62: 122. 2007. *Dianthera paniculata* Forssk., Fl. Aeg.-Arab. 7. 1775. *Peristriophe paniculata* (Forssk.) Brummitt, Kew Bull. 38: 451. 1983. *P. bicalyculata* (Retz.) Nees in Wall., Pl. Asiat. Rar. 3: 113. 1832; C.B.Clarke in Hook.f., Fl. Brit. India 4: 534. 1885; Gamble, Fl. Madras 2: 1084. 1924. *Dianthera bicalyculata* Retz. Acta Holm. 1775. 297. t. 9. 1776.

Vern.: *Chebeera, Oosukada, Sannapulla, Atrial.*

Herb, laxly branched, 1.5 m high; stem 6-angled, hirsute, grooved. Leaves ovate, 2-9 × 1.5-3.5 cm, pubescent, more dense beneath, dense lineolate, base rounded, margin entire, ciliate, apex acuminate; lateral nerves 4-6 pairs; petiole 2 cm long. Flowers small in large divaricate lax pubescent panicles; pedicel 0.6-1.6 cm long; bracts 2, unequal, larger linear-obovate, 8-9 mm long, smaller 6 mm, margin scarious, bracteoles

4 as 2 unequal pairs, linear-acuminate; calyx lobes 5, 4 mm, linear-acuminate, connate, valvate, ciliate; corolla pink, 1.5cm long, tube cylindric, 4-6mm, twisted, 2-lipped, upper obtuse, lower 3-lobed, pubescent without; stamens 2, filaments dilated, anther cells 2, slightly higher, muticous; ovary pubescent above, style filiform, 1 cm, glabrous; stigma bifid. Capsule 1 cm, clavate with solid basal beak, tip acuminate; seeds 4, flattened, orbicular, 11.5-20 mm across, capillate.

Very common, spreading throughout in hedges scrubs and waste lands in all districts.

Fl. & Fr.: November- March.

Sone (ADB), *TP* & *PVP* 4065; Ibrahimpet RF (NZB), *TP* & *BR* 6119; Velutla RF (NZB), *BR* 9517; Choutkoor (MDK), *TP* & *CP* 11902; Narsapur RF (MDK), *TP* & *CP* 11972; Dharmapur RF (RR), *MSM* & *KH* 10548; Rampur (MBNR), *BSS* & *SKB* 32635; Yerragondupenta (MBNR), *TDCK* 13695; Khammam hills (KMM), *RR* 113802 (BSID); Mudigonda (NLG), *BVRS* 619 (NU); Kodimial (KRN), *GVS* 21856 (MH).

DIPTERACANTHUS Nees emend Bremek.

Dipteracanthus prostratus (Poiret) Nees in Wall., Pl. Asiat. Rar. 3: 81. 1832. *Ruellia prostrata* Poir. in Lam., Encycl. 6: 349. 1804; C.B.Clarke in Hook.f., Fl. Brit. India 4: 411. 1884; Gamble, Fl. Madras 2: 1017. 1924. *Dipteracanthus dejectus* Nees in Wall., Pl. Asiat. Rar. 3: 82. 1832.

Vern.: *Nela neelambaramu.*

A prostrate diffuse undershrub, straggling, much branched, terete to subquadrangular, internodes long, nodes purple. Leaves ovate or elliptic, 3-5 × 1-2 cm, glabrescent, base acute, margin crenulate, apex obtuse or subacute; petiole 6-10 mm. Flowers solitary, axillary, subsessile; bracteoles elliptic, 1 cm long; calyx 8 mm long, divided from the base, segments linear-subulate, hairy; corolla blue to violet, 3 cm long, tube narrowly cylindrical below, infundibuliform above, 2 cm, lobes 5, 6-9 mm, obovate-oblong; stamens 4, didynamous, filaments glabrous, 0.5-0.9 cm; ovary pubescent, 3 mm long, style 2 cm. Capsule slightly pubescent, 1.8 cm long, clavate, beaked, seeds 16-20, orbicular, 3 mm diam, flattened, glabrous, hairs on the margins.

Frequent, in most districts in bushes, on banks, in dry places and also in gardens, often in hedges.

Fl. & Fr.: July – December.

Gandhari RF (NZB), *BR* & *CPR* 7231; Manchippa (NZB), *BR* & *GO* 9215; Narsingi (MDK), *BR* & *CP* 11534; Aklaspur (KRN), *GVS* 20139, 22501 (MH); Jallada – Kothagudem (KMM), *RR* 106096 (BSID); Cherukupally (NLG), *BVRS* 667 (NU).

DYSCHORISTE Nees

Dyschoriste vagans (Wight) Kuntze, Revis. Gen. Pl. 2: 486. 1891; Gamble, Fl. Madras 2: 718. 1924. *Calophanes vagans* Wight, Icon. Pl. Ind. Orient. t. 1526. 1850; C.B.Clarke in Hook.f., Fl. Brit. India 4: 411. 1884.

Vern.: *Adavi neelambaramu.*

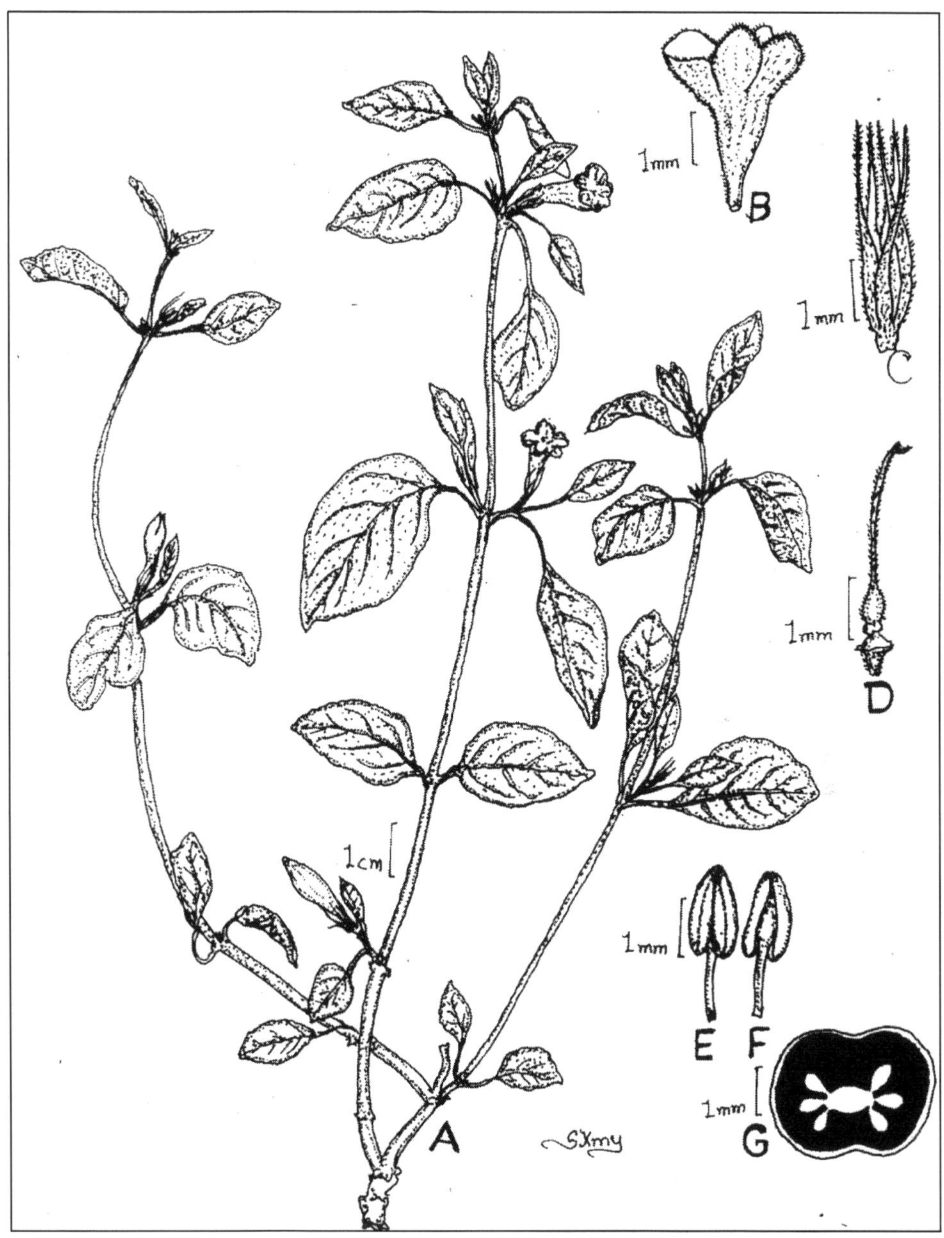

Figure 78: Dipteracanthus prostratus (Poiret) Nees

A. Twig, B. Corolla, C. Calyx, D. Pistil, E&F. Stamens, G. Ovary c.s.

Straggling herb, up to 45 cm high; branches obtusely quadrangular, puberulous, pale. Leaves ovate-lanceolate, 3-7.3 × 1.6-4 cm, glabrous, puberulous or more or less pubescent, pale beneath, base acute or sub acuminate or tapering, margin entire, apex acute to shortly acuminate; lateral nerves ca 8 pairs; petiole to 2.5 cm. Cymes axillary, shortly peduncled, 2-8-flowered, bracts and bracteoles small, linear; calyx 1.2 cm long, pubescent, tube 4 mm long, lobes subequal, lanceolate-subulate; corolla lobes bluish-violet, tubular-ventricose, pubescent without, to 19.7 mm, tube not exceeding the calyx, lobes subequal; stamens 4, filament pairs 4 and 5 mm long, anther 1.5 mm long, lobes with capillary white spur; ovary 2 mm, style filiform, 1.3 cm long. Capsule oblong; seeds 4, broadly ellipsoid, densely hairy.

Occasional, found in almost all hill ranges of dry and moist deciduous forests in Hyderabad, Khammam and Mahabubnagar districts.

Fl. & Fr.: September - February.

Mannanur RF (MBNR), *STR* 239 (MH); Hyderabad (HYD), *MRS* 81282 (DD); Ramavaram (KMM), *RR* 112618 (BSID); Pullaechelmal – Farhabad (MBNR), *SRS* 109160 (BSID).

ECBOLIUM Kurz

Ecbolium viride (Forssk.) Alston in Trimen, Handb. Fl. Ceylon 6 (Suppl.) 229. 1931, var. **viride**. *Justicia viridis* Forssk., Fl. Aeg. – Arab. 5. 1775. *Ecbolium linneanum* Kurz., Journ. Asiat. Soc. Bengal 40: 75. 1871; C.B.Clarke in Hook.f., Fl. Brit. India 4: 544. 1885; Gamble, Fl. Madras 2: 1074.1924. *Justicia ecbolium* L., Sp. Pl. 15. 1753.

Vern.: *Nakka thoka, Pacha vadambaramu*.

Undershrub, 50-120 cm high; branches terete, striate, young ones pubescent. Leaves coriaceous, elliptic-lanceolate, 3.6 –12 × 1.8-6 cm, glabrous above, pubescent and pale beneath, base tapering and decurrent, margin entire, apex obtusely acute-acuminate; lateral nerves ca 5 pairs, prominent; petiole slightly winged, 2 cm long. Spikes terminal, densely flowered, 4-7 cm long; bracts foliaceous, elliptic-oblong, pale green, to 2.1 × 1.4 cm, pubescent; bracteoles 2; calyx small, 3.5 mm, lobes subequal, linear-subulate, glandular pubescent; corolla bluish-green, tube very slender, 3 cm long, lobes 5, 2-lipped, upper lip 1.2 cm, shortly forked, lower lip with 2 wing like spreading lateral lobes, 1.5 cm; stamens 2, attached at the base of upper lip, exserted, filaments 2 mm, anther cells oblong, parallel, 2.5 mm, unequal, muticous; ovary 3 × 1 mm, style filiform, 3.5 cm long, stigma capitate. Capsule ovoid with basal beak, to 1.4 (2) × 0.5 mm, glabrous; seeds 2, ovate, orbicular, tuberculate.

Occasional in moist places of forests and along stream banks in Karimnagar and Mahabubnagar districts.

Fl. & Fr.: August - March.

ELYTRARIA Michaux *nom. cons.*

Elytraria acaulis (L.f.) Lindau in Engl. & Prantl, Nat. Pflanzenf. Nachtr. 1: 304. 1897; Gamble, Fl. Madras 2: 1009. 1924. *Justicia acaulis* L.f., Suppl. Pl. 84. 1781. *Elytraria crenata* Vahl, Enum Pl. 1: 106. 1804; C.B.Clarke in Hook.f., Fl. Brit. India 4: 394. 1884.

Vern.: *Yeddu adugu, Nela venu, Kukkapan, Pogetiaku.*

Herb to 25 cm high, stem 0. Leaves radical or sub radical, obovate-oblanceolate, or rarely oblong-obovate, 5.5-17 × 2.5-6.5 cm, villous or pubescent beneath, especially on the nerves, base cuneate to attenuate, margin crenate with rounded tips, apex obtuse; lateral nerves ca 12 pairs; petiole to 1.5 cm. Spike scapigerous, several, to 22 cm long, unbranched, rarely 2-branched; bracts 6 mm, rigid, ovate, acuminate or spinous tipped, spirally imbricate, lower sterile, upper fertile; bracteoles 2, linear, small, 4 mm; calyx lobes 4-5, lanceolate, sepals penicillate at apex, margins scarious; corolla pale blue or white, 7 mm long, tube 4 mm long, bilipped, 2+3; stamens 2; ovary 1.5 mm, style filiform, 8 mm long, stigma lanceolate. Capsule oblong, 1 cm long, valves recurved; seeds many, flattened, retinacula inconspicuous.

Common in dry districts, in grassy lands and open forest areas in most districts.

Fl. & Fr.: October - December.

Kagaznagar (ADB), *GO* 4365; Kandakurthi (NZB), *BR* & *GO* 9219; Gourammasari (MBNR), *TDCK* 13694; Mudigonda (NLG), *BVRS* 616; Pakhal (WGL), *ANH* 15981 (MH); Aklaspur (KRN), *GVS* 20220; Bhupatipur (KRN), *GVS* 22482 (MH); 8 KM towards north from Pasra (WGL), *RKP* 108230 (BSID); Yetapaka (KMM), *RR* 106081 (BSID).

ERANTHEMUM L.

1. Bracts white with green nerves .... **E. purpurascens**
1. Bracts green .... **E. pulchellum**

Eranthemum pulchellum Andrews, Bot. Repos. 2: t. 88. 1800. *E. nervosum* (Vahl) R. Br. ex Roem. & Schult., Syst. Veg. 1: 174. 1817; Gamble, Fl. Madras 2: 1024. 1924. *Justicia nervosa* Vahl, Enum. Pl. 1: 164. 1804. *Daedalcanthus nervosus* (Vahl) T. Anderson, J. Linn. Soc. London 9: 487. 1867; C.B.Clarke in Hook.f., Fl. Brit. India 4: 418. 1884.

Small shrub, up to 2 m high, stems 4-angled, puberulous. Leaves large, elliptic to ovate, 5-25 × 2.5-9 cm, pubescent or puberulous, base decurrent, margin obscurely dentate, or crenate-dentate, apex acuminate; petiole 1.2-6 cm. Flowers in terminal, ternate, paniculate spikes, 2.5-7.5 cm long; bracts white with green nerves. oblong-obovate, 1.2-1.8 cm long; calyx scarious; corolla blue, lobes broadly elliptic; stamens 4, 2 perfect and 2 staminodes, anther oblong, 2-celled. Capsule 1.3-1.5 cm long with a short point, glabrous.

Occassional in Hyderabad district.

Fl. & Fr.: January – May.

Eranthemum purpurascens Wight ex Nees in Wall., Pl. Asiat. Rar. 3: 106. 1832; Gamble, Fl. Madras 2: 1025. 1924. *Daedalacanthus purpurascens* (Nees) T. Anderson, J. Linn. Soc. London 9: 488. 1867; C.B.Clarke in Hook.f., Fl. Brit. India 4: 420. 1884. *Justicia pulchella* Roxb., Pl. Coromandel t. 177. 1805. non *Eranthemum pulchellum* T. Anderson 1800.

Herb to undershrub, to 1 m high, stem 4-angled, striate, shortly pubescent. Leaves ovate or elliptic, 4-10 × 2.5-8 cm, glabrescent, base rounded and then decurrent, sub cordate in upper sessile ones, margin crenulate, apex acuminate; lateral nerves ca 8 pairs, prominent; petiole decurrent, 3.5 cm, upper leaves much smaller than the lower ones, cordate, rounded or narrowed at base, sessile or shortly petiolate. Spikes terminal and axillary, 7 cm long, peduncle to 10 cm long; bracts green, ovate-lanceolate, to 2.5 cm long, long acuminate, conspicuously nerved, reticulate, margin ciliate; calyx scarious, 12 × 1 mm, oblong-lanceolate with long acumen, lobes 5; corolla blue, 4 cm long, tube long and slender, lobes obovate, twisted to the left in bud, spreading in flower; perfect stamens 2, staminodes 2; style long and slender, sparsely hairy. Capsule 1-1.2 cm long, clavate-oblong, pointed, glabrous.

Occasional along hilly areas of deciduous forests of Khammam, Warangal and Mahabubnagar districts.

Fl. & Fr.: November- February.

Pakhal (WGL), *ANH* 15923 (MH); Gangaram (WGL), *RKP* 110904 (BSID); Lakshmipuram RF (KMM), *RCS* 99009 (BSID); Perantapally (KMM), *RCS* 98990 (BSID).

GANTELBUA Bremek.

Gantelbua urens (B. Heyne ex Roth) Bremek., Verh. Kon. Ned. Akad. Wetensch., Afd. Natuurk. Sec. 2.148, 1944; Sant., Univ. Bomay Bot. Mem. 2: 28. 1952. *Ruellia urens* B. Heyne ex Roth, Nov. Pl. Sp. 302. 1821. *R. dura* Nees in Hook. Comp. Bot. Mag. 2: 311. 1836. *Hemigraphis dura* (Nees) T. Anderson, J. Linn. Soc. London 9: 461. 1867; C.B.Clarke in Hook.f., Fl. Brit. India 4: 422. 1884; Gamble, Fl. Madras 2: 1018. 1924.

A procumbent, hard hispid herb, bulbous based hairs, very prominent, 1-2 mm long, jointed, branches flexuose. Leaves oblong-elliptic or spathulate, 1.2-3 × 0.6-1.2 cm, base acute to attenuate, margin crenate, apex obtuse, lateral nerves 5-6 pairs; petiole 1-3 mm. Flowers in heads, outer bracts 4, linear-oblong, 1.8 × 4 mm long, inner bracts linear, 9 mm, densely hirsute, hairs brown at apical parts; calyx lobes divided from the base, subequal, to 9 mm long, linear, hirsute; corolla blue, 2.3 cm long, tubular, ventricose, lobes 5, sub equal, 1.3 × 1.8 cm, tube 8 mm long; stamens didynamous, filaments pairs 3 and 1.5 mm long, anther 2 mm long, divaricate, longer filaments hairy at the lower half; ovary 4 mm, style 1.7 cm long, sparsely hairy. Capsule linear-oblong, glabrous, grooved; seeds 6-8.

Occasional in wastelands and forests in Nizamabad district.

Fl. & Fr.: April – June.

Kandakurthi (NZB), *BR* 7261.

Note: The genus *Gantelbua* was separated from *Hemigraphis* on account of the peculiar structure of the inflorescence, but it distinguishes itself also from that genus by the deeply divided calyx and by the small size of the areola. The name is a latinization of the vernacular name of the type species (cf. Cooke, Fl. Bombay 2: 357. 1905).

GENDARUSSA Nees

Gendarussa vulgaris Nees in Wall., Pl. Asiat. Rar. 3: 104. 1832. *Justicia gendarussa* Burm. f., Fl. Ind. 10. 1768; C.B.Clarke in Hook.f., Fl. Brit. India 4: 532. 1885; Gamble, Fl. Madras 2: 1079. 1924.

Vern.: *Nela vavili, Gandha rasamu, Nallanochali.*

An erect undershrub, strong scented; branches terete, pale, striped. Leaves narrowly lanceolate, 6-15 × 2.5-6 cm, glabrous, base narrowed, margin entire, apex obtuse, lateral nerves 5 pairs; petiole to 4 mm. Flowers clustered in terminal, 10 cm long spikes; bracts linear, 4 mm, bracteoles 0; calyx lobes 5, free, linear-lanceolate, to 7 mm, acuminate; corolla white with purple streaked, 1.2 cm long, tube cylindric; stamens 2, lower anther cells spurred. Capsule glabrous.

Planted in hedges in shady position; frequently run wild in most districts.

Fl.: February- May.

Nirmal (ADB), *GO* & *BR* 9404; Pochampadu (NZB), *BR* 7300; Ramayampet (MDK), *BR* & *CP* 11570.

HEMIGRAPHIS Nees

1. Prostrate herb; longer filaments glabrous in the upper half, densely hairy in the lower half:
   2. Herb covered densely with white woolly hairs .... **H. hirta**
   2. Herb covered with scabrous hairs .... **H. venosa**
1. Erect or straggling herb; longer filaments densely hairy in the upper half, less so in the lower half .... **H. latebrosa**

Hemigraphis hirta T. Anderson, J. Linn. Soc. London 9: 462. 1867; C.B.Clarke in Hook.f., Fl. Brit. India 4: 42. 1884; Gamble, Fl. Madras 2: 1018. 1924. *Ruellia hirta* Vahl, Symb. Bot. 3: 84. t. 67. 1794.

Prostrate-decumbent, much branched, softly white – woolly herb, up to 15 cm long. Leaves ovate, chartaceous, 3-3.5 × 1-2 cm, base attenuate, margin crenate, apex obtuse to acute, pubescent. Flowers in small close terminal heads; bracts oblong or oblanceolate, 1-1.2 cm, calyx to 9 mm long, hairy without, tube to 5 mm long, lobes linear or spathulate in fruit; corolla purplish, pale blue or white, 1-1.2 cm diameter, tube up to 0.65 cm long, dilated in the upper half, lobes 5, up to 0.25 mm long; stamens 4, didynamous, short filaments to 0.5 mm, thick, longer ones laxly hairy, glabrous above, to 2 mm long, anthers oblong, 1.5 mm long, 2-celled. Capsule linear, 7.5-10 mm long, glandular-hairy at apex, 12-seeded.

Common, in grassy open places in Adilabad, Nizamabad, Karimnagar, Khammam and Mahabubnagar districts.

Fl. & Fr.: March – May.

Birsaipet (ADB), *GO* & *PVP* 4597; Gandhari RF (NZB), *TP* & *BR* 6336; Ibrahimpet (NZB), *BR* & *KH* 9046; Jalapenta (MBNR), *BSS* & *SKB* 32612; Rachapalli (KRN), *GVS* 22272 (MH); Rollapadu (KMM), *RR* 108034 (BSID).

Hemigraphis latebrosa (Roth) Nees in DC., Prodr. 11: 723. 1847 var. **latebrosa**; C.B.Clarke in Hook.f., Fl. Brit. India 4: 423. 1884; Gamble, Fl. Madras 2: 1018. 1924. *Ruellia latebrosa* Roth, Nov. Pl. Sp. 307. 1821.

Erect or straggling herbs, up to 35 cm high, stems faintly angled, ribbed, hairy. Leaves up to 7 × 4 cm, ovate, margin crenate-serrate, apex acuminate, base rounded or obtuse, hairy on both surfaces. Flowers pale blue on axillary, shortly peduncled cymes, bracts foliaceous, hairy; calyx hairy, lobes linear, strigose at first then scarious, lobes 5, unequal, tube 3 mm long, lobes unequal, to 6 × 1 mm; corolla blue, funnel-shaped; stamens 4, didynamous, shorter filaments glabrous, to 0.5 mm long, longer filaments densely bearded in the upper half with long hairs, to 2 mm long. Capsules 0.6 cm long, tetragonous, pubescent above, 6-seeded.

Common in undergrowth of deciduous forests in most districts.

Fl. & Fr.: November-March.

Potchera (ADB), *DAM* & *GO* 5060, 5067; Kunthala (ADB), *DAM* 5132; Pochammaralu (MDK), *BR* & *CP* 11471; Gangapur (MDK), *TP* & *CP* 12068; Kusumasamudram (RR), *MSM* 11028; Mannanur range (MBNR), *STR* 241 (MH); Kodimial (KRN), *GVS* 21863 (MH); Pakhal (WGL), *ANH* 15944 (MH), *RKP* 108258 (BSID); Kunavaram RF (KMM), *RCS* 98787 (BSID); Umamaheswaram (MBNR), *SRS* 109147 (BSID).

Hemigraphis latebrosa (Heyne ex Roth) Nees var. **heyneana** Bremek.

On laterite soil, common in Karimnagar district.

Fl. : December

Kodimial (KRN), *GVS* 21863 (MH).

Hemigraphis latebrosa (Heyne ex Roth) Nees var. **incana** Gamble, Fl. Madras 2: 1019. 1924.

Leaves very softly white-tomentose and thicker.

Occasional as an undergrowth in Warangal district.

Pakhal (WGL), *ANH* 15924 (MH).

Hemigraphis venosa C.B.Clarke in Hook.f., Fl. Brit. India 4: 423. 1884. Gamble, Fl. Madras 2: 1018. 1924.

A small, prostrate, hirsute herb, 5-30 cm, rooting at lower nodes. Leaves opposite, ovate, 2.5-3.2 × 1.2-1.8 cm, hispid, base acuminate, margin crenate, apex acute-acuminate; petiole to 1.8 cm. Flowers deep blue, in dense condensed spikes, axillary and terminal; bracts obovate, 1.2 cm, obovate; bracteoles 0.4 cm; sepals 5, 0.4 cm, subequal, lanceolate; corolla 1.6 cm, tube 0.6 cm, cylindric, mouth subequal; anthers subequal; ovary glabrous, tip glandular pubescent.

Occasional in Warangal district

Fl. & Fr.: November – April.

Pakhal (WGL), *RKP* 108263 (BSID).

HYGROPHILA R.Br.

1. Procumbent unarmed herbs; leaves opposite, entire; flowers in short spikes; calyx 5-lobed; capsule more than 15-seeded .... **H. polysperma**
1. Erect armed herb; leaves in pseudowhorls of 6, minutely dentate on margins; flowers in axillary whorled fascicles; calyx 4-lobed; capsule 4-8-seeded .... **H. schullii**

Hygrophila polysperma (Roxb.) T. Anderson, J. Linn. Soc. London 9: 456. 1867; C.B.Clarke in Hook.f., Fl. Brit. India 4: 406. 1884; Gamble, Fl. Madras 2: 1015. 1924. *Justicia polysperma* Roxb., Fl. India 1: 119. 1820. *Hemiadelphis polysperma* (Heyne ex Roth) Nees in Wall., Pl. Asiat. Rar. 3: 75. 80. 1832. *Ruellia polysperma* Heyne ex Roth, Nov. Pl. Sp. 305. 1821.

Procumbent herb with creeping stems, rooting at nodes, shoots 5-20 cm long, terete. Leaves narrowly oblong to ovate, to 20 × 6 mm, glabrous or slightly pubescent, base tapering, margin subentire to crenate, apex subobtuse; lateral nerves 4 pairs, not conspicuous; petiole 0-3 mm. Spikes terminal, to 9 cm long, bracts elliptic-oblong, obtuse, puberulous, bracteoles linear; calyx tubular, pubescent, divided from below the middle, teeth 5, narrowly linear, hairy, margin ciliate; corolla pale blue, 2-lipped, pubescent; stamens 2, staminodes teeth like, filaments filiform, anther cells parallel, muticous. Capsule narrowly oblong, 6-8 × 1.5 mm, tip shortly apiculate, glabrous, valves recurved after dehiscence; seeds 20-32, orbicular, hairy, 0.5 mm across.

Occasional in wet places along river banks, tanks, rice fields in Adilabad, Nizamabad, Karimnagar, Warangal and Hyderabad districts.

Fl. & Fr. September- October.

Kadam dam (ADB), *DAM* & *GO* 5401; Jannaram (ADB), *TP* & *PVP* 4193; Salura (NZB), *BR* 7277; Rechapalli (KRN), *GVS* 22262 (MH); Pakhal (WGL), *ANH* 15954 (MH).

Hygrophila schulli (Buch.-Ham.) M.R. Almeida & S.M. Almeida, J. Bombay Nat. Hist. Soc. (Suppl.) 221. 1986. *Behel schulli* Buch-Ham., Trans. Linn. Soc. Lond. 14: 289. 1825. *Asteracantha longifolia* (L.) Nees in Wall., Asiat. Rar. 3: 90. 1832; Gamble, Fl. Madras 2: 1015. 1924. *Barleria longifolia* L., Cent. Pl. 2: t. 22. 1756. *Hygrophila spinosa* T. Anderson in Thwaites, Enum. Pl. Zeyl. 225.1860; C.B.Clarke in Hook.f. Fl. Brit. India 4: 408. 1884. *H. auriculata* (Schumach.,) Heine, Kew Bull. 16: 172. 1962; Moulali in Pullaiah & Moulali, Fl. Andhra Pradesh 2: 713. 1997. *Barleria auriculata* Schumm. in Schumm. & Thonn., Beskr. Guin. Pl. 285. 1827.

Vern.: *Neerugubbi, Mulla gobbi, Kokilachi.*

Erect, armed unbranched, rarely less branched herb, up to 75 cm high; stem subquadrangular, reddish, strigose hairy, nodes thick, 6-spined. Leaves 6 at nodes, sometimes 8, the outer ones large, oblong-lanceolate or oblanceolate, 5.5-14 × 1.5-2.6 cm, sparsely hispid on both surfaces mainly on midvein, base tapering or cuneate, margin entire to minutely dentate, hispidular, apex subacute; lateral nerves ca. 10 pairs, prominent beneath, subsessile, thorns to 2.5 cm long, yellow. Flowers in whorls,

8 at each node; bracts leafy, lanceolate, hairy, bracteoles linear-lanceolate; calyx 4-partite, lobes unequal, upper one broader than other, to 1 cm long; corolla bluish purple to 3.2 cm, widely bilipped; stamens 4, filaments pairs 5 and 10 mm long, filaments of each pair united at the base, anther cells unequal; ovary oblong, 2.5 mm long, apically ciliate, style 2.5 cm long, slightly puberulous. Capsule 1 cm long, linear-oblong or subquadrangular, 4-8-seeded; seeds orbicular, 3 mm across.

Common throughout the state in swampy places, along water courses.

Fl.: November-April. Fruit through the year.

Khanapur (ADB), *TP* & *PVP* 4121; Kammarpally (NZB), *TP* & *BR* 6232; Mohammadabad RF (RR), *MSM* 10499; Molachintapally (MBNR), *TDCK* 15305; Pakhal (WGL), *GVS* 11657 (MH; Kodimial (KRN), *GVS* 21863 (MH); Narsapur (MDK), *KMS* 6808 (MH); Mannanur (MBNR), *SRS* 109112 (BSID); Rollapadu (KMM), *RR* 108037 (BSID); Narayanpur (NLG), *BVRS* 446 (NU).

INDONEESIELLA Sreem.

Note: C.P.Sreemadhavan (in Phytologia 15: 270-271. 1967; 16: 466. 1968) has described a new genus *Indoneesiella* Sreem. (=*Neesiella* Sreem.) based on the number of seeds as 4 (as found in *Andrographis echioides* (L.) Nees) as against many seeds in others.

1. Racemes longer than leaves, usually branched .... **I. longipedunculata**
1. Racemes shorter than leaves, not or scarcely branched .... **I. echioides**

Indoneesiella echioides (L.) Sreem., Phytologia 16: 466. 1968. *Justicia echioides* L., Sp. Pl. 16. 1753. *Andrographis echioides* (L.) Nees in Wall., Pl. Asiat. Rar. 3: 117. 1832; C.B.Clarke in Hook.f., Fl. Brit. India 4: 505. 1884; Gamble, Fl. Madras 2: 1061. 1924.

Vern.: *Aku pootha, Pottinelavemu, Sanyasi.*

Annual erect herb, to 50 cm high, stem quadrangular, grooved, pubescent. Leaves oblong to oblanceolate, 3-9.5 × 0.7-2.1 cm, sparsely white-villous hairy, base acute or cuneate, margin entire, apex obtuse or rounded; ciliate, lateral nerves ca 4 pairs; petiole 0 or very short. Racemes axillary, shorter than leaves, rachis pubescent; bracts lanceolate, minute; calyx lobes 5, linear, 6 mm, glandular pubescent; corolla white with brown tinge, tube 8 mm; stamens 2, adnate to the throat, filaments 8 mm long, sparsely hairy, anthers 2.5 mm, ovary elliptic, 1.5 mm, puberulous, style 15 mm, hairy. Capsule ellipsoid or ellipsoid-oblong, pointed at both ends, 1 cm long; seeds 4, ovoid-oblong, reticulately pitted.

Common, up to 500 m altitude in hill areas and plains in all districts.

Fl. & Fr.: August- December.

Ankusapuram (ADB), *GO* 4327; Manchippa RF (NZB), *TP* & *BR* 6036; Siddipet (MDK), *CP* 11373; Lingampalle (RR), *MSM* 12757; Irlapenta (MBNR), *BSS* 32384; Mannanur range (MBNR), *STR* 123 (MH); Durachipalli (NLG), *BVRS* 33 (NU); Pakhal (WGL), *KMS* 11653 (MH); Kodimial (KRN), *GVS* 21850 (MH); Krishnarajasagar (KMM), *RCS* 102462 (BSID); Garlapaadu RF (MBNR), *SRS* 109097 (BSID).

Indoneesiella longipedunculata (Sreem.) Sreem., Phytologia 16: 466. 1968.
Neesiella longipedunculata Sreem., Phytologia 15: 271. 1967.

An annual, erect, villous herb, up to 40 cm high, stem quadrangular, simple or branching from base. Leaves oblong to oblanceolate, 4-5 × 1.5-2 cm, sparsely white-villous, base acute to shortly auriculate, margin entire, apex subacute to apiculate. Flowers many in axillary, long internodal racemes; racemes longer than leaves; bracts elliptic-lanceolate, pubescent; calyx-lobes 5, glandular hairy without, enlarging in fruit; corolla pink or white with purple tinge, 5.5. mm long, unequal, 2 + 3, imbricate, slightly connate at base, linear, obtuse, hairy within; stamens 2. Capsule elliptic, lanceolate, 1.3 × 0.4 cm, glandular ciliate; seeds 4, reticulate pitted.

Occasional in foot hills in Medak, Nalgonda and Mahabubnagar districts .

Fl. & Fr.: December – March.

Sangareddy tank (MDK), *BR* 11297; Mannanur RF (MBNR), *GS* 18222; Rangapur RF (MBNR), *AJR* 20476; Nagarjuna Sagar (NLG), *KMS* 9743 (MH).

JUSTICIA L.

1. Flowers more than 2.5 cm long, shrubs:
   2. Plants more then 2 m high; bracts leafy .... **J. adhathoda**
   2. Plants less than 2 m high; bracts scarious with green nerves .... **J. betonica**
1. Flowers less than 2.5 cm long, herbs or undershrubs
   3. Bracts as long as calyx or smaller than calyx .... **J. oerstedii**
   3. Bracts longer than calyx:
      4. Corolla white with purple spots .... **J. glauca**
      4. Corolla pale pink:
         5. Bracts 5-8 mm long .... **J. vahlii**
         5. Bracts up to 4 mm long .... **J. japonica**

Justicia adhatoda L., Sp. Pl. 15. 1753; Stearn, J. Arn. Arb. 52: 640. 1971. *Adhatoda zeylanica* Medic. in Commentat. Acad. Elect. Sci. Theod.- Palat. 6: 393. 1790. *A. vasica* Nees in Wall., Pl. Asiat. Rar. 3: 103. 1832; C.B.Clarke in Hook.f., Fl. Brit. India 4: 540. 1885; Gamble, Fl. Madras 2: 1082. 1924.

Vern.: *Addasaram, Vasakam.*

A dense shrub, up to 2.5 m high, branches many, ascending, terete, lenticellate, grey pubescent. Leaves thin-coriaceous, elliptic-lanceolate, 8-24 × 2-7 cm, glabrous, puberulous when young, pale beneath, base acute or tapering, margin entire or slightly crenulate, apex acuminate, lateral nerves 12-14 pairs, prominent; petiole stout, to 3.8 cm, canaliculate. Spikes axillary or terminal, to 10 cm long; peduncle to 8 cm long; bracts leafy, to 3.3 x 2 cm, elliptic-ovate, 5-9-nerved, closely reticulately veined; bracteoles oblong-lanceolate, 1.8 x 0.7 cm, ciliate along margins; calyx lobes 5, slightly connate, 0.7-1.2 cm long, lobes imbricate, oblong-lanceolate, 3-nerved; corolla white,

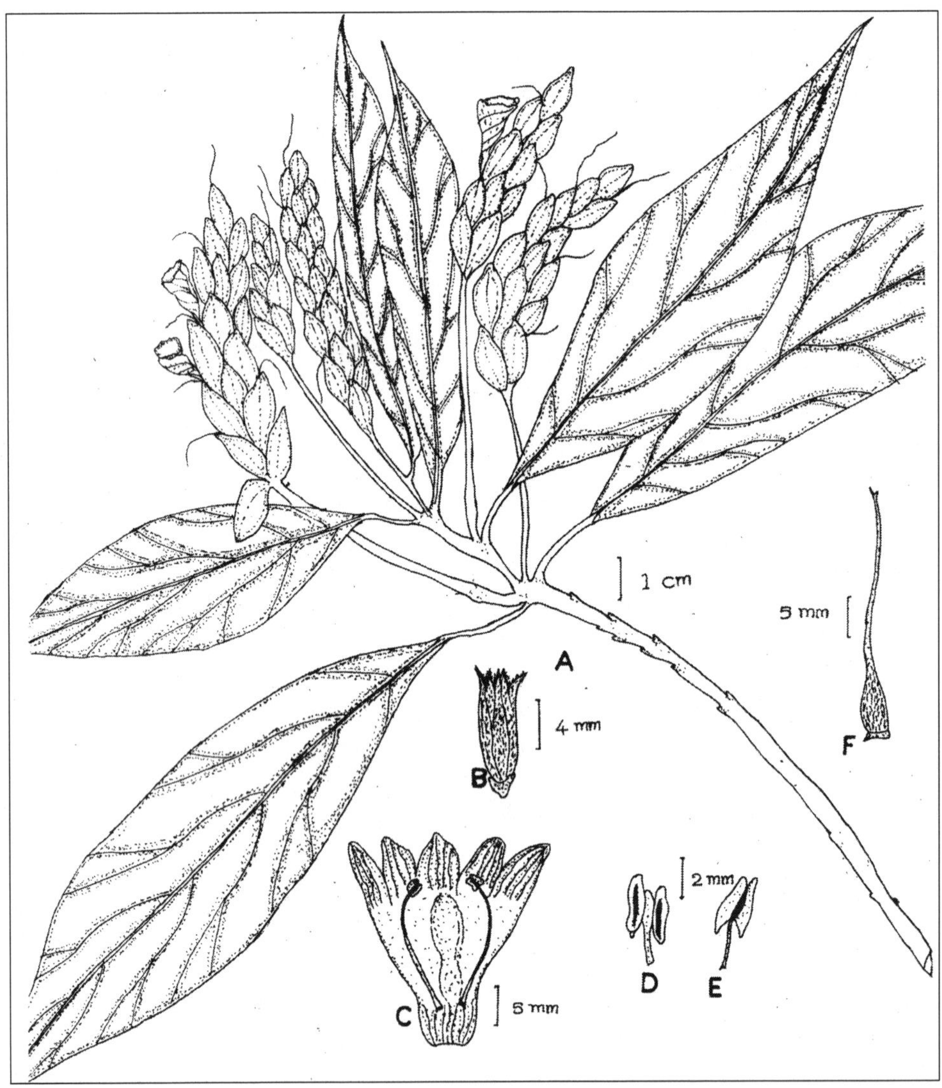

Figure 79: Justicia adhatoda L.

A. Twig, B. Calyx, C. Corolla split open, D&E. Stamens, F. Pistil.

puberulous without, 2-lipped, 2+3, upper lip ovate-oblong, curved, obtuse, notched, lower lip oblong, rounded; stamens 2, stout, curved, exserted, filaments to 2 cm, hairy at the base, anther cells parallel, apiculate at base; ovary oblong, 3 mm, pubescent; style 2.5 cm long, stigma slightly bifid. Capsule clavate, basally beaked, 1.8 cm long, pubescent; seeds 4, orbicular-oblong, tubercled.

Common, often found in or near villages in Adilabad, Nizamabad, Karimnagar, Ranga Reddi and Nalgonda districts, some times planted along fences.

Fl. & Fr.: October- March.

Kotapalli (ADB), *DAM* & *GO* 5145; Palwancha (NZB), *BR* & *KH* 9683; Mohammadabad (RR), *MSM* & *PRSR* 15123; Bhongir (NLG), *BVRS* 9 (NU).

Justicia betonica L., Sp. P1. 15. 1753. var. **betonica;** C.B.Clarke in Hook.f., Fl. Brit. India 4: 525. 1885; Gamble, Fl. Madras 2: 1078. 1925. *J. trinervia* Vahl, Enum. Pl. 1: 156. 1804; C.B.Clarke in Hook.f., Fl. Brit. India 4: 526. 1885; Gamble, Fl. Madras 2: 1079. 1924.

Vern.: *Telarantu, Tella-addasaramu. Velimungilu.*

Shrub to 1 m high, branches terete, striate, puberulous, nodes swollen. Leaves ovate or lanceolate, linear-elliptic, 2-11.5 × 1-4 cm, glabrous, base abruptly attenuate or cuneate, margin entire to sub crenate, apex acute or shortly acuminate; lateral nerves ca 5 pairs, conspicuous beneath; petiole 1-2 cm long. Flowers white, > 2.5 cm long in terminal, 4-15 cm long spikes, often panicled; bracts closely imbricate, ovate-mucronate, pubescent, 3-nerved from base, scarious with green nerves; hairy, 1.2-15 cm long, bracteoles similar to bracts; calyx lobes 4 mm long, linear-lanceolate, scarious, pubescent, 3-ribbed; corolla 2 cm long, 2-lipped, upper 2-lobed, lower 3-lobed, tube cylindric; stamens 2, slightly exserted, 6 mm long, curved, anther cells spurred; ovary pubescent. Capsule clavate, 0.7-1.2 × 0.3-0.5 cm, pubescent; seeds 2-4, corrugated or rugose.

Common, spread throughout in forests and by stream banks in Nizamabad, Medak, Ranga Reddi, Mahabubnagar and Warangal districts.

Fl. & Fr.: Through the year.

Kamareddy (NZB), *BR* & *KH* 9629; Narsapur RF (MDK), *BR* & *CP* 14110; Kanmankalva RF (RR), *MSM* 10457; Anantasagar RF (RR), *MSM* 10503; Irlapenta (MBNR), *BSS* & *SKB* 32373; Mannanur Range (MBNR), *STR* 52 (HH); Pakhal (WGL), *KMS* 11645 (MH); Hanmakonda (WGL), *RKP* 105369 (BSID).

Justicia glauca Rottl., Ges. Naturf. Freunde Berlin Neue Schriften 4: 219. 1802; C.B.Clarke in Hook.f., Fl. Brit. India 4: 529. 1885; Gamble, Fl. Madras 2: 1079. 1924.

Vern.: *Addesaram.*

Under shrub, 30-60 cm high, branchlets terete, grooved, puberulous, internodes long. Leaves broadly ovate, elliptic-ovate, 4-12 × 3-6 cm, glabrous, base rounded, to acute or obtuse, margin entire, apex acute, lateral nerves 4-6 pairs, prominent beneath; petiole to 2.4 cm. Flowers 0.7 cm long, in terminal and axillary 2.5-5 cm long spikes;

bracts orbicular or broadly ovate, 1 × 0.8 cm, ciliate along margins, acute, bracteoles linear, lanceolate, 4 × 1 mm; calyx lobes 5, lanceolate, 6 mm, acuminate, corolla white with purple spots, 8 × 3 mm, puberulous along the throat without; stamens 2, to 4 mm, slightly exserted, anther cells unequal, lower cell spurred at base; ovary 1.5 mm, style 7 mm, pubescent below. Capsule oblong, 1 × 0.5 mm, apically pointed; seeds 4, tubercled.

Common throughout in dry localities of waste lands and in forests.

Fl. & Fr.: October - January.

Sirpur (ADB), *GO* & *PVP* 4907; Manchippa RF (NZB), *TP* & *BR* 6024; Mingaram RF (NZB), *BR* 9292; Medak (MDK), *CP* 11721; Lingampalli (RR), *MSM* & *PRSR* 12752; Rangapur RF (MBNR), *AJR* 20482; Pakhal (WGL), *KMS* 13130 (MH); Rechapalli (KRN), *GVS* 20150 (MH); Hyderabad (HYD), *MRS* 83537 (DD); Yenkanapalem (KMM), *RR* 113828 (BSID); Nagarjunakonda (NLG), *BVRS* 93 (NU).

Justicia japonica Thunb., Fl. Jap. 20. 1784. *J. simplex* D.Don, Prodr. Fl. Nep. 118. 1825, non *Rostellularia simplex* Wight 1850; C.B.Clarke in Hook.f., Fl. Brit. India 4: 539. 1885; Gamble, Fl. Madras 2: 1080. 1924. *Rostellularia mollissima* Nees in Wall., Pl. Asiat. Rar. 3: 101. 1832. *Rostellularia japonica* (Thunb.) J.L. Ellis, Bull. Bot. Surv. India 22: 196. 1980 (1982).

A softly hairy diffuse herb, up to 35 cm high; branches often angled, zig-zag. Leaves oblong or linear, 2-4.2 × 0.7-1.8 cm, base narrowed, margin entire, apex acute, strigose on both surfaces, petiole 8 mm long, pubescent. Flowers in terminal or axillary spikes, bracts obovate, to 4 x 2 mm, pubescent without, margin scarious, long ciliate, bracteoles as long as calyx; calyx 3.8 mm long, lobes 4, subequal, unequal lanceolate, ciliate and hairy except at base, acute; corolla pale pink or lilac, to 5 mm long, tube 2.5 mm long, 2-lipped, upper lip slightly shorter than the lower one, obscurely 2-lobed, lower lip 2.5 mm long, 3-lobed, middle lobe rounded and hairy without, lateral ones acute, slightly longer than the middle lobe; stamens 2, adnate to the throat of corolla tube, filaments 2 mm long, flattened, 0.5 mm long, basal one white-appendaged at base. Capsule 3.5 × 1.5 mm, oblong, papery, villous at apex; seeds ovate oblong.

Occasional in open forests in Ranga Reddi district.

Fl. & Fr.: July - January

Justicia oerstedii Leonard, Publ., Field Mus. Nat. Hist., Bot. Ser. 18(4): 1230. 1938. *J. glabra* Koen. ex Roxb., Fl. Ind. 1: 132. 1820; C.B.Clarke in Hook.f., Fl. Brit. India 4: 535. 1885; Gamble, Fl. Madras 2: 1079. 1924.

Slender erect herb, up to 1 m high; branchlets streaked, glabrous or puberulous. Leaves ovate or elliptic-ovate, 4-12 × 3-6 cm, glabrous, base rounded, sometimes acute, rarely truncate, margin entire, apex acute to acuminate; lateral nerves ca 8 pairs, midvein raised beneath; petiole to 4.5 cm to 2 mm going upwards. Flowers in terminal or axillary lax dichotomous cymose panicles, rachis 5 cm long, pedicel slender, to 2 cm, bracts and bracteoles very small, to 1.5 mm, linear; calyx lobes 5, setaceous, to 3 mm long; corolla white, pinkish or pale yellow with purple spots;

stamens 2, to 4 mm, twisted; ovary 1.5 mm, style 6 mm long. Capsule clavate, 4-sided, 1 × 0.4 cm, pubescent, apex pointed; seeds 1.5 mm across, echinate, retinacula jointed.

Rare in open forests in Warangal district.

Fl. & Fr.: September - April.

Pakhal (WGL), *CSR* 317 (KU).

Justicia vahlii Roth, Nov. Pl. Sp. 14. 1821; Gamble, Fl. Madras 2: 1081. 1924. *J. diffusa* Roth var. *vahlii* (Roth) C.B.Clarke in Hook.f., Fl. Brit. India 3: 538. 1885. *Rostellularia vahlii* (Roth) Nees in Wall., Pl. Asiat. Rar. 3: 102. 1832.

A straggling herb with stout root stock, up to 50 cm high, branches slender, angular, grooved, scabrous; internodes long, nodes with short hairs. Leaves chartaceous, oblong-elliptic, glabrous, 2.7 × 0.8-1.5 cm, base cuneate, margin entire, apex acute or acuminate; lateral nerves ca 4 pairs; petiole 2-6 mm. Spikes terminal or axillary, narrow, to 17.5 cm long; bracts ovate-oblong, 2.5 mm, midrib and margins ciliate; calyx lobes 4, subequal, 4.4 mm, linear, midrib scarious, ciliate; corolla 6.6 mm long, pale pink with reddish streaks, sparsely hairy without; stamens 2, exserted, adnate at the throat of corolla, anther cells 0.8 mm long, lower anther cell spurred, filaments 3 mm long, flattened; ovary puberulous, style slender, 5 mm, stigma slightly bifid. Capsule 4 mm, clavate, pubescent; seeds brown, 1.2 mm across, papillose, flattened.

Common in open forests of most districts.

Fl. & Fr.: July – January.

Sattenapalli (ADB), *TP* & *PVP* 4095; Manchippa RF (NZB), *TP* & *BR* 6041; Pocharam tank (NZB), *BR* 9507; Kodimial (KRN), *GVS* 21876 (MH); Rollapadu (KMM), *RR* 108012 (BSID); Kambalpally (NLG), *BVRS* 156 (NU).

LEPIDAGATHIS Willd.

1. Capsule 2-seeded:
    2. Stem reddish .... **L. cristata**
    2. Stem not reddish:
        3. Stem faintly winged at nodes; leaves 5-11 cm long; corolla sericeous outside and at throat .... **L. hamiltoniana**
        3. Stem not as above; leaves 2- 5 cm long; corolla sericeous outside only .... **L. mitis**
1. Capsule 4-seeded .... **L. fasciculata**

Lepidagathis cristata Willd., Sp. Pl. 3: 400. 1800; C.B.Clarke in Hook.f., Fl. Brit. India 4: 516. 1885; Gamble, Fl. Madras 2:1066. 1924.

Vern.: *Mullabanti, Nakkapintuka, Karapani, Karapanipoudu.*

Perennial, procumbent herb, branches numerous, spreading, quadrangular, glabrous. Leaves linear-oblong or lanceolate-oblong, 3-6 × 0.5-1 cm, hairy beneath.

Spikes condensed, ovoid, on the lower part of the leafy branches; bracts ovate or elliptic, spinescent, bracteoles lanceolate, membranous hairy, spinescent; calyx 5-partite, densely hairy, 8 mm long, long spinous; corolla pale pink or pinkish, hairy without, tube 4 mm, narrow, upper lip oblong, obtuse, notched at apex, 5 mm, lower lip 3-lobed, divided nearly to the base; stamens 4, exserted, filament pairs 1.5 and 2.5 mm, anther cells unequal; ovary ellipsoid-globose, glabrous, style 7mm long, slightly hairy. Capsule ovoid-oblong, compressed; seeds 2, ovate, densely hairy.

Common throughout in waste lands, gravelly ground and also grows in poor soils in most districts.

Fl. & Fr.: September - March.

Boadh (ADB), *GO* 4632; Ibrahimpet RF (NZB), *TP* & *BR* 6127; Gandhari RF (NZB), *BR* & *CPR* 7233; Medak (MDK), *BR* & *CP* 11414; Dharmapur RF (RR), *MSM* & *KH* 10544; Edigonipally (MBNR), *DV* & *MV* 40713 (BSID); Nagarjunakonda (NLG), *DAM* 5904; Chandra Sagar (MBNR), *STR* 461 (HH); Pakhal (WGL), *KMS* 11650 (MH); Ramavaram (KMM), *RR* 112673 (BSID).

Lepidagathis fasciculata (Retz.) Nees in Wall., Pl. Asiat. Rar. 3: 95. 1832; C.B.Clarke in Hook.f., Fl. Brit. India 4: 522. 1885; Gamble, Fl. Madras 2: 1068. 1924. *Ruellia fasciculata* Retz., Observ. Bot. 4: 28. 1786.

An aromatic, softly villous straggling herb, 20-30 cm high; much branched from the creeping root stock; branches glandular pubescent with raised green stripes. Leaves membranous, often in unequal pairs, ovate or elliptic-laceolate, 5-8.5 × 1.4-3 cm, sparsely hairy above, densely beneath, base acute or cuneate, suddenly decurrent into the petiole, margin coarsly serrate or dentate, apex acute or shortly acuminate, lateral nerves 6-8 pairs; petiole up to 5 cm long. Flowers small, faintly fragrant, in sub capitate axillary spikes, ca 1.3 cm long; bracts foliaceous, lanceolate-oblong, ca 5 mm long; calyx hairy with simple glandular hairs, 5-partite nearly to the base, outer upper sepal the longest, oblong or lanceolate, ca 3-5 mm long, acute, 3-nerved, the two lower or anterior sepals linear, lateral one almost setaceous; corolla white to lilac, to 7 mm long, pubescent without, lower part tubular, to 3 mm long, ventricose part 2 mm; 2-lipped, lower lip 3-lobed, lobes obtuse, middle the longest, upper lip shortly 2-lobed; stamens 4, filaments 2-3 mm long, anthers 1 mm long, arising at the mouth of corolla tube; ovary globose, 1mm, disk 0.25 mm, style 3.5 mm long, hairy ¾ part from ovary, stigma 1 mm, capitellate. Capsule linear-oblong, 4-6 mm long, shortly beaked; seeds 4, orbicular, 1.5 mm across, hairy when wetted

Rare in rocky places and as an under growth in hilly forests in Karimnagar (Naqvi, 2001), Khammam and Mahabubnagar districts.

Fl. & Fr.: Novembr - April.

Nellippakka RF (KMM), *RCS* 99020 (BSID); Nilgiri view point (MBNR), *SRS* 109718 (BSID).

Lepidagathis hamiltoniana Wall., Pl. Asiat. Rar. 3: 96. 1832; C.B.Clarke in Hook.f., Fl. Brit. India 4: 516. 1984; Gamble, Fl. Madras 2: 1067. 1924.

Erect or diffuse herb with perennial root stock; stems 15-45 cm. Leaves linear, 5-11 × 0.4-1 cm, nerves 5-7 pairs, distant, base obtuse, margin minutely scabrid ciliate, apex acute. Flowers in heads both at the base of the stem and on the branches; bracts ovate or lanceolate, acuminate, and ending in a spine, glabrous but ciliate, with long stiff cilia; bracteoles 9 mm long, membranous below, ciliate upwards; calyx lobes lanceolate, acuminate and ending in a spine, glabrous but with long cilia; corolla silky hairy; stamens 4. Capsule narrowly conical or lanceolate, 6.5 mm long, acute, glabrous, some what compressed, 2-seeded.

Common in the forests in Khammam district.

Fl. & Fr.: October - February.

Edrallapalli RF (KMM), *RCS* 98757 (BSID).

Lepidagathis mitis Dalzell, Hooker's J. Bot. Kew Gard. Misc. 3: 226. 1851; C.B.Clarke in Hook.f., Fl. Brit. India 4: 516. 1885; Gamble, Fl. Madras 2: 1067. 1924.

A stiff undershrub, branches spreading, procumbent from woody root stock, glabrous, often purple when old. Leaves oblong or obovate, linear-oblong, 2-5 × 1-1.5 cm, lineolate above, more or less hairy on the nerves beneath, base narrowed, margin entire, ciliate, apex sub acute or obtuse. Flower heads globose, 12.5-18 cm diam., at the base of the stem; bracts chaffy, broad, obtuse, shortly spinous mucronate; the lower calyx lobes jointed to near the top; corolla white or pale pink, spotted with brown or purple; stamens 4. Capsule ovoid, 2-seeded.

Occasional in dry sandy places in Medak, Ranga Reddi, Khammam, Nalgonda and Mahabubnagar districts.

Fl. & Fr.: December - March.

Way to Dahhachelama (MBNR), *SRS* 107444; Billyguttalu (KMM), *RR* 106002; Nagarjuna Konda (NLG), *KMS* 9727 (CAL & MH).

MEYENIA Nees

Meyenia hawtayneana (Wall.,) Nees in Wall., Pl. Asiat. Rar. 3: 78. 1832; Gamble, Fl. Madras 2: 1009. 1924. *Thunbergia hawtayneana* Wall., Tent. Fl. Nap. 49: 1826; C.B.Clarke in Hook.f., Fl. Brit. India 4: 381. 1884.

Vern: *Podaparedu.*

A small climbing shrub, branches slender, many, stem quadrangular, glabrous. Leaves coriaceous, elliptic-ovate, 4-8.5 × 1.8-4.2 cm, glabrous, smooth and shining above, pale beneath, base cordate, margin entire, apex acute with apicule, 3-nerved from base, sessile. Flowers solitary or in pairs, axillary, bracteoles ovate-oblong, reticulate, leafy, 2.5 × 0.9 cm; calyx reduced, obscurely lobed, 2.5 mm; corolla bluish-purple yellow within, tube 2.5 cm, lobes 5, spreading, villous at the base; stamens 4, inserted at the mouth, included, filament pairs 1.2 and 1.8 cm long, swollen at base, acutely conical above, glabrous, valves woody; seeds 4, orbicular, 4 mm across.

Occasional in dry deciduous forests of Mahabubnagar district.

Fl. & Fr.: March - September.

NELSONIA R. Br.

Nelsonia canescens (Lam.) Spreng., Syst. Veg. 1: 42. 1824. *Justicia canescens* Lam., Encycl. 1: 41. 1791. *Nelsonia compestris* R. Br., Prodr. Fl. Nov. Holl. 481. 1810; C.B.Clarke in Hook.f., Fl. Brit. India 4: 394. 1884, Gamble, Fl. Madras 2: 1010. 1924.

Diffuse softly villous herb, 20-50 cm high with thick fleshy roots. Leaves elliptic, obovate-elliptic or elliptic-oblong, pilose on both surfaces, intermixed with glandular hairs, variable in size, lower large and sometimes 20 × 7.5 cm or more, upper few or one pair of smaller, base cuneate, apex obtuse or subacute, sometimes unequal, subsessile, 1.2-5 cm long. Flowers in dense short spikes, bracts foliaceous, imbricate; calyx lobes 5, unequal; corolla blue or pink, 2 – lipped, spreading; stamens 2, adnate to the middle of the carolla tube. Capsule ovoid, conical; seeds 1-5 in the lower part of each cell, closely minute granulate.

Frequent in moist and shady places and sides of streams in Nallamalais in Mahabubnagar district.

Fl. & Fr.: November – May.

PETALIDIUM Nees

Petalidium barlereioides (Roth) Nees in Wall., Pl. Asiat. Rar. 3: 82. 1832; C.B.Clarke in Hook.f., Fl. Brit. India 4: 416. 1884; Gamble, Fl. Madras 2: 1021. 1924. *Ruellia barlerioides* Roth, Nov. Pl. Sp. 310. 1821.

Straggling or scandant shrub, 1-2 m high; branches terete, stems numerous, subterete, glabrous, young ones pubescent. Leaves ovate, 4-10 × 3-6 cm, glabrous or nearly so except on the nerves beneath, base rounded, margin entire, apex acuminate; lateral nerves 6-8 pairs, prominent; petiole up to 5 cm long, canaliculate. Flowers subsolitary in opposite, decussate pairs or in subcapitate spikes on abbreviated lateral branches; bracteoles foliaceous, 2.2 x 1.5 cm, ovate, slightly acuminate, strongly nerved and reticulately veined, green, becoming pale brown and scarious; pedicel to 7 mm; calyx glandular hairy, tube 3 mm, lobes 8 mm, lanceolate-subulate, one is larger; corolla white, 3.3 cm long, pubescent without, prominently nerved, the palate with long deflexed hairs within; stamens 4, anther cells subsimilar, parallel, shortly spurred at base; ovary glabrous, style pubescent. Capsule clavate, cuspidate, 1.2-1.6 cm long, glabrous; seeds 2, orbicular, softly villous, mainly at the margins.

Rare in forests in Nizamabad and Khammam districts.

Fl. & Fr. January - May.

Mudheli RF (NZB), *BR* 6363; way to Tekuloddi (KMM), *PVS* 84095 (BSID); Thiruppumetta RF (KMM), *RCS* 99016.

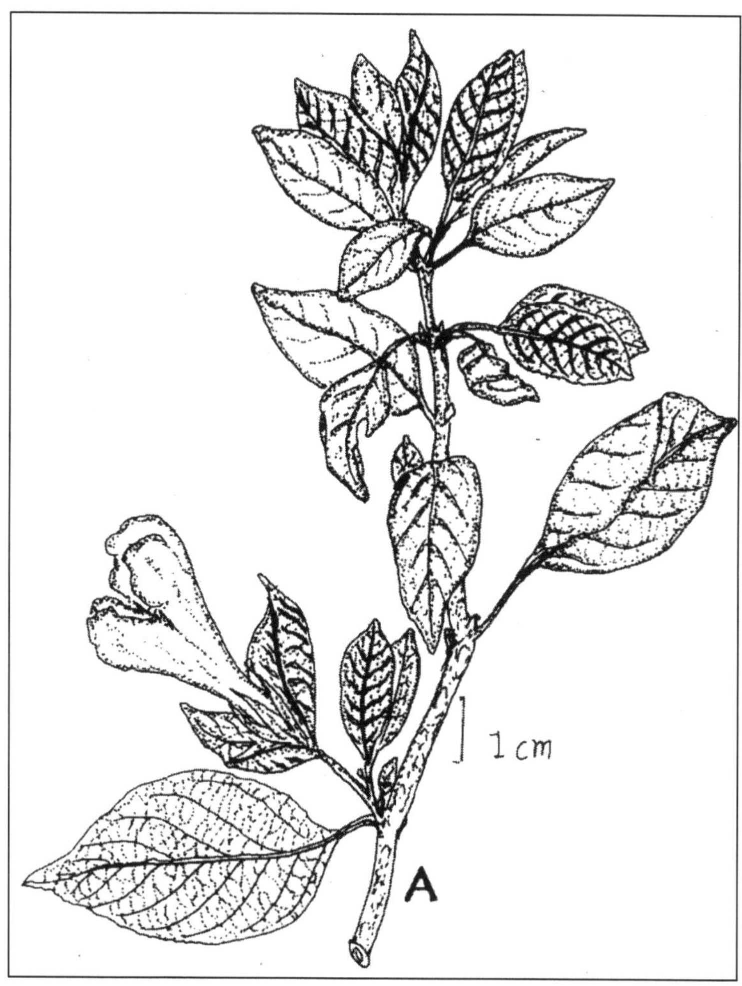

Figure 80: Petalidium barlereioides (Roth) Nees
A. Twig

RHINACANTHUS Nees

Rhinacanthus nasutus (L.) Kurz., J. Asiat. Soc. Bengal, 30: 79. 1870. *Justicia nasuta* L., Sp. Pl. 16. 1753. *Rhinacanthus communis* Nees in Wall., Pl. Asiat. Rar. 3: 109. 1832; C.B.Clarke in Hook.f., Fl. Brit. India 4: 541. 1885, Gamble, Fl. Madras 2: 1083. 1924.

Vern.: *Naga malli, Vontikommu rhinachettu, Pamu malli.*

Erect herb or subshrub, divaricately branched, to 1.5 m high; branches striate, subterete or obscurely angled, young shoot pubescent. Leaves elliptic or elliptic-lanceolate, 4-20 × 2-10 cm, puberulous, base acute to narrowed, margin entire or rarely very scarcely shallowly serrate, apex acute or shortly acuminate; lateral nerves 6-10 pairs, clearly impressed beneath; petiole 1-2 cm long, pubescent. Flowers sessile or shortly pedicellate, solitary or 2-3 together, distant on the divaricate branches of large, lax, terminal panicles; bracts and bracteoles ovate-lanceolate, 2 mm, glandular pubescent; calyx deeply partite, 3 mm, glandular pubescent, sepal lobes subequal, linear, subacute; corolla white, 3.2 cm, tube narrowly cylindric, 2.5 cm long, imbricate, 2-lipped, 2+3, upper lip linear-oblong, recurved, 9 × 1.5 mm, lower lip 1.3 cm long, lobes broadly oblong, obtuse, hispid without; stamens 2, attached at the throat, exserted, filaments 2 mm long, anther 2-celled, cells superposed, 2 mm; ovary 2 mm long, style 2.3 cm long, hairy, stigma minutely 2-fid. Capsule linear-ellipsoid, 1.8-2 cm, narrowed, pointed with a solid basal beak; seeds 4, orbicular, 4.3 mm diam, tuberculate, black.

Common, throughout among crevices in Ranga Reddi and Mahabubnagar districts.

Fl. & Fr. Most part of the year.

Anantagiri RF (RR), *MSM* & *PRSR* 12785.

ROSTELLULARIA Reichenb.

Note: The determination of various taxa belonging to the genus *Justicia* L. has been a problem, as the limits between them are very obscure and are based on intergrading characters. Santapau and other earlier workers also expressed similar opinion (Univ. Bombay Bot. Mem. 2; 83. 1952).

Nees in Wall., Pl. Asiat. Rar. 3: 76-100. 1832 proposed to split the genus *Justicia* L. into different genera like *Adhatoda, Gendarussa, Justicia, Rostellularia* based on the significant variations in the vegetative as well as floral features. Later Bremekamp (Verh. Nadul. Kad. Wet. 45(2): 1-78. 1948) resurrected these genera. Bremekamp is followed here.

1. Slender under shrubs .... **R. quinqueangularis**
1. Erect or prostrate or straggling herbs:
    2. Bracts and bracteoles usually shorter than calyx lobes:
        3. Raphides irregularly scattered; bracts and bracteoles generally glabrous .... **R. diffusa**
        3. Raphides ascendingly scattered; bracts and bracteoles more or less scabrous .... **R. prostrata**

2. Bracteoles as long as or slightly longer than calyx lobes:
   4. Branchlets and spikes conspicuously hairy .... **R. procumbens**
   4. Branchlets and spikes sparsely hairy:
      5. Leaves thin, ovate or elliptic; capsule oblong, pubescent ..... **R. crinita**
      5. Leaves linear-oblong; capsule ellipsoid .... **R. simplex**

Rostellularia crinita (Nees) Nees in DC., Prodr. 11: 373. 1847. *Rostellaria crinita* Nees in Wall., Pl. Asiat. Rar. 3: 101. 1832. *Justicia micrantha* Wall., ex. C.B.Clarke in Hook.f. Fl. Brit. India 4: 536. 1885; Gamble, Fl. Madras 2: 1080. 1924.

An erect herb, up to 60 cm high, branches trichotomous, pale, zigzag, striate and grooved, thinly pubescent. Leaves ovate or elliptic, 2.5-5 × 1.5-2.5 cm, sparsely hairy above, more or less glabrous beneath, lateral nerves 4-6 pairs, base acute to attenuate, margin entire, apex acute or subobtuse; petiole to 2 cm. Spikes terminal, bracts and bracteoles linear, blunt with bulbous based hairs, margins scarious near base, 4 mm; calyx lobes 4, unequal, scarious, midrib green, apex with blunt tip, 4-5 mm; corolla 2-lipped, pale purple, 6.2 mm, deltoid-oblong; stamens 2, filaments glabrous, 2.5 mm long, anther cells 0.6 mm; ovary 1.5 mm, pubescent at the tip, style 5 mm long, lower part hairy. Capsule 1.5 mm long, oblong, 2-4 mm, pubescent, 4-seeded; seeds 1 mm, very scarcely papillose.

Occasional in forests and waste lands of Adilabad, Ranga Reddi, Hyderabad, Karimnagar and Medak districts.

Fl. & Fr.: June-December.

Wankidi (ADB), *GO* 4472; East Narsapur RF (MDK), *BR* 11208; Mohammadabad RF (RR), *MSM* 10435; Saleswaram (MBNR), *BSS* & *SKB* 33338; Jalapenta (MBNR), *BSS* & *SKB* 32620 (BSID); Vempalli (KRN), *GVS* 20186 (MH); Pagarikutta (MDK), *KMS* 6650 (MH).

Rostellularia diffusa (Willd.) Nees in Wall. in DC., Prodr. 11: 371. 1847. var. **diffusa**. *Justicia diffusa* Willd., Sp. Pl. 1: 57. 1797; C.B. Clarke in Hook.f., Fl. Brit. India 4: 538. 1885; Gamble, Fl. Madras 2: 1081.1924.

A slender but sometimes woody herb, up to 40 cm high; branched, branchlets striate. Leaves elliptic-lanceolate, 2-4 × 2-3 cm, glabrous, raphides irregularly scattered, drying dark green, base truncate to rounded, margin entire, apex subacute, petiole to 1cm. Spikes linear, terminal or at the forks of dichotomy, peduncle to 6 cm; bracts and bracteoles ovate, 2 mm, midrib green, acute; calyx lobes 4, subequal, lanceolate, minutely scabrous at the tip, 3.5 mm long; corolla 3.5 mm long, pale purple, 2-lipped, upper lip slightly notched at apex, 2 mm, lower lip as broad as long, 3.5 mm, slightly 3-lobed at apex; stamens 2, to 2.5 mm long, anthers 0.6 mm; ovary 1 mm, style 3 mm long, pubescent at lower part. Capsule oblong, shortly pointed, glabrous, 4-4.8 mm long; seeds ovate, 0.9 mm.

Occasional in Adilabad district.

Fl. & Fr.: July-December.

Jaipur RF (ADB), *GO* & *PVP* 4826.

Rostellularia procumbens (L.) Nees in DC., Prodr. 11: 371. 1847. *Justicia procumbens* L., Sp. Pl. 15: 1753; C.B. Clarke in Hook.f., Fl. Brit. India 4: 539. 1885; Gamble, Fl. Madras 2: 1080. 1924.

A slender, diffuse herb, to 30 cm high; branchlets villous. Leaves elliptic-lanceolate or elliptic-oblong, 1-3 × 0.5-1.5 cm, sparsely pubescent, raphides transversely curved, base obtuse, margin entire, apex obtuse or acute, lateral nerves ca 4 pairs; petiole 0.8 cm long. Spikes up to 3 cm long, densely flowered; bracts and bracteoles linear, hirsute, margins scarious; bracts 4 mm long, bracteoles slightly longer than calyx; calyx 4-partite, to 3 mm long, linear, hisute, margin scarious, pale purple; corolla 2-lipped, 3.5 mm long; stamens 2, filaments 1-5 mm long, anthers 1mm, basal lobe acutely appendaged; ovary 1mm, pubescent, style 2.5 mm long, sparsely hirsute. Capsule 3 x 1.2 mm, sparsely hairy; seeds 0.8 mm across.

Common in open forests in most districts.

Fl. & Fr.: September - March.

Peddagutta (NZB), *BR* & *GO* 9201; Narsapur (MDK), *BR* & *CP* 11234; Rangammagudem RF (RR), *MSM* & *KH* 10576; Mudigonda (NLG), *BVRS* 618 (NU).

Rostellularia prostrata (Roxb. ex C.B.Clarke) R.B.Majumdar, Bull. Bot. Soc. Bengal 25: 75. 1971. *J. diffusa* Willd. var. *prostrata* Roxb. ex C.B.Clarke in Hook. f., Fl. Brit. India 4: 538. 1885. *J. prostrata* (Roxb. ex C.B.Clarke) Gamble, Fl. Madras 2: 1081. 1924.

Small, slender, branched, prostrate herb, up to 35 cm long, branches slender, angular. Leaves chartaceous, 2-4 × 0.5-2.1 cm, ovate or elliptic, glabrescent, base rounded to acute, margin entire, apex obtuse to acute, petiole 5 mm long, pubescent. Flowers in terminal and axillary pubescent spikes; bracts and bracteoles lanceolate; calyx with green midrib and scarious margins, corolla tube to 1 mm, 2-lipped, upper lip 2.5 mm long, 2-toothed at apex, pubescent without, lower lip to 3 mm long, 3-lobed, pubescent without; stamens 2, adnate to the throat of corolla-tube, filaments 2 mm long, flattened, villous within below the attachment, anther cells separate, 0.5 mm long, oblong, lower anther cell appendiculate. Capsule linear, 1cm long; seeds 4, ovoid, compressed, sub-concentrically ridged.

Common weed in fields, waste lands and road sides in all districts.

Fl. & Fr.: June – November.

Sirpur (ADB), *GO* 4391; Janakampalli (NZB), *BR* 6319; Sardhana (MDK), *BR* & *CP* 11626; Amrabad (MBNR), *BSS* & *SKB* 32533, *STR* 113 (MH); Koriguttalu (KMM), *RR* 106035 (BSID).

Rostellularia quinqueangularis (Koen. ex Roxb.) Nees in DC., Prodr. 11: 375. 1847. *Justicia quinqueangularis* Koen. ex Roxb. Fl. Ind. 1: 134. 1820; C.B.Clarke in Hook.f., Fl. Brit. India 4: 536.1885; Gamble, Fl. Madras 2: 1081. 1924.

A slender under shrub, with curving angular branches, up to 50 cm high, rooting at the nodes. Leaves chartaceous, elliptic-oblong or lanceolate, 2.5-7.5 × 0.3-0.9 cm, glabrous, base narrowed, margin entire, apex obtuse. Flowers in slender terminal spikes; calyx 5 mm long, glabrous without, 4-lobed, lobes subequal, nearly divided to the base, 1 mm broad, margins scarious; corolla rose, to 1 cm long, 2-lobed, lower lip to 5.5 mm long, 3-lobed, middle lobe hairy without, rounded, lateral lobe acute; stamens 2, adnate a little below the throat of the corolla-tube, villous below the attachment of the stamens; stamens 2, filaments 3.5 mm long, flattened, anther cells separate, 1 mm long; ovary 1.2 mm long, style 6 mm long. Capsule obovoid, glabrous, 7 × 2.2 mm, obtuse, apiculate; seeds with wavy, subconcentric ridges.

Common in forests of Hyderabad district (Ramana, 2010).

Fl. & Fr.: July – November.

Rostellularia simplex Wight, Icon. Pl. Ind. Orient. t. 1542. 1850, non *Justicia simplex* D.Don 1825. *Justicia notha* C.B.Clarke in Hook. f., Fl. Brit. India 4: 537. 1885; Gamble, Fl. Madras 2: 1080. 1924.

Erect herb, up to 40 cm high, branches striate, angled, zig-zag. Leaves sessile, 5-7.5 × 0.4-0.7 cm, linear-oblong, acute at both ends. Flowers in terminal, solitary, dense spikes; bracts ovate, scarious, with green midrib; sepals linear-oblong; corolla pink. Capsule ellipsoid.

Occasional in Karimnagar (Naqvi, 2001), Mahabubnagar and Ranga Reddi districts.

Fl. & Fr.: Aug. – March.

On the way to Koilkonda (MBNR), *SRS* 109108 (BSID).

RUELLIA L.

1. Roots tuberous .... **R. tuberosa**
1. Roots not tuberous:
    2. Leaves linear, to 16 x 1 cm .... **R. sindica**
    2. Leaves elliptic or ovate, to 4 x 2.5 cm .... **R. patula**

Ruellia patula Jacq., Misc. Austriac. 2: 358. 1781; C.B.Clarke in Hook.f., Fl. Brit. India 4: 412. 1884; Gamble, Fl. Madras 2: 1017.1924. *Dipteracanthus patulus* (Jacq.) Nees in Wall., Pl. Asiat. Rar. 3: 82. 1832.

Erect or suberect herb, to 50 cm high, young branches subquadrangular, hoary pubescent. Leaves elliptic or ovate, 1.5-4 × 0.6-2.5 cm, appressed pubescent, densely beneath, base rounded or acute, margin coarsely crenulate, apex obtuse or sub acute; lateral nerves 4 pairs, petiole ca 1.8 cm long. Flowers solitary, 2-3 together, axillary; bracteoles foliaceous, ovate or elliptic, 1cm long, densely pubescent; calyx 6 mm long, divided from the base, segments 5, linear-lanceolate; corolla blue to violet, 2.5-3 cm long, pubescent without, tube 1.5 cm, narrowly cylindric in the lower, upper part campanulate-infundibuliform, lobes 6 mm, broadly elliptic; stamens 4, filaments included, glabrous, 4 and 5.8 mm long, anthers 2 mm; ovary glabrous, 1.5 mm long, style pubescent, 1.4 cm long. Capsule 1.8 cm long, clavate, beak basal, solid, glabrous; seeds ca 12, suborbicular, flattened.

Common in most districts on waste lands and road sides.

Fl. & Fr.: July-December.

Bijjur RF (ADB), *GO* & *PVP* 4986; Zaheerabad (MDK), *CP* 11590; Achampeta Range (MBNR), *STR* 309 (MH); Venkatapuram (MBNR), *BR* & *MV* 41361; Nagarjunasagar (NLG), *KMS* 9747 (MH).

Ruellia sindica Ghafoor & Heine, Willdneowia 16: 122. 1986. *R. longifolia* T. Anderson, J. Linn. Soc. 9: 460. 1867; Cooke, Fl. Bombay 2: 432. 1967.

Ascending herb, up to 50 cm high, clothed all over with glandular hoary pubescence, stems numerous, terete, woolly at the base. Leaves linear, 3-9 cm long, margin entire, apex acute; petioles continuous with the attenuated leaf blade. Flowers solitary, axillary, very shortly pedicellate; calyx densely strigose, segments lanceolate-subulate; corolla pubescent outside, lilac, lobes 5, subequal; stamens 4, included; ovary pubescent. Capsule pubescent, clavate, pointed; seeds broadly ellipsoid, much compressed, clothed all over with glistening white hairs, when dry appear mucronate by the hairs converging at the apex.

Rare, occurs among bushes in shade in Karimnagar district

Fl. & Fr.: September – December.

Karimnagar, *Naqvi* 1786 (KU), *E.N. Murthy* 472 (KU).

Ruellia tuberosa L., Sp. Pl. 635. 1753; Gamble, Fl. Madras 2: 1017. 1924.

Vern.: *Jurbula gadda.*

Erect herb, up to 50 cm high, much branched, branches quadrangular, tender parts hispid. Leaves elliptic-obovate or elliptic-oblong, 4-12 × 3-4.5 cm, lineolate, base cuneate to tapering, margin entire to undulate or scarcely crenate, apex obtuse; lateral nerves 8-10 pairs; petiole 1-2.5 cm. Flowers in axillary dichasial cymes; bracts and bracteoles linear-lanceolate; sepals 5, subequal, 1.5 cm long, setaceous, puberulous; corolla blue to violet, to 5 cm long, lobes subequal; stamens 4, included, filament pairs 4 and 8 mm long, hairy at base, anthers 4 mm, sagittate, staminode narrowed; ovary 4 mm long, style 2 cm long, stigma bilamellate, ciliate. Capsule oblong, 2.8 cm long, linear-oblong; seeds flattened, brown, hairy, more dense on margins.

Common throughout, in wastelands, waysides near habitations.

Fl. & Fr.: Throughout the year.

Gacchibowli (RR), *MSM* 15200; Edigonipally (MBNR), *DV* & *MV* 40708; Kazipet (WGL), *RKP* 105204 (BSID); Yerraguntapalli (NLG), *BVRS* 629 (NU).

RUNGIA Nees

1. Flowering bracts similar .... **R. repens**
1. Flowering bracts not similar:
    2. Floral bracts villous .... **R. pectinata**
    2. Floral bracts glabrous .... **R. parviflora**

Rungia parviflora Nees in Wall., Pl. Asiat. Rar. 3: 110. 1832; C.B.Clarke in Hook.f., Fl. Brit. India 4: 550. 1885; Gamble, Fl. Madras 2: 1071. 1924. *Justicia parviflora* Retz., Observ. Bot. 5: 9. 1788.

A small, soft annual herb. Leaves membranous, variable, ovate, lanceolate or linear-spathulate, to 4 × 3 cm, apex obtuse. Flowers in small spikes; barren and flowering bracts very different, the former in 3 rows on one side of the spike, thick, elliptic, mucronate; the later with the flowers on the other side, conspicuously hyaline-margined, ovate or orbicular-ciliate; calyx-lobes 5; corolla pinkish blue; stamens 2, filaments glabrous, lower anther cell appendaged. Capsule elliptic-oblong, compressed, 4-seeded.

Occasional in dry forests in Karimnagar and Warangal districts.

Fl. & Fr.: November – March.

Aklaspur (KRN), *GVS* 25662 (MH); Pakhal (WGL), *KMS* 11652 (MH, CAL).

Rungia pectinata (L.) Nees in DC., Prodr. 11: 469. 1847 *p.p. Justicia pectinata* L., Amoen. Acad. 4: 299. 1759. *Rungia parviflora* (Retz.) Nees var. *pectinata* (L.) C.B. Clarke in Hook. f., Fl. Brit. India 4: 550. 1885; Gamble, Fl. Madras 2: 1071. 1924.

Vern.: *Pindi kunda.*

Annual prostrate to erect herb, 6-30 cm high, diffusely branched, branchlets hairy. Leaves chartaceous, leaf pairs unequal; 2.5-7.5 × 0.5-2.6 cm, sparsely hirsute, variable, ovate-lanceolate or linear-spathulate, base rounded or acute, margin subentire, apex acute, lateral nerves 3-5 pairs, raised beneath; petiole to 2 cm. Spikes to 1.8 cm long, axillary; sterile and fertile bracts different, sterile bracts thick, elliptic-oblong in two rows on one side of the spike, to 4 x 1.75 mm, arranged on the other side of the flowers; fertile bracts obovate, conspicuously hyaline margined, 4 x 3 mm, ciliate, bracteoles narrowed than floral bracts, 3.2 × 1.5 mm, hyaline, ciliate along margins; calyx 5-partite, long-hirsute, 3 mm long, linear-lanceolate; corolla 2-lipped, 5 mm long, upper lip shortly acuminate; stamens 2, filaments 2 mm long, anthers 1 x 0.25 mm; style 3 mm long, stigma bifid. Capsule ovate-oblong, 2.5 × 1.5 mm, acute and hairy at apex; seeds 4, orbicular, 1 mm across, tuberculate.

Frequent along hilly slopes in moist localities on banks of streams and forest undergrowth in most districts.

Fl.: November-April; Fr. December-May.

Potchera (ADB), *DAM* 5056, 5074; Alisagar (NZB), *BR* 7279; Bodhan (NZB), *BR* 7271; Choutkoor (MDK), *TP* & *CP* 11913; Kondapur RF (RR), *MSM* & *KH* 10322; Borapuram (MBNR), *BSS* & *SKB* 32388; Jalapenta (MBNR), *BSS* & *SKB* 32613; Aklaspur (KRN), *GVS* 22483 (MH); Musalivera beat, Ramavaram range (KMM), *RR* 112614 (BSID); Perantapally RF (KMM), *RCS* 98992 (BSID).

Rungia repens (L.) Nees in Wall., Pl. Asiat. Rar. 3: 110. 1832; C.B.Clarke in Hook.f., Fl. Brit. India 4: 549. 1885; Gamble, Fl. Madras 2: 1070.1924. *Justicia repens* L., Sp. Pl. 15. 1753.

Vern: *Palakavelli, Kharmor.*

Erect or decumbent herb, up to 30 cm high; stem slender, sub terete, strigose, often rooting at nodes. Leaves elliptic-oblong or oblong-lanceolate, 1.8-7 × 0.2-1.4 cm, sparsely spreading strigose, base obtuse or rounded, margin entire, apex acute; lateral nerves 6 pairs; petiole 1-2 mm. Spikes erect, 2-8 cm long, terminal; bracts closely imbricate, 4-farious, orbicular, 7 x 6 mm, margins conspicuously whitish, scarious, apex acuminate or cuspidate; bracteoles 2; 3.8 x 1 mm, lanceolate with scarious margins, with glandular cilia; calyx 3.2 mm long, ciliate, glandular hairs; corolla pinkish with darker spots, pubescent without, 1.2 cm long, limb 2-lipped, upper lip oblong, emarginate, lower lip larger, shortly 3-lobed; stamens 2, filaments 2.8 mm, anther 1.8 mm, oblong, cells superposed, lower cells with short appendage at base. Capsule 5 mm long, ovoid-oblong, compressed, membranous; seeds 4, ridged, pale brown.

Common throughout, among grasses in plains, bunds of paddy fields, sometimes along roadsides and in forests.

Fl. & Fr.: September - March.

Jannaram (ADB), *TP* & *PVP* 4183; Manchippa RF (NZB), *TP* & *BR* 6042; Gandhari (NZB), *TP* & *BR* 6348; Narsapur (MDK), *BR* & *CP* 14100; Molachintapally (MBNR), *TDCK* 15316; Saleswaram to Rampur (MBNR), *BSS* & *SKB* 3339; Durachipalli (NLG), *BVRS* 33 (NU); Kodimial (KRN), *GVS* 21838 (MH); Yenkur (KMM), *RR* 108563 (BSID).

THUNBERGIA Retz. *nom. cons.*

Note: Bremekamp (Verh. Kon. Ned. Akad., Wetensch. Afd. Natuur. Sect. 2. 50: 51. 1955), Airy Shaw (Willis, Dict. Fl. Pl. & Ferns. Rev. Ed. 1966) and Santapau (1967) treated *Thunbergia* Retz. under the separate family Thunbergiaceae. However, we retain this genus under Acanthaceae only.

1. Flowers axillary, solitary or in pairs on long pedicels; calyx 10-16 toothed; corolla white or yellow:
   2. Corolla white; petiole not winged .... **T. fragrans**
   2. Corolla yellow; petiole winged .... **T. alata**
1. Flowers in racemes; calyx entire or crenulated ring; corolla pale blue .... **T. grandiflora**

Thunbergia alata Bojer ex Sims. in Bot. Mag. t. 2591. 1825; C.B.Clarke in Hook.f., Fl. Brit. India 4: 391. 1884; Gamble, Fl. Madras 2: 1008. 1924.

English name: Black-Eyed Susan.

Slender branched vine, stem hairy, slightly winged. Leaves sagittate, 4-9.5 x 3-6.8 cm, pubescent, more dense beneath, base hastate, margin entire to scarcely dentate, ciliate, apex mucronate, prominent, 5-nerved from base, petiole to 9 cm, winged. Flowers 1-2, axillary, bracts 2, 2.3 × 1.6 cm, pubescent, hastate, pedicel 4.8 cm; calyx divided from base, lobes 15, subulate, hairy, unequal, 3 mm; corolla yellow with purple throat, 3.5 cm long, lobes 5, spreading; stamens 4, attached at the throat of the tube, filament pairs 5 and 4 mm, base curved, anther lobes bearded, unequal, spurred,

4 mm; ovary 2 mm, style erect, 1.2 cm, stigma 2-fid, lobes lamellate, one above the other. Capsule depressed globose, pubescent, beak apical, long.

Cultivated, often as an escape.

Fl. & Fr.: Through the year.

Thunbergia fragrans Roxb., Pl. Coromandel t. 47. 1796 var. **fragrans**; C.B.Clarke in Hook.f. Fl. Brit. India 4: 390. 1884; Gamble, Fl. Madras 2: 1007. 1924. *T. fragrans* Roxb. var. *laevis* (Nees) C.B.Clarke in Hook f. Fl. Brit. India 4: 391. 1884; Gamble, Fl. Madras 2: 1007. 1924. *T. laevis* Nees in Wall., Pl. Asiat. Rar. 3: 77. 1832.

Eng.: The White Thunbergia.

Twining perennial, stem slender, slightly hairy, longitudinally 2-grooved. Leaves deltoid-ovate, 4-12 × 3-5.5 cm, thinly puberulous, base truncate, with 1 or 2 broad teeth on either side, margin entire, apex acute with apicule; basally 3-5-nerved, petiole to 3 cm, scabrid, canaliculate. Flowers axillary, solitary or rarely 2-together, pedicel to 5 cm long, bracteoles 1.6 cm long, lanceolate; calyx appressed hairy without, tube 1.5 mm long, deeply lobed, teeth 12-15, 3-4 mm, subulate, glandular pubescent; corolla white, 3.7 cm across, up to 5 cm long, tube cylindric, 2.6 cm long, slightly contracted, limb 5-lobed, spreading; stamens 4, didynamous, inserted near the base of the corolla tube, filament pairs 4 and 6 mm, anther oblong, unequal, 3 × 3.5 mm, linear-oblong, beaked at apex; ovary 2 mm, style 2 cm, stigma 2-lipped. Capsule globose below, 9 mm, suddenly narrowed into a sword-shaped beak, to 1.8 cm long; seeds 4, blackish-brown, 5 mm diam, excavated on the plane face, reticulate.

Common on moist deciduous and dry deciduous forests on thickets, fences in Mahabubnagar district.

Fl. & Fr.: Throughout the year.

Pochammaralu (MDK), *BR* & *CP* 11487; Naskal (RR), *MSM* & *PRSR* 15108.

Thunbergia grandiflora Roxb. ex Edw., Bot. Reg. 6 : t. 495. 1820; Hook.f., Fl. Brit. India 4 : 392. 1884; Gamble, Fl. Madras 2 : 1008. 1924. *Flemingia grandiflora* Roxb. ex Rottl., Ges. Naturf. Fr. Berlin Neue Schr. 4 : 202. 1803.

Large, perennial, woody climbers. Leaves deltoid to ovate-lanceolate, 8-15 x 4-7 cm, acute to subacuminate at apex, subcordate to deeply cordate at base with angular lobes, pubescent on both surfaces; petiole to 7 cm long. Flowers in terminal racemes, rarely axillary, solitary or in clusters; bract small, deciduous; inner bract foliaceous; bracteoles falcate–elliptic; calyx reduced to a subentire, velvety ring, to 1.5 mm long; corolla light-blue, 5-7 cm long; lobes subequal, spreading, 2-3 cm long. Capsule upcurved, with 4-quetrous beak, 3-5 cm long.

Grown as garden ornamental in hedges and on arches.

Fl. & Fr.: September– December.

Alisagar (NZB), *BR* 7289.

VERBENACEAE

1. Inflorescence cymose or paniculate:
   2. Cymes capitate, flowers few, sessile .... **Symphorema**
   2. Cymes paniculate, flowers many, pedicellate:
      3. Flowers essentially actinomorphic .... **Tectona**
      3. Flowers more or less zygomorphic:
         4. Fruits composed of 1-celled pyrenes; calyx accrescent in fruit:
            5. Flower bud symmetrical or, if asymmetrical, usually expanding abruptly on upper side due to resupination; anterior corolla lobe only slightly (if at al) larger than the other; anthers versatile; stigma lobes equal...... **Clerodendrum**
            5. Flower bud markedly asymmetrical, corolla expanding abruptly on lower side only; anterior corolla lobe frequently much larger than the other four; anthers usually basifixed (occasionally approaching versatile); stigma lobes frequently unequal .... **Rotheca**
         4. Fruits composed of one 4-celled and 3 – celled pyrenes; calyx usually not accrescent in fruits:
            6. Leaves simple; corolla 4 – lobed .... **Premna**
            6. Leaves compound; corolla 5 – lobed
               7. Corolla-tube small, campanulate .... **Vitex**
               7. Corolla-tube large, funnel-shaped .... **Gmelina**
1. Inflorescence racemose or spicate:
   8. Inflorescence spicate:
      9. Pyrenes normally 1-celled and 1-seeded:
         10. Perfect stamens 2 .... **Stachytarpheta**
         10. Perfect stamens 4:
            11. Fruits drupaceous, succulent .... **Lantana**
            11. Fruits dry, hard:
               12. Procumbent or creeping herbs; spikes usually elongating in fruit; bractlets not 4-ranked .... **Phyla**
               12. Erect bushy under-shrubs or shrubs; spikes not elongating in fruit; bracts or bractlets + 4 ranked .... **Lippia**
      9. Pyrenes normally 2 – celled and 2-4-seeded .... **Priva**

8. Inflorescence racemose; flowers pedicellate:
    13. Shrubs:
        14. Calyx broadly saucer-shaped, petaloid; inflorescence axillary and terminal; flowers red or yellow .... **Holmskioldia**
        14. Calyx tubular, not petaloid; inflorescence terminal; flowers bluish or white .... **Duranta**
    13. Trees; flowers fragrant .... **Citherexylum**

CITHEREXYLUM L.

Citherexylum spinosum L., Sp. Pl. 625. 1753 *C. subserratum* Sw., Nov. Gen. Pl. 91. 1788. *C. fruticosum* L. var. *subserratum* (Sw.) Mold., Phytologia 1: 17. 1933 & 6: 383. 1958. Rajendran & Daniel, Indian Verbenaceae 77. 2002.

Trees, up to 5 m high; branches and branchlets obtusely 4-angular, ribbed, glabrous, sparsely lenticellate. Leaves 8-12 × 4-6 cm, elliptic-ovate, chartaceous, base acute, margin entire, apex obtuse, glabrous; lateral nerves 4-6 pairs. Flowers white in terminal drooping racemes; peduncles 2-4 cm long; bracts elliptic-linear, pubescent; calyx cupular, 5-toothed; corolla white, tubular-campanulate, 5-lobed; stamens 2; stigma dilated. Fruit not seen.

Occasional and planted in house compounds for its flowers.

Fl. & Fr.: May – October.

Yadapalli (NZB), *BR* 6468.

CLERODENDRUM L.

1. Inflorescence a cyme:
    2. Erect shrub; leaves sinuate or serrate, pubescent; calyx deeply toothed .... **C. phlomides**
    2. Scandent or scrambling shrub; leaves entire, glabrous; calyx subtruncate or shallowly toothed .... **C. inerme**
1. Inflorescence a corymb .... **C. chinense**

Clerodendrum chinense (Osbeck) Mabb., Plant-book Repr. 707. 1989. *C. philippinum* Schauer in DC., Prodr. 11: 667. 1847 (*Clerodendron*); Howard & Powell, Taxon 17: 54. 1968; Rajendran & Daniel, Indian Verbenaceae 137. 2002. *C. fragrans* Hort. ex Vent., Tard. Mal. 2: t. 70. 1804 pro. syn. Gamble, Fl. Madras 2: 1100. 1924. *C. japonicum* (Thunb.) Sweet var. *planiflorum* (Schauer) Maheshwari, Taxon 15: 44. 1966. *C. fragrans* var. *planiflorum* Schauer in DC., Prodr. 11: 666. 1847.

Vern.: *Dilbahar, Mysore malli.*

Shrub to 2 m high, branchlets pubescent. Leaves broadly ovate, 20 × 10 cm, glabrous above, pubescent beneath, base subcordate or rounded, margin coarsely prominently toothed, apex acute, lateral nerves ca 5 pairs; petiole to 7 cm long. Flowers fragrant, compact in terminal, corymbose panicles; calyx reddish, deeply 5-lobed; corolla white, double. Drupe blue, seated on large accrescent red calyx.

It is a Chinese shrub with sweet scented double flower often met in gardens and run wild in topes and on road sides.

Fl. & Fr.: Throughout the year.

Vikarabad (RR), *TP* & *MSM* 11089.

Clerodendrum inerme (L.) Gaertn., Fruct. Sem. Pl. 1: 271. t. 75. 1788; C.B.Clarke in Hook.f., Fl. Brit. India 4: 589. 1885; Gamble, Fl. Madras 2: 1099. 1929; Rajendran & Daniel, Indian Verbenaceae 112. 2002. *Volkameria inermis* L., Sp. Pl. 637. 1753.

Vern.: *Pisangi, Nilla vuppi, Pisinka.*

Straggling shrub, up to 2 m high, much branched. Leaves coriaceous, elliptic, broadly elliptic or obovate, 1.5-4 × 1.5–2.5 cm, base cuneate, margin entire, apex rounded, obtuse, acute, lateral nerves ca 3 pairs, not conspicuous, petiole to 1.5 cm. Flowers in 3-flowered or sometimes 7-flowered cymes; bracts linear, 2 mm; calyx cupular, 5 mm, minutely toothed; corolla white with pink streaks; corolla-tube 2-3 cm long, slender, lobes oblong, to 0.7 cm long; stamens 4, inserted in the upper half of tube, filaments filiform, 1.8 cm long, anthers oblong, 2.5 mm; ovary oblong, 2 mm, imperfectly 4-celled, style filiform, 3.2 cm. Drupe globose or pyriform.

Common along the river banks. Often planted in gardens used as hedge and perhaps run wild.

Fl.: July - April; Fr.: October – June.

Yedapally (NZB), *BR* & *KH* 6471; Central University campus (RR), *MSM* 15197.

Clerodendrum phlomides L.f., Suppl. Pl. 292. 1782; C.B.Clarke in Hook.f., Fl. Brit. India 4: 590. 1885; Gamble, Fl. Madras 2: 1099. 1924; Rajendran & Daniel, Indian Verbenaceae 136. 2002.

Vern.: *Thakkalaku, Konda takai, Thakkali.*

Large bush or small tree, to 7 m high; branches pale yellowish pubescent. Leaves ovate-rhomboid or deltoid-ovoid, chartaceous, 2-6 × 1-3.5 cm, base truncate or subcordate, margin coarsely sinuate or serrate, apex acute or obtusely acute; lateral nerves ca 4 pairs, clearly marked beneath; petiole to 2.3 cm. Flowers moderate sized, scented in dichotomous, axillary or terminal cymose panicles; bracts lanceolate, leafy, 5 mm; calyx campanulate, ovate-lanceolate, to 1 cm, acuminate; corolla white or purplish, tube to 2.2 cm long, hairy without, limb 5-lobed, lobes elliptic, obtuse, prominently nerved, subequal, to 8 mm long; stamens 4, exserted, inserted on the corolla tube, filament pairs 2 and 2.3 cm long, anthers oblong, 2.5 mm, greyish; ovary globose, 2 mm, style to 3.7 cm long, stigma 2-fid. Drupe obovoid, 4-lobed, black, to 1 cm long, with accrescent calyx lobes.

Common throughout, in hedges near villages.

Fl. & Fr.: February – May. Flower fragrant at night.

Khanapur (ADB), *TP* & *PVP* 4122; Kunthala (ADB), *DAM* & *GO* 5117; Sirnapally RF (NZB), *BR* & *PSPB* 9572; Domakonda (NZB), *BR* 9690; Mekavaripalli (RR), *TP* & *MSM* 11049; Kodimial (KRN), *GVS* 21861 (MH); Pakhal (WGL), *KMS* 115466 (MH);

Kanapur RF (WGL), *KMS* 11566 (MH); Nallagonda (NLG), *Cleghorn* 39981 (MH); Khammam to Dhaniyaigudem (KMM), *RR* 108088 (BSID); Edupayalu RF (MDK), *RG* 113434 (BSID).

DURANTA L.

Duranta erecta L., Sp. Pl. 637. 1753; Mathew, Rec. Bot. Surv. India 20: 185. 1969; Rajendran & Daniel, Indian Verbenaceae 149. 2002. *D. plumieri* Jacq., Select. Am. 186. t. 176. f. 76.1763; Hook.f., Fl. Brit. India 4: 560. 1885; Gamble, Fl. Madras 2: 1106. 1924.

Shrubs, up to 2 m high; young shoots pubescent or sub-glabrous; branchlets slightly angular. Leaves 2.4-5 × 1.2-2.3 cm, obovate-elliptic, base cuneate, margin dentate or slightly serrate, apex acute. Racemes usually terminal, rarely axillary or panicled, drooping. Flowers blue or violet-blue; calyx 5-ribbed, 5-toothed, teeth small; corolla salver-form, lobes 5, unequal; stamens 4; stigma capitate. Drupe 4-5 mm across, globose, 4-furrowed, enclosed with an accrescent calyx.

Commonly cultivated as a hedge plant, often runs wild.

Fl. & Fr.: Through out the year.

GMELINA L.

1. Armed shrub; bark brownish-white; leaves ovate-lanceolate, glaucous beneath .... **G. asiatica**
1. Unarmed tree; bark yellowish-grey; leaves ovate-cordate, tomentose beneath .... **G. arborea**

Gmelina arborea Roxb., Pl. Coromandel t. 246. 1815; Hook.f., Fl. Brit. India 4: 581. 1885; Gamble, Fl. Madras 2: 1097. 1924; Rajendran & Daniel, Indian Verbenaceae 155. 2002.

Vern.: *Gamar teku, Gummadi teku, Pasini, Pedda gumudu.*

Deciduous tree, up to 15 m high, bark yellowish-grey, rather corky; wood greyish-white, glossy, smooth and even-grained, medium coarse textured; young branches yellowish or tawny tomentose. Leaves sub-coriaceous, broadly ovate-cordate, 3-20 × 3.6-15.5 cm, glabrous above when mature, stellately fulvous tomentose beneath, base cordate, margin entire, apex acuminate; secondary nerves 5-6 pairs; petiole to 10.5 cm. Flowers in terminal or axillary panicles, rachis yellow tomentose; bracts subulate, 5-7.5 mm long, caducous; calyx campanulate, 5-6.2 mm, velutinous without, 5-lobed, lobes deltoid, fulvous tomentose without; corolla yellow or yellow-brown, 2.5-3.7 cm long, tomentose hairy without, upper lip shorter than the lower, lobes 5, to 2.5 cm; stamens 4, inserted on the throat of the corolla, filament pairs 1.3 and 1.7 cm, anthers 2 mm; ovary 4 mm, pubescent, style 2 cm long. Drupe ovoid, 2.5 × 1 cm, orange-yellow when ripe, smooth, fleshy.

Common in deciduous forests in most districts.

Fl.: February- April; Fr.: May – June.

Wankidi RF (ADB), *TP* & *GO* 5482; Gundaram RF (NZB), *BR* & *KH* 9677; Thornal (MDK), *CP* 11707; Tandur (RR), *MSM* 19241; Pakhal (WGL), *ANH* 15958 (MH).

Gmelina asiatica L., Sp. Pl. 626. 1753; C.B.Clarke in Hook.f. Fl. Brit. India 4: 582. 1885; Gamble, Fl. Madras 2: 1098. 1924; Rajendran & Daniel, Indian Verbenaceae 161. 2002.

Vern.: *Kavva gummidi, Chiru gummudu, Salla gummudu, Chiru nelli, Nela gummudu.*

Armed shrub, armed with axillary divaricate spines; bark brownish-white, thin; wood grey. Leaves chartaceous, ovate-lanceolate, 1.6-3.5 × 1.4-2.5 cm, glabrous, glaucous beneath, base acute, margin entire, apex obtuse to subacute; secondary nerves 3-4 pairs, impressed beneath; petiole to 1.4 cm. Flowers in axillary and terminal, 8 cm long racemes; calyx 4-lobed, 6 mm, puberulous without, lower part of the tube narrow, upper portion cap-shaped, hairy and glandular without; corolla yellow, 4 cm long, tube slender, lobes 5, some what glandular pubescent; stamens 4, filament pairs 1 and 1.7 cm long, anthers 2.5 mm; ovary 4 mm, style slender, 3 cm long, stigma shortly bifid. Drupe yellow, obovate, 1.8-2.5 cm long; seeds 2.

Common, in scrub jungles in Adilabad, Karimnagar, Nalgonda and Mahabubnagar districts.

Fl.: May –August. Fr.: November – February.

Jallada – Kothagudem (KMM), *RR* 106089 (BSID); Yenkannapalem (KMM), *RR* 113827 (BSID); Chityal (NLG), *BVRS* 413 (NU).

HOLMSKIOLDIA Retz.

Holmskioldia sanguinea Retz., Observ. Bot. 6: 31.1791. Mold., Phytologia 8 : 58. 1961. f. **citrina** Mold., Sixth Summ. Verbac. Etc. 550. 1980 & Phytologia 48 : 349. 1981; Rajendran & Daniel, Indian Verbenaceae 174. 2002; Bahadur & *al.*, Proc. A.P.Akademi Sci. 11 : 132. 2007.

Straggling woody shrub, up to 6 m high; branches and branchlets acutely 4-angular, young part pubescent, glabrous when mature. Leaves ovate or elliptic-ovate, 4-10 × 3-7 cm, base cuneate, margin crenate-serrate, apex acuminate–caudate, chartaceous, glabrous above, lateral nerves 4-5 pairs, gland dotted beneath, petiole slender, 1-3 cm long, pubescent. Flowers yellow, in axillary or terminal, few-flowered racemes, showy, pedicel ca 2 mm long; calyx campanulate to saucer-shaped, faintly 5-lobed or entire, lemon yellow coloured, membranous, glabrous, older calyx becoming greenish-yellow and light brown when dry; corolla infundibuliform, 2-lipped, 5-lobed, corolla-tube narrow, cylindric, 1 cm long, bright greenish lemon, membranous, pubescent out side; stamens 4, didynamous, filaments slender, 1-2 cm long, hairy, exserted, anthers ovoid, white, 2-celled; ovary conical, glabrous, style slender, 2.5 cm long, stigma 2-lobed. Drupes subglobose, deeply 4-lobed, splitting into 4 pyrenes; seeds 1-4, rugose, fuiting calyx accrescent, rotate, glabrous through out.

Grown in Karimnagar and Hyderabad as ornamental and naturalised.

Fl. & Fr. : Round the year.

Uppal Ring Road, Snehpuri, Hyderabad (HYD), *Bir Bahadur* 1 (MH).

LANTANA L.

1. Plants with recurved prickles; leaves scaberulous beneath; bracts lanceolate .... **L. camara**
1. Plants without prickles; leaves densely soft pubescent beneath; bracts broadly ovate:
   2. Leaf margin crenulate; corolla white .... **L. wightiana**
   2. Leaf margin crenate-serrate; corolla pink or red .... **L. indica**

Lantana camara L., Sp. Pl. 627. 1753; var. **camara**; Rajendran & Daniel, Indian Verbenaceae 180. 2002. L. *camara* L. var. *aculeata* (L.) Mold., Torreya 34: 9. 1934 & Lilloa 4: 289. 1939. *L. aculeata* L., Sp. Pl. 627. 1753; C.B.Clarke in Hook.f., Fl. Brit. India 4: 562. 1885; Gamble, Fl. Madras 2: 1087. 1924.

Vern.: *Gabbuseeki.*

Straggling or subscandant, prickly aromatic shrub, up to 2.5 m high; branches brownish, armed, prickles recurved, 1 to 2 mm long. Leaves ovate or ovate-oblong, 5-8 × 2-5 cm, scabrid above, pubescent beneath, base truncate, rounded, cuneate or acute, margin crenate-serrate, apex acute to shortly acuminate; lateral nerves ca 5 pairs; petiole to 2.2 cm. Flowers in short capitate umbellate spikes, scented, axillary and terminal; bracts lanceolate, hispidulous; calyx membranous, truncate, shortly or obscurely toothed; corolla white, yellow, dark crimson or variously coloured, hypocrateriform, tube slender, 7 mm long; limb 7.3 mm across, obscurely 2-lipped, lobes 5, spreading; stamens 4, didynamous, included, anthers oblong, cells parallel; ovary 2 mm, style short, stigma oblique. Drupe fleshy, blue, globose, enclosing 2, 1-seeded pyrenes.

Common in scrub forests in all districts.

Fl. & Fr.: Throughout the year.

Nirmal (ADB), *DAM* 5937; Manchippa RF (NZB), *TP* & *BR* 6005; Narsapur – Hyderabad road (MDK), *BR* & *CP* 11211; Mohammadabad RF (RR), *MSM* & *KH* 10207; Yerraguntapalli (NLG), *BVRS* 630 (NU); Moosi river bank (HYD), *KMS* 5940 (MH); Ramappa temple (WGL), *RKP* 10535; Ramanagutta (KMM), *RR* 107997 (BSID).

Key to Varieties

1. Plant usually armed, minutely pubescent; leaves ovate or ovate-oblong .... var. **camara**
1. Plant usually unarmed, densely hirsute; leaves lanceolate .... var. **splendens**

Lantana camara L. var. **splendens** (Medik.) Moldenke, Phytologia 33: 130. 1976 & Phytologia Mem. 2: 553; Rajendran & Daniel, Indian Verbenaceae 184. 2002.

Introduced, grown in gardens, now naturalized.

Fl. & Fr.: July - December .

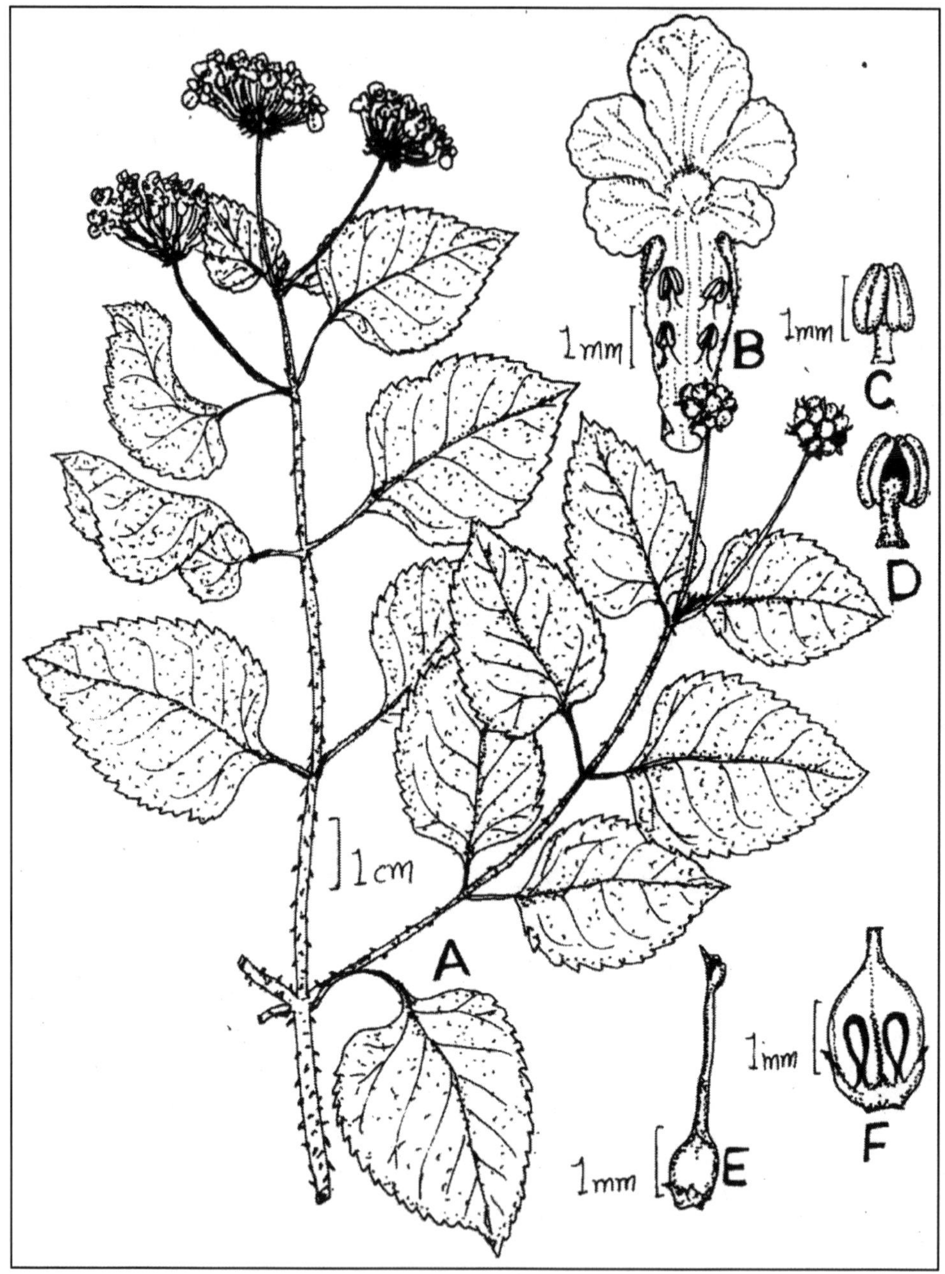

Figure 81: Lantana camara L.

A. Twig, B. Corolla open, C&D. Stamens, E. Pistil, F. Ovary l.s.

Lantana indica Roxb. Fl. Ind. 3 : 89. 1832; C.B.Clarke in Hook.f., Fl. Brit. India 4 : 562. 1885; Gamble, Fl. Madras 2 : 1087. 1924.

Scandent shrubs, 1-2 m high; branches 4-angular. Leaves ovate, 4-6 x 2-3 cm, base narrowed, rounded, margin creneate-serrate, apex acute or subobtuse, rugose and finely pubescent above, softly pubescent beneath; petioles 6-10 mm long. Flowers sessile, in axillary, pedunculate, ovoid heads 1.5-2 cm long, lengthening into spikes; calyx 1.2 mm long, long ciliate; corolla pink with a yellowish tube, hairy outside; tube 5-6 mm long; limb 6-8 mm across; lobes 4, rounded, spreading. Drupes globose, 4-5 mm in diam., dark purple.

Frequent, in hedges and also naturalized (Ramana, 2010).

Fl. & Fr.: August-November.

Lantana wightiana Wallich ex Gamble, Fl. Madras 2 : 1087. 1924. *L. indica* var. *albiflora* Wight ex C. B. Clarke in Hook.f., Fl. Brit. India 4 : 562. 1885.

Unarmed sub shrub; tender parts scabrous. Leaves elliptic-ovate, 2-4 x 1.5-3 cm, base rounded to acute, margins crenulate, apex subacute, softly appressed pubescent above; petiole to 1 cm long. Flowers white in umbellate spikes; spikes 3 cm long; calyx 2 mm long, margins obscurely 5-lobed, minutely ciliate; corolla 4 mm across, 5-lobed; lobes 2 mm long. Drupe globose.

In scrub jungles along waysides in Hyderabad (Ramana, 2010).

Fl. & Fr.: September-February.

LIPPIA L.

Lippia javanica (Burm.f.) Spreng., Syst. Veg. 2 : 752. 1825; Meeuse, Blumea 5 : 68. 1942. *Verbena javanica* Burm.f., Fl. Indica 12. 1768. *Lippia alba* (Mill) N.E. Br. ex Britt. & Wils. Bot. Porto Rico 6 : 141. 1925.

Aromatic straggling shrubs with erect branches. Leaves broadly elliptic-oblong, 7-8 x 3-3.5 cm. base round to acute, margins serrate-crenate, apex acute, sparsely hairy above, softly pubescent below; petiole 1 cm long. Flowers in axillary spikes, 2.5 cm long; calyx cupular, 4 mm long, lobes 2, lanceolate, concave; corolla pinkish, salver form, 6 mm long; lobes 5, bilabiate. Fruits not seen.

Occasional in Khammam district.

Fl. & Fr.: November-January.

Bhadrachalam (KMM), *RCS* 98800 (BSID); Parnasala RF (KMM), *RCS* 98772 (BSID); Perantapally RF (KMM), *RCS* 98998 (BSID).

PHYLA Lour.

Phyla nodiflora (L.) Greene, Pittonia 4: 46. 1899. *Verbena nodiflora* L., Sp. Pl. 20. 1753; Rajendran & Daniel, Indian Verbenaceae 206. 2002. *Lippia nodiflora* (L.) A. Rich. in Michaux, Fl. Bor. Amer. 2: 15. 1803; C.B.Clarke in Hook.f., Fl. Brit. India 4: 563. 1885; Gamble, Fl. Madras 2: 1088. 1924.

Vern.: *Bokkena, Bokkenaku, Mosalipappu.*

Prostrate herb, stems rooting at nodes, much branched, sub glandular, grooved, glabrous. Leaves thick-chartaceous, obovate-spathulate, some times orbicular, 5-18 × 5-8 mm, thinly pubescent, base decurrent, margin serrate at upper half, apex obtuse or rounded, 2-4-nerved. Flowers numerous, sessile, aggregated in condensed, cylindric axillary spikes, peduncle 3-6 cm, bracts broadly elliptic or obovate, 2.4 mm; calyx membranous, 2-lobed, lanceolate, closely covering the fruit, lobes acuminate, scarious; corolla white or pale pink, salver form, tube 1.5 mm, lobes 5, spreading, 2-lipped, 2 + 3; stamens 4, included, attached in pairs at the throat of the tube, filaments curved, 0.4 mm, anthers ovate, lobes parallel; ovary glabrous, 2-celled, ovule 1 per cell, style short, 0.5 mm long; stigma oblique, subcapitate. Drupe subglobose, 1 mm long, dry, splitting into two 1-seeded pyrenes.

Common on river banks, margins of ponds, ditches, canals and moist places of cultivated fields in almost all districts.

Fl. & Fr.: Throughout the year.

Sone (ADB), *TP & PVP* 4046; Nirmal (ADB), *MHR* 14423; Mudheli (NZB), *TP & BR* 6365; Medak vagu (MDK), *BR & CP* 11456; Gadirayal RF (RR), *MSM* 10524; Kodimial (KRN), *GVS* 20098 (MH); Khammam to Wyra (KMM), *RR* 108082 (BSID); Uppal (MBNR), *BR & MV* 35844; Near Dindi project (NLG), *BVRS* 673 (NU).

PREMNA L. *nom. cons.*

1. Plants stellate-tomentose; leaf apex acuminate, secondary nerves 10 pairs .... **P. tomentosa**
1. Plants not stellate-tomentose; leaf apex obtuse to acute, secondary nerves 3-4 pairs .... **P. mollissima**

Premna mollisima Roth, Nov. Pl. Sp. 286. 1821. *Premna latifolia* Roxb. var. *latifolia;* C.B.Clarke in Hook.f., Fl. Brit. India 4: 577. 1885; Gamble, Fl. Madras 2: 1096. 1924. *Premna latifolia* Roxb. var. *mollissima* (Roth) Hook. f., Fl. Brit. India 4: 578. 1885; Gamble, Fl. Madras 2: 1096. 1924.

Vern.: *Tekli, Nelli, Pedda nelli kura.*

Tree, up to 10 m high; bark thin, light-grey or brown, smooth. Leaves chartaceous, ovate to sub-orbicular, 3-8 × 2.5 cm, glabrous to pubescent, black when dry, base rounded, margin entire, apex obtuse to acute, sometimes shortly acuminate; secondary nerves 3-4 pairs; petiole to 1.7 cm. Flowers small, in terminal corymbs; calyx obscurely 4-lobed, pubescent or strigose, 2.5 mm; corolla 2-lipped, 3.5-5 mm long, cream-coloured, densely villous within, upper lip oblong, lower lip longer with 3 spreading, obtuse or rounded lobes; stamens 4, exserted, inserted in the middle of the tube, filament pairs 2 and 3 mm long, filiform, anther subglobose, 0.5 mm; ovary 1 mm, style filiform or subulate, 4.6 mm long, stigma shortly bifid. Drupe depressed globose, 7 mm long, with subentire, persistent, saucer-shaped calyx.

Common, in foot hills, scrubs or open forest of Adilabad, Karimnagar, Ranga Reddi, Hyderabad and Mahabubnagar districts.

Fl.: April- May, Fr.: May – June.

Mallampeta (ADB), *TP* & *GO* 5445; Rudraram RF (RR), *TP* & *MSM* 12710; Anantagiri (RR), *MSM* 12710; Nagarjunakonda (NLG), *BVRS* 100 (NU).

Premna tomentosa Willd., Sp. Pl. 3: 314. 1800; C.B.Clarke in Hook.f., Fl. Brit. India 4: 576. 1885; Gamble, Fl. Madras 2: 1095 .1924. *Cornutia corymbosa* Lam., Encycl. 1: 54. 1783 (auct non Burm.f. 1768).

Vern.: *Pedda narava, Kampu gummudu, Kokkiti chettu, Nagoora, Pomanti, Nela neredu, Adavinela.*

Deciduous tree, up to 10 m high; bark grey or light brown, smooth; wood light brown, smooth and close and even-grained; branchlets with densely soft, tawny tomentum of stellate hairs. Leaves thick, chartaceous, 2.5-15.5 × 2-9 cm, ovate, densely tawny yellow, stellate tomentose, base cordate, margin entire, apex acuminate; secondary nerves of 10 pairs; petiole to 3.2 cm, tawny-tomentose, canaliculate. Flowers small, scented in terminal, corymbose cymes; calyx sub-globose, tomentose, 5-lobed, slightly 2-lipped, obtuse; corolla cream, 2-lipped, tube 3 mm long, villous along the throat within, lobes 5, pubescent and glandular without; stamens 4, inserted at the throat, anthers rounded, 0.5 mm; ovary truncate, stellate-pubescent, style 2.5 mm long, stigma unequally 2-fid. Drupe ovoid or subglobose, 5 mm diam, endocarp smooth.

Frequent in forest hilly areas, foot hills in Adilabad, Ranga Reddi, Hyderabad, Nalgonda and Warangal districts.

Fl.: February - April, Fr.: March – May.

Vempalli RF (ADB), *GO* 4442; Mohammadabad RF (RR), *MSM* & *KH* 10212; Mallelatheertham (MBNR), *TDCK* 15350; Cherukupally (NLG), *BVRS* 345 (NU); Pakhal (WGL), *SSR* & *KSM* 18520, *KMS* 13157 (MH); Kacharam (KMM), *PVS* 84056 (BSID).

PRIVA Adans.

Priva cordifolia (L.f.) Druce, Bot. Soc. Exch. Club. Brit. Isles 1916: 641. 1917. *Buchnera cordifolia* L.f., Suppl. Pl. 287. 1781. *Priva leptostachya* Juss., Ann. Mus. Hist. Nat. 7: 70. 1806; C.B.Clarke in Hook.f., Fl. Brit. India 4: 565. 1885; Gamble, Fl. Madras 2: 1091. 1924.

Vern.: *Pachareni.*

Herb, branchlets 4-angular, pubescent with hooked hairs. Leaves ovate or elliptic, chartaceous, 4-7.5 × 3-5 cm, fine hooked hairy on both surfaces, base subcordate or truncate, margin coarsely crenate – serrate, apex acute; lateral nerves ca 5 pairs, prominently reticulately veined; petiole to 3.8 cm long. Flowers distant, subsessile, in terminal, elongate, 15-30 cm long spikes; bracts linear-lanceolate, persistent, hairy; calyx tubular, 5-ribbed, 5-toothed, 6 mm long, clothed with fine spreading hooked hairs; corolla white, tube 8 mm long, twisted, scarcely enlarged above, limb 2-lipped, upper lip 2-lobed, lower lip larger than the upper, deeply 3-lobed; stamens 4, filament pairs 1.5 and 2 mm long, inserted in the middle of the corolla tube, included, anther cells parallel or slightly divergent; ovary oblong-cylindric, 1.5 mm, 2-locular, each 2-

ovuled, style 4 mm long, filiform, persistent, stigma oblique. Drupe enclosed in the dilated calyx-tube, separating when ripe into two 2-seeded pyrenes, sticky pubescent.

Occasional on rocks, old walls, hedges of fields among bushes and also hill slopes of Karimnagar, Nalgonda and Mahabubnagar districts.

Fl.: November – February; Fr.: December - March.

Kambalpally (NLG), *BVRS* 216 (NU).

Note: The echinate pyrenes sticking to the clothes is a good guide in detecting the presence of the plant in the field.

ROTHECA Rafin.

Rotheca serrata (L.) Steane & Mabb., Novn 8: 206. 1998. *Clerodendrum serratum* (L.) Moon., Cat. Pl. Ceylon 46: n. 382. 1824; C.B.Clarke in Hook.f., Fl. Brit. India 4: 592. 1885; Gamble, Fl. Madras 2: 1100. 1924. *Volkameria serrata* L., Mant. Pl. 90. 1767.

Vern.: *Bharangi, Gantu bharangi, Bommalamarri, Angaravalli, Ciruteku.*

Slender shrub, up to 2 m high, not much branched, stems quadrangular, grooved and also striate, pale brown. Leaves subcoriaceous, ternate and opposite passing into bracts in inflorescence, elliptic-oblong, obovate, obovate-oblong or oblanceolate, to 27 × 10 cm, glabrous, dark brown when dry, base acuminate; lateral nerves 10 pairs, prominent; petiole to 6 mm, stout. Flowers numerous, showy in large, terminal, to 20 cm long, pyramidal panicles; bracts leafy, 1.2-3.2 cm, obovate-lanceolate, pubescent; calyx cup-shaped, 5 mm, hairy, teeth 5, triangular; corolla pale blue or violet, to 2 cm long, tube to 8 mm, cylindric, limb 2-lipped, upper and lateral lobes spreading, lower lobe deflexed, lip like, with a large nectary gland at base, to 1.5 cm long; stamens 4, didynamous, fully exserted, filaments inserted at the tube, 2.8 cm long, curved, basally hairy, anthers oblong, 2 mm, brown; ovary globose, 2.5 mm, style 3 cm long. Drupe broadly obovoid, 6 mm long, lobed, somewhat succulent, black when ripe.

Occasional near the streams in moist forests and grasslands, hills above 1000m. especially on cleared slopes in most districts.

Fl. & Fr.: August – December.

Chilapalli (ADB), *GO* & *PVP* 4904; Mudheli RF (NZB), *BR* & *KH* 7124, 7125; Narsapur RF (MDK), *TP* & *CP* 14031; Rudraram RF (RR), *MSM* & *TP* 12731; Manchippa Range (MBNR), *STR* 285 (HH); Thattlinga RF (KMM), *RCS* 102413 (BSID).

STACHYTARPHETA Vahl

1. Leaves sharply acute at the apex, spreading dentate, the teeth sharply acute with apices central; calyx with 1-2 peltate glands near the apex .... **S. urticifolia**
1. Leaves usually obtuse at the apex, crenate-dentate, the teeth short and blunt with apices excentric; calyx without glands .... **S. jamaicensis**

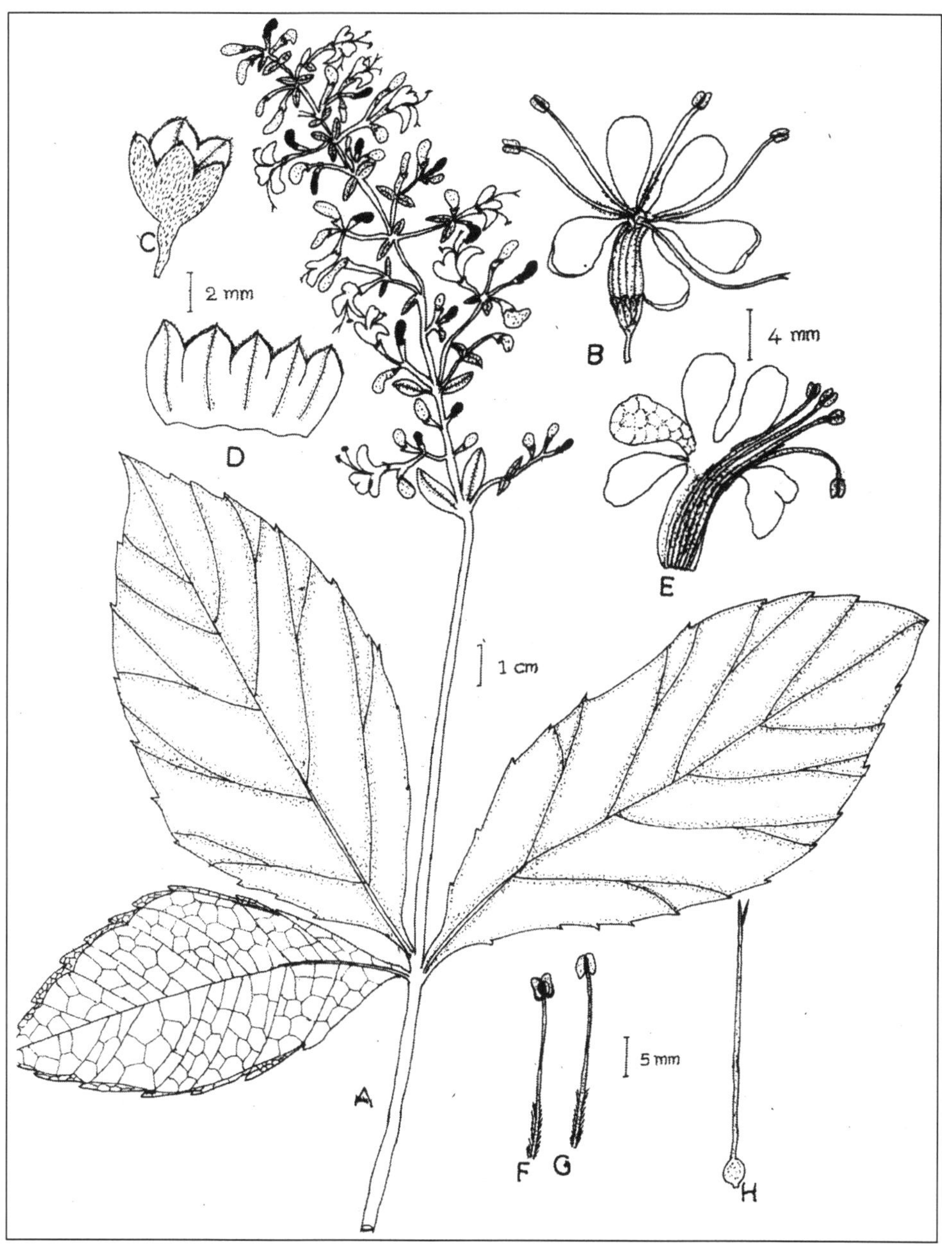

Figure 82: Rotheca serrata (L.) Steane & Mabb., [Syn.: *Clerodendrum serratum* (L.) Moon] A. Twig, B. Flower, C. Calyx, D. Calyx split open, E. Corolla split open, F&G. Stamens, H. Pistil.

Stachytarpheta jamaicensis (L.) Vahl, Enum. Pl. 1: 206. 1804; Meeuse, Blumea 5: 70. 1942; Mold., Ann Missouri Bot. Gard. (Fl. Panama) 60: 74. 1975; Rajendran & Daniel, Indian Verbenaceae 307. 2002. *Verbena jamaicensis* L., Sp. Pl. 19. 1753. *Stachytarpheta indica* var. *jamaicensis* (L.) Trimen, Cat. Pl. Ceylon 68. 1885 & Handb., Fl. Ceylon 3: 348. 1895; Gamble, Fl. Madras 2: 1090. 1924. *S. indica* sensu, C.B.Clarke in Hook.f. Fl. Brit. India 4: 564; *p.p.* (auct non (L.) Vahl 1804).

Erect woody herb, about 1 m high, dichotomously branched, nearly glabrous; branches sub quadrangular. Leaves variable, elliptic, obovate-elliptic, 5-10 × 2-3.5 cm, base tapering and decurrent, crenate-dentate, apex obtuse or acute; lateral nerves ca 6 pairs, prominently veined; petiole to 1cm long. Flowers sessile, in 30 cm long spikes, the rachis glabrous, hollowed out beneath each flower; bract 4.8 mm long, lanceolate-subulate; calyx 5-6 mm long, membranous, tubular, teeth 4, small; corolla blue, 7 mm long, tube slender, cylindric, 5 mm long, limb oblique, lobes 5, sub equal, flat, spreading, 5 mm; stamens 2, staminode 2, minute, filaments white-hairy, slender, 1.5 mm long, hairy, anthers 0.8 mm long; ovary glabrous, 2 mm, style 7 mm long, filiform. Fruit oblong, ribbed, 3-4 mm long, splitting into 2 pyrenes, dark chocoloate-brown.

Common in waste lands and road sides in Nizamabad, Medak, Karimnagar, Ranga Reddi and Hyderabad districts.

Fl. & Fr.: April – November.

Mosra forest nursery (NZB), *TP* & *BR* 6058; Manjira barrage (MDK), *BR* & *CP* 14094; Medak (MDK), *RG* 104110 (BSID); Central University (RR), *MSM* 19227; Katasintha (HYD), *KMS* 6005 (MH).

Note: The plant hitherto known as *S. indica* (L.) Vahl correspond to the type of *S. jamaicensis* (L.) Vahl and be cited as such (From Notes of J.P.H.Brenan, cited in Mathew, Fl. Tamil Nadu Carnatic 3(2): 1232. 1983).

Stachytarpheta urticifolia (Salisb.) Sims, Bot. Mag. t. 1848. 1816. *Cymburus urticifolius* Salisb., Parad. Lond. t. 53. 1806. *Stachytarpheta indica* auct. non (L.) Vahl 1804; C.B.Clarke in Hook.f., Fl. Brit. India 4: 564. 1885. *p. p.*; Gamble, Fl. Madras 2: 1090. 1924.

A tall herb, up to 1 m high; dichotomously branched, nearly glabrous. Leaves 3.5-8 × 2-3.5 cm, ovate-elliptic, base rounded, margin sharply dentate, apex sharply acute, glabrous or nearly so. Flowers in spikes which sometimes reach 30 cm in length; calyx narrowly tubular; corolla white or light bluish-violet, lever-shaped, tube slender, cylindric, usually curved, 5-lobed; stamens 2, perfect, with a posterior pair of staminodes, ovary 2-celled. Fruit 0.3-0.4 cm long, oblong, ribbed.

Rare in waste lands and open patches in the forests in Warangal district.

Fl. & Fr.: Throughout the year.

Pakhal (WGL), *KMS* 13199 (MH).

SYMPHOREMA Roxb.

1. Calyx lobes 6; corolla lobes 6-8 .... **S. involucratum**
1. Calyx lobes 6-7; corolla lobes 12-16 .... **S. polyandrum**

Symphorema involucratum Roxb., Pl. Coromandel t. 186. 1805 & Fl. Ind. 2: 262. 1832; C.B.Clarke in Hook.f., Fl. Brit. India 4: 599. 1885; Gamble, Fl. Madras 2: 1104. 1924; Mold. & Mold. in Dassanayake, Rev. Handb. Fl. Ceylon 4: 189. 1983.

Vern.: *Konda tekkali, Surudu, Buddamarri, Tella pippili.*

Scandent shrub, 3 to 10 m high, bark pale, lenticellate, stems slender, branchlets stellately tomentose. Leaves ovate or elliptic or rhomboidly elliptic, 3-12.2 × 2-7 cm, glabrous above or nearly so when mature, pale and stellate tomentose beneath, base rounded, margin toothed, crenate – serrate; lateral nerves ca 4 pairs, reticulately veined prominent above; petiole to 1.3 cm long. Flowers in capitate cymes, pedunculate, 7-flowered, involucral bracts foliaceous, scarious, ca 6, 0.6-1.8 cm; calyx 4.8 mm, tubular, 7-8-toothed, triangular – oblong, 1.5 mm long, stellately tomentose, slightly enlarged in fruit; corolla white, glabrous, 8 mm long, tube narrow-cylindric, 6-8-lobed, lobe 3 mm long, linear – oblong; stamens 6-16, inserted at the corolla tube, exserted, filaments glabrous, filiform, 3mm long, anthers parallel, oblong, ovary 2 or imperfectly 4-celled, style filiform, stigma shortly 2-fid. Fruit subglobose, dry, 4 mm across, enclosed in persistent calyx, bracts enlarged in fruit, 3.5 cm long, spreading, spathulate-elliptic, papery, densely pubescent, prominently reticulately veined.

Common in dry deciduous forests on trees in Ranga Reddi and Mahabubnagar districts.

Fl.: February – April. when more or less leafless. Fr.: April - May.

Kothapalli RF (RR), *MSM* & *KH* 10299; Sarlapally (MBNR), *TDCK* 13609.

Symphorema polyandrum Wight, Icon. Pl. Ind. Orient. t. 363. 1840 (*polyandra*); C.B.Clarke in Hook.f., Fl. Brit. India 4: 599. 1885; Gamble, Fl. Madras 2: 1104. 1924.

Large sub scandent shrub, stems terete, tawny tomentose, devoid of leaves during flowering time. Leaves 5-10 × 2.5-4 cm ovate, suborbicular, base rounded, margin crenate, dentate, apex acute. Flowers crowded at the end of peduncles into a head. Involucral bracts elliptic, elliptic-obovate, in fruit 3.7 cm long, obovate; calyx tube 7 mm long, 6-7-lobed, enlarging in fruit; corolla white, 12-16-lobed, glabrous, lanceolate, acute at apex; stamens equal in number to the corolla lobes, exserted, filaments 1 cm long, filiform; ovary ovoid, glabrous, glandular at apex, style glabrous, much longer than the stamens, filiform, few stalked glands present at the base of the style, stigma bifid. Drupe enclosed in persistent calyx.

Occasional in Warangal and Mahabubnagar districts.

Fl. & Fr.: March – July.

Pakhal (WGL), *RKP* 108289 (BSID); Mannanur Range (MBNR), *STR* 260 (HH).

TECTONA L. f. *nom. cons.*

Tectona grandis L.f., Suppl. Pl. 151. 1782; C.B.Clarke in Hook.f., Fl. Brit. India 4: 570. 1885; Gamble, Fl. Madras 2: 1092.1924; Rajendran & Daniel, Indian Verbenaceae 320. 2002.

Vern.: *Teku, Pedda teku*; Eng.: Teak.

Large deciduous tree, up to 20 m high; bark grey or light greyish brown, fibrous with shallow longitudinal fissures, exfoliating in long, thin, narrow some what corky flakes, wood golden yellow, turning brown, scented, and oily to the touch, coarse and uneven textured, branchlets quadrangular, stellately tomentose. Leaves broadly elliptic, 15-30 × 9-25 cm, rough above, dense grey tomentose beneath, base rounded or cuneate, margin entire, apex acute or obtuse; secondary nerves 8-9 pairs; petiole to 3 cm long. Flowers 6-8-merous, shortly pedicellate, in terminal, large, dichotomously branched panicles, 10-30 cm across; calyx stellately tomentose, campanulate, lobes 5-7, spreading, sub equal, 2-3.5 mm long; corolla yellowish white, glabrous, 1cm across, sub-rotate, lobes 5-7, spreading, broadly oblong, 5-6 mm long; stamens 5-7, inserted near the base of the corolla, equal, exserted, filaments 3 mm long, anthers ovate or oblong with distinct parallel lobes; ovary globose, 2 mm, hairy without, style linear, 4 mm, stigma shortly bifid. Drupe globose, 1.2 cm in diam, slightly 4-lobed, epicarp stellate-pubescent, spongy, endocarp bony, accrescent calyx to 2.5 cm, bladder like, enclosing the fruit.

Found throughout the state in dry and moist deciduous forests.

Fl.: April – June, Fl.: July – August.

Vempalli (ADB), *GO* & *PVP* 4779; Ghanapur RF (NZB), *BR* & *KH* 7114; Mutaypally (MDK), *CP* 11783; Anantasagar RF (RR), *MSM* & *KH* 10407; Pakhal (WGL), *KMS* 13155 (MH); Seetharamapuram (KMM), *AJR* & *PAN Reddy* 20465 A; Edrallapalli RF (KMM), *RCS* 102403 (BSID).

VITEX L.

1. Large trees; leaves glabrescent; inflorescence axillary .... **V. leucoxylon**
1. Shrubs to small trees; leaves pubescent or tomentose; inflorescence terminal:
    2. Leaves 1-3-foliolate, leaflets all sessile .... **V. trifolia**
    2. Leaves 3-5-foliolate, central leaflets petioled .... **V. negundo**

Vitex leucoxylon L. f., Suppl. Pl. 293. 1781; C.B.Clarke in Hook.f., Fl. Brit. India 4: 587. 1885; Gamble, Fl. Madras 2: 1103.1924; Rajendran & Daniel, Indian Verbenaceae 356. 2002. *Wallrothia leucoxylon* (L.f.) Roth, Nov. Pl. Sp. 391. 1821.

Vern.: *Luki, Kondavavili, Gajavavili, Nakkaneredu, Nevaledi.*

Large, deciduous tree, up to 20 m high; crown spreading, bark grey, smooth; wood dark purple, straight grained, fine-medium textured; young branches slightly puberulous. Leaves 3-5-foliolate, petiole to 7 cm long, leaflets thin, coriaceous, narrowly elliptic-lanceolate, elliptic-oblong, obovate-oblanceolate, 7-15 × 3-4 cm, glabrescent,

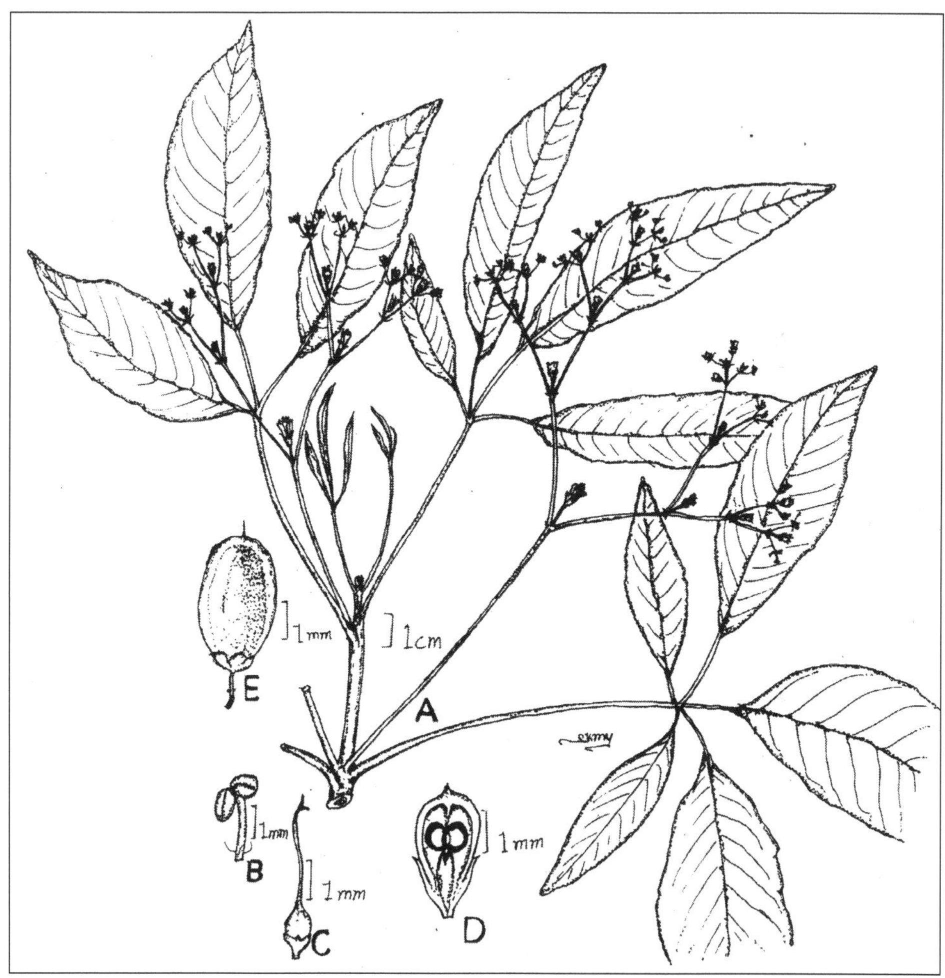

Figure 83: Vitex leucoxylon L.f.

A. Twig, B. Stamens, C. Pistil, D. Ovary l.s. E. Fruit

base acute, margin entire and toothed, apex sub obtuse; secondary nerves 8-10 pairs, closely reticulated; petiolule 2 cm long. Flowers mildly fragrant, in axillary, corymbiform, dichotomous cymes in 13 cm long panicles; peduncle 4-7 cm long, slender, glabrous; bract 2 mm long; calyx 5 mm long, puberulous, 5-toothed, corolla white, densely appressed silky without, 7 mm long, 2-lipped, upper lip 3 mm, 2-lobed, lower lip 3-lobed, 8 mm long, middle lobe obovate, purplish hairy within, margin crisped; stamens 4, filament pairs 4 and 6 mm long, hairy at the base, anthers 0.5 mm; ovary 2 mm, densely hairy at apex, style 7.8 mm long, stigma 2-lobed. Drupe ellipsoid-obovoid, dark purple, 1.5-2.4 cm long; seeds 4, oblong.

Common in forests in Nizamabad, Khammam and Mahabubnagar districts.

Fl.: February – April, Fr.: Through the year.

Sirnapally (NZB), *BR* & *GO* 9085; Farahabad (MBNR), *TDCK* 15346, *SSR* 15536; Appapur (MBNR), *SSR* 15536; Saleswaram (MBNR), *BR* & *SKB* 30731; Perantapalli (KMM), *PVS* 84065 (BSID).

Vitex negundo L., Sp. Pl. 638. 1753; C.B.Clarke in Hook.f., Fl. Brit. India 4: 583. 1885, incl. var. *incisa*; Gamble, Fl. Madras 2:1102. 1924; Rajendran & Daniel, Indian Verbenaceae 361. 2002.

Vern.: *Vavili.*

Shrub or small tree, to 5 m high, grey-pubescent, young branches 4-angled. Leaves 3-5-foliolate, petiole to 7.5 cm; leaflets chartaceous, oblanceolate or narrowly lanceolate, 1.5-11 × 2-3.2 cm, central larger, base acute, margin entire or crenate, glabrous above, grey-pubescent or finely tomentose beneath, apex acuminate, lateral nerves 14-16 pairs, clearly marked above; petiolule 2-3 cm long in terminal ones. Flowers in terminal, 10-25 cm long cymose panicles; calyx 2.5–3 mm, tomentose without, 5-toothed, teeth triangular, acute; corolla bluish purple to violet, 6-7 mm across, hairy, inside at the inception of the stamens, upper lip 2 mm, divided to the base into 2 obtuse lobes, lower lip 5 mm long, lateral lobes oblong-obtuse, terminal lobe 3 mm, crenulate, obovate; stamens 4, filaments hairy at the base, 3.5 and 5 mm long, anthers 1 mm; ovary glabrous, stigma forked. Drupe globose, black, 3 mm diam.

Common in all districts on the river banks or fencing near villages, on road side, and along the streams.

Fl. & Fr.: May – February.

Sadarmat (ADB), *TP* & *PVP* 4103; Mosra (NZB), *TP* & *BR* 6083; Siddipet (MDK), *CP* 11377; Medak (MDK), *BR* & *CP* 11412; Mohammadabad RF (RR), *MSM* & *KH* 10232; Talupenta (MBNR), *TDCK* 15318; Yetapakka (KMM), *RR* 106067 (BSID); Palempet (WGL), *RKP* 105347 (BSID).

Key to Varieties

1. Leaves grey- or white-pubescent; corolla blue or violet with pink tinge; filaments white .... var. **negundo**
1. Leaves purple-pubescent; corolla and filaments dark purple .... var. **purpurascens**

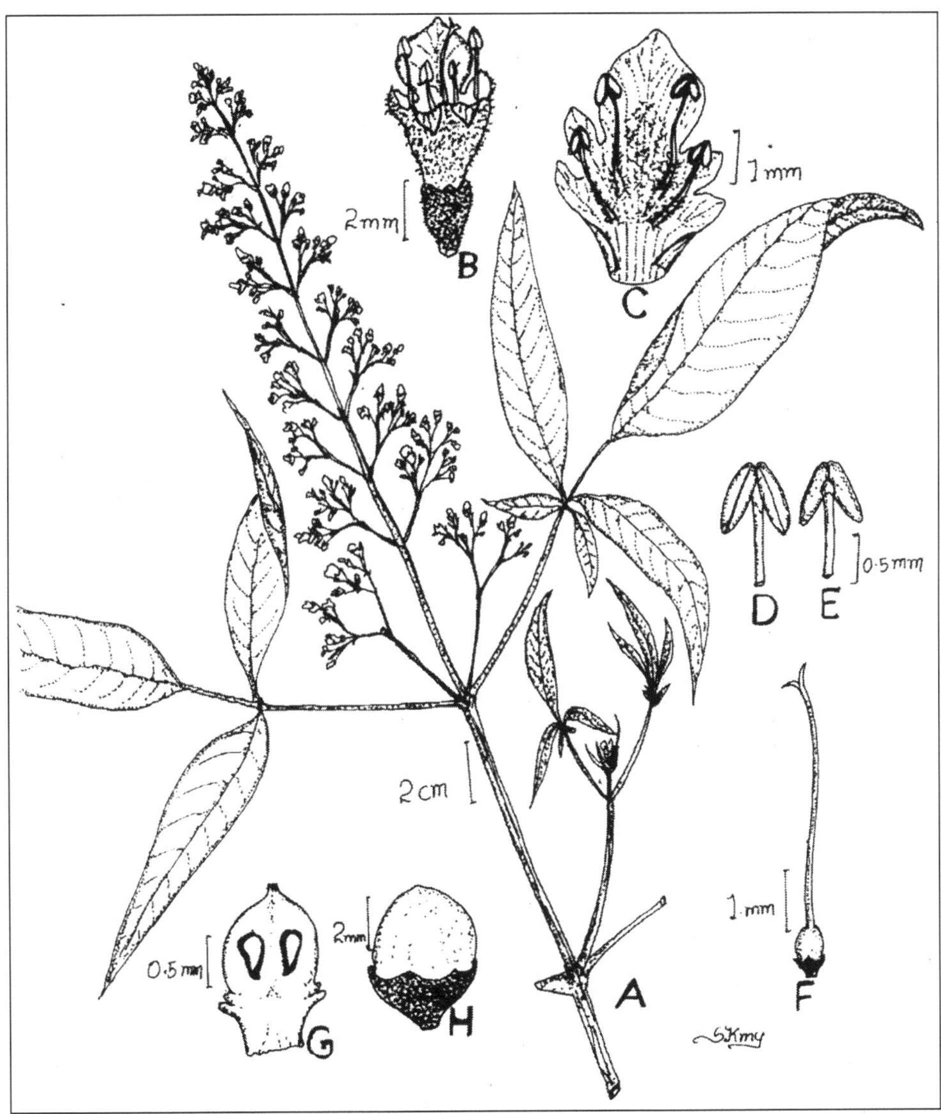

Figure 84: Vitex negundo L.

A. Twig, B. Flower, C. Corolla open, D&E. Stamens, F. Pistil, G. Ovary l.s., H. Fruit.

Vitex negundo L. var. **purpurascens** Sivar. & Mold., Phytologia 28: 404. 1974; Rajendran & Daniel, Indian Verbenaceae 366. 2002

Vern.: *Nalla vavili, Neeru vavili.*

An aromatic large shrub of about 3 m high. It differs from typical variety mainly in the presence of purplish black tomentum in the flowers, fruits and lower surface of leaves.

Rare in Kongarigidda of Warangal district (Sudhakar Reddy & *al.*, 2001).

Fl. & Fr.: January – July.

Vitex trifolia L., Sp. Pl. 638. 1753 (*trifolius*); C.B.Clarke in Hook.f., Fl. Brit. India 4: 583. 1885; Gamble, Fl. Madras 2: 1102.1924; Rajendran & Daniel, Indian Verbenaceae 377. 2002.

A slender bushy shrub or small tree, up to 4 m high, tender parts softly pubescent. Leaves trifoliolate, leaflets sessile, 1-6 × 0.6-3 cm, obovate, obovate-oblong, base tapering, margin entire or sometimes toothed, tip subacute or sometime acute. Flowers in terminal panicles; pedicelled; bracts minute; calyx teeth 5, minutely toothed, white-tomentose, corolla pale-purple or blue, tomentose outside, upper lip larger, lower lip with 3 rounded lobes; stamens 4, didynamous; filaments hairy at base. Drupe ovoid, purplish-black when ripe, enclosed more than half at the base of the enlarged calyx.

Occasional in Adilabad, Nizamabad and Mahabubnagar districts in forests.

Fl. & Fr.: Throughtout the year.

Sriramsagar project township (NZB), *BR* 9003.

Gmelina philippensis Chasn., *Petrea volubilis* L., *Verbena bipinnatifida* Schauer. are cultivated as ornamentals.

LAMIACEAE (*nom. alter.* LABIATAE)

1. Stamens declinate:
    2. Lower lip of the corolla one-lobed; stamens exserted:
        3. Lower lip of the corolla flat:
            4. Calyx deflexed in fruit, upper lobe curved, margin decurrent on the tube:
                5. Corolla tube exceeding the calyx; stigma entire .... **Orthosiphon**
                5. Corolla tube not exceeding the calyx; stigma bifurcate ... **Ocimum**
            4. Calyx suberect or declinate in fruit, upper lobe not recurved, margins not decurrent:
                6. Flowers in paniculate racemes .... **Basilicum**
                6. Flowers in globose heads or in spikes or racemes .... **Platostoma**
        3. Lower lip of corolla conspicuously concave or boat-shaped:
            7. Corolla distinctly two-lipped, lower lip larger than the upper:

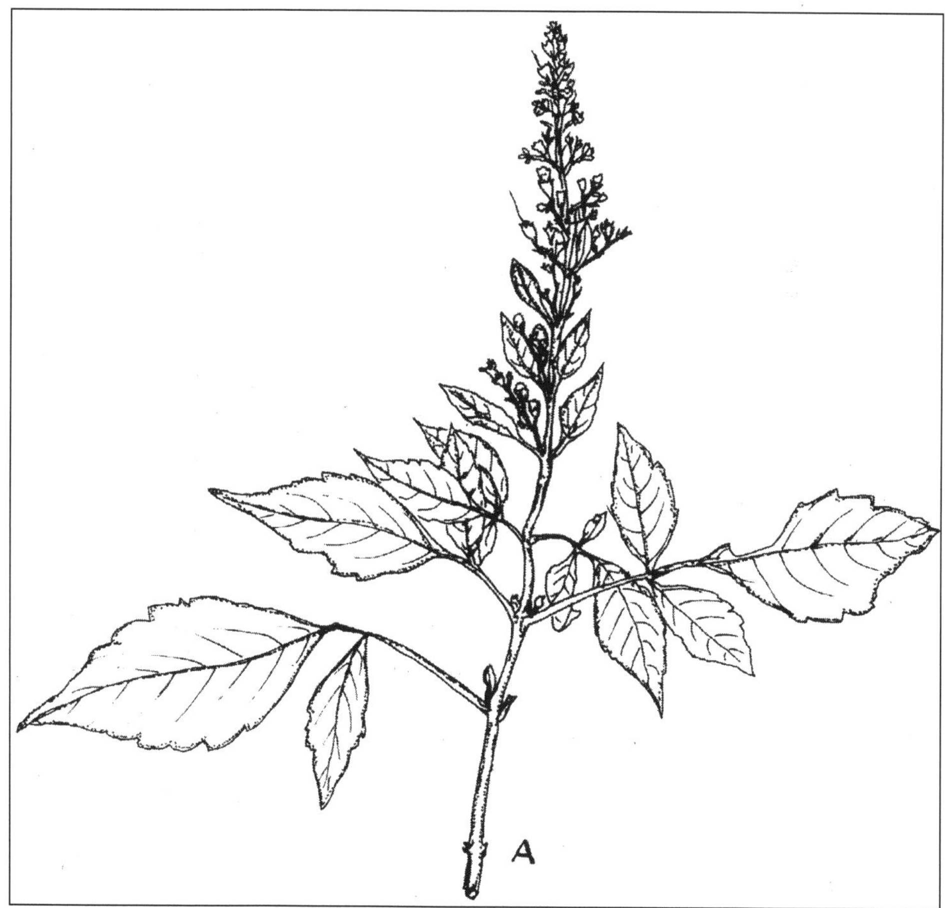

Figure 85: Vitex trifoliata L.

A. Twig

8. Calyx accrescent in fruit:
    9. Corolla pinkish white .... **Plectranthus**
    9. Corolla blue-violet .... **Isodon**
8. Calyx not accrescent in fruit .... **Anisochilus**

7. Corolla 5-lobed, lowest lobe shorter or equaling the other lobes:
    10. Calyx-lobes linear .... **Hyptis**
    10. Calyx-lobes lanceolate .... **Anisomeles**

2. Lower lip of the corolla 2-fid; stamens included .... **Lavandula**

1. Stamens straight or descending:
    11. Stamens 4, perfect:
        12. Stamens subequal, not didynamous:
            13. Teeth of fruiting calyx plumose and up to 6 mm long; tall sturdy shrubs .... **Colebrookea**
            13. Teeth of fruiting calyx not plumose and less than 5 mm long; annual herbs .... **Mentha**
        12. Stamens didynamous:
            14. Upper lip of corolla not woolly .... **Scutellaria**
            14. Upper lip of corolla densely woolly:
                15. Lower lip of corolla shorter than hood .... **Leonotis**
                15. Lower lip of corolla longer than hood .... **Leucas**
    11. Stamens 2, anterior pair only perfect:
        16. Connective elongated, anther cells unequal .... **Salvia**
        16. Connective not elongated, anther cells equal .... **Meriandra**

ANISOCHILUS Wall.

1. Shrubs .... **A. scaber**
1. Herbs:
    2. Leaves sessile oblong-lanceolate or oblancoolate .... **A. verticillatus**
    2. Leaves petiolate ovate-obovate .... **A. carnosus**

Anisochilus carnosus (L.f.) Wall., ex Benth. in Wall., Pl. Asiat. Rar. 2: 18. 1830; Hook.f., Fl. Brit. India 4: 627. 1885; Gamble, Fl. Madras 2: 1126. 1924. *Lavandula carnosa* L.f., Suppl. Pl. 273. 1781.

Vern.: *Ritchu – rodda, Sarugudu ganapa, Roga chettu, Karpuravalli, Omammaku.*

Aromatic, erect herb, 15-60 cm high, stem stout, 4-gonous, bluntly quadrangular, finely pubescent, often tinged with red. Leaves fleshy, ovate-obovate, to 9 × 6 cm, pubescent above, tomentose beneath, glandular when mature, base rounded to sub cordate, margin coarsely crenate, apex obtuse, acute; lateral nerves 3-5 pairs; petiole to 5 cm long. Spikes 1- 4.8 cm long, dense, compact, quadrangular, elongating and cylindric in fruit, peduncled, terminal; bracts broadly ovate, 15 mm; calyx 3.5 mm long, pubescent, tube 1.5 mm, upper lip 1.5 – 2 mm, lower lip reduced, truncate, membranous; corolla pale purple, to 7 mm long, bilipped, 4+1, upper lip with short erect, very shallow rounded lobes, lower lip concave, 3 - 4.5 mm, hairy without, tube narrowed below, inflated above; stamens 4, didynamous, declinate in lower lip, filament pairs of 2 and 3 mm, anthers 0.5 mm; ovary 0.6 mm, style 3.2 mm long, filiform, stigma bifid. Nutlets suborbiculer, compressed, smooth, inner face angular, shining brown.

Common throughout the state on exposed slopes especially along crevices of rocky boulders in all districts.

Fl.: October – January; Fr.: December – May.

Vempalli (ADB), *GO* 4437; Pocharam tank (NZB), *BR* 9504; way to Pathakutta (MDK), *KMS* 6761 (CAL); Cherukupally (NLG), *BVRS* 177 (NU); Aklaspur (KRN), *GVS* 25603 (MH).

Anisochilus scaber Benth. in DC., Prodr. 12: 81. 1848; Hook.f. Fl. Brit. India 4: 627. 1885; Gamble, Fl. Madras 2: 1127. 1924.

Shrub, branchlets pubescent. Leaves ovate, 3-5 × 3-4 cm, sub coriaceous, puberulous on either side, base acute-attenuate, margin minutely crenate, apex acute to obtuse. Flowers in spikes; calyx tube 1.5 mm, lobes 5, bilipped, 1+4, upper lip 2.5 mm, pubescent without, acute, lower lip reduced; corolla white, tinged with pink, 5×1 mm, lobes 5, bilabiate, 4+1, upper lip entire, lower lip 3.5 mm, pubescent without, acute; stamens 4, filament pairs 2 and 3 mm; anthers 0.5 mm; ovary 0.5 mm, style 6 mm.

Common in moist areas in Nizamabad and Ranga Reddi districts.

Fl. & Fr.: December – February.

Annaram RF (NZB), *BR* 7180; Rangammagudem RF (RR), *MSM* 10473.

Anisochilus verticillatus Hook.f., Fl. Brit. India 4: 629. 1885; Gamble, Fl. Madras 2: 1127. 1924.

An erect herb, up to 1.5 m high; stem stout, silky villous, grooved when old. Leaves in whorls of 4-6, each 5-8 cm long, sessile, oblong-lancelolate or oblanceolate, margin entire or crenulate, apex acute. Flowers in terminal spikes; corolla pale blue and bilabiate; stamens 4, didynamous. Nutlets sub-orbicular, compressed. Fruiting calyx with upper lip short, entire or notched; lower short, truncate, densely pubescent and glandular.

Rare in Mahabubnagar district.

Fl. & Fr.: October-November.

ANISOMELES R.Br.

Note: *Epimeredi* Adan. 1763 and *Anisomeles* R.Br. 1810 were considered congeneric by some authors. A proposal to conserve *Anisomeles* R.Br. against *Epimeredi* Adan. was rejected by the nomenclatural committee (Taxon 20: 388. 1971) since there is no evidence that they are congeneric.

1. Stems and branches obtusely tetragonous, white tomentose; leaves lanceolate .... **A. malabarica**
1. Stems and branches acutely tetragonous, pubescent; leaves ovate .... **A. indica**

Anisomeles indica (L.) Kuntze, Revis. Gen. Pl. 2: 512. 1891; Gamble, Fl. Madras 2: 1140. 1924. *Nepeta indica* L. Sp. Pl. 571. 1753. *Anisomeles ovata* W.T. Aiton, Hort. Kew.ed. 2. 2: 364. 1811; Hook.f. Fl. Brit. India 4: 672. 1885. *Epimeredi indicus* (L.) Rothm., Feddes Repert. 53: 12. 1944.

Vern.: *Adabeera, Chinna ranaberi.*

Suffruticose herb or under shrub, scarcely aromatic, softly pubescent, 0.6-2 m high, stem and branches acutely tetragonous. Leaves chartaceous, ovate, 3-9.5 (15) × 2-5 (7.5) cm, softly pubescent, base subtruncate or cuneate, margin crenate-serrate, apex acute to acuminate; petiole to 4.3 cm (6.3 cm) long. Spikes 3.5-20 (30) cm long, axillary, flowers in dense whorls; calyx lobes 5, equal, 1 cm long, villous without; corolla pale or pink-violet, 1.5 cm long, tube annulate within, upper lip oblong, rounded at the apex, lower lip much longer, 3-lobed, the lateral lobes small, rounded, mid one deeply divided into 2 oblong lobes; stamens 4, filaments to 1.2 cm long, hirsute, anther 0.4 mm; ovary 0.5 mm, style 1.6 cm long. Nutlets ovoid, 1.8-2 mm, black, polished, smooth, inner face obtusely angled, dorsal face rounded.

Common throughout in wastelands, waysides of all districts.

Fl. & Fr. September – March.

Sone (ADB), *TP* & *PVP* 4053; Manchippa RF (NZB), *TP* & *BR* 6051; Mosra – Malkapur road (NZB), *TP* & *BR* 6099; Pillutla RF (MDK), *BR* & *CP* 11267; Kanmankalva RF (RR), *MSM* 10460; Irlapenta (MBNR), *BR* & *BSS* 32372; Aklaspur (KRN), *GVS* 25611 (MH); Yerraguntapalli (NLG), *BVRS* 631 (NU); Pakhal (WGL), *KMS* 13174 (MH); Rollapadu (KMM), *RR* 108015 (BSID).

Anisomeles malabarica (L.) R. Br. ex Sims, Bot. Mag. 46: t. 2071. 1819; Hook.f., Fl. Brit. India 4: 673. 1085; Gamble, Fl. Madras 2: 1140. 1924. *Nepeta malabarica* L., Mant. Pl. 566. 1771.

Vern.: *Magabeera, Mogabheri, Pedda ranabheri.*

Under shrub, up to 2 m high, clothed with dense tomentose hairs or thick whitish woolly, stem obtusely quadrangular, slightly branched. Leaves lanceolate, thick, sub coriaceous, 4-9 × 1.6-2 cm, softly tomentose, base rounded or shortly cuneate, apex acute, margin crenate-serrate; lateral nerves 6-8 pairs, prominent beneath, petiole to 1.7 cm long. Spikes to 18 cm long, interrupted, bracts to 1.8 cm long, linear, soft

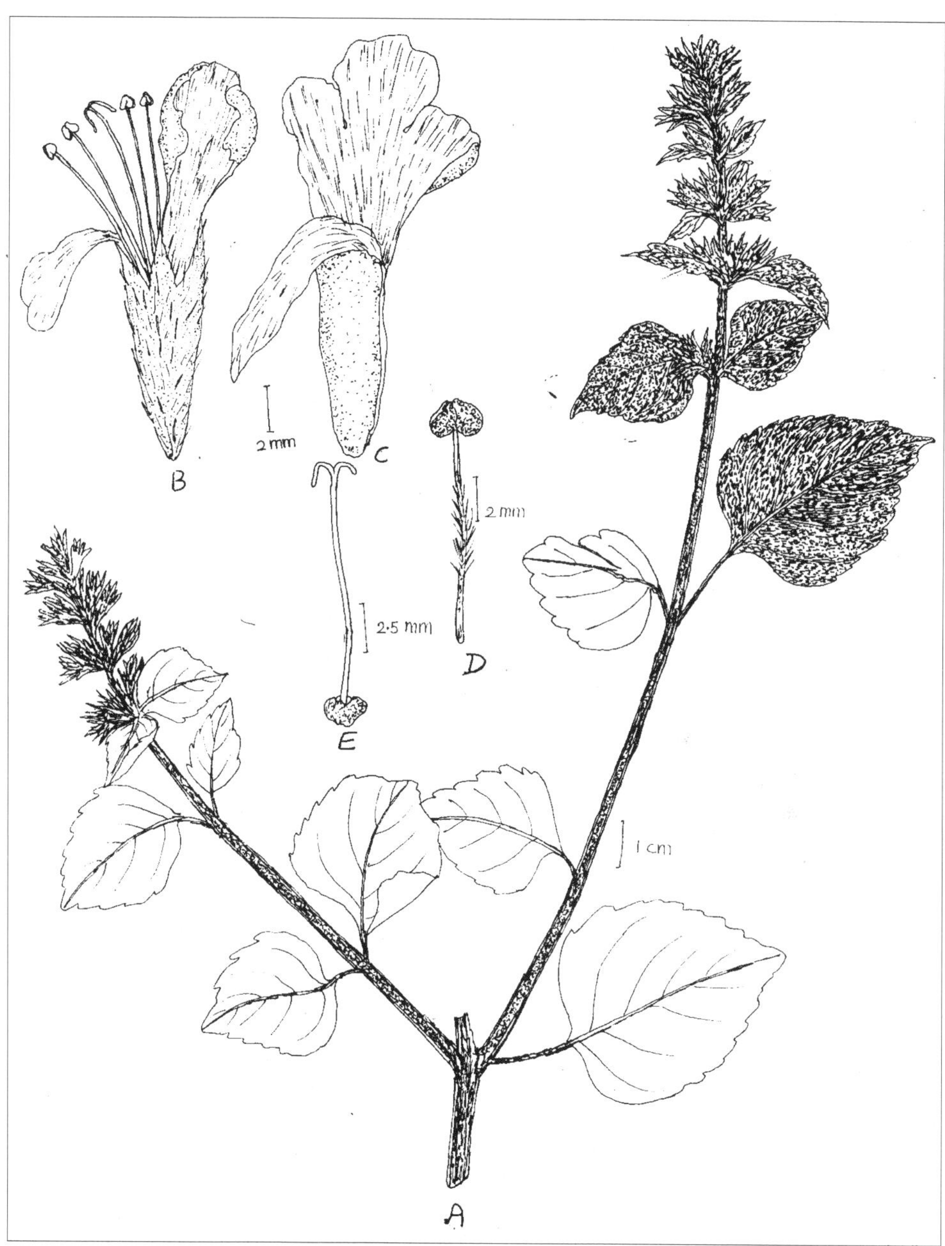

Figure 86: Anisomeles indica (L.) Kuntze

A. Twig, B. Flower, C. Corolla, D. Stamen, E. Pistil.

woolly, bracteoles to 9 mm long; calyx lobes 5, 8.5-10 mm long, densely villous without, teeth narrowly lanceolate; corolla purple, hairy without, 1.4 cm long, bilipped, 2+3, upper lip flat, obtuse, lower mid lobe marginate, lateral lobes short; stamens 4, didynamous, ascending to the upper lip, exserted, filaments hairy, to 1.2 cm, anther 0.7 mm; ovary 0.5 mm, style 3 cm long, strigose, bifid, to 0.4 mm. Nutlets pale, to 4 mm, ellipsoid, compressed, smooth, polished, brown.

Common in wastelands, along roadsides of the hills in Karimnagar, Nalgonda, Hyderabad and Mahabubnagar districts.

Fl. & Fr.: Throughout the year.

Nagarjunasagar (NLG), *KMS* 19180 (MH), *BVRS* 301 (NU).

BASILICUM Moench

Basilicum polystachyon (L.) Moench, Suppl. Meth. 143. 1802. *Ocimum polystachyon* L., Mant. Pl. 567. 1771. *Moschosma polystachyum* (L.) Benth. in Wall., Pl. Asiat. Rar. 2: 13. 1830 – 1831; Hook.f., Fl. Brit. India 4: 612. 1885; Gamble, Fl. Madras 2: 1117. 1924.

Slender, annual, erect herb or subshrub, to 1.5 m high; branches slender, sharply 4-angled, glabrous, brownish. Leaves ovate-deltoid, 1-5 × 1-3 cm, glabrous, base cuneate or truncate, margin serrate, apex acuminate, lateral nerves ca 5 pairs; petiole to 4.5 cm. Thyrses paniculate, terminal and axillary, rachis slender, flowers very small in whorls of 6, bracts ovate, pedicel 0.5mm; calyx tubular, ovoid, 2-lipped, 1+4, 0.5 – 1 mm; corolla pale pinkish, tube cylindric, 2 mm, limb obscurely 2-lipped, 1mm; stamens 4, didynamous, descending into the lower lip, filaments slightly exserted, to 1 mm long, anthers 2-celled, 0.2 mm, ovary 0.1 mm; style gynobasic, 2.3 mm long, stigma forked, disc anteriorly developed. Nutlets ellipsoid, 2 mm, erect, basilar, smooth.

Common, in moist localities in waste lands, river banks near the forests in Ranga Reddi district.

Fl. & Fr.: November – April.

Dharmapur (RR), *MSM* 10484.

COLEBROOKEA Sm.

Colebrookea oppositifolia Sm., Exot. Bot. 2: 111. t. 115. 1806 "*Colebrookia*"; Hook.f., Fl. Brit. India 4: 612. 1885; Gamble, Fl. Madras 2: 1138. 1924.

Vern.: *Maipeeth, Mandrabutta, Jolidi, Chadumbal.*

A functionally dioecious, much branched shrub, 1-3 m high, trunk stout, pale; branches pale, quadrangular, sulcate, young branches tomentose. Leaves sub coriaceous, elliptic or elliptic-oblong, 3-22 × 1.5-9 cm, glabrous above, pubescent beneath, more densely pubescent when young, base acute or cuneate, margin crenulate, apex acuminate; lateral nerves ca 15 pairs, ascending, prominent, petiole 1-4 cm. Flowers very small, dioecious, in bracteate, pedunculate (1 mm) clusters on elongate, trichotomous, 10 cm long spikes arranged in panicles; bracts 1.3 mm, linear-oblong, pubescent, bracteoles many, connate at base; calyx 5-partite, 1.5 mm long, covered with light-brown glands without on the tube, almost to the base, lobes subulate, plumose, elongating in fruit; corolla white, 1-1.5 mm, tube short, limb 2 + 3-lobed, upper lip emarginate, lower lip lobes subequal; stamens 4, exserted in male flowers, staminodes 4, included in female ones, filaments 2.5 mm (male) long, anthers orbicular, cells confluent, style slender, bifid at apex, the lobes subulate. Nutlets oblong-obovoid, pubescent, often solitary, villous at apex, basal scar small.

Occasional in hills of Mahabubnagar district.

Fl. & Fr.: January – May.

Saleswaram (MBNR), *BSS* & *SKB* 30583.

HYPTIS Jacq. *nom. cons.*

Hyptis suaveolens (L.) Poit., Ann. Mus. Hist. Nat. 7: 472. 1806; Hook.f., Fl. Brit. India 4: 630.1885; Gamble, Fl. Madras 2: 1129. 1924. *Ballota suaveolens* L., Syst. Nat. ed. 10. 1100. 1759.

Vern: *Seemathulasi, Danthi tulasi.*

Herb, up to 2 m high, aromatic; branches terete, to 4-angled, grooved, strigose hairy. Leaves chartaceous, ovate, 1.5-6.8 × 1-4.8 cm, densely hairy beneath, less above, base obliquely truncate, acute or sub cordate, margin irregularly serrate, apex acute; lateral nerves 6-8 pairs, prominent beneath; petiole to 4 cm. Flowers often 2-6 together on axillary peduncles or in bracteate axillary racemiform cymes, pedicel to 1cm; calyx 4-4.6 mm, campanulate, 10-ribbed, lobes 5, subequal, linear-acuminate, hairy; corolla blue, 6 mm long, bilipped, upper lip shortly 2-lobed, lower 3-lobed, mid lobe folded, short and bifid, finely hairy; stamens 4, didynamous, descending into the lower lip, partly exserted, filaments 2-3 and 1.5-2 mm long, strigose; ovary 0.7 mm, style gynobasic, 4.7 mm long, stigma very shortly bifid. Nutlets compressed, oblong, 3 mm long, emarginate at top, ribbed.

Fairly common in wastelands, waysides of hills and often forming gregarious patches in all districts. It is common in regeneration forest plots and along streams

Fl. & Fr. : November - April.

Khadam dam (ADB), *TP* & *PVP* 4141; Laxmapur (NZB), *TP* & *BR* 6267; Adluryellareddy (NZB), *BR* & *KH* 9056; Toopran (MDK), *BR* & *CP* 11284; Mohammadabad (RR), *MSM* & *KH* 10556; Pakhal (WGL), *KMS* 11563 (MH); Parvathapur, Appanapally (MBNR), *SRS* 104467 (BSID); on the way to Wyra from Khammam (KMM), *RR* 108058 (BSID); Boddupalle (NLG), *BVRS* 642 (NU).

ISODON (Schrad. ex Benth.) Spach

Isodon coetsa (Buch.-Ham. ex D. Don) Kudô, Mem. Fac. Sci. Taihoku Imp. Univ. 2: 131. 1929. *Plectranthus coetsa* Buch.-Ham. ex D. Don, Prodr. Fl. Nep. 117. 1825; Hook.f., Fl. Brit. India 4: 619. 1885; Gamble, Fl. Madras 2: 1121. 1924.

Erect, robust, aromatic undershrub, up to 2.5 m high; usually much branched; stems pubescent. Leaves opposite, ovate, to 13 × 5 cm, pubescent, base cuneate, margin serrate, apex acute to acuminate, petiole to 3 cm long. Flowers in cymes in axillary and terminal panicles, pedicels slender, pubescent; calyx campanulate, 2 mm, grey-pubescent, tube recurved in the middle, upper lip 3-lobed, lower 2-lobed, teeth subequal, acute; corolla blue or pale violet, 3 mm long, pubescent outside, tube declinate, suddenly swollen above the very short base, gibbous above, lower lip boat-shaped, upper with 4 short rounded lobes; stamens included in the lower lip. Nutlets oblong, inner face slightly angled, outer face rounded, tuberculate.

Common in edges of forests of Adilabad and Nizamabad districts mostly along road sides and along edges of the forests.

Fl. & Fr.: October - January.

Potchera (ADB), *DAM* & *GO* 5061; Adluryellareddy (NZB), *TP* & *BR* 6412; Devanpally (NZB), *BR* & *KH* 9631.

Note: In Flora of Nizamabad district this plant was given under *Plectranthus japonicas* (Burm.f.) Koidz. and *P. coetsa* Buch.- Ham. has been given as synonym. These two are different species. The differences are given below

1. Fruiting calyx with 5, equal or subequal straight teeth .... **Isodon japonicas**
1. Fruiting calyx with subequal teeth or 2-lipped, recurved .... **Isodon coetsa**

LAVANDULA L.

Lavandula bipinnata Kuntze, Revis. Gen. Pl. 2: 521. 1891; Gamble, Fl. Madras 2: 1129. 1924. *L. burmanni* Benth., Lab. Gen. et Sp. 151. 1861; Hook.f., Fl. Brit. India 4: 631. 1885.

A slender erect herb, up to 1.5 cm high; stem simple, thinly villous, quadrangular. Leaves 3-10 cm long, pinnatipartite or very deeply pinnatisect, the lobes linear, entire, cut or toothed, whorled, 1.5-3 mm broad, glabrescent, pale beneath. Flowers in simple or branched 5 cm long spikes or arranged umbellately; bracts finely pubescent, persistent, 6-8 mm long, broadly ovate, strongly nerved; calyx 5 mm, tubular, slightly curved with body pubescence, teeth equal, lanceolate, acute tips, 1mm long; corolla blue or white, pubescent without, 8 mm long, lower half slender, upper half slightly dilated, oblique, limb 2-lipped, upper lip 2-fid, lower lip 3-partite, the middle lobe oblong, rounded, more than twice as long as the obtuse lateral lobes; stamens 4, didynamous, declinate, included, filaments free, anther cells confluent, rounded, pubescent, disc equally 4-lobed. Nutlets 2 mm, oblong-ellipsoid, smooth, very mucilaginous when wetted, black with large white areole on the dorsal surface.

Frequent in open places in the hills of Ranga Reddi and Mahabubnagar districts.

Fl. & Fr.: September – February.

Anantagiri (RR), *MSM* 12793.

LEONOTIS (Pers.) R.Br.

Leonotis nepetiifolia (L.) R.Br., Hortus Kew. 3: 409. 1811 (*nepetaefolia*); Hook.f., Fl. Brit. India 4: 691. 1885; Gamble, Fl. Madras 2: 1155. 1924 var. **nepetiifolia.** *Phlomis nepetifolia* L., Sp. Pl. 586. 1753 (*nepetaefolia*).

Vern.: *Ranabheri, Mulla golimidi, Hanumantha beera, Rama chettu, Errathummi.*

Erect herb, 1-2 m high, stem stout, obtusely quadrangular, shallowly grooved, finely pubescent, pale. Leaves membranous or chartaceous, ovate, 3-8 × 2-5 cm, finely pubescent, base truncate to cuneate, running down into petiole, margin coarsely crenate, serrate, apex acute to shortly acuminate; petiole 2.5-6.5 cm long. Flowers in axillary dense globose, many-flowered, 3 to 7 cm diam. whorls, bracts numerous, linear, apex spinescent; calyx tubular and curved, 1.8 cm long, upper part densely pubescent without, glabrous within, teeth 8-9, unequal, upper most the largest; corolla orange scarlet, 2.5 cm long, tube slender, exserted, annulate within, upper lip 1.2 cm, densely woolly with orange-scarlet hairs, lower lip 4.5 mm, deeply 3-lobed; stamens 4, didynamous, lower pair larger, ascending, exserted, anthers connate, divaricate; ovary 1 mm, 4-partite; style gynobasic, 2 cm, stigma 2-fid, lobes unequal, capitellate, minutely bearded, disc cupular, anteriorly developed. Nutlets erect, oblong-obovoid, trigonous, basilar, 4 mm.

Common throughout the state, frequent in wastelands, way sides, scrub forest and near villages.

Fl.: November- April. Fr: Through the Year.

Neredugonda (ADB), *DAM* & *GO* 5100; Mudheli (NZB), *BR* & *CPR* 7212; Pocharam tank (MDK), *TP* & *CP* 12054; Narsapur (MDK), *KMS* 6789 (MH, CAL); Irlapenta (MBNR), *BSS* & *SKB* 32374; Rechapalli (KRN), *GVS* 22281 (MH); Pargi (RR), *MSM* & *PRSR* 15101; Kambalpally (NLG), *BVRS* 217 (NU).

Leonotis nepetifolia (L.) R.Br. var. **africana** (P.Beauv.) J.K.Morton, J. Linn. Soc. Bot. 58: 275. 1962. *Phlomis africana* P. Beauv., Fl. Oware 2: 82. t. 111. 1819; Ramana & *al.*, Taiwania 58: 254. 2011.

This variety differes from the typical variety by its yellowish cream flowers.

Occasional in open fallow lands and along road sides in Hyderabad.

Fl. & Fr.: August – November.

Osmania University campus (HYD), *MVR* 2078 (CAL & BSID).

LEUCAS R.Br.

Note: Singh (2001) brought out an excellent "Monograph on Indian *Leucas* R.Br. (Dronapushpi) Lamiaceae".

1. Calyx-mouth villous within, with a definite ring of villous hairs present:
    2. Mouth of the calyx truncate:
        3. Flower-whorls terminal, usually solitary:
            4. Bracts and calyx-tube 7-10 mm long, calyx-tube almost straight, not turbinate at mouth .... **L. stricta**
            4. Bracts and calyx–tube 4-6 mm long, calyx-tube decurved, turbinate at mouth .... **L. clarkei**
        3. Flowers in more than two axillary whorls .... **L. lanata**
            5. Hairs on the stem deflexed, often mixed with few scattered hairs .... **L. decemdentata**

5. Hairs on the stem erect or spreading, never deflexed .... **L. lanata**

2. Mouth of the calyx oblique or suboblique:

6. Tall erect herbs, 25-90 cm tall; bracts 3-4 mm wide .... **L. cephalotes**

6. Dwarf diffuse herbs, up to 40 cm tall; bracts 1.5-2 mm wide ...... **L. nutans**

1. Calyx-mouth glabrous, pubescent or tomentose within, without definite ring of villous hairs:

7. Mouth of the calyx truncate:

8. Leaves glabrate to pubescent or scabrous on both sides .... **L. biflora**

8. Leaves tomentose, velvety or densely hirsute on both sides or at least beneath .... **L. montana**

7. Mouth of the calyx oblique:

9. Calyx-mouth split above, produced below .... **L. urticaefolia**

9. Calyx-mouth split below, produced above:

10. Calyx-tube sigmoidly curved .... **L. martinicensis**

10. Calyx-tube straight or slightly curved:

11. Bracts ciliate with long marginal bristle like or bulbous-based stiff hairs:

12. Calyx-tube membranous in lower region, distinctly ribbed with raised veins, veins on pericarp of nutlets indistinct or absent .... **L. aspera**

12. Calyx-tube membranous throughout, veins visible, glabrous, veins on nutlets distinctly three .... **L. lavandulifolia**

11. Bracts ciliate with ordinary hairs, neither bristly nor bulbous based hairs .... **L. zeylanica**

Leucas aspera (Willd.) Link, Enum. Pl. Hort. Berol. Alt. 2: 113. 1822; Hook.f., Fl. Brit. India 4: 690. 1885; Gamble, Fl. Madras 2: 1150. 1924; V. Singh, Monograph on Indian *Leucas* 43. 2001. *Phlomis aspera* Willd., Enum. Pl. Hort. Berol. 2: 621. 1809.

Vern: *Thummi, Tella thummi.*

An erect or diffuse much branched annual herb, up to 50 cm high; branchlets thinly hairy. Leaves linear or narrowly oblong-lanceolate, 2-8 × 0.3-1.5 cm, more or less hispid on both surfaces, base acute or cuneate, margin entire to obscurely serrate, apex acute or obtuse, petiole to 5 mm long. Verticils axillary and terminal, distantly arranged, rarely superposed, 1-2.5 cm diam across, very dense, many-flowered, bracts-linear, ciliate, equalling to the calyx; calyx 8-11 mm long, tubular, oblique, slightly curved, smooth, membranous and glabrous in lower part, irregularly toothed, ribbed and hispid at upper, teeth 1-2 mm, narrow, spinescent, ciliate; corolla ca 10-18 mm long, lower lip twice as long as the upper one. Nutlets 3 mm long, smooth, brown, truncate at apex.

Very common in all districts, in fields, road sides and waste lands up to 1200 m.

Fl. & Fr.: Throughout the year.

Keslapur (ADB), *GO* & *PVP* 4568; Kammarpally (NZB), *TP* & *BR* 6244; Kusumasamudram RF (RR), *MSM* & *KH* 11026; Mannanur (MBNR), *TDCK* 13604; Nagarjunasagar (NLG), *KMS* 19064 (MH); Kudurpalli (KRN), *SLK* 70757 (NBG); Narsingpur (WGL), *RKP* 110864 (BSID); Musalivera beat, Ramavaram range (KMM), *RR* 112624 (BSID).

Leucas biflora (Vahl) R. Br. ex Sm., Cycl. 20 (2): 7. 1812. var. **biflora**; Hook.f., Fl. Brit. India 4: 683. 1885; Gamble, Fl. Madras 2: 1150. 1924; V. Singh, Monograph on Indian *Leucas* 49. 2001. *Phlomis biflora* Vahl, Symb. Bot. 3: 77. 1784.

Vern: *Gobbi.*

Slender procumbent pubescent herb, with woody root stock, 50-70 cm high, stem terete, pubescent with deflexed hairs. Leaves variable, ovate or elliptic-oblong, 1-4.3 × 0.5-2.5 cm, pubescent on both surfaces, glands conspicuous beneath, base slightly truncate to rounded, margin crenate-dentate or crenate-serrate, apex obtuse or sub acute; lateral nerves 3-4 pairs, petiole 0.2-1.2 cm long. Verticils 1-4-flowered, laxly arranged, axillary; pedicel to 2 mm long, bracts setaceous; calyx 8.5 mm long, mouth straight, 10-12-toothed and ribbed, glandular, hispid at the outer surface, especially on the ribs, teeth triangular-subulate, alternately subequal; corolla tube as long as calyx, annulate within. Nutlets 1.5 mm long, truncate at apex, tuberculate, brown.

Occasional in foothills, hill slopes, under growth in the forest in moist habitat, hedges, along road sides and cultivated fields and also in river banks in Adilabad, Nizamabad, Medak, Karimnagar, Nalgonda, Hyderabad and Mahabubnagar districts.

Fl.: July - April. Fruits through the year.

Sone (ADB), *GO* & *BR* 9421; Mondisadak (NZB), *TP* & *BR* 6367; Nagaram (NZB), *BR* 9707; Gangapur RF (MDK), *TP* & *CP* 12091; Amrabad (MBNR), *BSS* & *SKB* 32531; Bandirevu RF (KMM), *RCS* 98902 (BSID); Venkatapur (NLG), *BVRS* 272 (NU).

Leucas cephalotes (Koen. ex Roth) Spreng., Syst. Veg. 2: 743. 1825; Hook.f., Fl. Brit. India 4: 689. 1885; Gamble, Fl. Madras 2: 1150. 1924; V. Singh, Monograph on Indian *Leucas* 53. 2001. *Phlomis cephalotes* Koen. ex Roth, Nov. Pl. Sp.1821. *Leucas capitata* Desf. in Mem. Mus. Hist. Nat. Paris 11: 8. t. 4. 1824.

Vern.: *Pedda tummi, Kedari, Tumbikura.*

Erect, stout annual herb, 25-90 cm high, branchlets scaberulous with spreading hairs. Leaves narrowly ovate-lanceolate, 3-7.5 × 1-3.5 cm, membranous, long spreading hirsute hairs above, soft pubescent and glandular punctate beneath, base cuneate, margin crenate-serrate, apex subacute; lateral nerves 3-5 pairs; petiole to 7 mm long. Verticils terminal, globose, ca 3.5 cm in diam, dense, many-flowered, bracts ovate-lanceolate foliaceous, conspicuous, lanceolate, 1.5 cm long, imbricate and forming

Figure 87: Leucas cephalotes (Koen. ex Roth) Spreng.
A. Twig

an involucre covering the floral heads, ciliate; calyx ca 1.4 cm long, tubular, glabrous at the nutlets part, hirsute above, mouth oblique produced posterior side, 10-toothed, teeth subulate, hispidulous; corolla lower lip not much longer than the upper bearded lip. Nutlets 3 mm long, dark brown, smooth.

Common in forest clearings, cultivated lands and along way sides in most districts.

Fl. & Fr.: August - May.

Dabba (ADB), *GO* & *PVP* 4834; Annaram RF (NZB), *BR* & *CPR* 7176; Kamareddy (NZB), *BR* 9270; Narsapur RF (MDK), *TP* & *CP* 14027; Mekavaripalli (RR), *TP* & *MSM* 11048; Irlapenta (MBNR), *BSS* & *SKB* 32377; Mannanur (MBNR), *KS Singh & Party* 8998 (CDRI); Mamillagudem (KMM), *RR* 105926 (BSID); Subedari (WGL), *RKP* 105257 (BSID); Nagarjunasagar (NLG), *KMS* 9725 (MH); Yerraguntlapally (NLG), *BVRS* 184 (NU).

Leucas clarkei Hook.f., Fl. Brit. India 4: 688. 1885; Mukerjee, Rec. Bot. Surv. India 14. 170. 1940; V. Singh, Monograph on Indian *Leucas* 65. 2001. *L. stricta* Benth. var. *clarkei* (Hook.f.) Paul, Bangladesh J. Bot. 6(1-2): 42. 1977.

Erect herb; 30-50 cm high, branched, stem quadrangular, ascending, sparsely hairy with deflexed or spreading hairs. Leaves linear-lanceolate or linear, 2.5-7.5 × 0.7-1.5 cm, strigose above, hirsute beneath, base narrowed, margin entire or remotely toothed, apex obtuse, petiole 0 or very short. Verticils terminal, subglobose, 2 cm in diam, solitary; bracts slender, 4-6 mm long. linear, rigid, ciliate; calyx 4-6 mm long, curved, closely densely pubescent, ribs 10, mouth truncate or suboblique, longer teeth produced above; villi shorter than teeth, teeth 10, 1.5-2.5 mm, pubescent and sparsely ciliate, acicular at tip; corolla white, tube included within the calyx, upper lip short, narrowly oblong, lower very wide, about 10 mm long. Nutlets oblong, 2-2.5 mm long, rounded at apex, smooth, light brown.

Rare in Hyderabad district (Rajagopal, 1973).

Fl. & Fr.: September - January.

Leucas decemdentata (Willd.) R. Br. ex Smith in Rees, Cycl. 20. 1812 var. **decemdentata;** Benth. Lab. Gen. Sp. 609. 1834; V. Singh, Monograph on Indian *Leucas* 68. 2001. *Phlomis decemdentata* Willd., Sp.Pl. 3: 124. 1800. *Leucas flaccida* R. Br., Prodr. Fl. Nov. Holl. 505. 1810; Benth. Lab. Gen. Sp. 609. 1834; Hook.f., Fl. Brit. India 4: 684. 1885; Mukerjee, Rec. Bot. Surv. India 14(1): 179. 1940. *Leucas flaccida* R. Br. var. *scaberula* (Hook.f.) Srinivasan in Henry (ed.) Flora of Tamil Nadu 2: 176. 1987; Moulali in Pullaiah & Moulali, Fl. Andhra Pradesh 2: 770. 1997. *L. mollisima* Wall., ex Benth. var. *scaberula* Hook.f. Fl. Brit. India. 4: 682. 1885; Gamble, Fl. Madras 2: 1151. 1924. *Leucas mollisma* Wall., ex Benth. in Wall., Pl. Asiat. Rar. 1: 62. 1830; Hook.f., Fl. Brit. India 4: 684. 1885; Gamble, Fl. Madras 2: 1151. 1924.

Vern.: *Pulatummi.*

Annual herbs, woody at base, up to 1m high; branches adpressed with thin to dense deflexed and spreading hairs. Leaves ovate-lanceolate, 2-9 × 0.6-5 cm, base acute or rounded, margin serrate to crenate, apex acute, strigose villous above, densely pubescent beneath, lateral nerves 4-5 pairs; petiole up to 2 cm long, densely pubescent. Verticils axillary with very short peduncle, distant, flowers not very much dense, pedicel 0-2 mm, bracts 5 mm long, ciliate, linear-subulate; calyx tube ca 6 mm long, pubescent, 10-ribbed, mouth truncate, teeth 10, 2 mm long, subulate; corolla tube exserted, upper lip shorter than lower. Nutlets 1.5-2 mm long, obovate, rugose or smooth, greyish brown.

Rare in waste places and outskirts Warangal district.

Fl. & Fr.: October - June.

Pakhal (WGL), *KMS* 11647 (MH).

Nomenclatural Note: Singh (2001) discussed at length about the nomenclature of this plant. It has been now widely accepted that *Leucas flaccida* R.Br. and *L. mollissima* Wall. ex Benth. are conspecific to *L. decemdentata* (Willd.) R.Br. ex Smith.

Leucas lanata Wall., ex Benth. in Wall., Pl. Asiat. Rar. 1: 61. 1830; var. **lanata;** Hook.f. Fl. Brit. India 4: 681. 1885; Gamble, Fl. Madras 2: 1151. 1924; Mukerjee, Rec. Bot. Surv. India 14(1) 178. 1940; V. Singh, Monograph on Indian *Leucas* 97. 2001.

Erect or spreading perennial herb, with many silky branches from a stout root stock, hairs erect or spreading, not deflexed. Leaves narrowly ovate-lanceolate, ovate or orbicular, rather thick, 1.5-6 × 0.6-2.5 cm, base rounded, margin crenate-serrate, apex acute or obtuse, veins marked above, petiole 0 or very short. Verticils axillary, many- flowered, densely hairy; bracts setaceous, densely hairy, much shorter than calyx; calyx tubular-campanulate, 6-8 mm long, straight, ribbed, densely tomentose outside, pubescent within on the upper part, mouth truncate, villi absent, teeth 10, alternately short and long, triangular- subulate; corolla white, 7.5- 8.5 mm long, tube included or little exserted, upper lip densely hairy, lower lip broader, longer than upper lip. Nutlets oblong, ca 1.5 x 1 mm, brown, obliquely truncate, tuberculate.

Occasional in Hyderabad district (Rajagopoal, 1973).

Fl. & Fr.: November – January.

Leucas lavandulifolia Smith in Rees. Cycl. 20. 1812; Mukerjee, Rec. Bot. Surv. India 14(1): 167. 1940; V. Singh, Monograph on Indian *Leucas* 104. 2001. *Leucas indica* (L.) R. Br. ex Vatke, Oesterr. Bot. Z. 25: 95.1975 var. *indica;* Moulali in Pullaiah & Moulali, Fl. Andhra Pradesh 2: 771. 1997. *Leonurus indicus* L., Syst. Nat. ed. 10. 1101. 1759. *Leucas linifolia* (Roth) Spreng., Syst. Veg. 2: 743. 1825; Hook.f. Fl. Brit. India 4: 690. 1885; Gamble, Fl. Madras 2: 1149. 1924. *Phlomis linifolia* Roth, Nov. Pl. Sp. 260. 1821.

Vern: *Pula pumni, Pula tummi.*

Erect, slender annual herb, 30-70 cm high; branches conspicuously 4-angled, puberulous. Leaves linear-lanceolate, 4.5-13 × 0.5-1 cm, pale, finely or minutely puberulous, more or less dense beneath, base tapering, margin entire to sparingly distantly serrate, apex sub acute; lateral nerves 3 - 4 pairs, ascending; petiole usually very short or obsolete, rarely as much as 1 cm long. Verticils axillary and terminal, 2 – 3, superposed and laxly arranged towards apical region of stems and branches, 1.5 cm in diam; bracts 8 mm long, narrowly linear; calyx 8-9 mm long behind, 5-6 mm long in front, glabrous, tube curved, mouth very oblique, rounded below and forming above a large acute lobe much produced forward and tipped by minute spinescent tooth, other one minute, distant, erect along the oblique margin; corolla 12 mm long, lower lip patent, upper lip oblong, white woolly. Nutlets 2.5 mm long, pale brown, adpressed, puberulous.

Occasional in moist shady hills or outskirts of the forests in Nizamabad, and Hyderabad districts.

Fl. & Fr.: August – June.

Rajakampet (NZB), *BR* & *KH* 9051.

Leucas martinicensis (Jacq.) R. Br., Prodr. Fl. Nov. Holl. 504. 1810 & in Aiton f., Hort. Kew. ed. 2. 3: 409. 1811; Hook.f. Fl. Brit. India 4: 688. 1885; Gamble, Fl. Madras 2: 1149. 1924; Mukerjee, Rec. Bot. Surv. 14(1): 169. 1940; V. Singh, Monograph on Indian *Leucas* 118. 2001. *Clinpodium martinicense* Jacq., Enum. Syst. Pl. Carib. 25. 1760 & Amer. Hist. t. 177. f. 75. 1763.

Vern. : *Pedda thummi.*

A tall stout annual herb, up to 1 m high, stem hirsute, with short deflexed hairs, angled, angles rounded, grooved. Leaves membranous, oblong or widely-lanceolate, base cuneate, rarely rounded, margin coarsely irregularly deeply serrate, apex shortly acuminate, or subacute or obtuse, lateral nerves 3-4 pairs, petiole to 1.6 cm long. Verticils 2.5 cm across, axillary, globose, many-flowered, flowers sessile, floral leaves decreasing upwards, bracts narrowly linear-lanceolate or subulate, ciliate spinescent at their tips; calyx 10 mm, tubular, sigmoidly curved, reticulate, slightly pubescent without, glabrous within, mouth oblique, teeth 10, narrowly linear-lanceolate, uppermost longest to 4 mm long, others alternately long and short; corolla small, upper lip as long as lower, 2.5 mm long, obovoid-oblong, obliquely truncate at apex, dark brown, shining. Nutlets *ca* 2 mm long, oblong-obovoid, dark brown, shining.

Common on laterite soils of Karimnagar district.

Fl. & Fr.: November - March.

Kodimial (KRN), *GVS* 218431 (MH).

Leucas montana Spreng., Syst. Veg. 2: 742. 1825; Hook.f. Fl. Brit. India 4: 687. 1885; Gamble, Fl. Madras 2: 1151. 1924; V. Singh, Monograph on Indian *Leucas* 121. 2001.

Vern.: *Konda thummi.*

Straggling under shrubs, branches 4-gonous, softly woolly or tomentose with deflexed hairs. Leaves ovate or suborbicular or ovate-rotundate, 2.5-5 × 0.8-3 cm, usually green and clothed with silky hairs above, velvety-villous or densely white or grey tomentose, base rounded, truncate or subcordate, margin crenate-serrate, serratures mucronate, apex obtuse or acute; lateral nerves 5-6 pairs; petiole 3-9 mm long. Verticils few-flowered, rarely many-flowered (8-10), axillary, distant; bracts linear, to 3 mm long, densely villous; calyx tube 5-8 mm long, campanulate, straight or slightly curved, ribs 10, distinct, produced into teeth, silky villous outside, the upper half finely pubescent inside glabrous throughout within, mouth truncate, villi absent, teeth 10, minute, 1-1.5 mm long, triangular-subulate, subequal or alternately longer and shorter; corolla white, tube exserted, annulate within, upper lip densely woolly, shorter than lower, middle lobe of lower lip broadly obovate, 2-fid. Nutlets trigonous.

Found in hills and waste lands in Mahabubnagar district.

Fl. & Fr.: November – March.

Mannanor Range (MBNR), *STR* 106 (MH).

Leucas nutans (Roth) Spreng., Syst. Veg. 2: 743. 1825; Hook.f., Fl. Brit. India 4: 688. 1885; Gamble, Fl. Madras 2: 1149. 1924; V. Singh, Monograph on Indian *Leucas* 127. 2001. *Phlomis nutans* Roth, Nov. Pl. Sp. 263. 1821.

Diffusely branched herb, up to 40 cm high, branches 4-angular, hairs both long spreading and shorter deflexed. Leaves ovate-oblong, or oblong-lanceolate, 2-6 × 1-2.6 cm, pubescent especially on nerves, gland-dotted beneath, base rounded, margin shallowly crenate, apex gradually acute; petiole short. Verticils axillary and terminal, often 2-3 on a branch, many-flowered, to 2 cm diam; bracts thin, foliaceous, lanceolate, finely spinous-tipped, ciliate, base narrowed; calyx tubular, 10 mm, abruptly falcately decurved in the middle, ribs prominent, mouth oblique, produced on the upper side, villi present, shorter than teeth, teeth 10, 1.5-3 mm, the upper most tooth larger than the others, lanceolate, spinous-tipped, ciliate; corolla white, to 1.3 cm long, tube 9 mm, little exserted, slightly annulate near the base, upper lip 2 mm, white woolly, lower lip 6 mm, middle lobe large, obovate, rounded. Nutlets 2-3mm long, obovoid oblong, 3-gonous, brown, smooth.

Occasional weed on exposed rocky habitats in Nizamabad, Medak, Karimnagar and Ranga Reddi districts.

Fl. & Fr.: July – January.

Mudheli RF (NZB), *BR* & *KH* 7121; Narsapur RF (MDK), *TP* & *CP* 14053; Lingampalle (RR), *MSM* 12765.

Leucas stricta Wall. ex Benth. in Wall., Pl. Asiat. Rar. 1: 61. 1830; Hook.f. Fl. Brit. India 4: 688. 1885; Gamble, Fl. Madras 2: 1150. 1924; Mukerjee, Rec. Bot. Surv. India 14(1): 170. 1940; V. Singh, Monograph on Indian *Leucas* 138. 2001.

Erect, some what diffusely branched herb, 15-30 cm high, hispidly hairy with bulbous base. Leaves linear or linear-oblong or elliptic, 1.2-4 × 0.2-0.9 cm, hispidulous on both surfaces, glandular beneath, base narrowed or acute, margin entire, or distantly serrulate, revolute in linear ones, apex obtuse or subacute; lateral nerves 3-4 pairs; petiole to 2 mm long. Verticils terminal, 2 cm in diam, bracts linear-filiform, ciliate, about as long as calyx; calyx 7-10 mm long, nearly straight, or tubular to funnel-shaped, hairy in upper part, glabrous in lower, mouth slightly oblique, villous within, shorter than teeth; corolla with large lower lip. Nutlets ca 3 mm long, oblong, apex rounded, smooth, brown with canescent surface.

Occasional in Nizamabad, Karimnagar, Medak, Khammam and Mahabubnagar districts.

Fl. & Fr. August – January.

Kamareddy (NZB), *BR* & *KH* 7148; Pathur (MDK), *TP* & *CP* 12120; Pegarikutta (MDK), *KMS* 6658 (CAL); Bhupatipur (KRN), *GVS* 20195 (MH); Murugampadu RF (KMM), *RCS* 102431 (BSID).

Leucas urticifolia (Vahl) R.Br., Prodr. 504. 1810; Hook.f. Fl. Brit. India 4: 680. 1885; Gamble, Fl. Madras 2: 1149. 1924; Mukerjee, Rec. Bot. Surv. India 14(1): 166. 1940; V. Singh, Monograph on Indian *Leucas* 144. 2001. *Phlomis urticifolia* Vahl, Symb. Bot. 2: 76. 1794.

An annual erect herb, 20-60 cm high; branches pale yellowish, adpressed with soft tomentum. Leaves ovate, 3-7 ×1.8-4 cm, thin, soft pubescent beneath, puberulous above, base rounded or slightly cuneate, margin crenate-serrate, apex obtuse; lateral nerves 4-5 pairs, petiole to 2 cm long. Verticils axillary, many-flowered, dense, globose; bract linear-lanceolate, ciliate, slightly equalling the calyx, foliaceous; calyx membranous, cylindric, to 1.3 cm long, pubescent, mouth oblique, produced on the anterior side, ribs 10, ending into spinulose teeth, anterior teeth longer, posterior one with minute teeth; corolla 6.5-12 mm long, tube slender, upper lip half the length of lower lip. Nutlets truncate at the top, 1.5-2 mm long, obovoid-oblong, trigonous, obliquely truncate and glandular at top.

Occasional weed in shady habitat in waste lands of Ranga Reddi and Nalgonda districts.

Fl. & Fr.:July - December.

Boddupalle (NLG), *BVRS* 183 (NU).

Leucas zeylanica (L.) W.T. Aiton, Hot. Kew. 3: 409. 1811. var. **zeylanica**; Hook.f. Fl. Brit. India 4: 689. 1885; Gamble, Fl. Madras 2: 1150. 1924; V. Singh, Monogr. Indian *Leucas* 158. 2001. *Phlomis zeylanica* L., Sp. Pl. 586. 1753.

Annual, erect herb, to 45 cm high, much branched near the base, branches quadrangular, hispid with erect, spreading or deflexed or mixed hairs. Leaves variable, linear, lanceolate or oblong to linear-lanceolate, 2-7.5 × 0.7-1.3 cm, hispidulous with bulbous based hairs, base tapering, margin entire or serrate, often slightly recurved, apex obtuse, midrib conspicuous beneath; petioles to 7 mm long. Verticils terminal, sometimes subterminal whorls formed, 1-2 cm in diam, many-flowered; bracts 6 mm long, linear, spinous tipped, ciliate with long bulbous based hairs; calyx turbinate, 4-6 mm long, slightly curved, lower part glabrous or slightly pubescent outside, the upper part with a few long scattered hairs, mouth oblique, villi absent, teeth 7-8, the upper most tooth longer than others, linear-subulate; corolla white, 1-8 cm long, tube exserted, 6 mm long, upper lip densely bearded, the middle lobe broadly obovate, cuneate, rounded or sub truncate, emargiate, lateral lobes small, lower lip small. Nutlets dark brown or black, 3-4 mm long, obovoid-oblong, subtruncate at apex.

Weed in the fields, wastelands, river beds and forest outskirts in Medak and Mahabubnagar districts.

Fl. & Fr.: August – February.

MENTHA L.

Mentha spicata L., Sp. Pl. 576.1753; Bailey, Man. Cult. Pl. 863. 1949. *M.viridis* L., Sp. Pl. ed. 2, 804. 1763; Hook.f., Fl. Brit. India 4 : 647. 1885.

Erect, annual herbs, 30-40 cm high; stem quadrangular. Leaves linear-lanceolate, 3-6.5 x 0.5-0.9 cm, narrowed at base, distantly serrate, acute; sessile. Flowers whorled in terminal thyrsoid panicles up to 30 cm long; calyx tubular-campanulate, 2-2.5 mm long, appressed hairy outside; teeth 5, subequal, lanceolate-subulate, ciliate; corolla purple or lilac, 4.5-5 mm long; lobes 4, obovate, emarginated, rather unequal. Nutlets globose, 1-2 mm in diam., brown.

Cultivated for its leaves which are used as condiment.

Fl. & Fr.: September-January.

MERIANDRA Benth.

Meriandra dianthera (Roth ex Roem. & Schult.) Briq., Nat. Pflanzenfam. 4 (3a): 288. 1896. *Salvia dianthera* Roth ex Roem. & Schult., Syst. Veg. 1: 263. 1817. *Meriandra bengalensis* Benth. in Edward's Bot. Reg. 15. Sub. t. 1282. 1829; Gamble, Fl. Madras 1: 1159. 1924.

A straggling shrub. Leaves strongly smelling like that of camphor. Calyx 2-lipped; corolla lobes 4, stamens 2, anterior pair only perfect, anthers 2-celled, cells pendulous from a long connective.

Cultivated in gardens and frequently found run wild.

Hyderabad, *MRS* 80729 (DD).

OCIMUM L.

1. Sub shrubs .... **O. gratissimum**
1. Herbs:
    2. Pedicels as long as calyx .... **O. tenuiflorum**
    2. Pedicels shorter than the calyx:
        3. Calyx glabrous within .... **O. filamentosum**
        3. Calyx hairy within:
            4. Corolla 4- 6 mm long .... **O. americanum**
            4. Corolla 7 - 12 mm long .... **O. basilicum**

Ocimum americanum L., Cent. Pl. 1: 15.1755; H. Keng in Steenis, Fl. Males. Ser. 1.8:376.1978. *O. canum* Sims, in Curtis Bot. Mag. t. 2452.1823; Hook.f., Fl. Brit. India 4:607.1885; Gamble, Fl. Madras 2:1111. 1924.

Vern. *Kukka thulasi, Bhuthulasi*. Eng.: Hoary basil.

An erect, much branched strongly aromatic herb, up to 50 cm high; stem quadrangular, grooved, striate, puberulous. Leaves elliptic-ovate or elliptic-lanceolate, 2-3.5 × 0.5–2 cm, glabrous or nearly so, gland dotted, acute at both ends, margin entire or shallowly serrate; lateral nerves 4-5 pairs, impressed beneath, petiole 1.2 – 2.5 cm. Thyrses 15-20 cm long, flowers in rather close whorls, ca 6 in a whorl; bract elliptic-lanceolate, stalked; pedicel 2 mm; calyx pubescent, 1+4, upper lip suborbicular, rather broader than long, flat, shortly mucronate, reticulately veined, ciliate; lower lip with lanceolate-subulate teeth, 2-2.5 mm, central tooth of lower lip longer than the lateral and exceeding the upper lip of the calyx; corolla white, 4-6 mm, hairy without, upper lip broadly oblong, 4-toothed at the subtruncate apex, lower lip longer than the upper, oblong, obtuse; stamens much exserted, filaments slender, 5 mm, anthers 0.8 mm; ovary 0.5 mm, style 5.5 mm, exserted beyond the filaments. Fruit 3-3.5 mm, calyx not much enlarged, pedicel recurved; nutlets ellipsoid, black, 1.4 mm.

Common throughout, in hills, near villages and waste lands to 1200 m altitudes.

Fl. & Fr.: September – February.

Alampalli (ADB), *TP* & *PVP* 4174; Sarvaipet (ADB), *GO* 4523; Mosra (NZB), *TP* & *BR* 6084; Armoor (NZB), *BR* 7291; Medak (MDK), *BR* & *CP* 11420, *KMS* 6667 (MH, CAL); Mohammadabad RF (RR), *MSM* & *KH* 10229; Yerraguntapalli (NLG), *BVRS* 628 (NU); Kodimial (KRN), *GVS* 20104 (MH); Kazipet (WGL), *RK* 105218 (BSID); Mukunur (KRN), *PVS* 84027 (BSID); Koriguttalu (KMM), RR 106037 (BSID).

Ocimum basilicum L., Sp., Pl., 597. 1753. var. **basilicum**; Hook.f., Fl. Brit. India 4: 608. 1885; Gamble, Fl. Madras 2: 1111.1924. *O. americanum auct non* L. 1755.

Vern.: *Kammagaggiraku, Bhoothulasi, Rudrajada, Vibhudi patri.* Eng.: Sweet basil;

An erect herb, scented, 60 cm high; stem quadrangular, striate, yellowish brown, pubescent in young branchlets. Leaves elliptic-lanceolate, 2-7 × 1-3cm, glabrescent, base cuneate, margin entire to shallowly serrate, apex acute; lateral nerves ca 5 pairs, raised beneath. Thyrses 10 – 15 cm long, peduncles stout, purplish green, bracts oblanceolate, petioled, 2-3 mm, pedicel 3 mm, shorter than calyx; calyx 2-3 mm long, in fruit 5-8 mm long, upper lip 3 mm, sub orbicular, ciliate, becoming horizontal in fruit from the deflection of the pedicel, lower lip 4 mm, hairy within, central pair of teeth longer than the upper lip; corolla white or slightly pale purple, 7-12 mm long, tube inflated below, 4 mm long, lower lip fimbriate; stamens 4, filament pairs 6 and 7 mm long, toothed at the base, anthers 0.6 mm; ovary 0.7 mm, style glabrous, 8.5 mm long; fruiting pedicel recurved. Nutlets oblong or ellipsoid, 1.8 mm, tuberculate; seeds mucilaginous when wet.

Common throughout the state, gregarious on wastelands, by arable lands etc in Adilabad, Nizamabad, Hyderabad, Karimnagar, Warangal, Khammam and Mahabubnagar districts.

Fl. & Fr.: August – January.

Sattenapalle (ADB), *TP & PVP* 4159; Janakampalli (NZB), *TP & BR* 6318; Jakaram (WGL), *RKP* 105285 (BSID); Moosi river bank (HYD), *KMS* 5948 (MH); Bhupatipur (KRN), *GVS* 20200 (MH); Umamaheswaram (MBNR), *SRS* 109133 (BSID); Ravannapadu (KMM), *RR* 113854 (BSID).

Ocimum filamentosum Forssk., Fl. Aegypt.- Arab. 108. 1775. *O. adscendens* Willd., Sp. Pl. 3: 166. 1800; Hook.f., Fl. Brit. India 4: 609. 1885; Gamble, Fl. Madras 2: 1112. 1924. *Becium filamentosum* (Forssk.) Chiov., Nuovo Giorn. Bot. Ital. (ns) 26: 162. 1919; Harley, Kew Bull. 38: 56. 1983.

Vern.: *Adavi-tulasi.*

An annual, prostrate herb, scarcely aromatic. Branches ascending, to 50 cm high, branchlets puberulous, subtetragonous. Leaves chartaceous, ovate, 1-4 × 1-3 cm, glabrous, base cuneate or attenuate, margin serrate or repand, apex obtuse; lateral nerves ca 4 pairs, impressed beneath; petiole 1-2 cm. Racemes lax, whorls few-flowered, distant; calyx campanulate, enlarged in fruit, 2-lipped, 1+4, upper lip obovate, 4 mm, lower lip 3 mm, mid lobes mucronate, lateral lobes indistinct, extends in the form of a sheath, puberulous without, glabrous within; corolla white, tube 5 mm, limb 2-lipped, 4+1, upper lip spreading, 5 mm, midlobes 2, lateral lobes obsolete, lower lip deflexed, 3-5 mm; stamens 4, didynamous, declinate, 1.5 cm, upper pair toothed above the base, anther 1-celled, oblong, 0.8 cm; ovary 0.7 mm, style 1.9cm long, stigma 2-fid. Nutlet orbicular, compressed, basiler, reddish brown.

Occasional in waste lands in Nizamabad, Medak, Ranga Reddi, Hyderabad and Karimnagar districts.

Fl. & Fr.: November – February.

Vempalli RF (ADB), *GO & PVP* 4792; Wankidi (ADB), *PVP* 9491; Nadipally (NZB), *BR & KH* 6450; Lingampet (NZB), *BR & CPR* 9241; Narsingi (MDK), *BR & CP* 11523; Pocharam sanctuary (MDK), *CP* 11752; Gacchibowli (RR), *MSM* 12769; Borapur to Appapur (MBNR), *BR & BSS* 32510; Hyderabad (HYD), *KMS* 5988 (MH); Kodimial (KRN), *GVS* 20042 (MH); Yenkur (KMM), *RR* 107904 (BSID); Chityal (NLG), *BVRS* 420 (NU).

Ocimum gratissimum L., Sp. Pl. 1197. 1753; Hook.f. Fl. Brit. India 4: 608. 1885; Gamble, Fl. Madras 2: 1111. 1924.

Vern.: *Nimma thulasi, Ramtulasi.*

A much branched aromatic sub shrub, up to 2 m high; branchlets glabrescent, acutely tetrangular, grooved. Leaves membranous, ovate or elliptic – obovate, 5–12 × 2.8 – 6 cm, pubescent on the nerves above, very sparsely hairy on the nerves beneath, base acute or cuneate, margin coarsely crenate-serrate, apex acute to acuminate; lateral nerves 7 pairs, prominent above; petiole to 6 cm. Thyrses to 20 cm long, verticillasters 6-flowered; bracts lanceolate, acuminate, 3 mm, reticulately nerved; calyx bilipped, 1+4, upper lip rounded, 3 x 1.5 m, prominently nerved, upcurved in fruit, lower lip 2 mm, lateral teeth shorter and broader than the 2 central subulate teeth; corolla white,

5 mm long, pubescent without, lobes acute; stamens 4, exserted, filaments 4-4.5 mm, upper pair of filaments with a bearded tooth at the base, anthers 0.7 mm; ovary 1.5 mm, style 6 mm long, stigma bifid, villous at the upper half. Fruit with calyx 5 x 2 mm; nutlet 1.5 mm diam., subglobose, yellowish-brown, minutely verrucose.

Common in waste lands, along road sides near villages on laterite soils in most districts.

Fl. & Fr.: August – April.

Lingampet (NZB), *BR* 9297; Narsapur – Hyderabad Road (MDK), *BR* & *CP* 11213; Madikonakutta (MDK), *KMS* 6733 (MH); Anantasagar RF (RR), *MSM* & *NV* 10406; Borapuram (MBNR), *BSS* & *SKB* 32360; Perantapally, Papi hills (KMM), *RCS* 98988 (BSID).

Ocimum tenuiflorum L., Sp. Pl. 597. 1753; H. Keng in Steenis, Fl. Males. Ser. 1. 8: 378. 1978. *O. sanctum* L., Mant. Pl. 85. 1767; Hook.f. Fl. Brit. India 4: 609. 1885; Gamble, Fl. Madras 2: 1111. 1924.

Vern: *Tulasi;* Eng.: Sacred Basil.

Erect, much branched aromatic herb, up to 1 m high, branches hispid, obtusely quadragular. Leaves elliptic-oblong, 2-4 cm, softly, minutely pubescent and glandular beneath, base acute, margin serrate, apex obtuse with mucro, lateral nerves 5 pairs, impressed above, petiole to 1.5 cm long. Thyrses to 13 cm long, bracts ovate, pedicels slender, 3 mm; calyx purplish, 2-2.5 mm, hispid without and glandular, upper lip ovate or suborbicular, reflexed, apiculate, lower lip larger than the upper, teeth 4, lanceolate, aristate; corolla white, purplish within, 3.5 mm long, tube 1 mm long, lobes 5, sericeous; stamens 4, exserted, filaments slender, villous at the base of anther, 0.5 mm; ovary 1mm, style 4-5 mm long. Fruit with calyx 3-3.5 × 1.5 mm; nutlets brown with minute tubercles, nearly smooth, ellipsoid, 1 mm diam.

Very common, cultivated in pots and pedestals at all Hindu home steads and temples for religious offering, frequently found run wild, abundant in fallow fields.

Fl. & Fr.: Throughout the year.

Penganga river (ADB), *GO* & *PVP* 4289; Nasrullabad (NZB), *TP* & *BR* 6168; Medak (MDK), *RG* 104116 (BSID); Central University campus (RR), *MSM* 19242; Bourapaur (MBNR), *TDCK* 16814; Rollapadu (KMM), *RR* 108076 (BSID); Kanthikonda (WGL), *RKP* 111540 (BSID); Boddupalle (NLG), *BVRS* 641 (NU).

ORTHOSIPHON Benth.

1. Flowers pale lilac in long racemes .... **O. thymiflorus**
1. Flowers white, in terminal racemes .... **O. rubicundus**

Orthosiphon rubicundus (D. Don) Benth. in Wall., Pl. Asiat. Rar. 2: 14. 1830. var. **rubicundus**; Hook.f., Fl. Brit. India 4: 614. 1885; Gamble, Fl. Madras 2: 1114. 1929. *Plectranthus rubicundus* D. Don, Prodr. Fl. Nepal 116. 1825 (*rubicunda*).

Vern.: *Nela thappidi, Rubicunda.*

An erect, much branched herb, up to 70 cm high; root stock woody, the roots often furnished with fusiform root tubers. Leaves ovate-oblong, elliptic-oblong, 4.5-11 × 2-3.5 cm, puberulous or glabrous above, paler and glabrous or hispidulous on the nerves beneath, base cuneate, margin coarsely serrate-dentate, apex obtuse; petiole 0.6-2 cm. Flowers white, in terminal 10-30 cm long racemes, verticils distant and 6-flowered; bracts ovate-oblong, reflexed, 1.2-2.5 mm; calyx 5 mm, minutely pubescent, upper lobe subpetaloid, purple, slightly decurrent on the tube, 2.5-3 mm broad, lower lip 2-awned, side lobes with short awns; corolla twice as long as calyx, up to 1.2 cm long, tube 8 mm long, longer than the calyx; stamens 4, declinate, filaments glabrous, anther cells confluent, disc gibbous, ovary 6 mm long, style slender, to 1 cm long, stigma capitate. Nutlets suborbicular, compressed, calyx enlarged in fruit, to 8 mm, 12-nerved.

Common as undergrowth of diciduous forests on sandy and laterite soils.

Fl. & Fr.: September - June.

Gupalapatnam (ADB), *GO* 4410; Manchippa RF (NZB), *BR* & *KH* 9030; Velutla RF (NZB), *BR* 9532; Mallalatheertham (MBNR), *TDCK* 15391; Kodimial (KRN), *GVS* 20136 (MH); Vempalli (KRN), *GVS* 20187 (MH).

Orthosiphon thymiflorus (Roth) Sleesen, Reinwardtia 5: 42. 1959. *Ocimum thymiflorum* Roth, Nov. Pl. Sp. 269. 1821. *Orthosiphon tomentosus* Benth. in Wall., Pl. Asiat. Rar 2: 4. 1830-1831; Hook.f., Fl. Brit. India 4: 613. 1885. *O. glabratus* Benth. in Wall., Pl. Asiat. Rar. 2: 14. 1830-1831; Gamble, Fl. Madras 2: 1114. 1924 incl. var. *parviflorus*. *O. viscosus* Benth. in Wall., Pl. Asiat. Rar. 2: 41. 1830-1831; Gamble, Fl. Madras 2: 1113. 1924 incl. var. *rubiginosus*. *O. diffusus* (Benth.) Benth. in DC., Prodr. 12: 50. 1893; Hook.f., Fl. Brit. India 4: 613. 1885; Gamble, Fl. Madras 2: 1113. 1924.

Vern.: *Adavi sajja.*

An erect, glandular viscidly-pubescent tomentose herb with spreading hairs, up to 50 cm high. Leaves ovate, 1.5-5 × 1-3.5 cm, base rounded or cordate, margin crenate, apex sub obtuse, viscous-pubescent. Flowers pale lilac in long racemes; bracts ovate, cuspidate, reflexed; rachis pubescent; pedicels 2-5 mm, sub erect; calyx 3-5 mm in flower, fruiting calyx deflexed, 7.5 mm long, more than 10-nerved; awns of lower lip projecting beyond upper, thinly pubescent, upper lobes 3.5 mm broad in fruit, some what coloured and apiculate, decurrent at sides, lateral scarcely aristate; corolla-tube 5-6.2 mm long, puberulous, longer than calyx, lobes 5, bilipped; stamens 4, included in the corolla tube. Nutlets sub globose, 1.5 × 1 mm, basilar, enclosed in persistent calyx, inner face obtusely angled, minutely tuberculate.

Occasional in Mahabubnagar and Nalgonda districts in dry open forests in red rocky soils.

Fl. & Fr.: June - December.

Pullajalamala (MBNR), *VBH* 84987 (BSID); Vatavarlapalli (MBNR), *VBH* 84965 (BSID); Tirmalgiri (NLG), *BVRS* 475 (NU).

PLATOSTOMA P.Beauv.

Platostoma hispidum (L.) A.J. Paton, Kew Bull. 52: 273. 1997. *Gomphrena hispida* L., Sp. Pl. ed. 2. 1: 326. 1762. *Acrocephalus hispidus* (L.) Nicolson & Sivad., Taxon 29: 324. 1980. *A. capitatus* (Roth) Benth. in Edward's Bot.-Reg. 15: sub t. 1282. 1829; Hook.f., Fl. Brit. India 4: 511. 1885. *Ocimum capitatum* Roth, Nov. Pl. Sp. 276. 1821. *Acrocephalus indicus* (Burm.f.) Kuntze, Revis. Gen. Pl. 2: 511. 1891; Gamble, Fl. Madras 2: 1115. 1924. *Brunella indica* Burm.f., Fl. Indica 130. 1768.

Erect simple or much-branched herb, up to 50 cm high, stem obtusely quadrangular, strigose with reflexed hairs. Leaves lanceolate, elliptic or oblanceolate, 1-3.8 × 0.5-0.7 cm, glabrous, base cuneate or narrowed into the petiole, margin coarsely serrate, apex acute, lateral nerves ca 3 pairs; petiole to 1.5 cm. Flowers very small in terminal or axillary globose-oblong heads, 6-13 mm long, often with 2 spreading leaves at the base, white with violet tinge; bracts to 4 mm long, imbricate, broad; calyx 2.5 mm long, ovoid in flower, tubular in fruit, 2-lipped, the upper lip flat, entire, the lower 4-lobed, lobes linear, long hispidulous; corolla small, 1-3 mm, pale purple, some what 2-lipped, the upper lip subequally 4-lobed, the lower entire, scarcely longer; stamens 4, didynamous, declinate, filaments filiform, anther 1-celled; ovary 4-partite, style slender, shortly bifid at apex. Nutlets smooth, oblong-elliptic, 2 mm long, compressed, brownish-black.

Occasional in moist forests and damp grassy places as an undergrowth of Nizamabad and Karimnagar districts especially on laterite soils.

Fl. & Fr.: September – December.

Ibrahimpet RF (NZB), *TP* & *BR* 6134; Aklaspur (KRN), *GVS* 25613 (MH).

PLECTRANTHUS L'Herit. *nom. cons.*

Note: There is yet no general agreement on the generic limits of *Plectranthus* L'Herit. Some species of the genus *Coleus* Lour. treated under *Plectranthus* L'Herit. by some authors (Morton, J. Linn. Soc. Bot. 58: 231-283. 1962; Codd, Bothalia 11: 371-442. 1975 and Keng in Fl. Males. Ser. 1. 8: 382-393. 1978). Cramer (Kew Bull. 32: 551-561. 1977) however disagreed with the above authors and retained the species of *Coleus* Lour. in Sri Lanka. Willemse (Blumea 25: 507-511. 1979) disagreed with Cramer and transferred all species of *Coleus* Lour. of Sri Lanka to *Plectranthus* L'Herit. *Plectranthus* L'Herit. as treated here includes three species of *Coleus* Lour.

1. Filaments connate below into a sheath around the style:
    2. Calyx throat villous with a ring of white hairs:
        3. Thyrses lax .... **P. barbatus**
        3. Thyrses compact .... **P. caninus**
    2. Calyx throat glabrous within .... **P. amboinicus**
1. Filaments free .... **P. mollis**

Plectranthus amboinicus (Lour.) Spreng., Syst. Veg. 2: 690. 1825; Willemse, Blumea 25. 509. 1979. *Coleus amboinicus* Lour., Fl. Cochinch. 2: 372. 1790; Gamble, Fl. Madras 2:1124. 1924. *C. aromaticus* Benth. in Wall., Pl. Asiat. Rar. 2: 15. 1830-1831; Hook.f., Fl. Brit. India 4: 625. 1885.

Vern.: *Vamu aku, Daggu aku, Karpura valli.*

Fleshy, strongly aromatic shrub, up to 1 m high. Leaves fleshy, broadly ovate, 3-6 cm long, base cordate, or rounded, margin crenate, very villous. Flowers pale purple, densely crowded in distant whorls; calyx-throat glabrous within; filaments connate below into a sheath around the style. Fruits not seen.

Cultivated in homesteads in all districts, occasionally as an escape in Karimnagar and Ranga Reddi districts.

Fl.: May-June.

Plectranthus barbatus Andr., Bot. Repos. 9: t. 594. 1810; Codd., Bothalia 11: 394. 1975. *Coleus forskohlii* (Willd.) Briq. in Engl. & Prantl, Pflanzenfam. 4(3a): 359. 1897. *Plectranthus forskohlii* Willd., Sp. Pl. 3: 169. 1800, non Vahl 1790. *Coleus barbatus* (Andr.) Benth. in Wall., Pl. Asiat. Rar. 2: 15. 1830-1831; Hook.f. Fl. Brit. India 4: 625. 1885; Gamble, Fl. Madras 2: 1129. 1924.

Perennial aromatic herb, 30-60 cm high; stem decumbent and ascending, stout, villous; roots thick, fleshy. Leaves fleshy, ovate or obovate 2.5-10 × 1.3-3.5 cm, base narrowed, margin crenate, apex obtuse, pubescent on both sides, gland dotted below. Flowers in 6-10-flowered whorls, in densely hairy, spiciform racemes; bracts conspicuous, broadly ovate, mucronate, pubescent and ciliate, caducous; calyx hairy, 5 mm long, glandular, throat villous, lower lip with 4 teeth, subulate-sub aristate, upper lip ovate, acuminate; corolla purple or pale blue, to be bent at a right angle, upper lip short, broadly ovate, subequally 4-lobed, lower lip boat - shaped, joined to the tube by a narrow neck, villous without; stamens 4, didynamous, filaments connate at base into a sheath, anthers blue, with red glands on the lower surface, style slender, long, stigma bifid. Nutlets 4, smooth, dark brown.

Cultivated and as escape.

Fl. & Fr.: October - December.

Plectranthus caninus Roth, Nov. Pl. Sp. 279. 1821. *Coleus spicatus*. Benth. in Wall., Pl. Asiat. Rar. 2: 15. 1830-1831; Hook.f. Fl. Brit. India 4: 624. 1885; Gamble, Fl. Madras 2: 1123. 1923.

Fleshy herb, up to 60 cm high; stem subquadrangular, pale yellowish, sulcate, clothed with long hirsute hairs. Leaves ovate or obovate, fleshy, 2-3 × 1.5-2 cm, pubescent, base narrowed, margin crenate, apex obtuse, petiole to 7 mm long. Spikes terminal, flower whorls close in flowering, some what separating in fruiting; bracts closely imbricate, ovate, 6 mm long; calyx 4 mm long, woolly without, rings of hairs within the throat, upper lip shorter than the lower, orbicular, ciliate, lower lip 6 mm long, 4-triangular toothed, teeth ciliate, the 2 middle lobes broader than the 2 lateral ones; corolla bluish or violet, ca 1.3 cm long, tube bent downwards almost at a right angle, upper lip broadly obovate, rounded, 4-lobed, the middle lobes broader than the 2 lateral ones; lower lip 6 mm long, stipitately narrowed into a neck at the base, boat-shaped part is nearly horizontal; stamens 4, filaments connate below, 5.5 mm long, staminal sheath 8 mm, anther 0.8 mm; ovary 0.5 mm, style 1.5 cm long with 2 acute rather deep lobes. Nutlets subglobose, 1.4 mm diam, smooth, black, shiny.

Occasional in dry places, exposed slopes among rock and bushes.

Fl. & Fr.: September - February.

Appapur (MBNR), *BSS* & *SKB* 30588.

Plectranthus mollis (Aiton) Spreng., Syst. Veg. 2: 690. 1825. *Ocimum molle* Aiton, Hort. Kew. ed. 1.2: 322. 1789. *Plectranthus incanus* Link, Enum. Pl. Hort. Berol 2: 120. 1822; Hook. f., Fl. Brit. India 4: 621. 1885; Gamble, Fl. Madras 2: 1121.1924.

An erect herb, to 80 cm high, branchlets purberulous, stem thick and fleshy, tumid below the nodes. Leaves chartaceous, ovate, 4-9 × 2.4-7 cm, pubescent beneath, nearly glabrous above, base truncate to sub cordate, margin double crenate, apex acute or shortly acuminate; lateral nerves 3-5 pairs; petiole 1.5-7 cm long. Flowers in few-flowered cymes in 30 cm long racemes, bracts ovate, 3.5 cm long; pedicel 3-4 mm long, 8 mm in fruit; calyx 3 mm long, glabrous within, 2-lipped, 1+4, upper lip broadly ovate, 2 mm, lower lip lobes lanceolate, the two lower larger than 2-laterals, acuminate; corolla pale lilac or light blue, 6.8 mm long, hairy without, tube gibbous, decurved about the middle, bilipped, 4+1, upper lip 3.5 mm long, 4-lobed, lobes shallow, rounded, lower lip boat-shaped, 5mm long; stamens 4, included without the lower lip, filaments 3.5 mm, anthers 0.6 mm; ovary 0.6 mm, style 5 mm long. Nutlets subglobose, reddish-brown, dotted with brown spots, smooth, 1.5 mm across.

Common in Nizamabad, Medak and Ranga Reddi districts, gregarious by wastelands, roadsides and near streams and also among rocks on exposed slopes.

Fl. & Fr.: December-March.

Lingampet (NZB), *BR* 9299; Pillutla RF (MDK), *BR* & *CP* 11270; Pegarikutta (MDK), *KMS* 6666 (CAL).

SALVIA L.

1. Leaves glabrous, apex obtuse; corolla 3-5 mm long .... **S. plebeia**
1. Leaves pubescent, apex acute-acuminate; corolla 1.7-2 cm long .... **S. coccinea**

Salvia coccinea P. J. Buc'hoz ex Etl., Salv. 23. 1777; Keng in Van Steenis, Fl. Malesiana Ser. I. 8 : 359. 1978.

Erect, pubescent, herbs, to 50 cm high; stems acutely 4-angled, covered with white short recurved and long patent hairs. Leaves ovate or ovate-oblong, 3-5 x 2-3 cm, base cordate and deeply sinuate or sometimes slightly cuneate, margin crenate-serrate, apex acute to acuminate, pubescent on both the surfaces; petioles 3 cm long. Flowers 4-6 in whorls arranged on 10-15 cm long axis, subtended by bractiform, deciduous leaves; calyx tubular, 15-16-ribbed, 2-lipped, reticulate; corolla tubular,

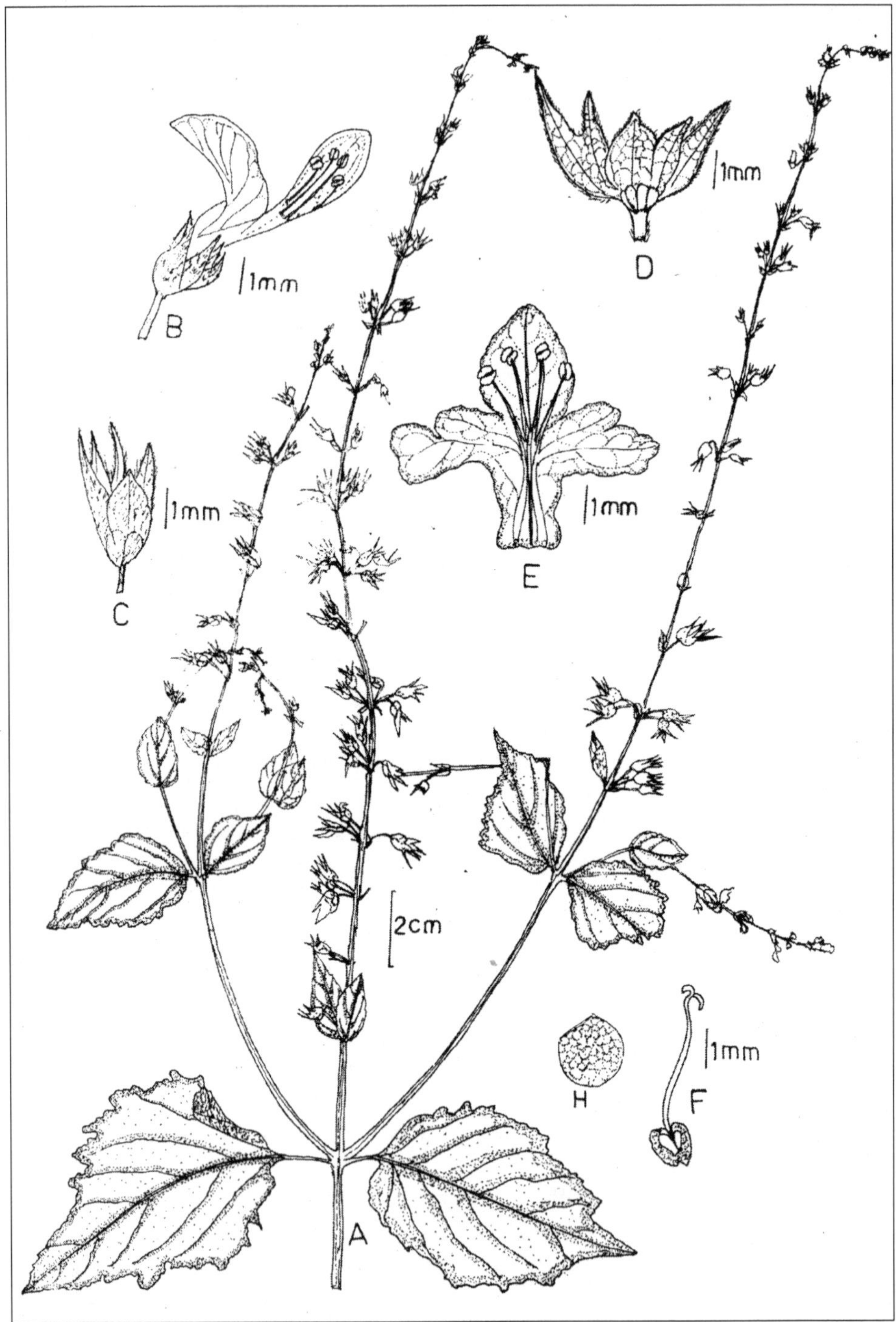

Figure 88: Plectranthus mollis (Aiton) Spreng

A. Twig, B. Flower, C. Calyx, D. Calyx split open, E. Corolla split open with stamen, F. Pistil, G. Stigma, H. Seed.

1.7-2 cm long, scarlet, pubescent outside; lower lip suborbicular, bifid. Nutlets ellipsoid, 1.5-2 mm long, rugose, blackish.

Grown in gardens as on ornamental, and also naturalized in Hyderabad (Ramana, 2010).

Fl. & Fr.: August-November.

Salvia plebeia R. Br., Prodr. Fl. Nov. Holl. 501. 1810; Hook.f., Fl. Brit. India 4: 655. 1885; Gamble, Fl. Madras 2: 1155. 1924.

Erect, pubescent herb, 30-60 cm high, stems stout, obtusely 4-angled, grooved on the sides, pubescent with short reflexed hairs. Leaves wrinkled, ovate, oblong or ovate-lanceolate, 2.5-7.5 cm long, base narrowed into long petiole, margin crenate, apex obtuse, glabrous. Flowers in terminal racemes, whorls 6-flowered, bracts oblong or lanceolate; calyx accrescent in fruit, upper lip entire, obtuse, decurrent on the tube, lower lip longer, obtusely 2-toothed; corolla 3-5 mm long; stamens ascending under the upper lip, anther cells widely separated by the lever like connective, one imperfect. Nutlets minute, ovoid, brown, mucilaginous when wetted.

Occasionally cultivated and run wild.

Fl. & Fr.: February - April.

Janakampalli (NZB), *TP* & *BR* 6318; Kandakurthi (NZB), *BR* 7250; Medak (MDK), *BR* & *CP* 11411; Rusthumpeta (MDK), *KMS* 80107 (MH).

SCUTELLARIA L.

Scutellaria violacea B. Heyne ex Benth. in Wall., Pl. Asiat. Rar. 1: 66. 1830; Hook.f., Fl. Brit. India 4: 668. 1855 *p.p*; Gamble, Fl. Madras 2: 1142. 1924.

A straggling herb, to 50 cm high, branchlets quadrangular, softly greyish pubescent. Leaves chartaceous, ovate, to 5 × 3.5 cm, hairy beneath, base truncate to subcordate, margin crenate, repand with dense hairs, apex obtuse or obtusely acute; lateral nerves 4-5 pairs; petiole to 3.4 cm long. Racemes terminal, to 17 cm, bracts ovate-lanceolate, 2.2 mm, pedicel to 0.6 cm long; calyx turbinate, 2-lipped, hairy without, 3 mm across, upper lip with outwardly curved scutellum, hairy; corolla violet, 1.8 cm long, 2-lipped, tube usually sharply recurved above the calyx, dilated at the throat, upper lip galeate, 3.5 mm, hairy within, lower lip 5 mm; stamens 4, didynamous, not fully exserted, filament pairs 4 and 7 mm, anthers hairy towards margin, upper pair anthers 2-celled, lower pair ones with one-celled anthers; ovary 4-partite, posterior ones larger than anterior, 3 mm, style slender, gynobasic, 1.7 cm long, stigma lobed. Nutlets basilar, subequal, spinulose, calyx lip closed in fruit, the upper lip deciduous, furnishing on the back with a large shield or pouch, the lower lip persistent.

Occasional in fields in Medak district.

Fl. & Fr.: Through the year. Flowers with a peak during October - January.

Venkatapuram (MDK), *TP* & *CP* 14007.

PLANTAGINACEAE

PLANTAGO L.

Plantago asiatica L., Sp. Pl. 113. 1753; Gamble, Fl. Madras 2: 1160. 1924.

Perennial herb. Leaves in rosettes, elliptic-ovate, 5-9 x 3-5 cm, chartaceous, 3-5-nerved from base, base rounded to acute, margin entire, decurrent, serrate or dentate, apex obtuse to acute; petiole to 10 cm. Flowers cream, in terminal spike; bracts 2, lanceolate; calyx lobes 4, ovate, 2.5 mm; corolla 3 mm across, tube 2 mm, lobes 4, ovate-lanceolate; stamens 4, inserted at the base of the tube, filaments filiform, 4 mm; style terete, 1.5 mm, upper half stigmatiferous. Capsule ellipsoid, gradually narrowed above, 4 mm; seeds about 30, mucilaginous.

Occasional in perennially moist places, by drainages in Medak and Hyderabad districts.

Fl. & Fr.: Through the year.

Ranjhole (MDK), *CP* 11600.

NYCTAGINACEAE

1. Annual or perennial, prostrate or erect herbs, up to 1.5 m high
    2. Flowers showy, with an involucre or prominent bracts .... **Mirabilis**
    2. Flowers small, bracts or involucel of bracts not prominent .... **Boerhavia**
1. Trees or shrubs, climbing or often straggling, up to 12 m or more long:
    3. Trees, unarmed .... **Pisonia**
    3. Climbing or straggling shrubs, armed .... **Bougainvillea**

BOERHAVIA L.

1. Anthocarp 10-ribbed with wart like appendages .... **B. chinensis**
1. Anthocarp 5-ribbed, glabrous:
    2. Diffuse herb, anthocarp glandular-hairy, top rounded, club-shaped .... **B. diffusa**
    2. Erect herb, anthocarp ribs undulate, top truncate, obconic shaped .... **B. erecta**

Boerhavia chinensis (L.) Rottb., Pl. Hort. Univ. Rar. Prog. 4. 1773; Stemmerik in Steenis, Fl. Malesiana ser. 1, 6: 455. 1964. *Valerina chinensis* L., Sp. Pl. 33. 1753. *Commicarpus chinensis* (L.) Heimerl, Nat. Pflanzenfam. ed. 2. 16c: 117. 1934. *Boerhavia repanda* Willd., Sp. Pl. 1: 22. 1797; Hook.f., Fl. Brit. India 4: 709. 1885; Gamble, Fl. Madras 2: 1163. 1925.

Vern. : *Teega mamidi*.

A creeper with ascending branches. Leaves triangular-ovate, 5-10 × 3-7 cm, base cordate to rounded, margin repand-sinuate or entire, apex acuminate, glandular dotted above, pubescent beneath; petiole 0.5-2 cm long. Flowers pink, in axillary and terminal, long-peduncled umbels; pedicel 1-1.5 cm long; bracts 0; bracteole 1; perianth 1.2 cm long, constricted below the middle; stamens 3-4. Anthocarp oblong-fusiform, 7 × 1.5 mm, black, 10-ribbed, viscid-glandular with wart like appendages; seed black, testa shiny, smooth.

Common in most districts in hedges and in forests.

Fl. & Fr.: Throughout the year.

Narsapur RF (MDK), *TP* & *CP* 12016; Domadugu (MDK), *BR* & *CP* 14124; Venkat Reddy palle (RR), *MSM* & *KH* 10310; Kambalpally (NLG), *BVRS* 219 (NU); Pakhal (WGL), *KMS* 11629 (MH); Kodimial (KRN), *GVS* 21881 (MH).

Note: This taxon is sometimes named *Commicarpus chinensis* (L.) Heimerl (1934). Stemmerik (*loc. cit.*) does not accept this generic segregation based on number of calyx ribs.

Boerhavia diffusa L., Sp. Pl. 3. 1753; Gamble, Fl. Madras 2: 1162. 1925. *B. repens* var. *diffusa* (L.) Hook. f., Fl. Brit. India 4: 709. 1885.

Vern. : *Atikamamidi.*

Perennial diffuse herb, hirsute, up to 1 m long; branches slender, pubescent. Leaves opposite, sub-equal, ovate, lanceolate, sub-orbicular, thick, 1–4 × 0.5–3 cm, base rounded or sub-cordate, margin some what undulate, apex obtuse, apiculate; greenish pubescent above, white beneath; lateral veins 4–5 pairs; petiole slender, up to 1.8 cm long, pubescent. Flowers in axillary and terminal umbels, glandular hairy; peduncle to 5 cm long, secondary peduncles 5, to 1.5 cm long; pedicel < 1 mm long; bracts 1 mm long; bracteole similar to bract; perianth tube 4 mm long; lower part prominently 5-ribbed; upper part deep pink, 4 mm across; 5-lobed; stamens 2 or 3; filaments 3 mm long; anthers 0.5 mm; ovary 0.6 mm long; style 3 mm long; stigma discoid. Anthocarp club-shaped, glandular-hairy, ellipsoid, 3 × 1 mm, rounded at apex.

Common weed of waste places, fields and road sides all over the State.

Fl. & Fr.: Throughout the year.

Sone (ADB), *TP* & *PVP* 4061; Bijjur RF (ADB), *GO* & *PVP* 4887; Kammarpally (NZB), *TP* & *BR* 6190; Mosra (NZB), *BR* 7287; Narsingi (MDK), *BR* & *CP* 11526; Mohammadabad (RR), *MSM* & *PRSR* 15124; Billyguttalu (KMM), *RR* 105971 (BSID); Amangal to Shadnagar (MBNR), *SRS* 104559 (BSID); Yerraguntapalli (NLG), *BVRS* 632 (NU).

Boerhavia erecta L., Sp. Pl. 3. 1753. *B. punarnava* Saha & Krishn., J. Sci. Industr. Res. 21c : 149. 1962; Rao & Hemadri, Sci. & Cult. 36: 218.1970

Vern.: *Punarnava, Tella atikamamidi.*

Annual erect herb, up to 70 cm high; stem pubescent, terete. Leaves opposite, sub-equal, ovate, 2–5 × 1.5–2 cm, glabrous above, pubescent below, base trunclate-obtuse, margin undulate, apex apiculate-acute; lateral veins 5–6 pairs; petiole 0.5–1.5 cm long. Flowers white, to 3 mm long, in axillary and terminal, 10–20 cm long, paniculate umbels, zigzag, slender, forked; pedicel to 0.6 cm long; bract and bracteole 1 mm long, lanceolate; perianth tube 2.5 mm long; lower parts with 5 faint ribs; upper part white, with pink tinge, 3 mm across; stamens 1 or 3; filaments 2.5 mm long; anthers 0.4 mm across; ovary 0.5 mm long; style 2.5 mm long. Anthocarp obconical, 2.5 mm long, glabrous, top truncate, the grooves between the ribs some what undulate

Occasional in waste lands in Adilabad, Karimnagar and Hyderabad districts.

Fl. & Fr.: September –January.

Sirpur (ADB), *GO* 4899.

BOUGAINVILLEA Comm. ex Juss.

1. Perianth-tube more or less pubescent; leaves and stems subglabrous; bracts magenta coloured .... **B. glabra**
1. Perianth-tube densely pubescent; leaves and stems pubescent; bracts deep rose, but varying to purple or greenish .... **B. spectabilis**

Bougainvillea glabra Choisy, Prodr. 13 (2): 437. 1849; Gamble, Fl. Madras 2: 1164. 1925.

Vern.: *Kagitha pulu, Kagithal chettu*

Large, thorny, climbing shrub, spines axillary, up to 1.5 cm long. Leaves ovate, 3–8 × 3–10 cm, base truncate-acute, margin entire, apex acute; lateral veins 5–7 pairs. Flowers axillary in dichotomous cyme; bracts 3, white or red, thin, reticulately veined, 4 × 3 cm; pedicels 3–4 cm long; calyx-tube 2 cm long, pink; lobes 5, pink below, white above, 0.5 cm long; corolla white, 5-lobed, each 2-fid, 0.3 cm across; stamens 8, in different hights, 1.5–2 cm long; anthers yellow, 0.3 mm across; ovary oblong, to 0.5 cm long; style hairy, curved; stigma linear, forked.

Cultivated in gardens. A native of Brazil.

Fl. & Fr.: January – April.

Bougainvillea spectabilis Willd., Sp. Pl. 2: 348. 1799.

Vern.: *Kagithala puvvu.*

Large scandent, thorny shrubs; spines stout with recurved tip. Leaves ovate, acuminate, pubescent on both the surfaces, veins 4-5 in alternate pairs. Flowers purple, deep rose or greenish supported by an involucres of 3 brilliantly coloured large bracts exceeding the flowers; perianth tubular, 5-lobed, limb wider, basal part looking like a sympetalous corolla; stamens 8, filaments unequal; ovary unilocular, ovule 1, basal. Fruit achene, seed erect.

Cultivated in gardens and habitations as an ornamental plant through out the state.

Fl. & Fr.: August – January.

MIRABILIS L.

Mirabilis jalapa L., Sp. Pl. 177. 1753; Gamble, Fl. Madras 2: 1164. 1925.

Perennial herb, to 75 cm high; root tuber black, cone-shaped, 20 × 6 cm; branches glabrous, succulent, swollen at nodes. Leaves opposite, triangular-ovate to elliptic, 4–10 × 3–6 cm, thick-chartaceous, glabrous, base truncate to shortly decurrent, margin entire, apex acute to acuminate; lateral veins 5 – 7 pairs; petiole to 3 cm long. Flowers sessile, axillary or terminal corymbose, subtended by an involucre of 5, ovate-elliptic, acute, shortly connate bracts, accrescent in fruit; perianth salver-shaped, 5-lobed, 5 cm long; lower part green, glabrous, 0.5 cm across; upper part pink or yellow or white, to 3 cm across; stamens 3–6, exserted; filaments to 6.5 cm long; anthers 1.5 mm across; ovary globose, 1.5 mm, sub-sessile; style 6 cm long; stigma shortly lobed, fimbriate. Anthocarp globose, to 6 mm across, ribbed and with knobs, black when mature, hard, shiny.

Cultivated as ornamental in gardens and often as an escape.

Fl. & Fr.: Throughout the year.

Vempalli (ADB), *GO* & *PVP* 4748.

PISONIA L.

Pisonia grandis R.Br., Prodr. 1: 422. 1810. *P. alba* Span., Linnaea 15: 342. 1841; Hook.f., Fl. Brit. India 4: 711. 1885 *p.p. P. morindifolia* R.Br. ex Wight, Icon. Pl. Ind. Orient. t. 1765. 1852; Gamble, Fl. Madras 2: 1164. 1925.

Evergreen tree, up to 12 m high. Leaves ovate, ca 11 × 5.5 cm, petiole 3 cm, base cuneate, entire, acute, glabrous, veins hairy below. Inflorescence terminal, corymbose cymes; male flower tubular, 5-lobed; stamens 8-9; male flower much smaller than female flower. Anthocarp clavate, 5-ribbed, angles with only 1 row of prickles.

Cultivated in gardens.

Fl. & Fr.: September – April.

AMARANTHACEAE

1. Bracteole wings entirely adnate to midrib, staminodes fimbriate, prominently appendaged within .... **Achyranthes**
1. Bracteoles not winged, staminodes not fimbriate:
    2. Stamens 2 .... **Nothosaerva**
    2. Stamens more than 2:
        3. Ovary containing 2- many ovules, style persistent, embedded ...... **Celosia**
        3. Ovary containing single ovule, style not persistent:
            4. Stems rooting at nodes; flowers spinescent; anthers 1-celled .... **Alternanthera**
            4. Stems lacking roots at the nodes; flowers not spinescent; anthers 2-celled:

5. Spikes tomentose:
    6. Perianth lobes 5 .... **Aerva**
    6. Perianth lobes 4 .... **Trichuriella**
5. Spikes not tomentose:
    7. Flowers unisexual or polygamous .... **Amaranthus**
    7. Flowers bisexual:
        8. Leave alternate:
            9. Flowers capitate; fruit a circumscissile utricle .... **Allmania**
            9. Flowers spicate; fruit indehiscent .... **Digera**
        8. Leaves opposite:
            10. Stigma bilobed .... **Gomphrena**
            10. Stigma capitate:
                11. Flowers all perfect .... **Psilotrichum**
                11. Perfect flowers surrounded by imperfect with hooked bristles .... **Pupalia**

ACHYRANTHES L.

Achyranthes aspera L., Sp. Pl. 204. 1753; Hook.f., Fl. Brit. India 4: 730. 1885; Gamble, Fl. Madras 2:1176. 1925.

Key to Varieties

1. Leaves silverily silky beneath .... var. **sicula**
1. Leaves other than silky .... var. **aspera**

Achyranthes aspera L. var. **aspera.**

Vern.: *Uttareni.*

Annual erect herb, up to 1.5 m high; branches erect to sub-scandent or straggling, much branched, 4-angular, sulcate, densely tomentose. Leaves opposite, elliptic-ovate or obovate, 2–6 × 2–4 cm, sparingly pubescent above, densely so below with white hairs, base rounded or acute, margin entire, apex obtusely acute; lateral veins 5–7 pairs; petiole 0.5–1.5 cm long. Spike terminal, 10–30 cm long, axis cylindrical, pubescent. Flower 0.3 cm long, 19–22 flowers per 5 cm length of spike axis, sessile, light pinkish-green; bracts lanceolate or narrowly deltoid-lanceolate, 3 × 1–1.5 mm, glabrous; bracteoles deltoid-lanceolate, 2–3 × 1–1.5 mm, glabrous, basal wings 0.2–0.3 mm long; spines 1 mm long; perianth segments 5; outer one longer than remaining, 3 × 1–1.5 mm, spiny acuminate, midrib distinct, lateral veins 2, distinct, margin pale; inner ones lanceolate, 2–3 × 1 mm, long acuminate, midrib distinct, margin pale; stamens 5, 1 mm long, alternating with subquadrate pseudo-staminodes, staminal sheath fimbriate at apex, prominently appendaged within; anthers 0.5 mm long;

ovary obconical, 0.7 mm across; style 1 mm long; stigma capitellate. Utricle oblong to ovoid, 1.5 mm long; operculum cup-shaped; seed oblong, testa coriaceous, white, 1–2 mm long.

Common weed on road sides, waste places in all districts.

Fl. & Fr.: July-December.

Sattenapally (ADB), *TP* & *PVP* 4081; Malkapur thanda (NZB), *TP* & *BR* 6101; Narsapur (MDK), *BR* & *CP* 11241; Central University campus (RR), *TP* & *MSM* 15128; Boddupalle (NLG), *BVRS* 640 (NU); Pakhal (WGL), *KMS* 11632 (MH); Aklaspur (KRN), *GVS* 22530 (MH); Rollapadu (KMM), *RR* 108077 (BSID); Kazipet (WGL), *RKP* 105209 (BSID).

Achyranthes aspera L. var. **sicula** L., Sp. Pl. 204. 1753. *Achyranthes aspera* L. var. *argentea* Hook. f., Fl. Brit. India 4: 730. 1885. *A. argentea* Thw., Enum. Pl. Zeyl. 249. 1861. non Lam. 1785.

Annual, erect herb, up to 1 m high; branches erect, quadrangular, glabrescent. Leaves opposite, ovate-elliptic, 3–4.5 × 2–3 cm, pubescent above, silky white villose below, base acute, margin entire, apex acuminate or acute; lateral veins 6–7 pairs; petiole to 1 cm long, pubescent. Spike terminal, to 40 cm long, glabrescent. Flowers 0.5 cm long, distantly arranged on spike axis, 10–12 flowers per 5 cm length of spike axis, sessile, white-green; bracts lanceolate, 3 × 1 mm, basally winged, wing 0.2–0.3 mm long, apically spiny; bracteoles 2, 4 × 1 mm, winged, wing 0.3–0.4 mm long; perianth segments 5, 5–6 mm across; outer one larger than remaining, 4–6 × 1–1.5 mm, lanceolate, acuminate, midrib distinct, spiny, margin thin; inner ones 3-4 × 1-1.2 mm, lanceolate, acuminate, midrib distinct, spiny, margin thin; stamens 5; filaments 1 mm long, alternating with thin pseudo-staminodes; staminal sheath filiform at top, 5 mm across, 1 mm height, prominently appendaged within; anthers white, 0.5 mm long; ovary obconical, 0.7 mm across; style 1 mm long; stigma capitate. Utricle oblong to elliptic, 2 mm across, operculum cap like; seed oblong, elliptic, 2.2 mm long, testa coriaceous, brown.

Rare, as undergrowth of trees in Karimnagar district (Naqvi, 2001).

Fl. & Fr.: June – December.

Bhupatipur (KRN), *GVS* 25597 (MH).

AERVA Forssk. *nom. cons.*

1. Leaves opposite-decussate, lanceolate; seed reniform .... **A. sanguinolenta**
1. Leaves alternate to subopposite, ovate or linear-oblong; seed annular:
    2. Leaves reniform to elliptic-ovate; inflorescence axillary dense, greenish-white spike; flowers hermaphrodite; stigma capitellate .... **A. lanata**
    2. Leaves oblancodate to spathulate; inflorescence terminal, white woolly spikes; flowers dioecious; stigmas 2, as long as the style .... **A. javanica**

Aerva javanica (Burm. f.) Juss. ex Schult., Ann. Mus. Hist. Nat. 11: 131. 1808; var. **javanica.** Hook.f., Fl. Brit. India 4: 727. 1885. *Iresine javanica* Burm. f., Fl. Ind. 212. t. 65. f. 2. 212. 1768. *Aerva tomentosa* Forssk., Fl. Aegypt-Arab. 122. 1775; Gamble, Fl. Madras 2: 1178. 1925.

Vern.: *Magabeera, Pindikonda, Pedda pindikonda.*

Annual or biannual, erect herb, to 1.5 m high; frequently woody; branches terete, striate, densely whitish tomentose. Leaves opposite at base, alternate towards apex, oblanceolate to spathulate, 5–9 × 1–2.5 cm, densely whitish tomentose, base cuneate to attenuate, margin entire, apex obtusely apiculate; lateral veins 5–6 pairs; petiole 2–3 mm long, tomentose. Spikes paniculate, terminal, 18–25 cm long. Flowers unisexual. Male flower: bracts and bracleoles deltoid-ovate, 0.75–2.25 × 0.5 mm, hyaline, acute or obtuse, midrib obscure, 3 mm long; stamens 5, 1.5 mm long, pseudostaminodes alternating the fertile stamens, gynoecium rudimentary. Female flower: bracts and bracteoles lanceolate, 1–2.5 × 0.5 mm; tepals 5, obovate-oblanceolate to spathulate, 2.5 mm long, tomentose without, obtuse; outer 2 tepals with faint midrib, 2–3 × 0.7–1 mm; inner 3 tepals with thick green midrib, 1–1.5 mm; ovary 0.5 mm across; style 1.8 mm long; stigma deeply forked; seed round, slightly compressed, black, shining, very faintly reticulate; testa coriaceous, 0.5–1 mm across.

Common in waste lands, open forests, hedges and fields in Hyderabad, Mahabubnagar and Nalgonda districts.

Fl. & Fr.: July – December.

Kambalpally (NLG), *BVRS* 196 (NU).

Aerva lanata (L.) Juss., Ann. Mus. Hist. Nat. 11: 131. 1808; Hook.f., Fl. Brit. India 4: 728. 1885; Gamble, Fl. Madras 2: 1178. 1925. *Achyranthes lanata* L., Sp. Pl. 204. 1753.

Vern.: *Telaga pindi konda.*

Erect or prostrate, sometimes straggling herb, to 2 m high; branches slender, 1–3 mm across, weak, white lanate or villous. Leaves alternate, variable, reniform to elliptic-ovate, basally large, to 2.5 × 2 cm and small towards apex, to 0.4 × 0.3 cm, pubescent above, tomentose below, base acute-attenuate, margin entire, apex orbicular or acute-apiculate; lateral veins 5–7 pairs; petiole sessile in apical leaves, 1–2 cm long. Flowers bisexual, in 1 cm long, axillary spikes; bracts deltoid-ovate to oblong-ovate, 0.7–1 × 0.5 mm; membranous with a short arista formed by the excurrent midrib, pilose, persistent; bracteoles similar to bracts; perianth lobes 5; outer 2 tepals ovate-oblong, 1.5–2 × 0.5–0.7 mm, hyaline, abruptly connate at apex to a distinct mucro formed by the excurrent mid vein; inner 3 tepals slightly shorter and narrow with a broad central green vitta along the midrib; stamens 5, 1 mm long, pseudostaminodes present at anthesis, attaining about half the length of the style; anthers 0.2 mm across; ovary 0.4 mm long; style 0.2 mm long; stigma capitellate. Utricle rotund, 1 mm across, compressed; seed reniform, 0.6–0.8 mm across, black, shining, testa smooth in the centre, faintly reticulate around the margin, coriaceous.

Common weed in all districts.

Fl. & Fr.: Throughout the year.

Sattenepalle RF (ADB), *TP* & *PVP* 4092; Kammarpally (NZB), *TP* & *BR* 6194; Ponnala (MDK), *CP* 11368; Mohammadabad RF (RR), *TP* & *MSM* 10419; Mannanur (MBNR), *TDCK* 13628; Peddavuru (NLG), *BVRS* 513 (NU); Pakhal (WGL), *KMS* 11567 (MH); Kodimial (KRN), *GVS* 20114 (MH); Tupakulagudem (WGL), *PVS* 76977 (BSID); Bandirevu RF (KMM), *RCS* 98933 (BSID).

Aerva sanguinolenta (L.) Blume, Bijdr. 547. 1826. *Achyranthes sanguinolenta* L., Sp. Pl. ed. 2. 294. 1762. *Aerva scandens* Wall. ex Moq., Prodr. 13: 302. 1849; Hook.f., Fl. Brit. India 4: 727. 1885; Gamble, Fl. Madras 2: 1178. 1925.

Erect herb, to 1 m high, stem hairy. Leaves opposite, decussate, lanceolate, 3-6 × 2-3 cm, base attenuate, entire, acute, glabrous above, pubescent beneath. Flowers in axillay, cylindric, white-woolly to 15 cm long spikes; flowers polygamous; tepals with midrib excurrent into tapering mucro. Fruit circumscissile; seed reniform, black, testa shiny.

Common along forest paths, in Mahabubnagar district (Raghava Rao, 1989).

Fl. & Fr.: July – April.

ALLMANIA R. Br. ex Wight

Allmania nodiflora (L.) R. Br. ex Wight, J. Bot. (Hooker) 1: 226. 1834. var. *nodiflora*; Hook.f., Fl. Brit. India 4: 716. 1885; Gamble, Fl. Madras 2: 1167. 1925. *Celosia nodiflora* L., Sp. Pl. 205. 1753. *A. nodiflora* (L.) R. Br. ex. Wight var. *aspera* (Roth) Hook. f., Fl. Brit. India 4: 717. 1885; Gamble, Fl. Madras 2: 1167. 1925. *Celosia aspera* Roth, Nov. Pl. Sp. 173. 1821. *Allmania nodiflora* (L.) R. Br. ex Wight var. *dichotoma* (Heyne ex Roth) Hook. f., Fl. Brit. India 4: 717. 1885; Gamble, Fl. Madras 2: 1167. 1925. *Celosia dichotoma* Heyne ex Roth, Nov. Pl. Sp. 172. 1821. *Allmania nodiflora* (L.) R. Br. ex Wight var. *longipedunculata* Trimen, Fl. Ceylon 3: 394. 1895. *A. longipedunculata* (Trimen) Gamble, Fl. Madras 2: 1168. 1925. *A. albida* R. Br. ex Hook. f., Fl. Brit. India 4: 717. 1885.

Vern.: *Pendibadde aku, Erra maddi aku.*

Annual ascending or prostrate herb, to 30 cm high; branches angular, pilose with whitish hairs, sulcate. Leaves alternate, variable, narrow spathulate or oblanceolate, 2–5 × 0.5–2 cm, thinly pilose below, base attenuate, margin entire, apex widely retuse to apiculate or obtuse; lateral veins 4–5 pairs; petiole to 1.5 cm long. Spike white or yellowish, axillary and terminal, to 2 cm long; sessile or peduncled; bracts deltoid-ovate, 2–3 × 1–2 mm, acuminate, aristate with excurrent dark midrib; bracteoles lanceolate, 2–3 × 1.5–2 mm, silvery membranous with a reddish or greenish midrib, excurrent in a very short mucronate tip, glabrous, or less pilose; perianth calycine, 5-lobed; lobes oblong-lanceolate, 0.5–0.7 × 0.1–0.2 cm, glabrous or pilose without; stamens 5, 3-3.5 mm long, subulate, filaments alternating the sheath; ovary sub-globose, 1 mm long; style 2 mm long; stigma capitate. Fruit subglobose or ovoid, membranous, utricle circumscissly dehiscent; seed lenticular, black, reticulate, more or less shining, areolate, dull, testa crustaceous.

Common in waste lands, along road sides, and fields in all districts.

Fl. & Fr.: July – December.

Vempalli (ADB), *GO* & *PVP* 4744; Bijjur (ADB), *GO* & *PVP* 4865; Arsapally (NZB), *BR* & *KH* 6481; Borgaun (NZB), *BR* 9282; Narsapur RF (MDK), *TP* & *CP* 14059; Kusumasamudram (RR), *MSM* & *NV* 10364; Amrabad (MBNR), *BSS* & *SKB* 32535; Kodimial (KRN), *GVS* 20029, 20043, 21881 (MH); Pakhal RF (WGL), *KMS* 11029 (MH); Way to Rusthempet-Narsapur Road (MDK), *KMS* 6786 (MH); Durachipally (NLG), *BVRS* 31 (NU).

ALTERNANTHERA Forssk.

1. Inflorescence pedunculate, axillary .... **A. philoxeroides**
1. Inflorscence sessile, axillary:
    2. Anthers 5, oblong .... **A. bicoidea**
    2. Anthers 3-5, round:
        3. Pseudostaminode margins fimbriate .... **A. maritima**
        3. Pseudostaminode margins laciniate, dentate or entire:
            4. Bracts and bracteoles glabrous:
                5. Tepals dissimilar, with a tuft of glochidiate bristles on the keel; leaves ovate to obovate:
                    6. Tepals 5-7 mm, sparsely villous; leaf blades usually as broad as long; pseudostaminode margins dentate .... **A. pungens**
                    6. Tepals 3-5 mm, densely villous; leaf blades longer than broad; pseudomale margins usually entire .... **A. caracasana**
                5. Tepals similar, without glochidiate bristles; leaves lanceolate to spathulate .... **A. sessilis**
            4. Bracts and bracteoles barbellate hairy:
                7. Pseudostaminodes much shorter than stamens; plant mat-forming with prostrate branches ..... **A. paronychioides**
                7. Pseudostaminodes as long as the stamens; plant erect or ascending (cultivated) .... **A. bettzickiana**

Altermanthera bettzickiana (Regel) G. Nicholson, Ill. Dict. Gard. 1: 59. 1884. *Telanthera bettzickiana* Regel, Index Sem. (St. Petersburg) 1862: 28. 1862. *Alternathera ficoides* (L.) R. Br. subsp. *bettzickiana* (Regel) Backer in Fl. Males. 4: 93. 1949. *Alternanthera tenella* Colla var. *bettzickiana* (Regel) Veldkamp, Taxon 27: 313. 1978.

Erect or semierect, much branched perennial herbs, up to 50 cm high. Leaves 1.2-7.6 × 0.5-4 cm, elliptic, oblanceolate, rhomboid-ovate, base attenuate, apex acute or acuminate; petioles hairy. Heads 1-5-together, globose, sessile; tepals 5, unequal. Utricle 1.5 mm long, obovoid-orbicular, 1.5-2 mm, margins winged or thickened; seed ovoid, lenticular.

Cultivated as a border ornamental in gardens.

Fl. & Fr.: Through out the year.

Public Gardens (HYD), *MHR* 6684.

Alternanthera caracasana Kunth, Nov. Gen. Sp. 2: 205. 1818. *Telanthera caracasana* (Kunth) Moq., Prodr. 13(2): 370. 1849; M.H.Reddy & R.R.V.Raju, J. Econ. Taxon. Bot. 21: 585. 1997.

Perennial prostrate herbs, rooting at nodes. Leaves opposite, elliptic, 1-3 × 0.75-1 cm, glabrous or thinly appressed – pilose below, base acute, obtuse. Flowers white or yellow in terminal and axillary globular heads. Utricle globular, radially symmetrical; seed ovoid, compressed.

Rare in Hyderabad

Fl. & Fr.: Through out the year.

Hussain Sagar (HYD), *MHR* 6700.

Alternanthera ficoidea (L.) Sm., Cycl. 39 (2. Suppl.): n. 8. 1819. *A. tenella* Colla in Mem. Reale Accad. Sci. Torino 33: 131. t.9. 1829. *A. tenella* Colla var. *versicolor* (Lem.) Veldkamp, Taxon 30: 208. 1981. *Telanthera ficoides* var. *versicolor* Lem., Ill. Hortic. 12, t. 440. 1865. *Alternanthera polygonoides* R.Br. var. *erecta* Mart., Nov. Gen. Sp. Pl. 2: 51. 1826.

Annual, decumbent or erect herb, to 1.5 m high; stem slightly angular, with appressed acroscopic hairs. Leaves opposite, elliptic-oblong or obovate-oblong 1.5–6.5 × 0.3–3.25 cm densely clothed with appressed long hairs below, less above, base attenuate, margin entire, apex acute-apiculate, dotted on both sides; lateral veins 5–7 pairs; petiole to 1 cm long. Flowers silvery or light yellowish in terminal spikes and near the top of the branches, 2–3 spikes together on common peduncle; bracteole lanceolate, 2–2.5 × 0.5–0.7 mm, very acute, densely hairy on dorsal side; perianth 5-lobed, oblong or ovate-oblong, 3–3.5 × 0.8–1 mm, 1-nerved, in lower half rather densely hairy on dorsal side; stamens 5, 1.5–2 mm long, alternating with pseudo-staminodes, strap-shaped, about as long as stamens, at top divided into a few narrow strips; ovary oblong, 2 mm across; style broader than long, 0.25–0.5 mm long; stigma capitellate. Utricle oblong, 1–1.5 mm, compressed, lenticular, testa crustaceous.

Rare in moist waste lands and roadsides in Medak, Karimnagar and Ranga Reddi districts.

Fl. & Fr.: Through out the year.

Patancheru (MDK), *BR* & *CP* 11319; Zaheerabad (MDK), *CP* 11595; Central University campus (RR), *MSM* 15149; RRL (HYD), *TR* 404 (HH); Dhaniyaigudem (KMM), *RR* 108526 (BSID).

Alternanthera maritima (Mart.) A. St. Hil., Voy. Distr. Diam. 2: 43. 1833. *Bucholzia maritima* Mart., Nov. Gen. Sp. Pl. 2: 50. 1826.

Perennial, erect herbs, up to 50 cm high. Leaves opposite, densely appressed – pilose on both sides when young, ovate, 3-8 × 1-3 cm, base ecuneate, apex acuminate, rarely acute. Flowers pale white to yellowish in terminal dense spikes. Utricle ellipsoid; seed ovoid, compressed, lenticular,.

Rare in moist shady places in Hyderabad.

Fl. & Fr.: July – December.

Public Gardens (HYD), *MHR* 6670.

Alternanthera paronychioides A. St. Hil., Voy. Distr. Diam. 2: 43. 1833; Veldk., Blumea 19: 167. 1971.

Prostrate creeping herb, stem with long crisped hairs on the younger parts. Leaves oblanceolate-elliptic, 1-1.5 × 0.5-0.7 cm, sessile, hairy. Flowers in numerous heads; tepals pilose on back, 3-nerved; stamens 3-5, alternating with sublacerate pseudo-staminodes. Utricle orbicular; seed discoid, 1 mm across, faintly reticulate.

Rare on bunds of paddy fields and along road sides in Mahabubnagar district. Native of Tropical Anerica, naturalized in India.

Fl. & Fr.: February – July.

Tekulapenta (MBNR), *BSS* & *SKB* 32622.

Alternanthera philoxeroides (Mart.) Griseb., Abh, Königl. Ges. Wiss. Göttingen 24: 36. 1879; Maheshwari, Bull. Bot. Surv. India 6: 313. 1964; V. S. Raju, Indian Bot. Reptr. 5: 207. 1986. *Bucholzia philoxeroides* Mart., Beitr. Amarantac. 107. 1825.

Perennial aquatic or marshy, prostrate herb, with many erect branches, lower nodes rooting, stem fistular, longitudinally striate. Leaves opposite, elliptic-lanceolate to obovate-lanceolate, 4-11 × 0.7-2 cm, base narrowed down to an indistinct petiole, glabrous. Flowers in axillary, globose, pedunculate heads; perianth segments shining white, glabrous.

Rare weed in waste lands in Warangal district. Introduced from America.

Fl. & Fr.: May – October.

Alternathera pungens Kunth, Nov. Gen. Sp. 2: 206. 1818. *Achyranthes repens* L., Sp. Pl. 205. 1753. *Alternanthera repens* (L.) Link, Enum. Pl. Hort. Berol. 1: 154. 1821; (non Gmelin 1791).

Vern. : *Mullu ponnaganti.*

Prostrate herb to 50 cm long; rooting at nodes; branches villous with white hairs. Leaves opposite, unequel, obliquely elliptic to orbicular, 1.5–3 × 1–2 cm, distinctly villous above, more below, base attenuate, margin entire, apex mucronate; lateral veins 6–7 pairs; petiole to 0.5 cm long. Flowers white, in axillary clustered heads, to 0.5 cm across; bracts oblanceolate, to 5 mm long, midnerve shortly awned; margin scarious; bracteoles 3 mm long, similar to bract; tepals 5; outer 2 lobes oblanceolate,

4–5 × 1.5–2 mm, very rigid, 5-nerved, mid nerve and outer pair stronger, jointed above into a pungent awn; inner one faintly visible, lobes oblong, 3–3.5 × 1–1.5 mm, 3-nerved, margin serrate, obtusely apiculate, with a tuft of glochidiate, barbellate bristles at base; inner 2 lobes linear, curved with a tuft of glochidiate, barbellate bristles at middle, apex acute; stamens 5; filaments 0.3 mm long; anthers oblong, 0.3 mm long; staminodes membranous, alternating the stamens, dentate; ovary depressed-globose, 2 mm long; style broader than long, 0.1 mm; seed discoid, brown shining, faintly reticulate; testa crustaceous.

Common weed in waste lands, along road sides and fields in most districts.

Fl. & Fr.: Throughout the year.

Varni nursery (NZB), *TP* & *BR* 6170; Narsapur (MDK), *BR* & *CP* 11235; Mokarlabad (RR), *MSM* & *KH* 10263; Farahabad (MBNR), *BR* & *BSS* 34439; Cherukupally (NLG), *BVRS* 668 (NU).

Alternanthera sessilis (L.) R. Br. ex DC., Cat. Pl. Horti. Monsp. 77. 1813; Hook.f., Fl. Brit. India 4: 731. 1885. *Gomphrena sessilis* L., Sp. Pl. 225. 1753. *Alternanthera triandra* Lam., Encycl. 1: 95. 1783 (*nom. illegit.)*; Gamble, Fl. Madras 2: 1179. 1925. *A. nodiflora* R.Br., Prodr. Fl. Nov. Holl. 417. 1810; Hook.f., Fl. Brit. India 4: 732. 1885.

Vern. : *Ponnagantikura*:

Annual, prostrate to erect herb, to 50 cm high, sometimes rooting at basal nodes; branches 5–6-angular, glabrescent, villous at nodes or sparsely hairy. Leaves opposite, variable in shape and size, linear-lanceolate or ovate-oblong or oblanceolate to spathulate, 1–9 × 0.2–2 cm, glabrous or thinly pilose below, base cuneate to attenuate, margin distinctly serrate, apex obtuse or acute; lateral veins 5–7 pairs; petiole to 2 mm long or sessile. Flowers white, in axillary, to 0.8 cm long spikes; bracts deltoid-ovate, 1 × 0.4 mm, scarious, white, mucronate with excurrent midrib, glabrous; bracteoles similar to bracts, 1–1.5 × 0.5–0.8 mm, persistent; perianth segments 5, ovate-elliptic, 2.5–3 × 0.5–1 mm, 1-nerved, acuminate to rather blunt, white, glabrous, apex shortly mucronate with midrib, margins obscurely denticulate; stamens 3; anthers fertile, staminodes 2, 1–1.5 mm long, alternating with pseudostaminodes resembling the filaments, shorter than fertile ones; ovary depressed-globose, 1 mm across; style 0.2 mm long; stigma simple, fimbriate. Utricle cordiform, strongly compressed, 1.2 × 2 mm; seed discoid-orbicular, 0.5–0.75 mm, brown shining, faintly reticulate, testa crustaceous

Common weed of damp places along ponds and streams in most districts.

Fl. & Fr.: Throughout the year.

Sattenaplle (ADB), *TP* & *PVP* 4071; Manchippa (NZB), *TP* & *BR* 6002; Mamidipally (NZB), *BR* & *KH* 6426; Narsapur tank (MDK), *BR* & *CP* 11246; Medak (MDK), *BR* & *CP* 11463; Dharmapur (RR), *MSM* & *KH* 10283; Gandipet (HYD) *KI* & *NVN* 18769; Pakhal (WGL), *KMS* 11690 (MH); Kodimial (KRN), *GVS* 20090 (MH); Billyguttalu (KMM), *RR* 105972 (BSID).

AMARANTHUS L.

1. Stamens and tepals 5 each:

2. Leaf axils with paired spines; styles 2 .... **A. spinosus**

2. Leaf axils without paired spines; styles 3:

3. Plants glabrous or nearly so:

4. Complex thyrsoid structures pendulous; bracts and bracteoles inconspicuously pointed at apex:

5. Inflorescence red .... **A. caudatus**

5. Inflorescence green .... **A. dubius**

4. Complex thyrsoid structures erect, bracts and bracteoles conspicuously pointed at apex:

6. Tepals as long as or longer than fruit; bracts twice as long as perianth .... **A. hypochondriacus**

6. Tepals distinctly shorter than fruit; bracts one and half times longer as perianth .... **A. cruentus**

3. Plants hairy:

7. Fruit longer than perianth .... **A. hybridus**

7. Fruit enveloped in persistent perianth .... **A. retroflexus**

1. Stamens and tepals 3 each:

8. Fruit dehiscent with a circumscissile lid:

9. Bracteoles exceeding or equaling the perianth lobes, awned .... **A. tricolor**

9. Bracteoles shorter than the perianth lobes, acute .... **A. graecizans**

8. Fruit indehiscent:

10. Stem erect, somewhat branched; utricle very rugose .... **A. viridis**

10. Stem ascending, much branched; utricles smooth or slightly rugose .... **A. blitum**

Amaranthus blitum L., Sp. Pl. 990. 1753. var. **blitum.** *A. lividus* L., Sp. Pl. 990. 1753. subsp. *lividus* var. *lividus*

Annual herb, erect, ascending or prostrate, 6-60 cm long. Leaves ovate-rhomboid ovate, 1-8 × 0.6-6 cm, base cuneate, margin entire or slightly undulate, apex emarginate; petiole 1-3.5 cm long. Flowers green, in slender to stout terminal spikes and axillary cymose clusters, male and female flowers intermixed; bracts and bracteoles deltoid-ovate to lanceolate; perianth segments 3, membranous-margined; stamens shorter than perianth; stigmas 2-3. Capsule subrotund to shortly pyriform, compressed, exceeding perianth; seeds round, compressed, dark brown to black.

Cultivated for leafy vegetable.

Fl. & Fr.: August – September.

Amaranthus caudatus L., Sp. Pl. 990. 1753; Hook.f., Fl. Brit. India 4: 719. 1885; Gamble, Fl. Madras 2: 1170. 1925.

Vern.: *Netakura.*

Erect herb, to 1.5 m high, stem hairy when young, glabrous when mature. Leaves alternate, green or red, rhomboid-ovate or ovate-elliptic, 4-10 × 1-5 cm, base cuneate, margin entire, apex acute, glabrous; petiole 1-15 cm long. Flowers in axillary and terminal, thyrsoid spikes, formed of increasingly approximated cymose clusters, apex pendulous; flowers unisexual, pink; bracts and bracteoles spathulate, elongate, awned; tepals red, shorter than fruit, 2-2.5 mm, transparent; stamens 5; stigmas 3. Utricle circumscissile; seed annular, 0.5 mm across.

Cultivated as leafy vegetable in most districts. Native to Neotropic.

Fl. & Fr. September – February.

Nagsanpally RF (MDK), *RG* 104179 (BSID).

Amaranthus cruentus L., Syst. Nat. ed. 10. 2: 1269. 1759. *A. paniculatus* L., Sp. Pl. ed. 2: 1406. 1763; Hook.f., Fl. Brit. India 4: 718. 1885; Gamble, Fl. Madras 2: 1170. 1925.

Vern. : *Dantu kura.*

Erect herb, to 80 cm high, stem (sub) succulent. Leaves alternate, lanceolate to rhomboid or ovate, 3-6 × 0.5-4 cm, petiole length gradually decreasing from bottom to top, base attenuate, entire, acute to obtuse, glabrous. Floweres in axillary clusters, yellowish green, reddish or purple, unisexual, intermixed; tepals shorter than fruit. Utricle 3-4 mm diam., circumscissile; seed annular, 1 mm across, testa shiny, black.

Commonly cultivated in gardens in most districts.

Fl. & Fr.: Throughout the year.

Kadthala (ADB), *GO* & *DAM* 5022.

Amaranthus dubius Mart. ex Thell., Fl. Adv. Montpellier 203. 1912.

Annual erect herb, up to 1 m high. Leaves alternate, glabrous, ovate-rhomboid, 2-8 × 0.8-4.5 cm, base cuneate, margin entire, apex retuse; petiole of proximal leaves equaling or longer than blade, becoming shorter distally. Flowers dioecious in axillary clusters in terminal panicles and axillary spikes. Utricle ovoid – subglobose, circumscissile; seed ovoid, black, compressed.

Introduced and naturalized. Occasional weed in waste places.

Fl. & Fr.: September – March.

Amaranthus graecizans L., Sp. Pl. 990. 1753. *A. graecizans* L. subsp. *thellungianus* (Nevski ex Vassilcz.) Gusev, Bot. Zhurn (Moscow & Leningrad) 57: 462. 1972. *A. thellungianus* Nevski ex Vassilcz in Komarov Fl. U.R.S.S. 6: 365. 1936. *A. polygonoides auct. non* L.; Roxb., Fl. Ind. 3: 602. 1832, Shivakumar & Prabhakar, J. Econ. Taxono. Bot. 37: 414. 2013. *A. polygamus* sensu: Hook. f., Fl. Brit. India 4: 721. 1885. *p.p.;* Gamble, Fl. Madras 2: 1171. 1925.

Vern. : *Chirikura.*

Diffuse, prostrate herb, stem red. Leaves alternate, elliptic-obovate, 1-2 × 0.5-1.5 cm, glabrous, base attenuate, margin entire, apex mucronate. Flowers in axillary fascicles, unisexual; tepals 3, equal, spreading, oblong-lanceolate; stamens 3, filaments 1 mm. Utricle ovoid, rugose; seed orbicular, 1 mm across.

Common weed of cultivated fields and waste places in Medak and Hyderabad districts.

Fl. & Fr.: July – December.

Vempalli (ADB), *GO* & *PVP* 4747.

Note: Linnaeus record of his *Amaranthus polygonoides,* mentioned by Trimen is clearly an error. *A. polygonoides* L. is confined to the Central regions of the American continent and West Indies. (Bull. Bot. Surv. India 8 : 88. 1966).

Amaranthus hybridus L., Sp. Pl. 990. 1753. subsp. **hybridus;** Towns. in Dassan. & Fosb., Rev. Handb. Fl. Ceylon 1: 13. 1980.

Vern. : *Dantu kura.*

Annual, erect herb, up to 2 m high; branches stout, angular, glabrous or thinly to moderately furnished with short or long muticellular hairs. Leaves alternate, broadly lanceolate to ovate, 3–20 × 1.5–10 cm, gradually narrowed, thinly pilose below on veins, base cordate, margin entire, apex blunt to mucronulate; lateral veins 5–15 pairs; petiole up to 15 cm long. Spikes red or purple, axillary and terminal, varying from a single spike to a broad, much-branched panicle, 10–30 × 3–5 cm; bracts and bracteoles unequal in size, deltoid-ovate 2.5–3 × 0.5–1 mm, pale membranous, scarious, acuminate with a long tip, arista erect and formed by the stout excurrent midrib, exceeding the perianth. Flowers androgynous, formed of cymose clusters, increasing upwards; perianth segments 5, lanceolate or oblong, 3–3.5 × 1–1.5 mm, acute-aristate or the inner sometimes blunt in the female flower, 3-nerved; stamens 5, 1.5 mm long, dorsifixed, X-shaped; ovary conic or pyriform; styles 3, fimbriate, 1 mm long, curved. Utricle subglobose to ovoid, 2–3 mm, capsule lid swollen with a distinct neck; seed ovoid, 0.75–1.25 mm, compressed, almost smooth, faintly reticulate around the margin, testa crustaceous.

Commonly cultivated in gardens in most districts.

Fl. & Fr.: Throughout the year.

Kadthala (ADB), *GO & DAM* 5022.

Amaranthus hypochondriacus L., Sp. Pl. 2: 991. 1753. *A. hybridus* L. var. *hypochondriacus* (L.) H. Rob., Rhodora 10: 32. 1908. *A. hybridus* subsp. *hypochondriacus* (L.) Thell.,Fl. Adv. Montpellier 204. 1912.

Annual erect herb, up to 70 cm high, glabrous or slightly pilose. Leaves rhombic-ovate or oblong-lanceolate, 3-8 × 1.5-3.5 cm, glabrous, base cuneate, margin entire or undulate, apex acute or acuminate; petiole 1- 6 cm long. Flowers green in terminal, thyrsoid spikes; bracts green or purple, ovate, subulate; tepals green or purple; stigmas 2 or 3. Utricle exceeding perianth; seeds white, subglobose.

Cultivated as leafy vegetable. Native of North America.

Fl. & Fr.: July – November.

Amaranthus retroflexus L., Sp. Pl. 2: 991. 1753.

Annual, erect herb, up to 70 cm high, densely pubescent. Leaves ovate-rhombic or elliptic, 5-10 × 2-5 cm, shortly hairy on both surfaces, base cuneate, margin entire and undulate, apex acute or notched, with a mucro. Flowers in axillary and terminal thyrsoid spikes; bracts and bracteoles white, subulate; tepals white, oblong or oblong-obovate, membranous, with a green midvein; stamens slightly longer than perianth; stigmas 3. Utricle light green, ovoid, compressed, shorter than perianth, circumscissile; seeds brown or black.

Occasional in waste lands, road sides and hedges of fields.

Fl. & Fr.: August – November.

Amaranthus spinosus L., Sp. Pl. 991. 1753; Hook. f., Fl. Brit. India 4: 718. 1885; Gamble, Fl. Madras 2: 1170. 1925.

Vern.: *Mulluthota kura.*

Armed, annual, erect herb, up to 1 m high; stem stout, angular, reddish, glabrous basally, multicellular long hairy apically. Spines axillary, 1.5 cm long, paired or clustered, subtending inflorescence. Leaves alternate, ovate to rhomboid-ovate, rarely oblong-lanceolate, 1.5–7 × 0.8–4 cm, glabrous or thinly pilose on the veins on lower side, base cuneate or attenuate, margin entire, apex obtuse-apiculate with fine mucro; lateral veins 5–7 pairs; petiole up to 9 cm long. Spikes clustered, axillary or terminal, up to 7 cm long, 0.8 cm diameter; lower clusters entirely female; upper flowers male and bisexual; bracts deltoid-ovate, 1–1.5 × 0.5–0.75 mm, pale scarious margin, pale or reddish awn formed by the excurrent green midrib, to 0.5 cm; bracteoles deltoid-ovate, 2–2.5 × 1–1.5 mm, sub-equaling or shorter than perianth lobes. Flowers pale greenish-white, up to 0.5 cm across; perianth segments 5; those of female flowers narrowly oblong or spathulate-oblong, 1.5–2.5 × 0.5–1 mm, obtuse or acute, mucronulate, frequently with a greenish dorsal midvein; those of male flowers broader than the female, oblong-lanceolate, 1.5–3 × 0.75–1 mm, acute or acuminate, midrib excurrent; stamens 5; filaments 1–1.5 mm long; anthers oblong, 0.2 mm long, sagittate; ovary 0.5 mm across; style 0.5 mm long; stigmas 3, flexuose or reflexed, 1–1.5 mm. Utricle ovoid, urceolate, 1.5–2 mm, with a short inflated neck below the style base, circumscissile, rarely indehiscent; seed globular, 0.75-1 mm, compressed, black, shining, very faintly reticulate, testa crustaceous.

Common in waste lands, fields, and hedges near villages in most districts.

Fl. & Fr.: Throughout the year.

Sattenapalle (ADB), *TP & PVP* 4156; Mosra (NZB), *TP & BR* 6085; Mombojipally (MDK), *BR & CP* 11637; Vikarabad (RR), *TP & MSM* 11090; Kodimial (KRN), *GVS* 20075 (MH); Laxmipuram (WGL), *PVS* 76983 (BSID); Karepally (KMM), *RR* 105964 (BSID).

Amaranthus tricolor L., Sp. Pl. 989. 1753. var. **tricolor.** *A. gangeticus* L., Syst. Nat. ed. 10. 1268. 1759 incl. var. *tristis*; Hook.f., Fl. Brit. India 4: 719; Gamble, Fl. Madras 2: 1171. 1925. *A. mangostanus* L., Cent. Pl. 1: 32. 1775; Gamble, Fl. Madras 2: 1175. 1925.

Vern.: *Perugu thotakura*

Annual erect or decumbent herb, to 50 cm high; stem angular, glabrescent, upper parts pilose with less crisped hairs. Leaves alternate, rhomboid-ovate, broadly elliptic, 1–4 × 0.5–3 cm, glabrous, white dotted below, base attenuate, margin entire, apex emarginated to apiculate; lateral veins 4–7 pairs; petiole to 7 cm long, longer than lamina. Flowers green to crimson in globose clusters on spike, male and female flowers intermixed; bracts deltoid ovate, 1–1.5 × 0.5–0.7 mm, bracteoles broadly ovate, 2.5–3 × 0.5–1 mm, sub-equal to the perianth, membranous, lanceolate, apex apiculate, midrib green; perianth segments 3, elliptic or oblong-elliptic, 3–3.5 × 1–1.5 mm, narrowed above, pale membranous, midrib excurrent into a long, pale-tipped awn, perianth segments slightly accrescent in fruit; stamens 3; filaments 2–2.5 mm long; anthers 0.3 mm long; ovary compressed, 1 mm across; style 3-fid, minute. Utricle ovoid-urceolate, 2.25–2.75 mm, with a short neck below, circumscissile, membranous; seed ovoid 1–1.5 mm, black or brown, shining, very faintly reticulate, lenticular, testa crustaceous.

Occasional weed in waste places, fields, and along road sides, most districts.

Fl. & Fr.: Throughout the year.

Surzapur (ADB), TP & PVP 4155; Kammarpally (NZB), *TP & BR* 6191; Rampur (MDK), *BR & CP* 11654; Thupakulagudem (WGL), *PVS* 76994 (BSID).

Amaranthus viridis L., Sp. Pl. ed. 2. 1405. 1763; Hook.f., Fl. Brit. India 4: 720. 1886; Gamble, Fl. Madras 2: 1171. 1925. *Amaranthus gracilis* Desf., Tabl., Ecole Bot. 43. 1804.

Vern.: *Chilakathota kura*.

Annual, erect herb, up to 75 cm high; stem glabrous when mature, pilose when young, with floccose multicellular hairs, angular. Leaves alternate, deltoid-ovate to rhomboid-oblong, 2–7 × 1.5–5.5 cm, glabrous or shortly to fairly long-pilose when young and on the veins of lower surface, base shortly cuneate or sub-truncate, margin entire, apex obtuse or emarginate-apiculate; lateral veins 7–9 pairs; petiole. 2.5–12 cm long. Flowers in axillary and terminal spike, often paniculate, intermixed but later more numerous; bracts and bracteoles deltoid-ovate to ovate-lanceolate, 1–2 × 0.5–0.7 mm, whitish membranous with a very short, pale awn formed by excurrent midrib; perianth lobes 3, very rarely 4; male flower segments ovate-oblong, 2–2.5 × 0.5–1 mm, margin scarious, apex mucronate, midrib green; stamens 3, filaments 1.5 mm long; anthers 0.3 mm long; ovary 0.7 mm across; style 0.2 mm, stigmas 2–3, erect. Utricle subglobose, 1.25–1.5 mm, not exceeding the perianth, indehiscent; seed ovoid, 1–1.25 mm, slightly compressed, dark brown to black, reticulate and verrucose, testa crustaceous.

Very common weed in waste lands, fields and along road sides, in all districts.

Fl. & Fr.: Throughout the year.

Bersaipet (ADB), *GO* 4619; Kadthala (ADB), *GO* & *DAM* 5020; Kammarpally (NZB), *TP* & *BR* 6195; NSF (MDK), *BR* & *CP* 1161; Anantasagar RF (RR), *MSM* & *KH* 11001; Bhupatipur (KRN), *GVS* 20207 (MH).

CELOSIA L.

Celosia argentea L., Sp. Pl. 205. 1753; Hook.f., Fl. Brit. India 4: 714. 1885; Gamble, Fl. Madras 2: 1166. 1925.

Key to Varieties

1. Inflorescence dissected, cocks comb like spike .... var. **cristata**
1. Inflorescence dense, cylindrical spike .... var. **argentea**

Celosia argentea L.var. **argentea**

Vern.: *Gurugu aku.*

Annual, erect herb, up to 2 m high; branches ascending, strongly ridged and sulcate, glabrous, reddish or greenish. Leaves alternate, oblong-lanceolate to narrowly linear, 2–15 × 1–3.5 cm, glabrous, sometimes reddish, base attenuate, margin entire, apex acuminate or obtuse, apiculate; lateral veins 7-9 pairs; petiole 2–10 mm. Flowers silvery to pink, in terminal, dense spike, 2.5–10 × 1.5–3 cm, conical at first but becoming cylindrical; peduncle sulcate, up to 20 cm long; bracts and bracteoles lanceolate, the lower deltoid, 3–4 × 2–3 mm, much shorter than tepals, hyaline, aristate with excurrent midrib, margins scarious, persistent; perianth segments 5, elliptic-oblong, 6–8 × 2–3 mm, acute to rather blunt, mucronate, lateral veins 2-4, margin widely hyaline; stamens 5, 6–7 mm long; filaments delicate; anthers creamy; ovary ovoid, 2–3 mm long; style filiform, 5–6 mm; stigmas 2–3. Urticle ovoid, 3–4 mm across; seeds lenticular, 1–1.5 mm diameter, black, shining; testa very finely reticulate, crustaceous.

Common weed in post harvested fields and waste lands.

Fl. & Fr.: August – December.

Sone (ADB), *TP* & *PVP* 4008; Kammarpally (NZB), *TP* & *BR* 6192; Narsapur (NZB), *BR* & *CP* 11216; Medak (MDK), *CP* 11796; Lingampalle (RR), *MSM* 12745; Nagarjunasagar (NLG), *KMS* 19101 (MH); Aklaspur (KRN), *GVS* 20219 (MH); Pakhal (WGL), *KMS* 11597 (MH); Dhaniyaigudem (KMM), *RR* 108507 (BSID).

Celosia argentea L. var. **cristata** (L.) Kuntze, Revis. Gen. Pl. 1: 541. 1891. *Celosia cristata* L., Sp. Pl. 205. 1753; Hook.f., Fl. Brit. India 4: 715. 1885; Gamble, Fl. Madras 2: 1166. 1925.

Vern.: *Kodijuttu chettu*. English: Cocks comb.

Annual sub-shrubs, to 1.5 m high. Leaves alternate, lanceolate, 4-12 × 0.5-4 cm, base acuminate, margin entire, apex acute, glabrous. Flowers in terminal dense spikes, cockscomb like, variously branched with sterile shoot, red or reddish-orange. Utricle ovoid, enveloped in persistent perianth; seeds compressed-reniform.

Commonly cultivated in gardens, in most districts.

Fl. & Fr.: November – March.

Jannaram (ADB), *TP* & *PVP* 4180.

DIGERA Forssk.

Digera muricata (L.) Mart., Nova Acta Phys.- Med. Acad. Caes. Leop.-Carol. Nat. Cur. 13: 285. 1826. *Achyranthes muricata* L., Sp. Pl. ed. 2. 295. 1762. *Digera arvensis* Forssk., Fl. Aegypt-Arab. 65. 1775; Hook.f., Fl. Brit. India 4: 717. 1885; Gamble, Fl. Madras 2: 1169. 1925.

Vern.: *Chenchali kura*.

Annual, prostrate or decumbent, sometimes erect herb, to 40 cm high; stem glabrous or pilose, angular. Leaves alternate, broadly ovate or rarely sub-elliptic, 2–5 × 0.5–3 cm, glabrous or hairy below, base attenuate or narrowed, margin entire, apex acute or acuminate-apiculate; lateral veins 7–10 pairs; petiole slender, 4–5 cm long. Flowers whitish-pink, in axillary, slender, 10 cm long spike, distantly arranged; fertile flowers each subtended by 2 lateral, sterile, modified flowers consisting of scales; bract obovate, 2.5–3 × 0.75–1 mm, midrib prominent; bracteole lanceolate, 2–2.5 × 0.5–0.75 mm, glabrous, membranous with a green or brownish percurrent midrib; perianth segments 5, ovate-oblong, 3–4.5 × 1–1.5 mm, acute, the inner perianth lobes slightly shorter, more delicate, blunt or erose, 1–3-nerved, margin hyaline with a darker central midrib; stamens 5, 3 mm long; ovary globose, 1 mm across; style slender, to 2 mm long; stigma bifid, finally recurved. Fruit subglobose, 2–2.5 mm, indehiscent, slightly compressed, bluntly keeld, furnished throughout with verrucose, testa crustaceous.

Common in irrigated dry fields in all districts.

Fl. & Fr.: Throughout the year.

Sone (ADB), *TP* & *PVP* 4004; Kammarpally (NZB), *TP* & *BR* 6239; Lingareddypally (MDK), *CP* 11717; Mekavaripalli (RR), *TP* & *MSM* 11038; Modimalakala (MBNR), *TDCK* 16837; Rampur (MBNR), *BR* & *BSS* 32641; Boddupalle (NLG), *BVRS* 644; Kodimial (KRN), *GVS* 20055 (MH); Billyguttalu (KMM), RR 105994; Eturunagaram (WGL), *RKP* 110819 (BSID).

GOMPHRENA L.

1. Heads globose or depressed-globose; dorsal crest of the bracteoles well-developed; flowers purplish or crimson .... **G. globosa**
1. Heads ovoid or spiciform; dorsal crest of the bracteoles very narrow; flowers white .... **G. serrata**

Gomphrena globosa L., Sp. Pl. 224. 1753; Hook.f., Fl. Brit. India 4: 732. 1885; Gamble, Fl. Madras 2: 1179. 1925.

Vern.: *Erra pagada banti, Rudrakshi, Bandumalli.*

Erect herbs, up to 40 cm high, hairy. Leaves elliptic to obovate-oblong, 2-6 × 1.5-3 cm, base attenuate, margin entire, apex acute; petiole to 1cm. Flowers sessile, above the uppermost pair of leaves, in globose or depressed-globose heads, purplish or crimson, hermaphrodite; seed annular, 2 × 1.5 mm, black, testa shiny.

Commonly cultivated and ornamental plant in all districts.

Fl. & Fr.: Throughout the year.

Gomphrena serrata L., Sp. Pl. 224. 1753; Mears, Taxon 29: 86. 1980. *G. decumbens* Jacq., Pl. Hort. Schoenbr. 4: t. 482. 1804; Gamble, Fl. Madras 2: 1179. 1925. *G. celosioides* auct. non Mart.: Bennet Fl. Howrah 142. 1979.

Vern.: *Tella pagada banthi.*

Annual, ascending or erect herb, to 20 cm high; much branched, spreading from base, densely furnished with long, white lanate hairs, villose. Leaves opposite, elliptic-oblong to oblanceolate 1.5–5 × 0.5–1.5 cm, pilose with lanate hairs, base attenuate, margin entire, apex obtuse-mucronate; lateral veins 5–7 pairs; petiole to 0.5 cm long, pilose. Flowers silvery white, in terminal, to 2.5 cm long spike; peduncle to 6 cm long with pilose hairs and having terminal pair of leaves. basal 1 cm long spike sterile; bracts deltoid-ovate, 2.5–3 × 0.5–1 mm, shortly mucronate with excurrent midrib; bracteoles lanceolate, 3–3.5 × 0.7–1 mm, laterally compressed, hyaline, villose; perianth segments 5, narrowly lanceoalte, 4.5–5 × 1–1.5 mm, 1-nerved; the outer 3 more or less flat, lanate at the base, apex mucronate; the inner ones densely lanate almost to the tip, slightly longer than the outer ones; stamens 5, staminal tube subequalling the perianth; anthers yellow; ovary compressed-globose, 1.3 mm; style and stigma together 1 mm, stigma divergent. Capsule shortly compressed pyriform, to 1.75 mm long; seed ovoid, compressed, brown, faintly reticulate, shining, testa coriaceous.

Common along road sides, waste lands and open forest in all districts.

Fl. & Fr.: Throughout the year.

Bijjur RF (ADB), *GO* & *PVP* 4891; Nasrullabad (NZB), *TP* & *BR* 6167; Sangareddy tank (MDK), *BR* & *CP* 11289; Central University campus (RR), *MSM* 15143; Cherukupally (NLG), *BVRS* 349 (NU); Kodimial (KRN), *GVS* 20099 (MH); Pakhal (WGL), *KMS* 11670; Rollapadu (KMM), *RR* 108045 (BSID); Gaarlapaadu RF (MBNR), *SRS* 109092 (BSID).

Note: The species, to which the name *G. serrata* L. was given, was renamed by Jacquin (1805) as *G. decumbens* without reference to Linnaeus' species name. *G. dispersa* Standl. is a name applied to some phenotypes of the same variable species. *G. serrata* L. is the most common and widely distributed non cultivated species of *Gomphrena*. It is frequently misidentified as *G. celosioides* Mart., a closely related species native to South America except in cultivation.

NOTHOSAERVA Wight

Nothosaerva brachiata (L.) Wight, Icon. Pl. Ind. Orient. 6: 1. 1853; Hook.f., Fl. Brit. India 4: 726. 1885; Gamble, Fl. Madras 2: 1175. 1925. *Acyranthes brachiata* L., Mant. Pl. 50. 1767.

Annual, erect herb, up to 50 cm high, branches spreading from base upwards; stem subterete, striate, sparsely hairy. Leaves decussate, lower leaves on main stem ovoid-elliptic, 1–3 × 0.5–2 cm, upper leaves on main stem and leaves of branches becoming shorter and narrower, reduced in flowering branches, glabrous to thinly hairy above, pubescent below, more on nerves, base acute to attenuate, margin entire, apex acutely apiculate; petiole about half the length of the lamina, to 1.5 cm long on main stem, shorter on branches, reduced towards apex and finally sessile. Flowers pale whitish woolly, bisexual, in axillary, dense paniculate spikes, subtended by reduced leaves; peduncles short, 3 mm long, axis thinly to rather densely pilose; bracts ovate, 0.5 mm, hyaline, acute or shortly acuminate, glabrous or very thinly hairy, nerveless; bracteoles 2, broadly ovate, concave, 1–1.25 mm, hyaline, acuminate, persistent; perianth segments 5, sub equal, oblong, deltoid-ovate, 1.25–1.5 × 0.5–0.75 mm, hyaline, membranous, 1-nerved, sub acute to shortly acuminate, villous on the outer suface, with a dark greenish midrib, extends up to $^2/_3$ of the segment, truncate, acute; stamens 2; filaments 1–2.5 mm long; anthers 2-celled, hemispheric, 0.1 mm across; ovary obconic to depressed-globose, 0.3 mm across, 1-locular; ovule 1 on short, erect funicle; style short, 0.5 mm long; stigma capitellate. Utricle pyriform, oblong, 0.5–0.75 mm, compressed, epicarp membranous, irregularly rupturing, falling off with the perianth-lobes; seed lenticular, 0.5 mm, brown, smooth and shining, testa crustaceous.

Rare, in moist low-lying grounds, along river banks in Karimnagar, Khammam, Medak, Warangal and Mahabubnagar districts.

Fl. & Fr.: September – March.

Rollapadu (KMM), *RR* 108042 (BSID); Kasumthumcheru (MDK), *RG* 104121 (BSID); Mangapet (WGL), *RKP* 110876 (BSID).

PSILOTRICHUM Blume

1. Inflorescence elongated spikes; tepals outer 2 membranous, inner 3 densely long-pilose .... **P. elliotii**
1. Inflorescence lax panicles; tepals outer 2 pungent, inner 3 silky ...... **P. sericeum**

Psilotrichum elliottii Baker, Fl. Trop. Afr. 6 (1): 58. 1909. *P. calceolatum* Moq. ex Hook. f., Fl. Brit. India 4: 725. 1885. *nom. illeg.*; Gamble, Fl. Madras 2: 1174. 1925.

Erect herb, up to 75 cm high; stem and branches in the older parts terete, striate, finally glabrescent at the base. Leaves ovate to broadly oblong, 2-4.5 × 1.5-3 cm, base truncate, acuminate, moderately furnished on both sides with long, sub appressed white hairs, sometimes gabrescent above with age; petiole 5 mm. Flowers in terminal or axillary elongated spikes. Utricle compressed; seed ovoid, 1 mm, compressed, brown, testa shiny, coriaceous.

Rare in swampy ground on the edges of forest in Mahabubnagar district.

Fl. & Fr.: October – January.

Psilotrichum sericeum (König ex Roxb.) Dalzell in Dalzell & A. Gibson, Bombay Fl. 216. 1861. *Achyranthes sericea* Koen. ex Roxb., Fl. Ind. 2: 502. 1824. *Psilostachys sericea* (Koen. ex Roxb.) Hook. f. in Benth. & Hook.f., Gen. Pl. 3: 32. 1880 & in Fl. Brit. India 4: 726. 1885; Gamble, Fl. Madras 2: 1174. 1925.

Slender dichotomous herbs. Leaves opposite, broadly ovate, to 2.5 × 2 cm, base rounded or cordate, entire; petiole very short. Flowers minute, in lax panicles. Utricle indehiscent; seeds inverse, testa coriaceous.

In Ranga Reddi district (Seshagiri Rao, 2012)

Fl. & Fr.: October – January.

PUPALIA Juss. *nom. cons.*

Pupalia lappacea (L.) Juss., Ann. Mus. Natl. Hist. Nat. 2: 132. 1803; Hook.f., Fl. Brit. India 4: 724. 1885; Gamble, Fl. Madras 2: 1173. 1925. *Achyranthes lappacea* L., Sp. Pl. 204. 1753. *P. atropurpurea* (Lam.) Moq. in DC., Prodr. 13(2): 331. 1849; Hook.f., Fl. Brit. India 4: 723. 1885; Gamble, Fl. Madras 2: 1173. 1925. *Achyranthes atropurpurea* Lam., Encycl. 1: 546. 1785.

Key to Varieties

1. Leaves glabrous or thinly pilose; spike unbranched .... var. **lappacea**
1. Leaves other than above; spike branched .... var. **orbiculata**

Pupalia lappacea (L.) Juss. var. **lappacea**.

Vern. : *Erra uttareni, Tella uttareni.*

Annual or perennial, erect herb, up to 2 m high; stem much branched and swollen at nodes; branches opposite, slender, obtusely 4-angled to almost terete, thinly pilose to tomentose. Leaves opposite, variable in shape and size, ovate-elliptic to oblong or orbicular, 2–7 × 1–4 cm, sparsely tomentose or pilose on the veins in lower side, base attenuate gradually, margin entire, apex acuminate to apiculate; lateral veins 7–9 pairs; petiole to 1 cm long, pilose. Flowers in terminal, unbranched, to 50 cm long spike; peduncle tomentose; flowers 3 in a cluster, central one 1-bracteate, lateral

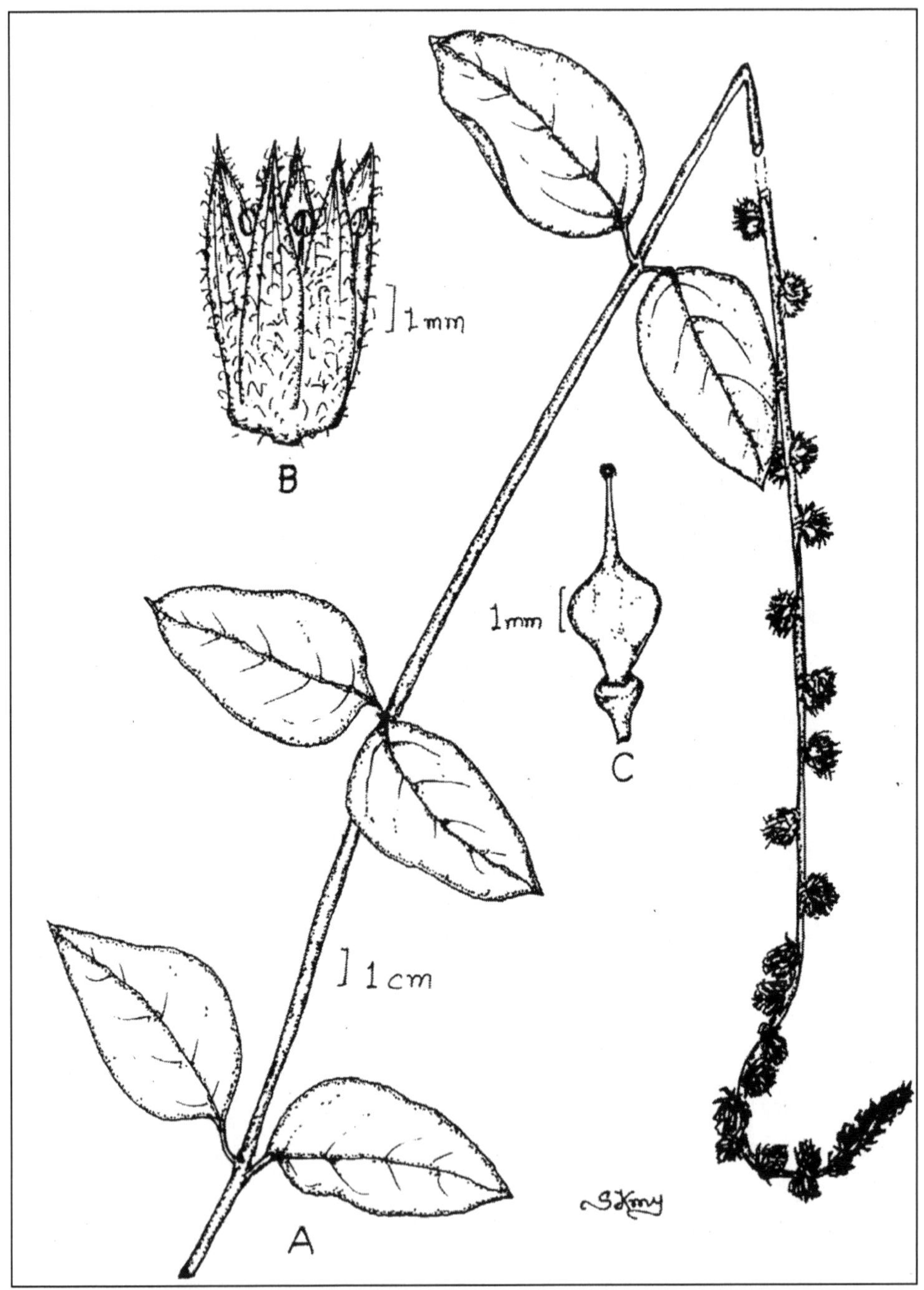

Figure 89: Pupalia lappacea (L.) Juss.
A. Twig, B. Flower, C. Pistil

flowers 2 bracteolate; sterile flowers modified into hooked bristles in fruit; bracts lanceolate, 1.5–2.5 × 0.5–0.7 mm, persistent, more or less deflexed after the fall of the fruit, sub-glabrous or pilose, sharply awned with percurrent midrib; bracteoles broad, subcordate-ovate, 3–5 × 0.5–1 mm, abruptly narrowed, dark arista formed by the excurrent midrib, membranous with pale margin, densely hairy; bracteoles of sterile flowers ovate-lanceolate, pilose; perianth segments oblong-ovate to lanceolate, 3–6 × 1–1.5 mm, narrowed to obtuse-mucronate at apex, aristate; outer tomentose, dorsally 3-nerved; inner 2-nerved, mucronate at apex with arista; inner most 3 slightly shorter and more densely pilose; stamens 5, 2.5–3 mm long, anthers yellowish; ovary depressed-globose, basally narrowed, 1 mm across; style erect, slender, 0.8–2 mm long; stigma capitate. Capsule depressed globose, circumscissile, 2.5 mm long, top areolate; seed ovoid-oblong, 1.5–2 mm, lenticular, dark brown, shining; testa at first faintly reticulate but finally smooth or punctulate, thinly crustaceous.

Common along road sides, out skirts of the forests and waste lands.

Fl. & Fr.: Throughout the year.

Sirpur (ADB), *GO* & *PVP* 4913; Manchippa RF (NZB), *TP* & *BR* 6056; Ibrahimpet RF (NZB), *TP* & *BR* 6123; Nathyanipally (MDK), *BR* & *CP* 11250; Rangammagudem RF (RR), *MSM* 10474; Ieeja (MBNR), *BR* & *MVS* 35803; Kambalpally (NLG), *BVRS* 209; Pakhal (WGL), *KMS* 11630 (MH); Kodimial (KRN), *GVS* 21879 (MH); Yelandu – Karepally (KMM), *RR* 105953 (BSID).

Pupalia lappacea (L.) Juss. var. **orbiculata** (Heyne ex Wall.) C. Towns., Kew Bull. 29: 469. 1974. *Achyranthes orbiculata* Heyne ex Wall. in Roxb., Fl. Ind. 2: 507. 1824. *Pupalia orbiculata* (Heyne ex Wall.) Wight, Icon. Pl. Ind. Orient. t. 1783. 1852; Hook.f., Fl. Brit. India 4: 724. 1885; Gamble, Fl. Madras 2: 1173. 1925.

Prostrate herb, hispidulous. Leaves orbicular-very broadly ovate, 2.5-3 × 4 cm, base rounded, margin entire, apex obtuse or retuse, densely pilose. Flowers in axillary and terminal to 10-20 cm long stout spikes, in clusters, distant, woolly-pubescent.

Occasional in Medak and Nalgonda districts.

Fl. & Fr.: Throughout the year.

Nagsanpally RF (MDK), *RG* 104176 (BSID).

TRICHURIELLA Bennet

Trichuriella monsoniae (L. f.) Bennet, Indian J. Forest. 8: 86. 1985. *Trichurus monsoniae* (L.f.) C.C. Towns., Kew Bull. 29: 466. 1974 & in Dassan & Fosb., Rev. Handb. Fl. Ceylon 1: 36. 1980; Venkata Raju & Padmavathi in Pullaiah & Moulali, Fl. Andhra Pradesh 2: 808. 1997. *Illecebrum monsoniae* L.f., Suppl. Pl. 161. 1781. *Celosia monsoniae* Retz., Observ. Bot. 2: 13. 1781. *Aerva monsoniae* (Retz.) Mart., Nova. Act. Phys.-Med. Acad. Caes. Leop.-Carol. Nat. Cur. 13: 291. 1825; Hook.f., Fl. Brit. India 4: 728. 1885; Gamble, Fl. Madras 2: 1178. 1925.

Annual, tufted, prostrate herb, to 90 cm long; stem wiry, slender with ridges, branches horizontally spreading, opposite or whorled, pilose with white hairs, villous. Leaves whorled, linear-filiform, up to 2.5 cm long, crowded at nodes, more commonly thinly pilose, with multicellular hairs, sessile, acuminate. Flowers pale pink in terminal and axillary spike; bracts and brateoles lanceolate, 1–1.5 × 0.5 mm, hyaline, with a green midrib; perianth lobes 4, lanceolate, sub-clavate, 2–2.5 × 0.7 mm, hyaline with a green midrib, dorsally lanate, curving towards the apex of the spike, persistent; stamens 4; filaments 1–1.5 mm long, alternating with triangular to sub-quadrate, pseudo-staminodes; ovary depressed globose, 0.2 mm across; style simple; stigma capitate. Utricle ovoid, to 1 mm, circumscissile, longitudinally split at maturity, lid much shorter than the base; seed ovoid, compressed, 0.6–0.7 mm, shining, reddish-brown, testa crustaceous.

Occasional in waste lands, fields and along road sides.

Fl. & Fr.: June – January.

Sirpur (ADB), *GO* & *PVP* 4915; Nasrullabad (NZB), *TP* & *BR* 6152; Sangareddy (MDK), *BR* & *CP* 11287; Pochammaralu (MDK), *BR* & *CP* 11479; Kothapalli RF (RR), *MSM* & *KH* 10565; Amrabad (MBNR), *BR* & *BSS* 32538; Nagarjunasagar (NLG), *BVRS* 67 (NU); Pakhal (WGL), *KMS* 11684 (MH); Kodimial (KRN), *GVS* 21875 (MH).

Note: *Aerva monsoniae* (Retz) Mart. is quite distinct from other species of *Aerva* Forssk. in several charaters of external morphology. Townsend (1974) erected a new genus, *Thichurus* to accommodate this species. *Trichurus* Townsend (1974) is a later homonym of *Trichurus* Clements (1896) which belongs to fungi. Bennet (1985) proposed *Trichuriella* as a substitute name for the later homonym.

CHENOPODIACEAE

1. Flowers mostly bisexual; perianth all alike on the same plant and merely enlarging in fruit:
   2. Flowers in dense glomerules, arranged in spikes or panicles; leaves coarsely serrate .... **Disphania**
   2. Flowers solitary or a few in lax cymes, arranged in panicles; leaves pinnatifid .... **Chenopodium**
1. Flowers unisexual; perianth of two kinds .... **Spinacia**

CHENOPODIUM L.

1. Branchlets green or with purple streaks; tepals ovate-rhomboid; cultivated .... **C. album**
1. Branchlets not streaked; tepals ovate or oblong; wild .... **C. murale**

Chenopodium album L., Sp. Pl. 219. 1753; Hook.f., Fl. Brit. India 5 : 3. 1886; Gamble, Fl. Madras 2: 1181. 1925.

Vern.: *Pappukura, Chakravarthi kura*.

Erect herbs, up to 50 cm high, branchlets with green or purple streaks, glabrous or thinly clothed with white vesicles. Leaves ovate-lanceolate, 2.5-6 × 1-2 cm. Flowers in axillary and terminal panicles, 2 mm across, hermaphrodite; tepals 5, ovate-rhomboid, 1.3 mm; stamens 5, filaments 0.25 mm, anthers 0.3 mm; stigmas 2, spreading. Utricle 1.5 mm across; seed horizontal obscurely keeled, 1.2 mm across.

Cultivated as leafy vegetable and found along the streams, often as an escape.

Fl. & Fr.: February – June.

Neredugonda (ADB), *GO* & *DAM* 5102; Gowraram (NZB), *TP* & *BR* 6387; Devanpally RF (NZB), *BR* & *KH* 9632; Rampur (MDK), *BR* & *CP* 11652.

Chenopodium murale L., Sp. Pl. 219. 1753; Hook.f., Fl. Brit. India 5: 4. 1886; Gamble, Fl. Madras 2: 1181. 1925.

Erect herbs, to 1 m high. Leaves alternate, rhomboid to deltoid-ovate, 2-7.5 × 1-4.5 cm, petiole 3 cm, base attenuate, margin sharply toothed, apex acute. Flowers minute, in axillary spikes in spreading cymes, polygamous; bract and bracteoles 0; tepals ovate-oblong. Utricle enclosed in the perianth; seed horizontal, testa coriaceous.

Occasional weed in cultivated fields in Nizamabad, Mahbubnagar and Karimnagar districts.

Fl. & Fr.: September – November.

Armoor (NZB), *BR* 7294; Aklaspur (KRN), *GVS* 22516 (MH).

DYSPHANIA R.Br.

Dysphania ambrosioides (L.) Mosyakin & Clements, Ukrayisk. Bot. Zhurn. 59: 382. 2002. *Chenopodium ambrosioides* L., Sp. Pl. 219. 1753; Hook.f., Fl. Brit. India 5: 4. 1886; Gamble, Fl. Madras 2: 1181. 1925.

Erect herbs, up to 60 cm high, stem striate, glandular-pubescent. Leaves alternate, oblong-lanceolate, 0.5-3.5 × 0.2-0.5 cm, base cuneate, margin sinuate-dentate, upper half entire, apex acute; petiole 2 mm. Flowers in axillary and terminal, simple or panicled spikes. Utricle enclosed in the perianth, seed horizontal, smooth, testa coriacious.

Common on hill slopes, along river banks in Hyderabad district.

Fl. & Fr.: Throughout the year.

SPINACEA L.

Spinacia oleracea L., Sp. Pl. 1027. 1753; Hook.f. Fl. Brit. India 5 : 4. 1886.

Erect, simple or branched, dioecious herbs, 20-60 cm high; stems angular-ribbed. Leaves radical and middle leaves long-petiolate, ovate-oblong or triangular-hastate, apex obtuse or acute, margin entire or with projecting lobes at base on either side, bright green, soft, glabrous; higher ones smaller and becoming lanceolate in inflorescence. Male flowers mostly in spikes or panicles. Tepals ovate, obtuse, 1.2-2 mm long. Pistillate perianth with 2 spines. Utricle enclosed in perianth and adnate to it, hard, compressed; seeds dull, obtusely margined.

Cultivated for leafy vegetable in kitchen gardens.

Fl. & Fr.: November-March.

Beta vulgaris L. (Beet root) is rarely cultivated for fleshy roots.

BASELLACEAE

BASELLA L.

Basella alba L., Sp. Pl. 272. 1753; Hook.f., Fl. Brit. India 5: 20. 1886; Gamble, Fl. Madras 2: 1185. 1925.

Vern.: *Bachali kura.*

Twining, succulent herb, stem and petioles red (var. *rubra*) or green (var. *alba*). Leaves broadly ovate, ovate or orbicular, 3.5 – 12 × 2.2-9 cm, base rounded to shortly decurrent, apex gradually acute; petiole to 1.5 cm. Flowers pinkish or purple in spikes; perianth urceolate, 3 mm across, tube 2.5 mm, lobes 5; stamens 5, inserted below the lobes, filaments 1.5 mm. Pseudoberry globose or lobed, 4-7 mm high, 5-10 mm wide, black with deep purple juice.

Commonly cultivated. Also found in wild state on rocky grassy hillsides.

Fl. & Fr.: October – February.

Nizamabad (NZB), *BR* & *KH* 9023; Choutkoor (MDK), *TP* & *CP* 11925; Mombojipally (MDK), *BR* & *CP* 11641; Peddavuru (NLG), *BVRS* 514 (NU).

POLYGONACEAE

1. Climber; inflorescence tip tendrillar .... **Antigonon**
1. Other than climbers; inflorescence tip not tendrillar:
   2. Branchlets modified into phylloclades; leaves deciduous .... **Muehlenbeckia**
   2. Branchlets otherwise; leaves persistent:
      3. Tepals 6; stigmas penicillate .... **Rumex**
      3. Tepals 5; stigmas capitate:
         4. Ocreae generally transluscent, glabrous, 2-lobed .... **Polygonum**
         4. Ocreae opaque, glabrous to scabrous, not 2-lobed .... **Persicaria**

ANTIGONON Endl.

Antigonon leptopus Hook. & Arn., Bot. Beechey Voy. 308. 1838; Gamble, Fl. Madras 2: 1192. 1925.

Vern.: *Teegarosa.*

Climbing shrub, with tuberous root. Leaves alternate, simple, elliptic-ovate, 6-12 × 3.5-8 cm, base hastate, margin entire, apex acute, stipules conspicuous, to 2 mm; petiole 1.5 cm. Flowers in axillary and terminal panicles, hermaphrodite, pink or white; perianth segments 5, 6-15× 3-7 mm; stamens 8; carpels 3, ovary ovoid, trigonous, styles 3, stigma capitate. Nut 5 mm long.

Cultivated in gardens as an ornamental, often as an escape.

Fl. & Fr.: Throughout the year.

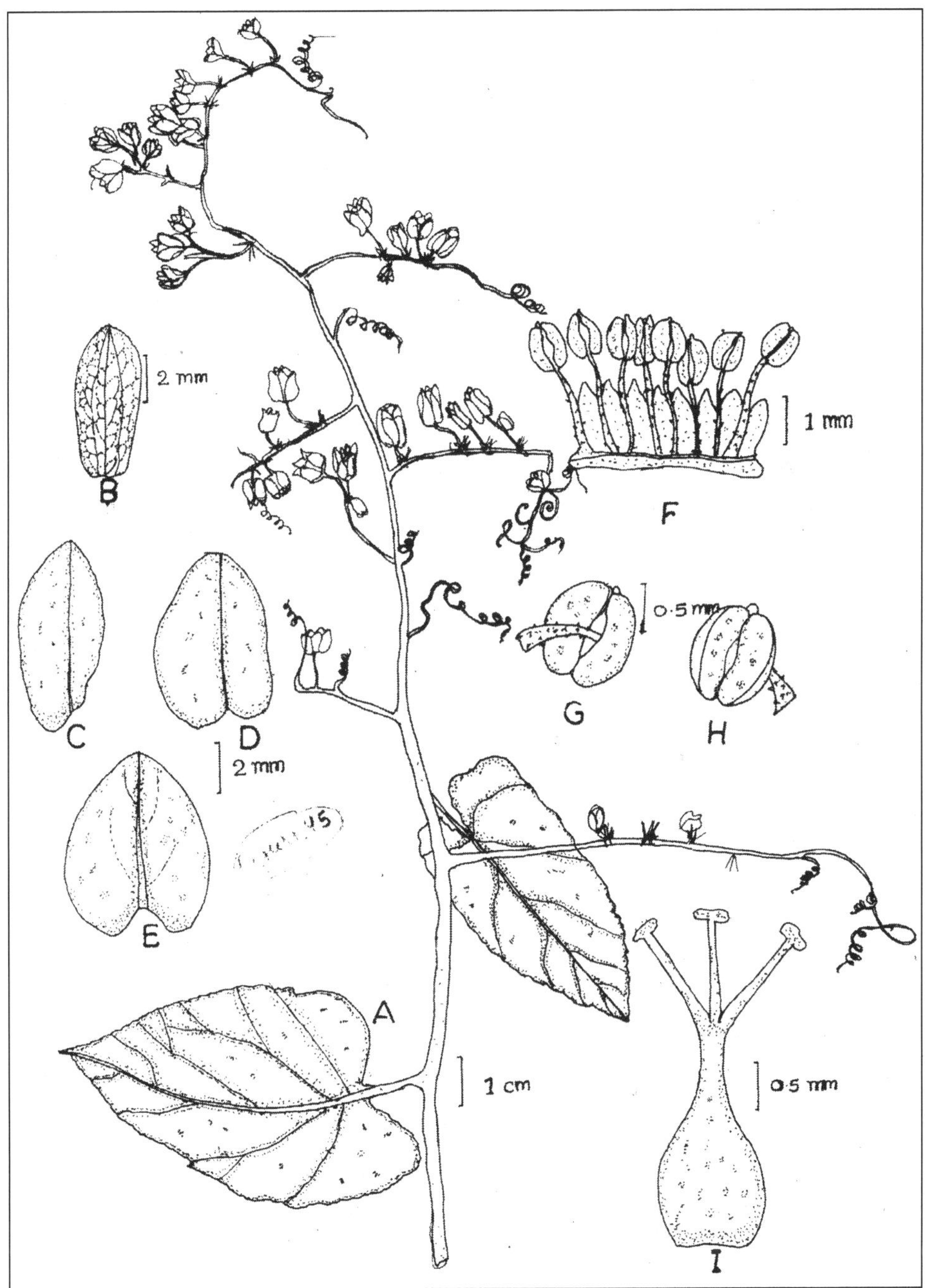

Figure 90: Antigonon leptopus Hook. & Arn.

A. Twig, B-E. Tepals, F. Staminal column, G-H. Stamens, I. Pistil.

Yellareddy (NZB), *TP* & *BR* 6298; Medak (MDK), *CP* 11744; Vikarabad (RR), *TP* & *MSM* 11091.

MUEHLENBECKIA Meissn. *nom. cons.*

Muehlenbeckia platyclada (F.J. Müll.) Meissn., Bot. Zeitung (Berlin) 22: 313. 1865; Gamble, Fl. Madras 2: 1192. 1925. *Polygonum platycladum* F.J. Müll., Trans. Philos. Soc. Victoria 2: 73. 1858.

Vern.: *Kurathutty.*

Erect herb or sub shrub, to 55 cm high; branchlets green, flattened, jointed, modified into phylloclades, nodes flattened. Leaves deciduous. Flowers greenish yellow, hermaphrodite, in axillary, short lateral spikes.

Cultivated in gardens as an ornamental.

Fl. & Fr.: Throughout the year.

PERSICARIA L.

1. Bracts and ocreae glabrous .... **P. glabra**
1. Bracts and ocreae ciliate or hispid:
   2. Nutlets biconvex .... **P. pulchra**
   2. Nutlets trigonous .... **P. barbata**

Persicaria barbata (L.) H. Hara, Fl. E. Himalaya 70: 1966. *Polygonum barbatum* L., Sp. Pl. 1: 362. 1753; Hook.f., Fl. Brit. India 5: 37. 1886; Gamble, Fl. Madras 2: 1189. 1925.

Erect herbs, to 1.5 m high, stem glabrescent; ocreae tubular, truncate, 1.5 cm, strigose, coarsely ciliate. Leaves lanceolate, 3.5-15 × 1-2.5 cm, glabrous, base acute to cuneate, margin entire, apex acuminate; petiole 1 cm. Flowers pink, in terminal and axillary, to 13 cm long spikes, hermaphrodite; bracts truncate, ciliate along margins; perianth hyaline, lobes 5, oblong, to 2.5 mm long; stamens 5-8; ovary trigonous; styles 5. Nutlet trigonous, 3 × 1.5 mm, brown.

Common along the streams, growing in association with *P. glabra* in Adilabad, Nizamabad and Karimnagar districts.

Fl. & Fr.: Throughout the year.

Sone (ADB), *TP* & *PVP* 4144; Potchera water falls (ADB), *MHR* 14492; Janakampally (NZB), *TP* & *BR* 6323; Dharmareddy (NZB), *BR* 9512.

Persicaria glabra (Willd.) M.Gomez, Anales Inst. Segunda Ensen 2: 278. 1896. *Polygonum glabrum* Willd., Sp. Pl. 2: 447. 1799. var. *glabrum*: Hook.f., Fl. Brit. India 5: 34. 1886; Gamble, Fl. Madras 2: 1189. 1925.

Vern.: *Buradagogu.*

Sub-shrub, up to 1 m high. Ocreae tubular, 2 cm, glabrous. Leaves alternate, narrowly lanceolate, 3.5-18 × 1-3 cm, base cuneate, margin entire, apex acuminate. Flowers hermaphrodite, in terminal lax spikes or from upper axils; bract 1, hyaline, subtending 2 flowers; bracteole hyaline, 2-lobed, subtending sessile floret; perianth lobes 5, hyaline, boat-shaped; anthers 7-8, dithecous; ovary compressed; style 2-fid, 2 mm, stigma capitate . Nutlet compressed 1.5 × 1 mm, faces flattended, black, shining.

Common near the streams and tanks or in water logged places in most districts.

Fl. & Fr.: Throughout the year.

Potchera (ADB), *GO* & *DAM* 5048; Laxmapur (NZB), *TP* & *BR* 6266; Salura (NZB), *BR* 7272; Medak (MDK), *BR* & *CP* 11465; Venkatammacheruvu (RR), *MSM* & *KH* 10259; Mannanur (MBNR), *TDCK* 13651; Jinugunta (MBNR), *KI* & *NVN* 18736; Bhupatipur (KRN), *GVS* 25705 (MH); Perantapalli RF (KMM), *RCS* 98986 (BSID).

Persicaria pulchra (Blume) Sojak, Preslia 46: 154. 1974. *Polygonum pulchrum* Blume, Bijdr. Fl. 530. 1826; R.A. Graham in Fl. Trop. East Africa (Polygonaceae) 20. 1958. *P. tomentosum* Willd., Sp. Pl. 2: 477. 1799. non Scharm 1789; Hook.f., Fl. Brit. India 5: 30. 1886; Gamble, Fl. Madras 2: 1189. 1925.

Erect herb, up to 1.5 m high. Ocreae tubular, 2-3 cm, margin truncate, chartaceous, softly ciliate. Leaves lanceolate or linearly obovate, 10-15 × 2-4 cm, silky tomentose, base cuneate, margin entire, apex gradually acuminate; petiole 2 cm,. Flowers pinkish, in terminal panicles; perianth 4 mm across, lobes 5, equal, ovate, 3.5 mm, shortly connate at base; stamens 5, filaments 2 mm; style 1 mm, stigmas capitellate. Nutlet biconvex, 3 mm across.

Occasional in wet places and ditches in Mahabubnagar district.

Fl.: December-February.; Fr.: Most part of the year.

POLYGONUM L. *nom. cons.*

1. Leaves up to 5 mm long; ocreae hyaline; flowers in fascicles .... **P. plebeium**
1. Leaves more than 2 cm long; ocreae membranous or chartaceous; flowers in racemes or capitate clusters .... **P. longisetum**

Polygonum longisetum Bruijn var. **rotundatum** A.J.Li, Bull. Soc. Res., Harbin. 15(4): 418. F. 6. 1995. *Polygonum barbatum* L. var. *gracile* (Danser) Steward, Contr. Gray Herb. 88: 55. 1930. *P. barbatum* L. subsp. *gracile* Danser, Bull. Jard. Bot. Buitenzorg, Ser, 3. 8: f. 2. 146 1927. *P. serrulatum sensu* Hook. f., Fl. Brit. India 5: 38. 1886. *p.p.* non Lagasca 1816.

Herb with woody base, rooting at nodes below. Leaves alternate, sub-sessile, drying green, linear to linear-lanceolate, 7-8 × 1-2.5 cm, acute to acuminate, glabrous or sparsely adpressed-hairy on midrib and margins. Flowers in paniculate dense racemes. Nutlet trigonous, 1.5 mm, smooth and shiny.

Occasional along banks of streams in Adilabad district.

Fl. & Fr.: Throughout the year.

Sone (ADB), *TP* & *PVP* 4044.

Polygonum plebeium R. Br., Prodr. Fl. Nov. Holl. 420. 1810 var. **plebeium;** Hook.f., Fl. Brit. India 5: 27. 1886. incl. var. *indica;* Gamble, Fl. Madras 2: 1188. 1925.

Vern.: *Chanti saga, Muthi saga.*

Prostrate herb. Ochreae tubular, 2 mm, hyaline, lacerate. Leaves alternate, oblong, 2-5 × 1-2 mm, subsessile, base attenuate, margin entire, apex acute to obtuse. Flowers hermaphrodite, in axillary fascicles, pink, pedicel to 1 mm; perianth lobes 5, lobes hooded at apex; stamens 8; ovary glabrous, styles 3. Nutlet strongly trigonous, 2 x 1 mm with persistent style.

Common weed in harvested rice fields, dried stream beds in most districts.

Fl. & Fr.: Throughout the year.

Indervelli (ADB), *GO* 4541; Mingaram RF (NZB), *TP* & *BR* 6314; Kandakurthi (NZB), *BR* 7256; Ramayampet (MDK), *BR* & *CP* 11517; Peddakunta (RR), *MSM* & *NV* 10422; Rayagir tank (NLG), *BVRS* 2 (NU); Kodimial (KRN), *GVS* 20085 (MH).

RUMEX L.

1. Flowers bisexual; styles terminal:
    2. Leaves narrowly panduriform, 5-20 × 2-5 cm .... **R. nigricans**
    2. Leaves oblong-lanceolate, 1-6 × 0.5-1.2 cm .... **R. dentatus**
1. Flowers unisexual or polygamous; styles arising from the angles of the ovary .... **R. vescicarius**

Rumex dentatus L., Mant. Pl. 2: 226. 1771; Hook.f., Fl. Brit. India 5 : 59. 1886.

Annual, erect, deep-rooted herb, to 50 cm high; branches all arising from the root stock, red, glabrous. Leaves radical and cauline, oblong-lanceolate, 1-6 × 0.5-1.2 cm, glabrous, base rounded or acute, margin entire or obscurely-crenate, apex rounded; petiole to 5 cm. Flowers in leafy or leafless whorls; perianth inner segments enlarged in fruit, reticulately veined, with an ovoid-oblong tubercle at back and with pectinate margins. Nuts trigonous, winged, to 0.2 × 0.1 cm.

Rare along river and canal banks in Nizamabad district.

Fl. & Fr.: September – January.

Kandakurthi (NZB), *BR* 7263

Rumex nigricans Hook.f., Fl. Brit. India 5 : 59. 1886; Gamble, Fl. Madras 2 : 1192. 1925.

Erect, slender annual herb, 20-60 cm high; stems angled and grooved. Leaves narrowly panduriform, 5-20 × 2-5 cm, contracted above cordate base, sometimes lower ones linear-oblong, thin, apex obtuse. Flowers in clusters along terminal leafy panicles; pedicels short, 0.12-0.15 cm long, recurved, jointed below the middle; perianth green; tepals 0.3-0.45 cm long, lanceolate, venose, turning black when dry, with very large oblong, dark tubercles and lanceolate tips; margins very narrow, entire or with 1-2 short, straight or spines. Nuts acutely trigonous, smooth, enclosed in accrescent perianth.

Near marshy localities, along margins of the lakes in Hyderabad (Ramana, 2010).

Fl. & Fr.: February-April.

Rumex vescicarius L., Sp. Pl. 336. 1753; Hook.f., Fl. Brit. India 5:61. 1886; Gamble, Fl. Madras 2: 1192. 1925.

Vern.: *Chukkakura.*

Erect succulent herb, up to 50 cm high, stem fleshy, stipules membranous. Leaves alternate, elliptic, 2-7 × 0.5-2 cm, base cordate to hastate, margin crisped, apex obtuse to acute, succulent; petiole 4 cm. Flowers pink, in axillary and terminal panicles, arranged in clusters, polygamous; seed ovoid, 1 x 0.75 mm, black.

Culivated in gardens as leafy vegetable.

Fl. & Fr.: July – December.

Kadthala (ADB), GO & DAM 5008.

ARISTOLOCHIACEAE

ARISTOLOCHIA L.

1. Cultigens; flowers large; perianth limb up to 7 cm across .... **A. littoralis**
1. Wild plants; flowers small; perianth limb up to 1 cm across:
   2. Prostrate herbs; leaves reniform .... **A. bracteolata**
   2. Twining herbs or shrubs; leaves oblong, to obovate-oblong spathulate .... **A. indica**

Aristolochia bracteolata Lam., Encycl. 1: 258. 1783. *A. bracteata* Retz., Observ. Bot. 5: 29. 1788; Hook.f., Fl. Brit. India 5: 75. 1886; Gamble, Fl. Madras 2: 1201. 1925.

Vern.: *Gadidagadapa*.

Perennial, slender, prostrate herb, to 50 cm high; branches angular with ridges, white glaucous. Leaves reniform-cordiform, 3–5 × 4–6 cm, glaucous below, glabrous above, base cordate to broadly sagittate, margin entire to undulate, apex obtuse to sub acute; lateral veins 5 at base, palmately divergent; petiole to 3 cm long. Flowers axillary, solitary, 5 cm long, purple; bract cordate-orbicular, 1 × 1 cm; pedicel to 0.5 cm long, tube 2 cm long, curved, glabrous without, hirsute within, base inflated, 0.5 cm across; lip dark purple, linear, 2 cm long, with revolute margin, apex obtuse; mouth 0.4 cm across; stamens 6; anthers oblong, 0.6 mm long; ovary 8 × 0.3 mm; style 2 mm long; stigmatic lobes 6, connivent, acute. Capsule about 2.5 cm long 1.5 cm across, glabrous, ellipsoid-oblong, 12-ribbed, septicidally opened; seeds cordate, 4 mm across, partly covered by the remains of the placenta, glands dotted; testa thin, albumen fleshy.

Common weed of cultivated fields and waste places in Adilabad, Nizamabad, Karimnagar, Hyderabad and Nalgonda districts.

Fl. & Fr.: June – December.

Tiryani RF (ADB), *PVP* 9808; Kandakurthi (NZB), *BR* 7266; Molachintapally (MBNR), *TDCK* 15304; Boddupalle (NLG), *BVRS* 645; Kodimial (KRN), *GVS* 20149.

Aristolochia indica L., Sp. Pl. 960. 1783; Hook.f., Fl. Brit. India 5: 75. 1886; Gamble, Fl. Madras 2: 1202. 1925.

Twining shrubs, stems long, slender, woody at the base, grooved, glabrous. Leaves variable, 4-7 × 2-4 cm, oblong to obovate – oblong, base rounded, truncate or sub-cordate, margin entire, apex obtusely acuminate, glabrous. Flowers greenish white, tube with globose inflated base, then bent at a right angle and suddenly narrowed into a cylindric tube, to 2 cm, with oblique trumpet-shaped mouth, to 4 cm across, gradually passing into a long narrow, linear oblong, obtuse brownish lip, lip to 1.5 cm; stamens 6; anthers oblong, 1 mm; ovary 1 × 2.5 cm, glabrous; style 3 mm; stigmatic lobes 6, erect, 1.5 mm, acute. Capsule 4 x 2 cm, oblong or globose – oblong, opening from below upwards, 6-valved; seeds many, flattened, laterally winged.

Occasional in forests in most districts.

Fl. & Fr.: September – March.

Kundram RF (ADB), *TP* & *GO* 5456; Gowraram RF (NZB), *TP* & *BR* 635; Nagulabanda (MDK), *CP* 11334; Dharmapur (RR), *MSM* & *KH* 10288; Kusumasamudram RF (RR), *MSM* & *NV* 10367; Vamulapaya vagu (MBNR), *TDCK* 13623; Amaragiri (NLG), *BR* & *TSS* 28884; Kodimial (KRN), *GVS* 21877 (MH); Yenkannapalem (KMM), *RR* 113830 (BSID); Kazipet (WGL), *RKP* 105213 (BSID).

Aristolochia littoralis Paroch., Anal. Soc. Cient. Argent. 5: 155.1878; Loigier, Rhodora 67: 328. 1965; Raizada, Suppl. Fl. Up. Gang. Pl. 241.1976. *A. elegans* Mast, in Card. Chron. 2 : 301.1885; Bailey, Man. Cult. Pl. 346. 1949.

Slender, perennial twiner with numerous striate branches. Leaves broadly ovate to sub-reniform, 5-10 cm across, cordate at base with broad sinus, with undulate margins, apex obtuse, green above, some what glaucous beneath; petioles 2-4 cm long; stipules foliaceous. Flowers solitary, axillary; perianth ash-grey or white with inflated base and yellowish-green cylindric tube; limb 6-7 cm across, blotched with violet-brown, orbicular. Capsules ovoid-oblong, 3-4 x 2-2.5 cm, 6-grooved, greenish-brown; seeds ovoid, 2-3 mm long, flat, black.

Cultivated in gardens for its showy flowers in Hyderabad (Ramana, 2010) and elsewhere.

Fl. & Fr.: September– March.

Mallelatheeratham (MBNR), *BR* & *TSS* 29860.

Aristolochia ringens Vahl, Symb. Bot. 3: 99. 1794. Native to Brazil, but now cultivated throughout the tropical world, including Telangana. Reported in Hyderabad and Mahabubnagar districts.

PIPERACEAE

1. Leaves alternate, not punctuate, nerves prominent; flowers unisexual ...... **Piper**
1. Leaves opposite, pellucid punctate, nerves obscure; flowers bisexual .... **Peperomia**

PEPEROMIA Ruiz & Pavon

Peperomia pellucida (L.) Kunth, Nov. Gen. Sp. 1: 64. 1816; Gamble, Fl. Madras 2: 1210. 1925. *Piper pellucidum* L., Sp. Pl. 30. 1753

Annual, succulent, delicate, glabrous herb, membranous when dry; stems weak, erect, 6-15 cm high. Leaves spiral, rarely opposite, greenish above, whitish beneath, pellucid, broadly ovate deltoid, distinctly 5-nerved, 1-2.5 cm wide and long, base slightly cordate, entire, apex subacute; petiole 0.5-1.5 cm long. Inflorescence leaf-opposed, terminal spike, 2-4.5 cm long, glabrous, peduncle *ca* 5 mm long; stamens 2, filaments very short, anther cells confluent; ovary superior, 1-carpellary with 1 basal orthotropous ovule; stigma penicillate, apical. Fruits 0.8 mm long, indehiscent nutlet, striate, with 10 vertical ridges and minute, transverse, raised lines across the pericarp; seed minute, testa membranous, albumen floury.

A shade and moist loving weed of gardens, pots and damp walls, particularly during monsoon period, till December (C.S. Reddy andV.S. Raju. 2002).

Fl. & Fr.: September – December.

Hanamkonda (WGL), *CSR* 1116 (KU).

PIPER L.

1. Leaves crispately pubescent beneath; ovary ellipsoid .... **P. hymenophyllum**
1. Leaves glabrous; ovary globose .... **P. attenuatum**

Piper attenuatum Buch.-Ham. ex Miq., Syst. Piperac. 306. 1844; Hook. f., Fl. Brit. India 5: 92. 1886; Gamble, Fl. Madras 2: 1207. 1925. *P. sylvestre* Wight, Icon. Pl. Ind. Orient. t. 1937. 1853; Hook.f., Fl. Brit. India 5: 93. 1886.

Slender climbers with often somewhat zig-zag branchlets, stems terete, striate. Leaves about 12 × 9 cm, ovate or broadly elliptic-ovate, alternate, base rounded to broadly acute, rarely sub-cordate, margin entire, apex acuminate, glabrous. Flowers unisexual, green, 0.5 mm long, in leaf-opposed spikes, about 11 cm long. Berry smooth, globose, 3.4 mm across.

Occasional in Nallamalais in Mahabubnagar district.

Fl. & Fr.: July – September.

Borapuram (MBNR), *BSS* & *SKB* 32362.

Piper hymenophyllum Miq., Hook. J. Bot. 5: 554. 1846; Hook.f., Fl. Brit. India 5: 93. 1886; Gamble, Fl. Madras 2: 1207. 1925.

Climbing herbs, stems terete, grooved, glabrous, nodes swollen, Leaves 7-13 × 4-6 cm, elliptic – lanceolate, alternate, base rounded, margin entire, apex shortly acuminate, coriaceous, thinly pubescent below. Flowers green in slender spikes, 10 cm long. Berry globose, 4 mm across, when dry with persistent styles.

Occasional in forests of Mahabubnagar district

Fl. Fr.: Febr.uary– December.

Mallelatheertham (MBNR), *M.Md. Bakshu* & *KVR* 24200; Saleswaram (MBNR), *BSS* & *KP* 33343.

LAURACEAE

1. Leafless twining parasite .... **Cassytha**
1. Leafy shrubs or trees .... **Litsea**

CASSYTHA L.

Cassytha filiformis L., Sp. Pl. 35. 1753; Hook. f., Fl. Brit. India 5: 188. 1886; Gamble, Fl. Madras 2: 1241. 1925.

Vern.: *Sitammavari jadalu, Antaravalli, Pachi teega, Nullu teega*.

Perennial, parasitic, green, glabrous, leafless, very long, wiry, twining herb; branches slender, green, attached to hosts by suckers or haustoria, younger ones yellow. Flowers in spikes, white, 3-merous, bisexual, 2 mm across; tepals 6, free, unequal, 3 + 3, truncate, obtuse at apex; outer lobes ovate-orbicular, 1 mm long, ciliate; inner lobes obovate, 1.5 mm long; fertile stamens 9; filaments 1.5 mm; glands sessile; anthers 2-celled, 1 mm long; staminodes 3, to 1 mm long; ovary 1.5 mm; style 0.5 mm long; stigma capitate. Drupe globose, enclosed within inflated perianth, crowned by lobes, white when mature, 0.5 across; seed 1, glabrous, globose; testa thin, hard, black, 3 mm across.

A common parasite on *Carissa spinarum, Vitex negundo, Acacia latronum, Ziziphus mauritiana, Azadirachta indica,* etc. Common in scrub jungles.

Fl. & Fr.: July-December.

Wankidi (ADB), *GO* 4481; Malkapur – Nizamabad road (NZB), *TP* & *BR* 6113; Ramareddy RF (NZB), *BR* & *PSPB* 9554; Pochammaralu (MDK), *BR* & *CP* 11481; Mohammadabad RF (RR), *MSM* & *KH* 10599; Mannanur (MBNR), *TDCK* 13624; Ramavaram (KMM), *RR* 112539 (BSID); Srinivaspur (NLG), *BVRS* 411 (NU).

LITSEA Lam. *nom. cons.*

Litsea glutinosa (Lour.) C.B. Rob., Philipp. J. Sci. 6(5): 321. 1911. *Sebifera glutinosa* Lour., Fl. Cochinch. 2: 638. 1790. *Litsea chinensis* Lam., Encycl. 3.574. 1789; Hook.f., Fl. Brit. India 5: 157. 1886; Gamble, Fl. Madras 2: 1235. 1925. *L. sebifera* Pers., Syn. 2:4 1807; Hook.f., Fl. Brit. India 5. 157. 1886.

Vern.: *Meda nara-naliki, Narva- alagi, Kanuju, Burra manu, Kanuja nalike*.

Evergreen trees, up to 12 m high; bark brown, somewhat corky; wood greyish-brown, fairly straight grained, medium and even textured; branchlets pubescent. Leaves coriaceous, very variable in shape, elliptic, obovate, or elliptic-oblong, glabrous, base acute, margin entire, apex obtuse or acute, secondary nerves 7-12 pairs; petiole 1-3.5 cm. Flowers white or yellowish, small in compound, few-flowered umblels. Fruit globose. black, 1.2 × 0.8 cm.

Occasional, in deciduous forests.

Fl.: June. Fr.: September.

Bheemaram RF (ADB), *TP* & *GO* 5462.

HERNANDIACEAE

GYROCARPUS Jacq.

Gyrocarpus americanus Jacq., Select. Strip. Amer. Hist. 282. 1763; Gamble, Fl. Madras 2: 1242. 1925. *G. jacquini* Roxb., Pl. Coromandel t. 1. 1795; Hook.f., Fl. Brit. India 2: 461. 1878.

Vern.: *Tella poniki, Kummara poliki, Thanaku, Thanaki, Poliki*.

Deciduous tree, to 20 m high; trunk cylindric, bark whitish; wood light, soft, white; young shoots pubescent. Leaves broadly ovate, 3-lobed, 5–10 × 8–15 cm, chartaceous, glabrous above, thinly pubescent below with appressed hairs, base truncate to sub-cordate, margin entire, apex acute to acuminate; lateral veins forking, extending to $^{2}/_{3}$ of the leaf; veins flattened to impressed above, slightly raised below; petiole 8–12 cm long. Flowers small, white or pale yellow, unisexual in large branched cymes; males numerous; females few. Male flowers 2–3 mm across; tepals 4–7, unequal, ovate, concave, 2 mm, thinly pubescent without, acute; stamens 4–7; filaments 3 mm, sparsely hairy; anthers 0.5 mm long, valves opening from base upwards; staminodes flat, obovate-spathulate, 1 mm across, obtuse. Female flowers 3 mm across; tepals 2, basally adnate to ovary, 2-winged above, appressed-pubescent, obtuse. A few of bisexual flowers scattered among a mass of male flowers, yellow or creamish. Gland white, disc-shaped, 1 mm across. Drupe obovoid-globose, longitudinally ribbed, crowned by 2, spathulate wings; wings 7 × 0.7 cm, apex obtuse, divaricate; seed 1; cotyledons large, convolute.

Common, in dry deciduous forests and along the roadsides in Adilabad, Karimnagar, Ranga Reddi, Nalgonda, Warangal, Hyderabad and Mahabubnagar districts.

Fl: January-March, Fr.: May-September.

Gadirayal RF (RR), *MSM* & *KH* 10260; Rayagir (NLG), *BVRS* 13 (NU); Pakhal (WGL), *ANH* 15947 (MH); Khammam hills (KMM), *RR* 113817 (BSID); Umamaheswaram (MBNR), *SRS* 109130 (BSID); Perantapalli (KMM), *PVS* 84066 (BSID); Pakhal (WGL), *RKP* 108279 (BSID).

PROTEACEAE

GREVILLEA R.Br. ex Salisb.

Grevillea robusta A. Cunn. ex R. Br., Suppl. Prodr. Fl. Nov. Holl. 24. 1830; Gamble, Fl. Madras 2: 1243.1925.

Vern.: *Parana, Perane*. Eng.: Silver Oak.

Evergreen tree, 10-25 m high; sparingly branched at the top; young branches hoary tomentose. Leaves 2-pinnatifid, fern like, the secondary lobes entire or again lobed, lanceolate, margins recurved, silvery beneath. Flowers orange with slender pedicelled in unilateral rcemes; pedicel 1-15 cm long; sepals 1.5-2 cm long, hooded, at first all fused together except on one side, later fused in twos basally and apically, free for greater length in the middle; stamens sessile; disc annular; gynophores about 2-3 mm long; ovary glabrous, styles 1-2.5 cm long; stigma conical. Fruit broad, very oblique, 2 to 2.5 cm long, dehiscing on one side; seeds 1-2, often winged, ca. 1 cm in diam.

Planted as avenue tree in parks in Hyderabad.

Fl.: May-June; Fr.: July - August.

LORANTHACEAE

1. Flowers in dichasial fascicles; flower buds lanceolate; corolla lobes twisted and interlocked after anthesis; filaments coiled .... **Helicanthes**
1. Flowers in centrifugal pseudoracemose or compound dichasium or simple dichasium or reduced dichasium; flower buds otherwise; corolla-lobes reflexed or obliquely erect after anthesis; filaments erect or recurved:
    2. Corolla 5-merous, slightly zygomorphic, not deeply split on one side .... **Dendrophthoe**
    2. Corolla mostly 4-merous, zygomorphic with a deep split on the inner side of the curved tube:
        3. Inflorescence usually 3-10-flowered, simple raceme, rarely reduced 2-flowered umbel; fruit pyriform, obovoid or clavate, distinctly stipitate, not woody .... **Scurrula**
        3. Inflorescence a few-flowered, simple, 2-5-rayed umbel; fruit urceolate or ellipsoid, not stipitate, warty .... **Taxillus**

DENDROPHTHOE Mart.

Dendrophthoe falcata (L.f.) Blume in Schult. f., Syst. Veg. 7: 1730. 1830; Rajasekaran in N.P.Balakr. & *al.*, Fl. India 23: 3. 2012. *Loranthus falcatus* L.f., Sp. Pl. Suppl. 211. 1781. *L. longiflorus* Desr. in Lam., Encycl. 3: 598. 1792; Hook.f., Fl. Brit. India 5: 214 1886. *L. longiflorus* Desr. var. *falcata* (L.f.) Kurz, Forest Fl. Burma 2 : 321. 1877; Hook.f., Fl. Brit. India 5: 215. 1886; Gamble, Fl. Madras 2: 253. 1925.

Vern.: *Badanika chettu, Bajinika, Jiddu, Yelinga.*

Large bushy, usually glabrous branch parasites; bark grey, smooth. Leaves very variable in shape and venation, 5-12 × 2-3.5, lanceolate or elliptic-ovate, alternate and opposite, base acute to cuneate, margin undulate, apex obtuse, thick, coriaceous. Flowers in racemes, racemes to 6 cm; pedicel short, to 0.6 cm; bracts broadly ovate, 1.5-2 × 0.8-1 mm; calyculus-tube subentire to denticulate, ca 1.2 cm long; petals 5, united, 2.5-5 cm long, yellow-scarlet at base, green at apex, tube terete, lobes 5, linear-elliptic, shorter than tube; stamens 5; ovary oblong, style exserted, stigma capitate. Fruit ellipsoid-ovoid to oblong, with persistent apical calyculus-tube, glabrous, bright pink.

Occasional on trees all over the State. A partial stem parasite on *Albizia lebbeck, Chloroxylon swietenia, Hardwickia binata* etc.

Fl. & Fr.: November– March.

Narsapur RF (MDK), *TP* & *CP* 11971; Lothuvagu (RR), *MSM* & *KH* 10296; Kothapalli RF (RR), *MSM* 10448; Srinivasapur (NLG), *BVRS* 409 (NU); Kodimial (KRN), *GVS* 20068 (MH).

Key to Varieties

1. Leaves linear, up to 2.5 cm wide .... var. **falcata**
1. Leaves broader, more than 2.5 cm wide:
    2. Leaves sessile or subsessile, amplexicaul at base; flowers glabrous throughout .... var. **amplexifolia**
    2. Leaves petiolate, not amplexicaul at base; flowers covered with malpighian hair when young .... var. **pubescens**

Dendrophthoe falcata (L.f.) Blume var. **amplexifolia** (DC.) Rajasekaran, J. Econ. Taxon. Bot. 31: 182. 2007 & in N.P.Balakr. & *al.*, Fl. India 23: 4. 2012. *Loranthus amplexifolius* DC., Prodr. 4: 305. 1830. *L. longiflorus* var. *amplexifolia* (DC.) Thwaites, Enum. Pl. Zeyl. 134. 1864; Hook.f., Fl. Brit. India 5: 215. 1886.

This variety is more robust than var. *falcata*. The leaves are larger, sessile, orbicular to oblong, 10-18 cm across and cordate-amplexicaul at base.

Common in Adilabad and Khammam districts.

Fl. & Fr. November - April.

Neelwai RF (ADB), *GO* & *DAM* 5165; way to Kunthala water falls (ADB), *GO* & *DAM* 5166; Sathiyampet RF (KMM), *RR* 112641 (BSID).

Dendrophthoe falcata (L.f.) Blume var. **falcata**; Rajasekaran in N.P.Balakr. & *al.*, Fl. India 23: 4. 2012.

Characterised by long, narrow leaves, with the calyx either glabrous or scurfy white tomentose.

Common throughout the state.

Fl. & Fr.: Almost throughout the year, mostly in November – March.

Laxmapur (NZB), *TP* & *BR* 6259; Lothuvagu (RR), *MSM* & *KH* 10296; Gangapur (MDK), *RG* 112496 (BSID); Narsapur (MDK), *RG* 116470.

Dendrophthoe falcata (L.f.) Blume var. **pubescens** (Hook.f.) V.Chandras. in A.N.Henry & *al.*, Fl. Tamil Nadu, Ser. I Analysis 2: 215. 1987; Rajasekaran in N.P.Balakr. & *al.*, Fl. India 23: 5. 2012. *Loranthus longiflorus* var. *pubescens* Hook.f., Fl. Brit. India 5: 215. 1886.

This variety is characterised by the presence of short hairs on young flowers.

In deciduous forests in Adilabad, Nizamabad, Medak, Ranga Reddi and Khammam districts.

Fl. & Fr.: January - December.

Indervelli (ADB), *GO* & *PVP* 4547; Ibrahimpet RF (NZB), *TP* & *BR* 6120; Gangapur RF (MDK), *BR* & *CP* 11442, 11443; Kusumasamudram (RR), *MSM* & *KH* 11022; way to Tekuloddi (KMM), *PVS* 84095 (BSID).

HELICANTHES Danser

Helicanthes elasticus (Desv.) Danser, Verh. Kon. Ned. Akad. Wetensch. Afd. Natuurk. Sect. II, 29(6): 55. 1933; Rajasekaran in N.P.Balakr. & *al.*, Fl. India 23: 10. 2012. *Loranthus elasticus* Desv. in Lam, Encycl. 3: 599. 1789; Hook.f., Fl. Brit. India 5: 216. 1886.

Shrubs, entirely glabrous, branches pseudo-dichotomous or trichotomous. Leaves opposite, decussate, obovate to elliptic-ovate or suborbicular, 5-12 × 2.5-6.5 cm, base cuneate, margin undulate, apex obtuse or subacute. Flowers in dichasial fascicles, sessile; bracts lunate; calyculus cupular, truncate; corolla-tube split equally up to the middle, 0.9-1.4 cm long, lobes 5, linear; stamens 5, red, filaments coiled; ovary subspherical with depressed sides, style curved, stigma narrowly elliptic-ovate. Fruit more or less spherical, pink when ripe.

Rare in Mahabubnagar district. Parasite on *Madhuca*.

Fl. & Fr.: December – April.

Mannanur range (MBNR), *simme coll.* 1131 (HH) on *Madhuca.*

SCURRULA L.

1. Leaves white tomentose on both surfaces .... **S. cordifolia**
1. Leaves glabrous above, rusty tomentose beneath .... **S. parasitica**

Scurrula cordifolia (Wall.) G. Don, Gen. Hist. 3: 421. 1834; Rajasekaran in N.P.Balakr. & *al.*, Fl. India 23: 23. 2012. *Loranthus cordifolius* Wall. in Roxb., Fl. Ind. (ed.) Carey & Wall. 2: 222. 1824; Hook.f., Fl. Brit. India 5: 209. 1886; Gamble, Fl. Madras 2: 1252. 1925.

Large parasitic shrubs, branchlets white tomentose. Leaves opposite, 9-11 × 5-7 cm, ovate, alternate, base cordate, margin entire, apex obtuse, coriaceous, white tomentose on either side. Flowers orange in fascicles, zygomorphic, variable in length; calyculus with annular hairy rim; petals united; tube with median split extending up to half the length, 1.5-1.8 cm long, lobes 4, narrowly elliptic-ovate; stamens recurved; ovary clavate, style 4-gonous, stigma conical. Fruit pyriform, tomentose with rusty brown hairs.

Occasional in forests in Karimnagar and Medak districts.

Fl. & Fr. : March – May.

Kodimial (KRN), *GVS* 21886 (MH); Gangapur to Chownpur (MDK), *RG* 106738 (BSID).

Scurrula parasitica L., Sp. Pl. 110. 1753; Rajasekaran in N.P.Balakr. & *al.*, Fl. India 23: 27. 2012. *Loranthus scurrula* L., Sp. Pl. (ed.2) 472. 1762; Hook.f., Fl. Brit. India 5: 208. 1886. *L. buddleioides* Desv. in Lam., Encycl. 3: 600. 1792; Gamble, Fl. Madras 2: 1251. 1925.

A perennial, much-branched, stem parasite, branches terete, branchelets rusty tomentose. Leaves 4-6.5 × 5-8 cm, elliptic – ovate, opposite, base rounded or sub-acute, margin entire, apex obtuse, coriaceous, glabrous above, rusty tomentose below. Flowers scarlet, in short axillary racemes, zygomorphic; variable in length, 1-2.4 cm long (including calyx and corolla), buds acute; bracts ovate or elliptic-ovate; pedicels 1-2 mm long; calyculous obsolete or up to 4 mm long; petals 4, united, tube gibbous, median split up to two-third the length, dark reddish brown outside, purplish inside, lobes sagittate; stamens 4, filaments 1.8-2.5 mm long; ovary clavate, style up to 1 mm long, stigma broadly conical. Fruit pyriform, pale green, rusty brown due to candelabra hairs.

Occasional in forests Nizamabad, Medak, Ranga Reddi and Nalgonda districts.

Fl. & Fr.: May – December.

Gandhari RF (NZB), *TP* & *BR* 6330; Mingaram RF (NZB), *BR* 9293; Gangapur RF (MDK), *TP* & *CP* 12074; Ponnala (MDK), *CP* 11322; Kothapalli RF (RR), *MSM* & *KH* 10298; Pullaichilmal, Farahabad (MBNR), *SRS* 109162; Srinivaspur (NLG), *BVRS* 410 (NU).

TAXILUS Tieghem

1. Leaves elliptic – sub-orbicular, plant parts and flowers tomentose .... **T. heyneanus**
1. Leaves obovate-spathulate; plant parts and flowers glabrous at maturity .... **T. cuneatus**

Taxillus cuneatus (Roth) Danser, Bull. Jard. Bot. Buitenzorg. Ser. 3. 10: 354. 1929; Rajasekaran in N.P.Balakr. & *al.*, Fl. India 23: 32. 2012. *Loranthus cuneatus* Heyne ex Roth, Nov. Pl. Spec. 193. 1821; Hook.f., Fl. Brit. India 5: 214. 1886; Gamble, Fl. Madras 2: 1253. 1925.

Bushy shrubs, bark grey, branches long, slender, numerous, with prominent nodes, parasitic on great variety of trees and shrubs. Leaves 1.5-3.5 × 1-2 cm, obovate-spathulate, alternate, base cuneate, decurrent into a short petiole, margin entire, apex rounded or obtuse, coriaceous, glabrous. Flowers bisexual, 2-5 in axillary, sessile or pedunculate cymes; bracts elliptic to ovate-elliptic, clasping ovary; calyculous tube subcylindrical, irregularly 5-toothed; petals 5, united; stamens 5; ovary oblong-cylindric, style tetragonous, stigma subcapitate. Fruit obconic, truncate, crowned with persistent calyculus, red.

Occasional in forests as stem parasite on various trees and shrubs.

Fl. & Fr.: July – October.

Vatavarlapalli (MBNR), *SKB* & *BSS* 32912.

Taxillus heyneanus (Schult.f.) Danser, Bull. Jard. Bot. Buitenzorg, ser. 3, 10: 355. 1929; Rajasekaran in N.P.Balakr. & *al.*, Fl. India 23: 33. 2012. *Loranthus heyneanus* Schult.f., Syst. Veg. 7: 106. 1829 (non Schult.f. 1830). *L. bracteatus* Heyne ex Wall. in Roxb., Fl. Ind. 2: 220. 1824; Hook.f., Fl. Brit. India 5: 213. 1886; Gamble, Fl. Madras 2: 1252. 1925. *Taxillus bracteatus* (Wall.) Tieghem, Bull. Soc. Bot. France 42: 256. 1895; Jayamma in Pullaiah & Moulali, Fl. Andhra Pradesh, 2: 831. 1997.

Vern.: *Jana-bhajanika.*

Large, semiparasitic shrubs with haustoriferous runners, to 1 m high on trees; stem brown tomentose, corky; branches whitish, tawny-tomentose. Leaves alternate, elliptic, sub-orbicular, 1.5–2.5 × 1–2 cm, pubescent, base truncate-obtuse, acute, margin entire, apex obtuse; lateral veins 3–5 pairs, mostly raised from base; petiole to 0.5 cm long, tometose. Inflorescence axillary umbellate raceme or cyme. Floweres 2.5 cm long, subtended by a lateral bract, bract oblanceolate, 7 × 3 mm; calyx 4 mm long, margin obscurely 2–4-toothed, white-pilose without; corolla greenish yellow, red inside, 0.4 cm across; tube 2.4 cm long, inflated above middle, pilose without; lobes 5, linear lanceolate, reflexed, 7 × 1.5 mm, subacute, red within; stamens 5; filaments 3 mm long; anthers 2 mm long, basifixed; ovary oblong-cylindric, 2 × 1.5 mm; style 2.2 cm long; stigma capitate, light yellow. Fruit oblong or ellipsoid, up to 2 cm long, red, with persistent bract, annulate above by persistent calyx; seed 1; yellowish, oblong, to 1.7 cm long, sticky nature.

Occasional in forests as stem parasite on various trees.

Fl. & Fr.: April – October.

Kanmankalva (RR), *MSM* & *KH* 10298; Borapuram (MBNR), *BR* & *BSS* 32516; Domalapenta (MBNR), *VBH* 84950 (BSID); Hamsanpalli RF (MDK), *RG* 104039 (BSID).

VISCACEAE

VISCUM L.

1. Internodes flattened, flattening in a plane at right angles to the preceding one; inflorescences laterally placed .... **V. articulatum**
1. Internodes terete; inflorescence lateral or around the node:
    2. Central flower male; young fruits warty .... **V. heyneanum**
    2. Central flower female; young fruits smooth .... **V. orientale**

Viscum articulatum Burm. f., Fl. Ind. 311. 1768; Hook.f., Fl. Brit. India 5: 226. 1886; Gamble, Fl. Madras 5: 1257. 1925; Sanjai & N.P.Balakr. in N.P.Balakr. & *al.*, Fl. India 23: 50. 2012. *V. nepalense* Spreng, Syst. Curi-post. 47. 1827.

Subshrub, moderate sized, semiparasite, evergreen, up to 1 m long, drooping; branches usually dichotomously branched, sometimes more than 2 branches at each node, yellowish green, internodes flattened, at right angle to the preceding one, gradually broadening from the base to tip, twisted near the base, longitudinally prominently grooved. Leaves absent. Flowers sessile triads in axillary fascicles, unisexual, actinomorphic, central flower of the triad female and lateral male. Male

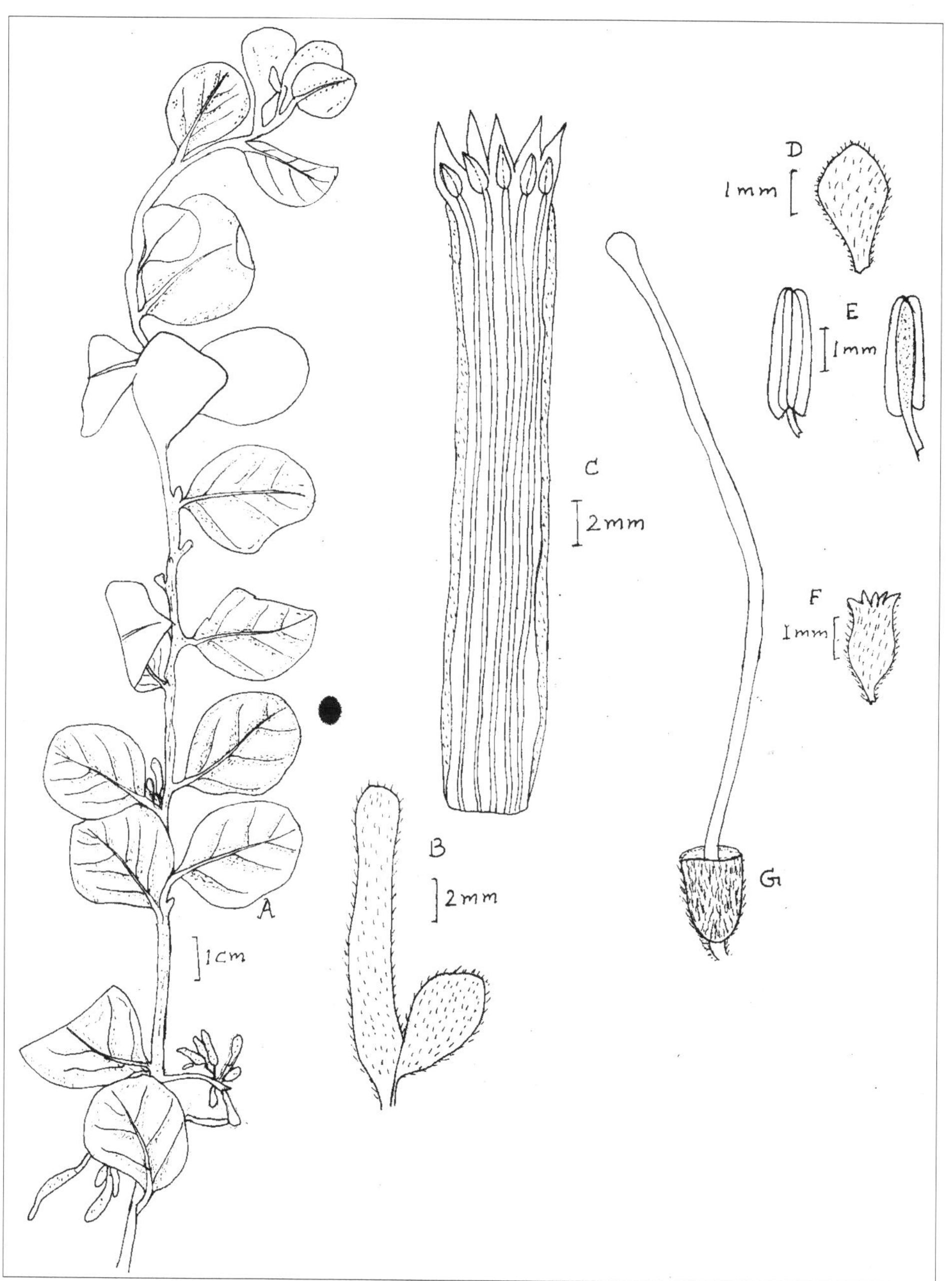

Figure 91: Taxillus heyneanus (Schult.f.) Danser, [Syn.: *Taxillus bracteatus* (Wall.) Tieghem] A. Twig, B. Flower, C. Corolla split open, D. Bract, E. Anthers, F. Calyx, G. Pistil.

flowers sessile, usually 2 in a triad on either side, up to 3 mm across, actinomophic, epigynous; perianth lobes 4, green, triangular, 1.5 mm long, valvate; stamens 4, 1 × 0.7 mm, opposite and adnate to the perianth lobes; filaments absent; pollen sacs many, dehisce introrsely. Female flower sessile, centre in the triad, up to 4 mm across, actinomorphic; perianth lobes 3 or 4, green, triangular, valvate, up to 2 mm across; ovary inferior; style absent; stigma nipple-shaped, truncate. Fruit berry, globose, green when young, yellowish when mature, up to 4 mm across, apex truncate; seed 1, globose, 2.5 mm across.

A partial stem parasite on spiny bushes such as *Ziziphus mauritiana, Z. oenoplia, Mangifera indica* etc. Occasionally found in the deciduous forests in most districts.

Fl. & Fr.: August – December.

Laxmapur (NZB), *TP* & *BR* 6283; Akkannapet (MDK), *BR* & *CP* 11558: Rangammagudem RF (RR), *MSM* & *KH* 10315; Pogila (NLG), *BVRS* 337 (NU); Pakhal (WGL), *KMS* 11699 (MH); Kodimial (KRN), *GVS* 20073 (MH); Lankapally hills (KMM), *RR* 10616 (BSID).

Viscum heyneanum DC., Prodr. 4: 2780. 1830; Sanjai & N.P.Balakr. in N.P.Balakr. & *al.*, Fl. India 23: 51. 2012. *V. verruculosum* Wight & Arn., Prodr. Fl. Ind. Orient. 379. 1834; Hook.f., Fl. Brit. India 5: 224. 1886; Gamble, Fl. Madras 2: 1258. 1925. *V. orbiculatum* Wight, Icon. Pl. Ind. Orient. t. 1016. 1845; Hook. f., Fl. Brit. India 5: 224. 1886; Gamble, Fl. Madras 2: 1258. 1925.

Vern.: *Badanika*.

Semiparasitic herbs, up to 40 cm long, much branched, branchlets terete. Leaves obovate-oblong, coriaceous, 1.5-3 x 0.6 - 1.5 cm, base attenuate, apex obtuse. Flowers axillary in fascicles. Male flowers: obovoid, perianth lobes 4, deltoid; stamens 4. Female flowers: slightly compressed, perianth lobes 4, ovary inferior, style short, stigma nipple shaped. Berries oblong, warty, with a crown of persistent or subpersistent perianth lobes.

Rare in forests in Medak district (Narasimha Rao, 1986).

Fl. & Fr.: November – May.

Viscum orientale Willd., Sp. Pl. 4: 737. 1806; Hook. f., Fl. Brit. India 5: 224. 1886; Gamble, Fl. Madras 2: 1258. 1925; Sanjai & N.P.Balakr. in N.P.Balakr. & *al.*, Fl. India 23: 54. 2012.

Vern.: *Bajinika*.

Large shrubs, parasitic on many kinds of trees, *Albizia, Pongamia* etc, stems much branched, internodes terete, faintly striate, thickened at the nodes; twigs slender, glabrous. Leaves 2.5-7 × 1-3 cm; obovate-elliptic or spathulate, base tapering into a short often obscure petiole, opposite, margin entire or undulate, apex obtuse or acute, coriaceous, glabrous. Flowers in short axillary dichasia, usually triads, green in colour. Male flowers: raised, perianth lobes 4 with anther attached on the inner side; stamens 4. Female flowers: sub sessile, perianth lobes 4; ovary inferior, style short, stigma capitate. Berry subglobose to oblong or ovoid

Common parasite on forest trees in Adilabad and Mahabubnagar districts.

Fl. & Fr.: February – March, and August – October.

Bheemaram (ADB), *GO* & *DAM* 5190; Gangapur to Chownpur (MDK), *RG* 106741 (BSID).

SANTALACEAE

SANTALUM L.

Santalum album L., Sp. Pl. 349. 1753; Hook.f., Fl. Brit. India 5: 231. 1886; Gamble, Fl. Madras 2: 1261. 1925; Sanjai & N.P.Balakr. in N.P.Balakr. & *al.*, Fl. India 23: 64. 2012.

Vern.: *Chandanam, Srigandham, Harichandanam*.

Small, evergreen, semiparasitic, glabrous tree with slender drooping branches, up to 12 m high, bark grey, cracked. The sap wood white and odorless, the heart wood yellowish brown, strongly scented. Leaves chartaceous, 4-6.5 × 2-3 cm, elliptic ovate to lanceolate, opposite, base acute, margin entire, apex sub-acute, sub-coriaceous. Flowers maroon or brownish-purple within, green without in 6 cm long axillary pedunculate cymes, shorter than leaves, trichotomous, terminal and axillary; perianth lobes 4 or 5, broadly triangular; stamens 4 or 5, filaments short, inserted at the mouth of perianth-tube; ovary at first free, later semi-inferior, 1-loculed, stigma 3-lobed. Drupe subglobose, 1 cm across, glabrous, purple black when ripe, endocarp hard.

Occasional in open dry deciduous and scrub forests in Nizamabd and Medak districts, frequently planted in gardens.

Fl. & Fr.: January – October.

Sone (ADB), *GO* & *BR* 9425; Tekrial (NZB), *TP* & *BR* 6424; Zaheerabad (MDK), *BR* & *CP* 11558; Medak (MDK), *CP* 11742; Kalmankalva (RR), *MSM* 10452; Amangal to Shadnagar (MBNR), *SRS* 104568 (BSID);

EUPHORBIACEAE

1. Inflorescence a cyathium; tepals 0 or minute; stamen 1, jointed:
    2. Branchlets zig-zag; cyathia zygomorphic, scarlet red .... **Pedilanthus**
    2. Branchlets not zig-zag; cyathia actinomorphic, green or reddish
        3. Involucral glands 4 or 5, separate and alterating with the involucral lobes of cyathium along the rim .... **Euphorbia**
        3. Involucral glands connate into an annular ring .... **Synadenium**
1. Inflorescence other than a cyathium; tepals conspicuous; stamens 2-10 or many, filaments not jointed:
    4. Stamens branched .... **Ricinus**
    4. Stamens not branched:
        5. Stamens alternating with tepals; cells of ovary 1-ovuled:

6. Perianth 2-seriate; inner tepals longer than outer, usually petaloid:
    7. Filaments inflexed in bud .... **Croton**
    7. Filaments short, straignt, elongating in flower:
        8. Stamens 1 or 2 bundles or raised on a receptacle:
            9. Shrub; ovary 3-locular; capsule hard, dehiscent .... **Jatropha**
            9. Tree; ovary 2-locular; drupe fleshy, indehiscent .... **Aleurites**
        8. Stamens free or on a column not raised on a receptacle:
            10. Monoecious herbs .... **Chrozophora**
            10. Dioecious trees .... **Givotia**
6. Perianth 1-seriate; tepals sepaloid (except *Manihot*)
    11. Outer tepals (sepals) valvate in male flowers:
        12. Inner tepals (petals) present in male flowers:
            13. Herbs .... **Chrozophora**
            13. Shrubs .... **Codiaeum**
        12. Inner tepals (petals) absent in male flowers:
            14. Climbers .... **Tragia**
            14. Erect herbs, shrubs or trees:
                15. Filaments much branched, staminal bundles indefinite:
                    16. Dioecious; capsule smooth .... **Homonoia**
                    16. Monoecious; capsule prickly .... **Ricinus**
                15. Filaments not branched:
                    17. Anthers vermiculiform or linear; styles laciniate .... **Acalypha**
                    17. Anthers not vermiculiform or linear:
                        18. Leves peltate .... **Macaranga**
                        18. Leaves not peltate:
                            19. Anther cells pendulous or adnate .... **Mallotus**
                            19. Anthers cells discrete, ascending, divaricate .... **Claoxylon**

11. Outer tepals (sepals) imbricate in male flowers:
    20. Inner tepals (petals) present in male flowers:
        21. Inner tepals free .... **Jatropha**
        21. Inner tepals (petals) connate .... **Givotia**
    20. Inner tepals (petals) in male flowers absent:
        22. Herbs or undershrubs:
            23. Leaves elliptic-ovate, 2-glandular at base .... **Baliospermum**
            23. Leaves linear-lanceolate, eglandular...... **Microstachys**
        22. Small trees .... **Falconeria**

4. Stamens opposite to tepals; cells of ovary 2-ovuled:
    24. Inner tepals (petals) present, small:
        25. Outer tepals (sepals) valvate:
            26. Ovary 2-celled; fruit a drupe with 2 pyrenes................ **Bridelia**
            26. Ovary 3-celled; fruit a capsule .... **Cleistanthus**
        25. Outer tepals (sepals) imbricate .... **Andrachne**
    24. Inner tepals (petals) 0; outer tepals (sepals) imbricate in bud:
        27. Disc wanting:
            28. Flowers in clusters, stamens connate by their connectives in a column, style also connate in a column .... **Glochidion**
            28. Flowers in male clusters, female sub-solitary, stamens free or rarely connate by their filaments.. **Putranjiva**
        27. Disc conspicuous:
            29. Flowers in spikes or recemes .... **Antidesma**
            29. Flowers in clusters:
                30. Stamens 5 .... **Flueggea**
                30. Stamens 3, rarely 5, filaments connate or sometimes not:
                    31. Tepals of male flower turbinate; ovary truncate .... **Breynia**
                    31. Tepals spreading; ovary globose .... **Phyllanthus**

ACALYPHA L.

1. Male and female flowers on separate spikes .... **A. alnifolia**
1. Male and female flowers on same spikes:
    2. Petiole shorter than the leaf blade; leaf apex gradually acuminate .... **A. fruticosa**
    2. Petiole as long as or longer than the leaf blade; leaf apex acute or abruptly acuminate:
        3. Spikes less than 4 cm long; bracts aciculate:
            4. Female bracts almost tubular, fimbriate along the rim with long subulate hispid teeth, not gland tipped .... **A. ciliata**
            4. Female bracts campanulate, not tubular, shortly toothed along margins, bordered with long hispid gland-tipped hairs .... **A. malabarica**
        3. Spikes more than 4 cm long; bracts dentate:
            5. Leaf base shortly attenuate .... **A. indica**
            5. Leaf base obtuse .... **A. lanceolata**

Acalypha alnifolia Klein ex Willd., Sp. Pl. 4: 525. 1805; Hook.f., Fl. Brit. India 5: 415. 1887; Gamble, Fl. Madras 2: 1330. 1925. N.P.Balakr. & Chakrab., Fam. Euphorb. India 124. 2007; Susila Rani & N.P.Balakr. in N.P.Balakr. & *al.*, Fl. India 23: 91. 2012.

Vern.: *Chiru kuppinta.*

Undershrub, up to 70 cm high, monoecious; branchlets terete, striate, tomentose. Leaves 1-2.5 × 0.6-2 cm, ovate, base rounded or cordate, margin crenate-serrate, apex acute; petiole 1.5 cm; stipules linear. Male flowers many, minute, in slender, elongate, cylindric, lower axillary spikes; flowers 0.7 mm across; pedicel 0.6 mm, hirsute; tepals 4, 0.6 × 0.4 mm, free, broadly ovate, concave, acute, hirsute without; stamens many, filaments short, anthers 0.35 mm, vermiculiform. Female flowers few, in short terminal spikes; flowers 1.3 × 5 mm; pedicel 1 mm; bracts 6.5 × 5 mm, cupular, dentate, ciliate; tepals 3, 1.5 × 0.6 mm, free, minute; pistil 1, ovary 2 mm across, hirsute, styles 3, to 1 cm, each with many filiform hairs, some further divided. Capsule 5 mm across, globose, hirsute; seeds globose, smooth.

Common in scrub jungles, in bushes and rocky hill areas in Karimnagar, Nalgonda, Hyderabad and Mahabubnagar districts.

Fl. & Fr.: April – September, Fr.: Throughout the year.

Amrabad (MBNR), *BSS* & *SKB* 32552; Mallelatheertham (MBNR), *SRS* 108989 (BSID); Bottichalami (NLG), *KMS* 19200 (MH); Kambalpally (NLG), *BVRS* 212 (NU).

Acalypha ciliata Forssk., Fl. Aegypt. – Arab, 162. 1775; Hook.f., Fl. Brit. India 5: 417. 1887; Gamble, Fl. Madras 2: 1330. 1925; N.P.Balakr. & Chakrab., Fam. Euphorb. India 125. 2007; Susila Rani & N.P.Balakr. in N.P.Balakr. & *al.*, Fl. India 23: 94. 2012.

Vern.: *Tellamurupindi.*

Undershrub, up to 80 cm high, monoecious; branchlets terete, striate. Leaves 3-9 × 2.5 cm, ovate, base rounded, margin serrate, apex abruptly acuminate; petiole to 5 cm. Flowers in axillary, short spikes. Male flowers many, minute, from above the middle, terminating in an allomorphic flower; flowers 0.5 mm across; pedicel 0.1 mm; tepals 4, 0.5 × 0.4 mm, free, broadly ovate, acute, muricate without; stamens 8, filaments short, anthers 0.3 mm, vermiculiform. Female flowers few, clustered at the lower half of the spikes; flowers 4 × 2.5 mm; pedicels 0.6 mm; bracts 4 × 2.6 mm, folded, campanulate, aciculate; tepals 4-5, 0.7 × 0.3 mm, lanceolate, acute, hairy without; pistils 2, one sterile, ovary 1.2 mm across, hispid, styles 3, to 1.6 mm, each with 3-4 filiform hairs. Capsule 2 mm across, globose, glabrous; seeds globose, smooth.

Occasional in waste lands and as forest undergrowth in scrub jungles in most districts.

Fl. & Fr.: August-November.

Sone (ADB), *GO* & *BR* 9426; Wankidi (ADB), *PSPB* 9153; Yedapalli (NZB), *BR* & *KH* 6474; Narsapur RF (MDK), *TP* & *CP* 14064; Anantagiri RF (RR), *TP* & *MSM* 11063; Pakhal RF (WGL), *KMS* 13170 (CAL); Kodimial (KRN), *GVS* 20053 A (MH).

Acalypha fruticosa Forssk., Fl. Aegypt.-Arab. 161. 1775; Hook.f., Fl. Brit. India 5: 415. 1887; Gamble, Fl. Madras 2: 1331. 1925; N.P.Balakr. & Chakrab., Fam. Euphorb. India 125. 2007; Susila Rani & N.P.Balakr. in N.P.Balakr. & *al.*, Fl. India 23: 96. 2012.

Vern.: *Mabbaku.*

Shrub, up to 90 cm high, monoecious; branchlets terete, striate. Leaves 1-4 × 0.5-2.5 cm, ovate, base truncate-rounded, margin crenate-dentate, apex gradually acuminate, thinly pubescent, glandular below; petiole to 1.5 cm. Flowers in axillary short spikes. Male flowers many, minute, from above the middle; flowers 0.4 mm across; pedicel 0.1 mm; tepals 4, 0.6 × 0.4 mm, free, broadly ovate, acute, concave, muricate without; stames 8-10, filaments short, anthers 0.2 mm, vermiculiform. Female flowers few, at the lower end of the spike; flowers 3 × 2 mm; pedicel 1 mm; bracts folded, campanulate, crenate; tepals 3, 1.1 × 0.45 mm, lanceolate, acute, hairy without; pistils 2, one sterile, ovary 0.7 mm across, hispid, styles 3, to 1.5 mm across, globose, glabrous. Capsule tomentose with large yellow glands in the sinuses; seeds globose, smooth.

In dry rocky situations in Hyderabad district (Ramana, 2010).

Fl. & Fr.: July - December.

Acalypha indica L., Sp. Pl. 1003. 1753; Hook.f., Fl. Brit. India 5: 416. 1887; Gamble, Fl. Madras 2: 1330. 1925; N.P.Balakr. & Chakrab., Fam. Euphorb. India 125. 2007; Susila Rani & N.P.Balakr. in N.P.Balakr. & *al.*, Fl. India 23: 96. 2012.

Vern.: *Muripinda, Kuppinta.*

Herb, up to 60 cm high, monoecious; branchlets terete, striate, thinly pubescent. Leaves 2-5 × 1-4 cm, broadly ovate, base shortly attenuate, margin crenate-serrate, apex acute; petiole to 6 cm; stipules small, caducous. Flowers in elongate, axillary spikes. Male flowers many, minute, from above the middle, terminating in an allomorphic flower; flowers 0.7 mm across; pedicel 0.1 mm; tepals 4, 0.4 × 0.3 mm, broadly ovate, concave, thinly pubescent without; stamens 8, filaments short, anthers 0.3 mm, vermiculiform. Female flowers 12-15, at the lower half of the spike; flowers 6 mm across; pedicel 0.1 mm; bracts 6.3 × 5 mm, folded, shortly dentate, ciliate; tepals 3, 0.7 × 0.35 mm, broadly ovate, acute, ciliate; pistils 2, one sterile, ovary 1.2 mm across, hispid, styles 3, to 3.2 mm, each with 2-3 filiform hairs. Fruit trilobite, 2 mm diam., sparingly pubescent; seeds ovoid, 1.3 mm long.

A common weed in wastelands in all districts.

Fl. & Fr.: Throughout the year.

Kowtal (ADB), *GO* & *PVP* 4829; Neela (NZB), *BR* & *GO* 9224; Sangareddy (MDK), *BR* & *CP* 11285; Rusthumpet (MDK), *KMS* 6787 (CAL); Mekavaripalli (RR), *TP* & *MSM* 11041; Bottichalami (NLG), *KMS* 19350 (MH); Kaleswaram (KRN), *SLK* 70882 (NBG); Gunavaram RF (KMM), *RCS* 98940 (BSID); Mannanur (MBNR), *SRS* 107445 (BSID).

Acalypha lanceolata Willd., Sp. Pl. 4: 524. 1805; Gamble, Fl. Madras 2: 1331. 1925; N.P.Balakr. & Chakrab., Fam. Euphorb. India 126. 2007; Susila Rani & N.P.Balakr. in N.P.Balakr. & *al.*, Fl. India 23: 94. 2012. *A. fallax* Müll.Arg., Linnaea 34: 43. 1865; Hook.f., Fl. Brit. India 5: 416. 1887.

Undershrub, up to 70 cm high, monoecious; branchlets terete, striate, glabrous, pubscent when young. Leaves 4.5 – 6.5 × 1.5-3 cm, ovate-lanceolate, base obtuse, margin crenate-serrate, apex acute; petiole to 3 cm; stipules small, lanceolate, ciliate. Flowers in axillary spikes. Male flowers few, minute at top; flowers 0.6 mm across; pedicel 0.1 mm; tepals 4, 0.6 × 4 mm, broadly ovate, concave, acute, muricate without; stamens 8, filaments short, anthers 0.25 mm, vermiculiform. Female flowers many; flowers 2.5 mm across; pedicel 0.05 mm; bracts 2 × 2.8 mm, dentate, teeth 10-12, ciliate, short, glandular tipped hairs without; tepals 3, 0.5 × 0.25 mm, free; pistil 1, ovary 1.5 mm across, hispid, styles 3, to 1.4 mm, each with 1-3 filiform hairs. Capsule 2 mm across, globose, hispid; seeds globose, smooth.

Occasional in wastelands, near moist places, shady moist places along road sides and fields in all districts.

Fl. & Fr.: April-August.

Bellampalli (ADB), *PSPB* 9152; Mamidipalli (NZB), *BR* & *KH* 6428; Nagulabanda (MDK), *CP* 11358; Kusumasamudram (RR), *MSM* & *NV* 10348; Cherukupally (NLG), *BVRS* 347; Aklaspur (KRN), *GVS* 25625; Vempalli (KRN), *GVS* 20191 (MH); Maraigudem RF (KMM), *RCS* 104249 (BSID).

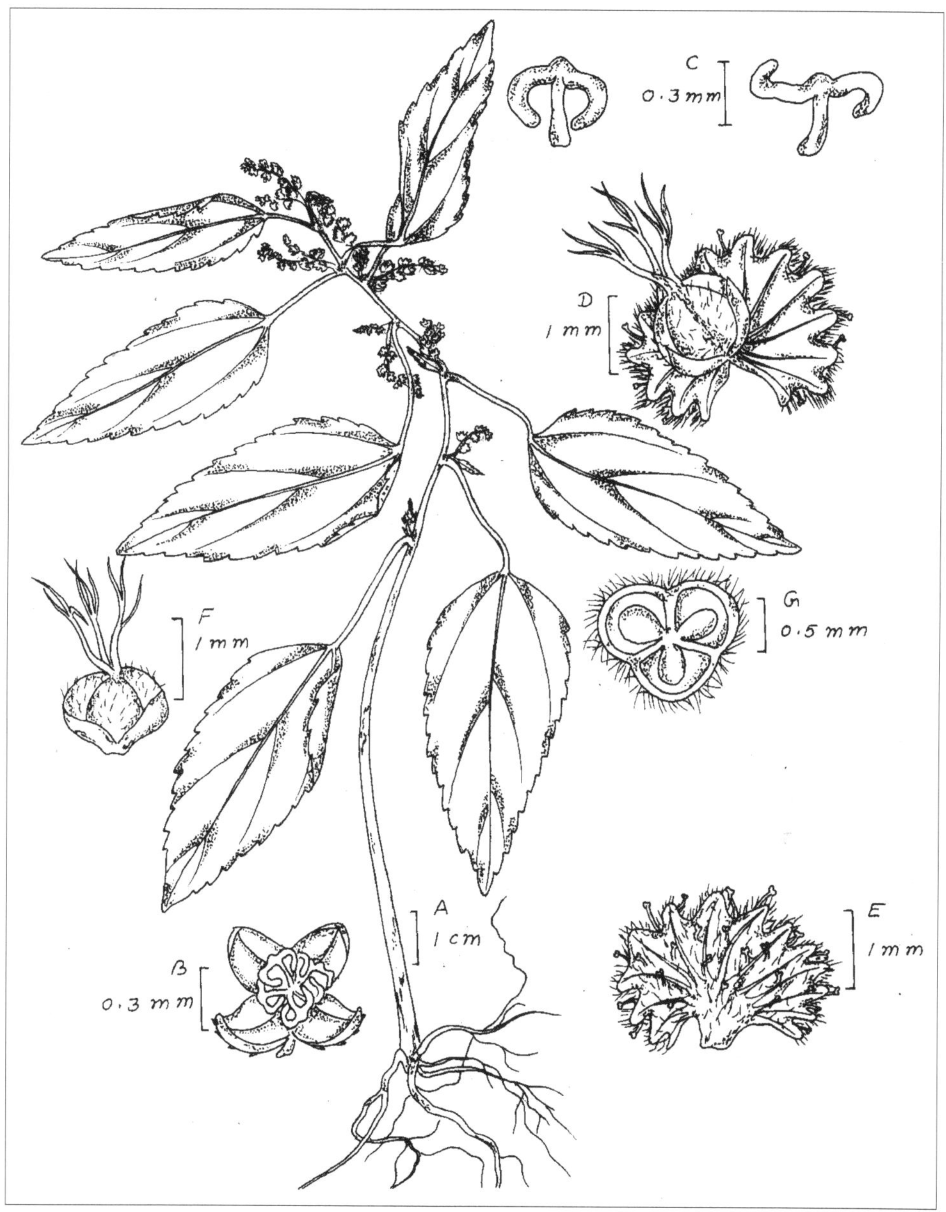

Figure 92: Acalypha lanceolata Willd.

A. Twig, B. Flower, C. Stamens, D. Flower, with bract, E. Bracts (dorsal view), F. Pistil, G. Ovary T.S.

Acalypha malabarica Müll.Arg., Linnaea 34: 42. 1865; Hook.f., Fl. Brit. India 5: 416. 1887; Gamble, Fl. Madras 2: 1330. 1925; N.P.Balakr. & Chakrab., Fam. Euphorb. India 126. 2007; Susila Rani & N.P.Balakr. in N.P.Balakr. & *al.*, Fl. India 23: 100. 2012.

Herb, up to 40 cm high, monoecious; branchlets striate, thinly pubescent. Leaves 2.5-4.5 × 1.5-3 cm, ovate or elliptic lanceolate, base cuneate-obtuse, margin crenate-serrate, apex abruptly acuminate; petiole to 4.5 cm. Flowers in axillary spikes. Male flowers few, at top; flowers 0.6 mm across; pedicel 0.4 mm; tepals 4, 0.6 × 0.4 mm, broadly ovate, acute, muricate without at middle; stamens 8, filaments short, anthers 0.2 mm, vermiculiform. Female flowers many, clustered at the base; flowers 3.5 mm across; pedicel 0.2 mm; bracts 3 × 4.2 mm, folded, dentate, teeth 12-13, ciliate, short gland tipped hairs within near the margins; tepals 3, 0.8 × 0.45 mm, broadly ovate, acute, hairy without; pistils 2, one sterile, ovary 1.5 mm across, thinly hispid at top, styles 3, to 2 mm, each with 4-5 filiform hairs. Capsule globose, glabrous or thinly hispid at top.

Occasional in moist shady places as forest undergrowth near the foothills, in Adilabad, Medak and Mahabubnagar districts

Fl. & Fr.: August-November.

Wankidi (ADB), *PPSPB* 11802.

Acalypha amentacea Roxb. subsp. **wilkesiana** Müll.Arg. is cultivated in gardens as ornamental.

ALEURITES J.R.Forst. & G. Forst.

Aleurites moluccana (L.) Willd., Sp. Pl. 4: 590. 1805; Hook.f., Fl. Brit. India 5: 384. 1887; Gamble, Fl. Madras 2: 1318; N.P.Balakr. & Chakrab., Fam. Euphorb. India 192. 2007; Chakrab. & N.P.Balakr. in N.P.Balakr. & *al.*, Fl. India 23: 207. 2012. *Jatropha moluccana* L., Sp. Pl. 1006. 1753.

Evergreen tree; young shoots densely clothed with brownish or grey stellate tomentum. Leaves subcoriaceous, broadly elliptic or ovate to deltoid, tomentose when young, base truncate, apex shortly acuminate; petiole 6-12 cm long, 2-glandular at apex. Flowers white in large, cymose, terminal panicles. Male flowers: calyx subglobose in bud, 6-8 mm long, lanceolate or ovate, disc urceolate, with 5 glands; stamens 15-20, verticillate in 3-5 rows, filaments hairy. Female flowers: sessile at the division of panicle; calyx cylindrical and conical, splitting into 2 or 3 segments; petals 5, narrowly oblong, 9-10 mm long; disc urceolate with 5 glands; ovary densely hispid, styles 2, each bifid with linear-subulate arms. Drupe globose, fleshy, 4 cm across with hard stone; seeds large.

Planted in gardens.

Fl.: December – February. Fr.: February-May.

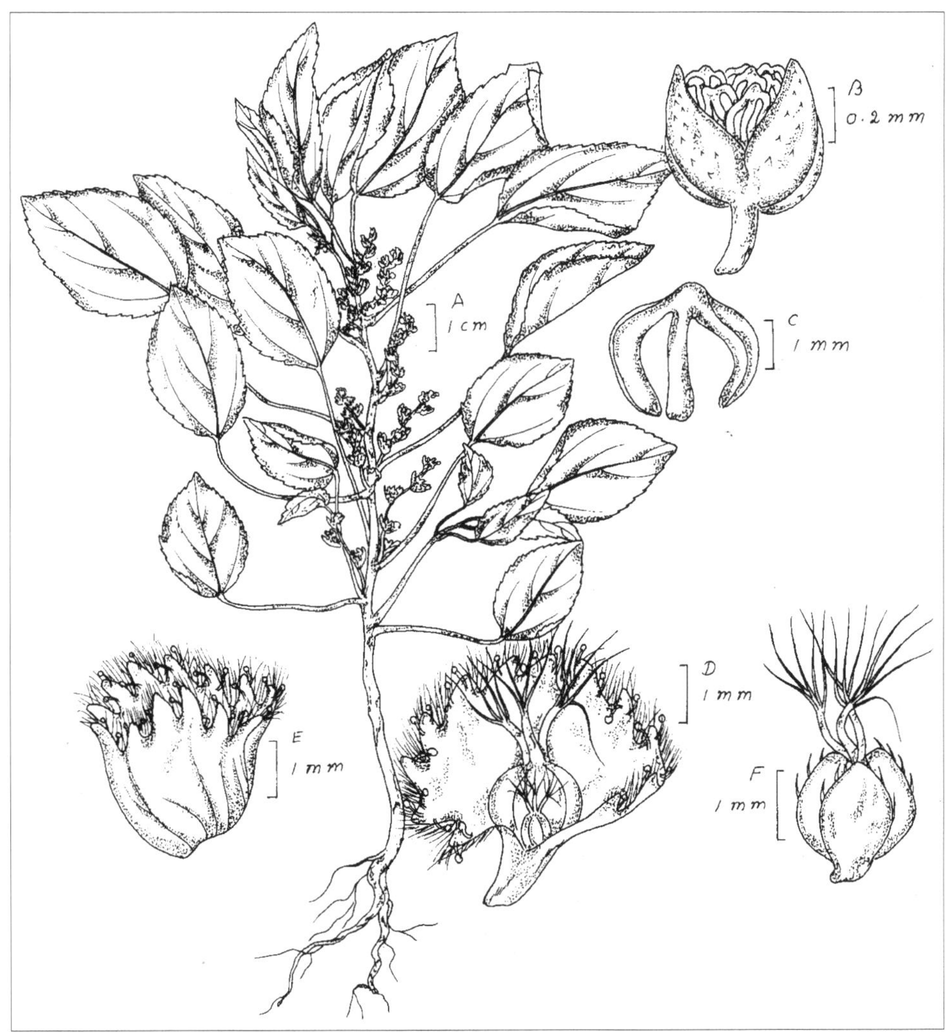

Figure 93: Acalypha malabarica Müll. Arg.

A. Twig, B. Flower, C. Stamens, D. Flower, with bract, E. Bracts (side view), F. Pistil.

ANDRACHNE L.

Andrachne telephoides L., Sp. Pl. 10114. 1753; Hook.f., Fl. Brit. India 5: 284. 1887; N.P.Balakr. & Chakrab., Fam. Euphorb. India 338. 2007; Chakrab. & *al.*, in N.P.Balakr. & *al.*, Fl. India 23: 408. 2012.

Glaucous, erect or diffuse monoecious herb; branches 7-15 cm long. Leaves oblong-obovate, 3-5 × 1-3 mm, base cuneate, margin entire, apex obtuse, penninerved, membranous, glabrous; petioles 1-2.5 mm long; stipules subpeltate, 1.5-2.0 mm long, irregularly incised. Flowers axillary, solitary or males often 2-3 together. Male flowers: pedicels 0.8-1.0 mm long; sepals 5-6, imbricate, connate at base, obovate, ca. 1.5 × 0.5 mm; petals 5 or 6, linear-oblong, deply notched at apex, ca 0.5×0.2 mm; disc glands 5 or 6, 0.15-0.18 mm long, 5-lobed; stamens 5 or 6, 0.4-0.6 mm long, filaments free or partially connate, anthers 4-lobed, ca. 0.15 × 0.2 mm, elliptic-oblong; pistillode present. Female flowers: pedicels ca 2 mm long; sepals larger than in male flowers; petals minute; disc glands as in male; ovary 3-loculed; style short, bifid to base. Capsules depressed globose, 2-3 mm in diam., glabrous, consisting of three, 2-valved cocci; endocarp thinly woody; seeds 2 per locule, triquetrous, with a convex, punctulate back, curved, rugose, sculptured, estrophiolate.

Occasional in open grounds, harvested fields, and dry mud in Nizamabad district (Sardesai and Chavan, 2010).

Fl. & Fr.: December – February.

Mirzapur (NZB), *Madhukar* 6348 (BMAU).

ANTIDESMA L.

1. Branchlets pubescent, spikes woolly .... **A. ghaesembilla**
1. Branchlets glabrous, spikes glabrous .... **A. alexitaria**

Antidesma alexiteria L., Sp. 1027. 1753 *p.p.*; Hook.f., Fl. Brit. India 5: 359. 1887; N.P.Balakr. & Chakrab., Fam. Euphorb. India 304. 2007; Chakrab. & M. Gangop. in N.P.Balakr. & *al.*, Fl. India 23: 357. 2012. *A. zeylanicum* Lam., Encycl. 1: 207. 1782; Gamble, Fl. Madras 2: 1297. 1925; Surya Prakash Babu in Moulali & Pullaiah, Fl. Andhra Pradesh 2: 843. 1997.

Shrub or small tree, to 6 m high, dioecious; branchlets terete, glabrous. Leaves 5-8 × 2.5-3.5 cm, elliptic to elliptic-lanceolate or oblong-elliptic, base acute-cuneate, margin entire, apex acutely apiculate, subcoriaceous; petiole to 5 mm; stipules lanceolate. Flowers in axillary and terminal spikes. Male flowers 2.5 mm across; tepals 4 or 5, free, concave, thinly pubescent without; staments 3, filaments 1.5 mm, anthers 0.5 mm across; pistillode acute. Female flowers 1.2 mm across, sessile; tepals 4, free, concave, thinly pubescent without; ovary 0.8 × 1.2 mm, globose, styles 2, one of them 2-fid. Drupe 4-5 × 3-4 mm, globose.

Occasional on either sides of the streams in the deciduous forests.

Fl. & Fr.: September-December.

Satnella river bed (ADB), *GO* & *PVP* 4218; Mudheli RF (NZB), *TP* & *BR* 6373.

Antidesma ghaesembilla Gaertn., Fruct. 1: 189, t. 39. 1788; Müll.Arg. . in DC., Prodr. 15: 215. 1866; Hook.f., Fl. Brit. India 5: 357. 1887; Gamble, Fl. Madras 2: 1298. 1925; N.P.Balakr. & Chakrab., Fam. Euphorb. India 305. 2007; Chakrab. & M. Gangop. in N.P.Balakr. & *al*., Fl. India 23: 359. 2012.

Ver.: *Polari, Puttam, Lona, Katha pulitsi, Janupulisaru.*

Small tree, to 10 m high, dioecious; branchlets terete, pubescent or fulvous tomentose. Leaves oblong, broadly oblong-elliptic to obovate, 5-9 × 2.5-5.5 cm, base rounded, margin entire, apex rounded or abruptly acute, often emarginated; petiole 1 cm, pubescent; stipules lanceolate. Flowers in slender paniculate axillary spikes, sessile. Male flowers 1.7 mm across; tepals 4 or 5, densely pubescent; stamens 4-5, filaments 1 mm, anthers 0.4 mm, curved; disc 5-lobed, pubescent. Female flowers 2.2 mm across; tepals as in the male flower; ovary 2 mm across, pubescent, styles 2, short, one of them 2-fid; disc as in the male flower, Drupe 3 mm across, globose, prominently rugose when dry; seeds covered with dried albumen.

Occasional near the streams in the deciduous forests in Adilabad, Karimnagar, Mahabubnagar and Warangal districts.

Fl. & Fr.: June-August.

Bijjur RF (ADB), *GO* & *PVP* 4881; Neelwai RF (ADB), *GO* 5437; Sirpur RF (ADB), *GO* & *PVP* 4923; Farahabad (MBNR), *MVS* & *MB* 38635; Mahadevpur to Kataram (KRN), *SLK* 70820 (NBG); Pakhal (WGL), *KMS* 13184 (MH); Gangaram (WGL), *PVS* 84017 (BSID).

BALIOSPERMUM Blume

Baliospermum solanifolium (J.Burm.) C.R.Suresh in Nicolson & *al*., Interpr. Van Rheede's Hort. Malab. 106. 1988; N.P.Balakr. & Chakrab., Fam. Euphorb. India 199. 2007; N.P.Balakr. in N.P.Balakr. & *al*., Fl. India 23: 196. 2012. *Croton solanifolius* J. Burm., Fl. Malab. 6. 1769; Geiseler, Croton Monogr. 74. 1807. *Jatropha montana* Willd., Sp. Pl. 4: 563. 1805. *Baliospermum axillare* B1ume, Bijdr. 604. 1825; Hook.f., Fl. Brit. India 5; 461. 1887. *Croton polyandrous* Roxb., Fl. Ind. 3: 682. 1832. *Baliospermum montanum* (Willd.) Müll.Arg. in DC., Prodr. 15: 1125. 1866; Gamble, Fl. Madras 2: 1342. 1925; Surya Prakash Babu in Pullaiah & Moulali, Fl. Andhra Pradesh 2: 843. 1997.

Vern.: *Kondamudam, Nelajidi.*

Herb or undershrub, to 1 m high, monoecious, often dioecious; branchlets terete, striate, tomentose. Leaves 3-21 × 1.5-8 cm, elliptic-ovate, often 3 to 5-lobed, base rounded or cuneate and 2-glandular, margin sinuate-toothed, apex acuminate, sparsely hirtellous; petiole to 6 cm; stipules lanceolate, 2-glandular. Flowers in axillary and terminal racemes; bracts 1.2 mm, widely ovate. Male flowers 2.5 mm across; tepals 5, 2.2 × 1.6 mm, slightly connate at base, suborbicular, concave, rounded; stamens numerous, filaments 1 mm long, anthers 0.5 × 0.8 mm; disc glands 6. Female flowers 3.5 mm across; tepals 6, connate at base, ovate-lanceolate, subacute, hirtillous without; ovary 2.5 mm across. Capsule 0.9 × 0.7 mm, obovate, pubescent; seeds elliptic-obovate, smooth.

Common along the roadsides in fields of Adilabad, Nizamabad, Karimnagar, Ranga Reddi and Mahabubnagar districts.

Fl. & Fr.: October-February.

Kadam river bank (ADB), *TP* & *PVP* 4127; Inderveli (ADB), *GO* & *PVP* 4530; Gandhari (NZB), *TP* & *BR* 6339; Manchippa (NZB), *PSPB* 9170, *PSPB* & *BR* 9608, 9609; Pargi (RR), *MSM* & *PRSR* 15103; Rechapalli (KRN), *GVS* 22237 (MH); Gangaram (WGL), *PVS* & *NRR* 76933 (BSID); Perantapally RF (KMM), *RCS* 98993 (BSID).

BREYNIA J.R. Forst.& G. Forst. *nom. cons*

1. Leaf apex apiculate; perianth greatly enlarged in fruit; capsule depressed globose .... **B. retusa**
1. Leaf apex obtuse-subacute; perianth not or scarcely enlarged in fruit; capsule globose .... **B. vitis-idaea**

Breynia retusa (Dennst.) Alston, Ann. Roy. Bot. Gard. (Peradeniya) 11: 204. 1929; N.P.Balakr. & Chakrab., Fam. Euphorb. India 340. 2007; Chakrab. & *al.* in N.P.Balakr. & *al.*, Fl. India 23: 410. 2012. *Phyllanthus retusus* Dennst., Schlussel Hortus Malab. 24. 1818. *P. patens* Roxb., Fl. Ind. 3: 667. 1832. *Breynia patens* (Roxb.) Rolfe, J. Bot. 20: 359. 1882; Hook.f., Fl. Brit. India 5: 329. 1887; Gamble, Fl. Madras 2: 1304. 1925. *B. angustifolia* Hook.f., Fl. Brit. India 5: 330. 1887.

Shrub, to 2 m high, monoecious; branchlets angular. Leaves ovate-elliptic or orbicular, base obtuse-rounded, margin entire, apex apiculate; petiole 2 mm. Flowers in axillary cymes; pedicels filiform. Male flowers 4 mm across; tepals 6, connate, lobes minute, triangular, acute; stamens 3, filaments connate, anthers oblong. Female flowers 5 mm across; tepals 6, obovate-orbicular, rounded; ovary cupular, styles 3, each 2-fid. Capsule globose or depressed globose, seated on the accrescent tepals cup; seeds wedge-shaped, smooth.

Common in dry localities near the forests in Karimnagar district.

Fl. & Fr.: February-July.

Kodimial (KRN), *GVS* 21892 (MH).

Breynia vitis-idaea (Burm.f.) Fischer, Bull. Misc. Inf. Kew 1932: 65.1932; N.P.Balakr. & Chakrab., Fam. Euphorb. India 340. 2007; Chakrab. & *al.* in N.P.Balakr. & *al.*, Fl. India 23: 411. 2012. *Rhamnus vitis-idaea* Burm. f., Fl. Ind. 61. 1768.*p.p. Phyllanthus rhamnoides* Retz., Observ. Bot. 5: 30. 1788. *nom. illeg. P. rhamnoides* Willd., Sp. Pl. 4: 580. 1805. *Breynia rhamnoides* (Willd.) Müll.Arg. in DC., Prodr. 15: 440. 1866 *nom. illeg.*; Hook.f., Fl. Brit. India 5: 330 1887; Gamble, Fl. Madras 2: 1304. 1925.

Vern.: *Errapurugudu, Erraballi.*

Shrub, to 3 m high, monoecious; with many horizontal branches, branchlets angular. Leaves 1.3-2.4 × 0.9-1.5 cm, base cuneate, margin entire, apex obtuse-subacute; petiole to 0.5 cm; stipules subulate from a triangular base, persistent. Flowers minute, in axillary cymes. Male flowers in small fascicles, pedicels filiform, variable in length;

flowers 2 mm across; tepals 6, turbinate or subcampanulate, entire or obscurely lobed; stames 3, filaments connate, staminal column to 6 mm, anthers 1 × 0.6 mm, linear. Female flowers solitary; flowers 2.5 mm across; pedicels filiform; tepals 6, campanulate, with 6 lobes, acute-obtuse; ovary 2 × 1.2 mm, truncate. Capsule 5 mm across, globose.

Common in open places in deciduous forests of Nizamabad and Medak districts.

Fl. & Fr.: September- January.

Rudroor (NZB), *BR & KH* 9642; Rusthempet (MDK), *KMS* 6790 (MH).

BRIDELIA Willd.

1. Trees up to 12 m high; branchlets tomentose; secondary nerves 15-25 pairs, parallel, straight .... **B. retusa**
1. Shrub to small trees up to 3 m high; branchlets glabrous; secondary nerves 6-10 pairs, not markedly straight or parallel .... **B. montana**

Bridelia montana Roxb., Willd. Sp. Pl. 4: 978. 1805; Hook.f., Fl. Brit. India 5: 269. 1887 *p.p.;* Gamble, Fl. Madras 2: 1281.1925; N.P.Balakr. & Chakrab., Fam. Euphorb. India 319. 2007; Chakrbarty & *al.* in N.P.Balakr. & *al.*, Fl. India 23: 379. 2012. *Clutia montana* Roxb., Pl. Coromandel 2: 36. t. 171. 1802. *Bridelia hamiltoniana* Müll.Arg., Linnaea 34. 77. 1865; Hook.f., Fl. Brit. India 5: 271.1887.

Vern.: *Panchavoni, Tella yegise, Pantangi, Balli, Kanchep, Sanna koctari, Fundu bigalu*.

Shrub to small tree, up to 3 m high; bark thin, rather rough; branchlets glabrous. Leaves coriaceous, obovate or oblanceolate, 4-10.5 × 2-5.8 cm, glabrous, green above, paler beneath, base cuneate, margin entire, apex acute or obtuse, secondary nerves 6-10 pairs, oblique; petiole to 5 mm. Flowers green, few in leaf axils. Male flowers: sepals triangular; petals flabellate; disc flat, entire; stamens 1.5-1.8 mm long. Female flowers: sepals rarely 6, triangular-oblong; petals 0.5-2 mm long; outer disc platelliform, inner disc urceolate, lobulate, covering the ovary; ovary subglobose or ovoid; styles ca 1 mm long, almost free, deeply bifid. Drupe globose, 0.5 cm in dia., pyrene furrows shallow, the ridge rounded, curving from a central one.

Common, in dry deciduous and degraded forests in most districts.

Fl.: September- October. Fr.: January - March.

Wankidi (ADB), *PSPB* 9154; Mothe RF (NZB), *BR & GO* 9068; Velutla RF (NZB), *BR* 9526; Nagulabanda (MDK), *CP* 11349; Gadirayal RF (RR), *MSM* 10518; Mannanur (MBNR), *TDCK* 13605; Kambalpally (NLG), *BVRS* 228; Bhupatipur (KRN), *GVS* 25706 (MH); Rangapur (WGL), *PVS & NRR* 76961 (BSID); Bandirevu RF (KMM), *RCS* 98914 (BSID).

Bridelia retusa (L.) A. Juss., Euphorb. Gen. 109. t.7, f. 22. 1824; Dressler, Blumea 41(2): 289. 1996; N.P.Balakr. & Chakrab., Fam. Euphorb. India 319. 2007; Chakrbarty & *al.* in N.P.Balakr. & *al.*, Fl. India 23: 380. 2012. *Clutia retusa* L., Sp. Pl. 1042. 1753 (*'Cluytia'*). *Bridelia crenulata* Roxb., Fl. Ind. 3: 734. 1832; Surya Prakash Babu in Pullaiah & Moulali, Fl. Andhra Pradesh 2: 845. 1997. *B. roxburghiana* (Müll.Arg.)

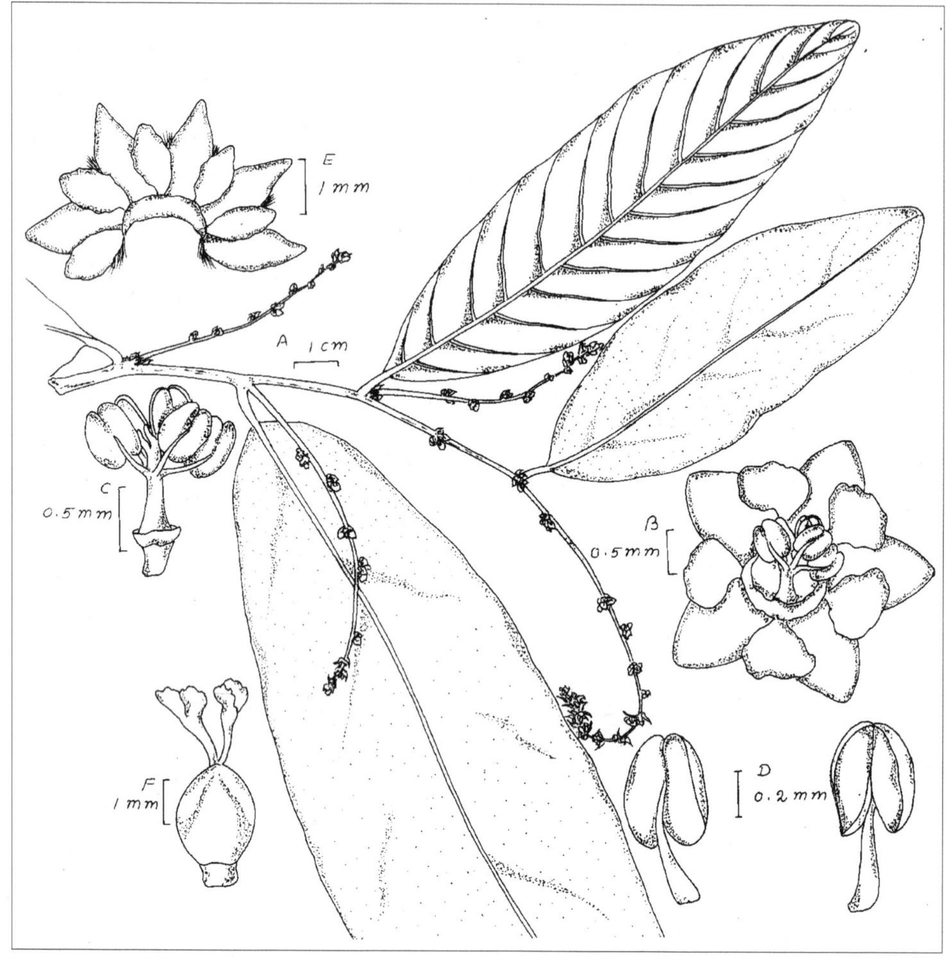

Figure 94: Bridelia retusa (L.) A. Juss.

A. Twig, B. Flower, C. Androecium with gynophores, D. Anthers, E. Perianth split open, F. Pistil.

Gehrm., Bot. Jahrb. Syst. 41: 30. 1908; Gamble, Fl. Madras 2: 1280. 1925. *B. retusa* var. *roxburghiana* Müll.Arg. in DC., Prodr. 15. 493. 1866; Hook.f., Fl. Brit. India 5: 268. 1887; Gamble, Fl. Madras 2: 1280. 1925. *B. retusa* var. *squamosa* (Lam.) Müll.Arg. in DC. Prodr. 493. 1866; Hook.f., Fl. Brit. India 5: 268. 1887. *B. retusa* var. *glauca* Hook.f., Fl. Brit. India 268. 1887. *B. cinerascens* Gehrm. *loc. cit.* 30. 1908; Gamble, *loc. cit.* 1280. 1925; Surya Prakash Babu, *loc. cit.* 2: 845. 1997.

Vern.: *Kora maddi, Mulu maddi, Puta karaka, Anemu, Koramanu, Koderu, Durga maddi, Pia maddi, Bonta yepi, Aanemchettu; Kosengi, Yerra sittuga.*

Deciduous tree, up to 12 m high; spiny when young; bark grey, longitudinally cracked; wood grey to crimson, shallowly or deeply interlocked grained in narrow bands, medium textured; branchlets tomentose. Leaves chartaceous, elliptic-oblong or obovate, 6-19.5 × 3-9 cm, glabrous above, chartaceous and glaucous beneath, base obtuse, margin entire, slightly crenulate, apex acute, or obtuse, secondary nerves 15-25 pairs, conspicuous with finely reticulate veins. Flowers greenish-yellow or purple in 5-10 cm long, terminal spikes or in axillary clusters. Male flowers: pedicels 1.5-2 mm long; sepals triangular; petals suborbicular, toothed or lobed; disc flat, entire; staminal column ca 1.5 mm long, filaments ca 1 mm long; pistillode conic. Female flowers: pedicels 0.5-2 mm long; sepals narrowly triangular; petals suborbiclar or oblong; outer disc disciform, inner disc conical-cupular; styles 0.8-1.5 mm long, shortly connate below and shortly bifid above. Drupe purple or black, seated on enlarged calyx, pyrene furrows curled and deep, 1-2 seeds with fairly thick bony shell, with brownish papery testa.

Common, in mixed and deciduous forests in Adilabad, Nizamabad, Karimnagar, Warangal and Nalgonda districts.

Fl.: May-July. Fr.: November-January.

Pulikunta (ADB), *PSBP* 9159; Sirpur RF (ADB), *GO* 4406; Varni (NZB), *TP* & *BR* 6172; Mudheli RF (NZB), *BR* & *CPR* 7138; Narsapur (MDK), *TP* & *CP* 11992, 14058, *RG* 113466 (BSID); Anantagiri RF (RR), *MSM* & *PRSR* 12782; Molachintapally (MBNR), *TDCK* 15397; Rechapalli (KRN), *GVS* 22267 (MH); Pakhal (WGL), *KMS* 13200 (MH); Pasra (WGL), *RKP* 105387 (BSID).

CHROZOPHORA Neck. ex A. Juss. *nom. cons.*

1. Leaves both radical and cauline, eglandular at base; stigmas orange; fruits grey or black, non tinctorial when wetted .... **C. prostrata**
1. Leaves cauline only, 2-glandular at base; stigmas red; fruits purple, tinctorial when wetted .... **C. rottleri**

Chrozophora prostrata Dalzell in Dalzell & A. Gibson, Bombay Fl. 233. 1861; N.P.Balakr. & Chakrab., Fam. Euphorb. India 166. 2007. var. **parvifolia** (Klotzsch ex Schweinf) N.P. Balakr., Bull. Bot. Surv. Ind. 15: 7: 1973 (1976); N.P.Balakr. in N.P.Balakr. & *al.*, Fl. India 23: 181. 2012. *C. parvifolia* Klotzsch ex Schweinf, Pl. Quaed. Nilot. 11. 1862; Gamble, Fl. Madras 2: 1316. 1925. *C. plicata* forma 2. Hook.f., Fl. Brit. India 5: 410. 1887.

Vern.: *Chinna-miriyam.*

Herb, to 30 cm high, monoecious; stem naked in the lower part, slightly branched above, branchlets stellate-tomentose, stellate hairs stalked. Leaves 1.5-3 × 0.8-2.3 cm, ovate-orbicular, base cuneate-truncate or obtuse, 2 glandular, margin sinuate, apex obtuse, bullate; petiole to 1.5 cm; stipules linear. Flowers in short, densely stellate-tomentose racemes; male flowers crowded above; female flowers at base, pedicellate; bracts linear. Male flowers 4 mm across; tepals in two whorls, 5 + 5; outer (sepals) 1.8 × 0.5 mm, free, oblanceolate, rounded, stellate hairy without; inner (petals) 1.6 × 0.6 mm, free, lanceolate, acute-obtuse; stamens 15, biseriate; filaments connate, outer short, inner longer, anthers 0.8 x 0.5 mm; disc glands 5, free. Female flowers 4.5 mm across; tepals in two whorls, 5 + 5; outer 1.5 × 0.5 mm, connate at base, narrowly triangular, acute, stellate hairy without; inner 0.9 × 0.5 mm, connate at base, narrowly triangular, acute, stellate hairy without; ovary 2.5 mm across, globose, stellate hairy, styles 3, slightly connate at base, each 2-fid, near the base, stigma orange in colour. Capsule 7 mm, densely stellate hairy, without silvery scales.

Occasional in dried up tanks in Medak and Nalgonda districts.

Fl. & Fr.: June-August.

Kambalpally (NLG), *BVRS* 229 (NU).

Key to Varieties

1. Branches erect or decumbent, woody; leaves oblong-lanceolate to elliptic, subacute, always with 2 glands at base beneath, near the midrib .... var. **parvifolia**
1. Branches prostrate to procumbent, less woody; leaves oblong-ovate to suborbicular, obtuse, usually eglandular, but sometimes with 2 glands at base near the midrib .... var. **prostrata**

Chrozophora prostrata Dalzell in Dalzell & A. Gibson, Bombay Fl. 233. 1861, var. **prostrata;** Gamble, Fl. Madras 2: 1316. 1925; N.P.Balakr. in N.P.Balakr. & *al.*, Fl. India 23: 181. 2012. *C. plicata forma* 3. Hook.f., Fl. Brit. India 5: 410. 1887.

Vern.: *Lingamiriyam.*

Procumbent humifuse herb, monoecious; branches spreadiang from the root, branchlets stellate-tomentose, stellate hairs sessile. Leaves 0.6-2 × 0.4-1 cm, broadly ovate – orbicular, base cordate-truncate, margin sinuate, apex obtuse, dark coloured, deeply bullate; petioles to 1.2 mm; stipules linear. Flowers in very short densely stellate-tomentose racemes; male flowers crowded at above; female flowers at base, pedicellate; bracts linear. Male flowers 4 mm across; tepals in two whorls, 5 + 5; outer (sepals) 2 × 0.6 mm, free, oblanceolate, acute, stellate hairy without; inner (petals) 2.9 x 0.8 mm, free, elliptic-lanceolate, obtuse, thinly stellate hairy without; stamens 15, biseriate, filaments connate, outer short, inner long, anthers 0.8 × 0.6 mm; disc glands 5, free. Female flowers 3.5 mm across; tepals in two whorls, 5 + 5; outer (sepals) 1.5 × 0.4 mm, connate at base, linear, acute, stellate hairy without; inner (petals) 1 × 0.25 mm, free, linear, acute; ovary 2.2 × 2.6 mm, globose, stellate-tomentose, styles 3, slightly connate at base, each 2-fid, near the base, stigma orange yellow in colour. Capsule 5 mm across, stellate hairy, with silvery scales.

Occasional in moist places of irrigated fields and on the edges of river beds in Karimnagar, Medak and Hyderabad districts.

Fl. & Fr: July-December.

Hussainsagar (HYD), *KMS* 5959; Kodimial (KRN), *GVS* 20100 (MH); Medak (MDK), *KMS* 7985 (MH, CAL).

Chrozophora rottleri (Geis.) Adr. Juss. ex Spreng., Syst. Veg. 3: 850. 1826; Gamble, Fl. Madras 2: 1317. 1925; N.P.Balakr. & Chakrab., Fam. Euphorb. India 167. 2007; N.P.Balakr. in N.P.Balakr. & *al.*, Fl. India 23: 181. 2012. *Croton rottleri* Geisel., Croton Monog. 57. 1807. *Chrozophora plicata* (non Adr. Juss.) Voigt, Hort. Sub. Calc. 156. 1846. *C. tinctoria* (non Adr. Juss.) Hook.f., Fl. Brit. India 5: 409. 1887. *p.p.*

Vern.: *Erra-meriyam.*

Erect herb, to 60 cm high, monoecious; branchlets fulvous-tomentose. Leaves 2-6 × 1.2-4 cm, broadly ovate, base acute-cordate, 2 glandular, margin sinuate, obtuse; petiole to 3 cm; stipules deciduous. Flowers in densely tomentose terminal racemes; male flowers crowded above; female flowers at base, pedicellate; bracts lanceolate. Male flowers 5 mm across; tepals in two whorls, 5 + 5; outer (sepals) 1.2 × 1 mm, connate at base, ovate-lanceolate, concave, acute-obtuse; inner (petals) 2.5 × 1.2 mm, free, lanceolate, obtuse; stamens 12-15, biseriate, filaments connate, outer short, inner long, anthers 1 mm across; disc glands 5, free. Female flowers 5 mm across; tepals in two whorls, 5 + 5; outer (sepals) 2 mm, connate at base, lanceolate-triangular, acute, stellate hairy wihtout; inner (petals) 1.5 mm, connate at base, lanceolate-triangular, acute, stellate hairy without; ovary 3 mm across, globose, stellate hairy, styles 3, slightly connate at base, each 2-fid more than halfway down, stigma red in colour. Capsule 9 mm across, depressed, 3-lobed, stellate-tomentose; seeds globose, smooth, shiny.

Common weed in cultivated fields especially in black soils in Adilabad, Nizamabad, Medak, Ranga Reddi, Khammam, Warangal and Nalgonda districts.

Fl. & Fr.: Throughout the year.

Satnella river bed (ADB), *GO* & *PVP* 4218; Sarvapur (NZB), *TP* & *BR* 6373; Ooracheru lake (MDK), *KMS* 7992; Mohammadabad RF (RR), *MSM* & *KH* 10234; Perantapalli (KMM), *PVS* 84063 (BSID); Eturnagaram (WGL), *RKP* 110822 (BSID); Rayagir (NLG), *BVRS* 15 (NU).

CLAOXYLON A. Juss.

Claoxylon mercurialis (L.) Thwaites, Enum. Pl. Zeyl. 271. 1861; Hook.f., Fl. Brit. India 5: 412. 1887; N.P.Balakr. & Chakrab., Fam. Euphorb. India 131. 2007; Susila Rani & N.P.Balakr. in N.P.Balakr. & *al.*, Fl. India 23: 114. 2012. *Micrococca mercurialis* (L.) Benth. in Hook. Niger. Fl. 503. 1849; Gamble, Fl. Madras 2: 1328. 1925; Surya Prakash Babu in Pullaiah & Moulali, Fl. Andhra Pradesh 2: 877. 1997. *Tragia mercurialis* L., Sp. Pl. 980. 1753.

Herb, up to 80 cm high, monoecious; branchlets sparsely hairy. Leaves 3-5 × 1-2.5 cm, ovate-lanceolate, base acute, margin serrate, apex subacuminate, puberulous; petiole to 3.5 cm; stipules minute. Flowers in long slender interrupted axillary spikes; bracts minute. Male flowers 1.8 mm across; tepals 3, 1.3 × 0.8 mm, slightly connate at base, ovate, acute; stamens 3-5, free, filaments 0.9 mm, anthers 0.3 × 0.25 mm. Female flowers 2.8 mm across; sepals 3, 1.8 × 1.2 mm, slightly connate at base, ovate-lanceolate, acute-acuminate, sparsely pubescent; styles 3, linear, horizontally spreading, plumose-laciniate; disc of glands linear. Capsule 4 mm, sparsely pubescent; seeds globose, pitted, minutely arillate.

Common along roadsides and wastelands in Karimnagar, Nalgonda and Mahabubnagar districts.

Fl & Fr.: Throughout the year.

Kambalpally (NLG), *BVRS* 220 (NU).

Note: Following Prain (1911), Sagun and van Welzen (2002) treat *Micrococca* and *Claoxylon* as independent genera. N.P.Balakrishnan and Chakraborty (2007) are of the the opinion that these two should be treated as single genus *Claoxyl*on.

CLEISTANTHUS Hook.f. ex Planch.

Cleistanthus collinus (Roxb.) Benth. ex Hook. f., Fl. Brit. India 5: 1887; Gamble, Fl. Madras 2: 1282; N.P.Balakr. & Chakrab., Fam. Euphorb. India 324. 2007; Chakrab. & *al.* in N.P.Balakr. & *al.*, Fl. India 23: 387. 2012. *Clutia collina* Roxb., Pl. Coromandel 2:37, t. 169. 1802 ('*Cluytia*').

Vern.: *Lodisa, Vodisa, Nalla-Kordtha*; Eng.: Star gooseberry.

Deciduous, monoecious tree, up to 8 m high; bark dark brown, rough; wood dark reddish brown; branchlets terete, glabrous. Leaves coriaceous, elliptic, orbicular or oblong-obovate, 4-12 × 3-7.5, base obtuse, margin entire, apex obtuse or retuse, glabrous, bright green; secondary nerves 5-6 pairs; petiole to 0.7 cm. Flowers yellowish-green in small, axillary, silky clusters; bracts broad-based subulate, 1.5-3 mm long. Male flowers: pedicel 1-2 mm long; sepals 5, occasionally 5, triangular-oblong or lanceolate-oblong, pilose outside; petals linear; disc annular; staminal column ca 1.5 mm long, filaments ca 1.2 mm long, anthers oblong; pistillode ovoid. Female flowers: pedicels 0.5-1.5 mm long; calyx-tube ca 1 mm long; sepals triangular-lanceolate; petals subulate; disc shortly cupular-annular; ovary subglobose to deeply bifid above. Capsule globose, 2 cm across, 3-valved, dark brown, shining; seeds 3, globose, testa smooth, shiny.

Very common in dry forests in Adilabad, Nizamabad, Medak, Warangal and Karimnagar districts.

Fl.: February, Fr.: May onwards dehiscing by next February.

Vempalli (ADB), *GO* & *PVP* 4794; Malkapur RF (NZB), *TP* & *BR* 6106; Ibrahimpet RF (NZB), *BR* & *KH* 9044; Gangapur RF (MDK), *BR* & *CP* 11430; Rayalatipenta (MBNR), *TDCK* 15321; Pakhal (WGL), *ANH* 15933; Kodimial (KRN), *GVS* 20144(MH); Parnasala (KMM), *RCS* 102472 (BSID).

CODIAEUM A. Juss. *nom. cons.*

Codiaeum variegatum (L.) Blume, Euphorb. Gen. Pl. Tent. 33.1824; Hook.f., F Brit. India 5 : 399.1887; T. Chakrab. & N.P.Balakr. in N.P.Balakr. & *al.*, Fl. India 23: 269. 2012. *Croton variegatus* L., Sp. Pl. (ed. 3) 1424. 1764.

Evergreen shrub, 1-2 m high. Leaves oblong-ovate, obovate, spathulate or linear, 20- 25 x 5-6 cm, rounded at base, entire, acute or obtuse, shining green or often variously spotted yellow or red above, pale beneath; petiole 5-6 cm long. Flowers in axillary raceme up to 25 cm long. Male flowers pedicels 1-3 mm long, usually fascicled; sepals ovate-suborbicular, 3-mm long; petals 5-6, minute, dentate at apex; disc glands 5-6, alternating with the petals; stamens 15-25, distinct. Female pedicels short, stout. Fruits not seen.

Grown in gardens as "croton" with several ornamental leaved varities.

Fl. & Fr .: March-May.

CROTON L.

Croton bonplandianus Baill., Adansonia 4: 339. 1864; N.P.Balakr. & Chakrab., Fam. Euphorb. India 212. 2007; T.Chakrab. & N.P.Balakr. in N.P.Balakr. & *al.*, Fl. India 23: 231. 2012. *C. sparsiflorus* Morong. in Ann. New York Acad. Sci. 7: 221. 1893; Gamble, Fl. Madras 2: 1316. 1925.

Vern.: *Galivana chettu, Nepalavemu.*

Herbs of subshrubs, to 60 cm high, monoecious; much branched, branchlets terete. Leaves 2-5.5 × 1-2.5 cm, ovate-lanceolate, base acute-obtuse, margin sinuate, apex acute, dotted stellate hairs below; petiole to 1.5 cm, 2 glandular at top. Flowers in elongated terminal racemes, pedicellate, bracteate; male flowers many, from above the middle; female flowers few, at the lower half. Male flowers 4 mm across; tepals in two whorls, 5 + 5, free; outer (sepals) ovate, acute; inner (petals) oblong, villous; stamens many, anthers 0.5 mm across. Female flowers 2.5 mm across; tepals 5, 1.5 mm, margin villous; stamens many, inserted in a hairy receptacle, filaments free, 0.7 mm, anthers 0.7 mm across; disc of 5 glands, opposite to outer tepals. Female flowers 5.5 mm across; tepals in two whorls, 5 + 5; outer (sepals) 1-2 × 0.4-1 mm, connate at base, elliptic, acute; inner (petals) 1 × 0.3 mm, free, linear, acute; ovary 1.5 mm across, globose, tomentose stellate hairy, styles 3, each 2-fid, near the base; disc of 5 glands. Capsule 6 mm across, subglobose, stellate hairy.

Occasional in waste lands and road sides in all districts.

Fl. & Fr.: August-January.

Kagaznagar (ADB), *GO* 4362; Mosra (NZB), *TP* & *BR* 6072; Sangareddy tank (MDK), *BR* & *CP* 11296; Central University campus (RR), *MSM* 19243; Kudichintalabailu (MBNR), *TDCK* 15374; Jallada – Kothagudem (KMM), *RR* 106086 (BSID); Subedari (WGL), *RKP* 105259 (BSID); Chityal (NLG), *BVRS* 421 (NU).

EUPHORBIA L.

1. Stems not developed above the ground; leaves in rosette at the ground level .... **E. fusiformis**

1. Stems well developed above the ground; leaves distributed along the stems and branches:

    2. Shrubs or small trees; stems and branches thick, fleshy:

        3. Branchlets unarmed, cylindric .... **E. tirucalli**

        3. Branchlets thick, fleshy, armed with stipular thorns:

            4. Branchlets cylindrical:

                5. Leaves 10-16 cm long, subspathulately obovate:

                    6. Shrubs, less than 50 cm tall .... **E. milli**

                    6. Trees, more than 5 m tall .... **E. nivulia**

                5. Leaves 2-4 cm, caducous .... **E. caducifolia**

            4. Branchlets angled or winged

                7. Branches crowded at apex; marbled white-coloured between the wings; flowers and fruits usually not developing .... **E. lactea**

                7. Branches spreading, not marbled white-coloured between the wings; flowers and fruits developing in every season:

                    8. Styles simple:

                        9. Branchlets exclusively three-winged .... **E. lacei**

                        9. Branchlets 3 to 5-angled .... **E.neriifolia**

                    8. Styles 2-fid .... **E. antiquorum**

    2. Herbs, stems and branches not fleshy:

        10. Stipules usually present, interpetiolar; involucral glands with petaloid limbs, though often very small or minute:

            11. Roots or at least root primordia present at all nodes; leaves retuse at apex; stiples connate .... **E. serpens**

            11. Roots or root primordia absent at all nodes; leaves rounded, obtuse, subacute or acute at apex; stipules free:

                12. Limbs of glands indistinct or of same size as the glands:

                    13. Plants prostrate, only branches or tips of branches decumbent; cyathia solitary, paired or 3 together:

                        14. Stems and branches pinkish grey; mature capsules remaining within the involucre ..... **E. thymifolia**

14. Stems, branches and leaves green, glaucous-green, purplish green or blackish green; mature capsules protruding out from the involucre:

15. Stems, branches and leaves blackish or purplish green; capsule prominently keeled, with ciliate hairs confined to keels only .... **E. prostrata**

15. Stems, branches and leaves green to glaucous green; capsule obtusely keeled, glabrous or sparsely hairy all over the surface........ **E. heyneana**

13. Plants erect, decumbent or ascending; cyathia more than 3 in glomerules:

16. Plants glabrous; leaves linear-lanceolate, 3-6 mm wide, glabrous to sparsely pilose .... **E. hyssopifolia**

16. Plants hispid hairy; leaves ovate, oblong-ovate to elliptic-lanceolate, more than 6 mm wide, appressed pubescent:

17. Plants with patent yellowish hairs, mixed with white crisped hairs; internodes ribbed; leaves often with a purple or pink patch in the middle; seeds transversely 3-4-furrowed ..... **E. hirta**

17. Plants with appressed flexuous or crispate white hairs; internodes terete, not ribbed; leaves without any such patch in the middle; seeds smooth or obscurely 1-2-furrowed .... **E. indica**

12. Limbs of glands distinct and a little to much larger than the glands:

18. Leaves obliquely obovate or linear-spathulate; 2 upper limbs of glands distinctly larger than the 2 lower limbs .... **E. rosea**

18. Leaves neither obliquely obovate nor linear-spathulate; limbs of glands equal in size:

19. Cyathia many in each cluster .... **E. corrigioloides**

19. Cyathia solitary, 2 or 3 in each cluster:

20. Floral leaves imbricate and concealing the cyathia:

21. Leaves hairy on both surfaces; limbs of glands laciniate .... **E. cristata**

21. Leaves glabrous on upper surface, hairy on lower surface, limbs of glands sinuate .... **E. elegans**

20. Floral leaves neither imbricate nor concealing cyathia .... **E. sebastinei**

10. Stipules absent or if present never interpetiolar; involucral glands without petaloid limbs:

22. Leaves usually of 3 types; cyathia always arranged in pseudoumbels:

23. Ray leaves connate at base; glands distinctly wavy-undulate at apex .... **E. perbracteata**

23. Ray leaves free; glands entire or slightly denticulate at apex:

24. All leaves almost equal in shape and size .... **E. dracunculoides**

24. Stem leaves, pseudoumbel leaves and ray leaves differing in shape and sizes .... **E. rothiana**

22. Leaves usually of 1 type, rarely 2 types; cyathia never in pseudoumbels:

25. Upper leaves sinuate, yellow-red at above; gland bilabiate, transversely oblong:

26. Floral leaves green with red or white base .... **E. cyathophora**

26. Floral leaves uniformly red .... **E. pulcherrima**

25. Upper leaves entire, uniformly green; gland cupular .... **E. heterophylla**

Euphorbia antiquorum L., Sp. Pl. 450. 1753; Hook.f., Fl. Brit. India 5: 255. 1887; Gamble, Fl. Madras 2: 1277. 1925; Binojk. & N.P. Balakr., Genus *Euphorbia* in India 294. 2010 & in N.P.Balakr. & *al.*, Fl. India 23: 319. 2012.

Vern.: *Bomma jemudu, Pedda jemudu*.

Succulent shrub or small tree, up to 5 m high; branchlets 3 to 5-winged, with stipular divaricate persistent sharp spines. Leaves small, obovate-oblong, 4-12 × 3-6 mm, base cuneate, margin entire, apex rounded or subtruncate, deciduous, subsessile. Cyathia in axillary, lax, dichasial cymes of 3, the central one sessile, the lateral two pedunculate, unisexual or bisexual; involucre 6.5 × 9.5 mm, cupular; glands 5, suborbicular, thick; appendages 0. Male florets in 5 groups, each of 6-8, few sterile; bracteole laciniate; stalk 3 mm; anther 0.7 mm across. Female floret erect; gynophores to 2 mm long, ovary 2.5 mm, bluntly 3-angled, styles 3, each 2-fid at above, recurved, stigma papillose. Capsule 1 cm, trigonous; seeds 2 mm, globose, minutely carunculate

Common in open plains usually in rocky places along foot hills in scrub jungles in Karimnagar, Ranga Reddi, Nalgonda and Mahabubnagar districts.

Fl. & Fr.: December-February.

Kambalpally (NLG), *BVRS* 620 (NU).

Euphorbia caducifolia Haines, Indian Forester 40: 154. 1914; Fischer, Bull. Misc. Inform. Kew 1925: 341. 1925; Gamble, Fl. Madras 2: 1276. 1925; Binojk. & N.P.Balakr., Genus *Euphorbia* in India 298. 2010 & in N.P. Balakr. & *al.*, Fl. India 23: 319. 2012.

Vern.: *Katte jemudu, Kola jamudu.*

Succulent shrub, up to 2 m high; much branched, branches cylindrical, 3-5 cm in diam., smooth, pale green, with silvery-white coating or peels when mature, with stipular divaricate persistent sharp spines. Leaves very small, oblong, 2-4 × 1-5 cm, fleshy, deciduous, subsessile. Cyathia in axillary cymes; solitary, paired or in triads, reddish pink, involucre 3 × 5 mm, cupular; glands 5, suborbicular, thick; appendages 0. Male florets in 5 groups, each of 4-6; bracteole laciniate; stalk 4 mm; anther 1mm across. Female floret laterally pendulous; gynophores to 1.3 mm long, perianth lobes present, 3, minute, triangular subacute, ovary 2.5 × 3.5 mm, sharply 3-angled; styles 3, connate at base, simple, recurved. Capsule 1 cm, trigonus, keeled; seeds ovoid to spherical, 2 mm, smooth, brownish.

Occasional in hill forests in Karimnagar district (Naqvi, 2001).

Fl. & Fr.: March – May.

Euphorbia corrigioloides Boiss. Cent. Euphrob. 11. 1860 & DC., Prodr. 15: 32. 1862; Hook.f., Fl. Brit. India 5: 251. 1887; Gamble, Fl. Madras 2: 1275. 1925; Binojk. & N.P. Balakr., Genus *Euphorbia* in India 153. 2010 & in N.P.Balakr. & *al.*, Fl. India 23: 282. 2012.

Vern.: *Errapala alumu.*

Prostrate herb; branches from a stout root stock. Leaves oblong to elliptic-oblong, 0.6-2 × 0.4-0.7 cm, base obliquely truncate, margin serrulate, apex obtuse, single nerved, blotched with red; petiole 1 mm; stipules triangular, laciniate. Cyathia 70-10, solitary in the upper axils or in terminal clusters; involucre 2.2 x 2 mm, broadly campanulate; lobes laciniate, thinly pubescent; glands 4, equal, suborbicular, in some involucres glands 0; appendages orbicular, subequal, white. Male florets 6-10; bracteole laciniate, pubescent; stalk 0.9 mm; anther 0.4 mm across. Female floret laterally pendulous; ovary 1.2 × 1mm, thinly pubescent, styles 3, each 2-fid near the base. Capsule 2.5 mm, thinly pubescent; seeds 1.5 mm, 4-angled, smooth

A common weed of cultivated fields especially in black cotton soils in Mahabubnagar district.

Fl. & Fr.: July – February.

Euphorbia cristata Heyne ex Roth, Nov. Pl. Sp. 226. 1821; Hook.f., Fl. Brit. India 5: 247. 1887; Gamble, Fl. Madras 2: 1274. 1925; Binojk. & N.P. Balakr., Genus *Euphorbia* in India 136. 2010 & in N.P.Balakr. & *al.*, Fl. India 23: 282. 2012. *Chamaesyce cristata* (Heyne ex Roth) Webster, J. Arnold Arbor. 48: 424. 1967.

Annual, erect, ascending or prostrate and spreading herb, up to 35 cm high, dichotomously branched; branches striate, pubescent. Leaves 0.5-1 × 0.4-0.8 cm, broadly ovate-oblong, base obliquely cordate, margin serrulate, apex acute, opposite, densely crowded at apex, sparsely hairy; petiole 1 mm; stipules ovate, obtuse; floral leaves distichous. Cyathia 1-3, in axillary and subterminal; involucre 2.2 × 4 mm, campanulate; lobes triangular, ciliate; glands 4, unequal, laciniate and ciliate. Male florets 6-8; bracteole ciliate; stalk 1.7 mm; anthers 0.3 × 0.5 mm. Female floret laterally pendulous; ovary 2 × 1.5 mm, pubescent; styles 3, each 2-fid, from near the base. Capsule 4 mm, pubescent; seeds 1.5 mm, 4-angled, furrowed, white.

Occasional in fallow fields, in crevices of exposed rocks and in forest areas in Medak district.

Fl. & Fr.: August-January.

Narsapur RF (MDK), *TP* & *CP* 11975.

Euphorbia cyathophora Murray, Comment. Soc. Regiae. Sci. Gott. 7: 81. 1786; Binojk. & N.P. Balakr., Genus *Euphorbia* in India 355. 2010 & in N.P.Balakr. & *al*., Fl. India 23: 329. 2012. *Poinsettia cyathophora* (Murr.) Klotzsch & Garcke, Monastber. Deutsch. Akad. Wiss. Berlin 1859: 253. 1859. *Euphorbia heterophylla* auct. non L.; Gamble, Fl. Madras 2: 1278. 1925.

Annual or biennial, erect herb, up to 1.5 m high, dichotomously branched; branchlets terete, striate. Leaves 5-7 × 1.5.3 cm, obovate-oblanceolate or panduriform, acute-attenuate at base, entire-serrate, acute, alternate below, opposite-whorled above, floral leaves with scarlet red or yellow patch at basal half; petiole to 1 cm. Cyathia 12-15, in terminal clustered pseudodichasia, each forking ca 4 times; involucre 3.5 × 3.8 mm, campanulate; lobes broadly triangular, laciniate; glands 1, bilabiate, transversely oblong; appendages 0. Male florets 20-30; bracteole laciniate, ciliate; stalk 3 mm; anther 0.8 × 1 mm. Female floret laterally subpendulous, ovary 2 mm; styles 3, each 2-fid, at above. Capsule trigonous subglobose, deeply 3-lobed, 3.5 mm, glabrous or sparsely puberulous; seeds 2.5mm, globose, tuberculate, greyish black.

Occasional weed in moist places, sometimes in the cultivated fields and gardens in most districts.

Fl. & Fr.: June-January.

Mosra (NZB), *TP* & *BR* 6066; Gowraram (NZB), *TP* & *BR* 6324; Narsapur (MDK), *BR* & *CP* 11240; Choutkoor (MDK), *TP* & *CP* 11920; Mekavaripalli (RR), *MSM* & *TP* 11042.

Note: *Euphorbia cyathophora* Murray superficially resembles *E. heterophylla* L. However, the two are distinguished by the former having floral leaves brightly coloured, involucral glands bilabiate and seeds sharply tuberculate and non-glandular, while the later has floral leaves green, involucral glands circular and seeds bluntly tuberculate and angular.

Euphorbia dracunculoides Lam., Encycl. 2: 428. 1788; Hook.f., Fl. Brit. India 5: 262. 1887; Gamble, Fl. Madras 2:1278. 1925; Binojk. & N.P. Balakr., Genus *Euphorbia* in India 355. 2010; & in N.P.Balakr. & *al*., Fl. India 23: 305. 2012.

Vern.: *Tillakada.*

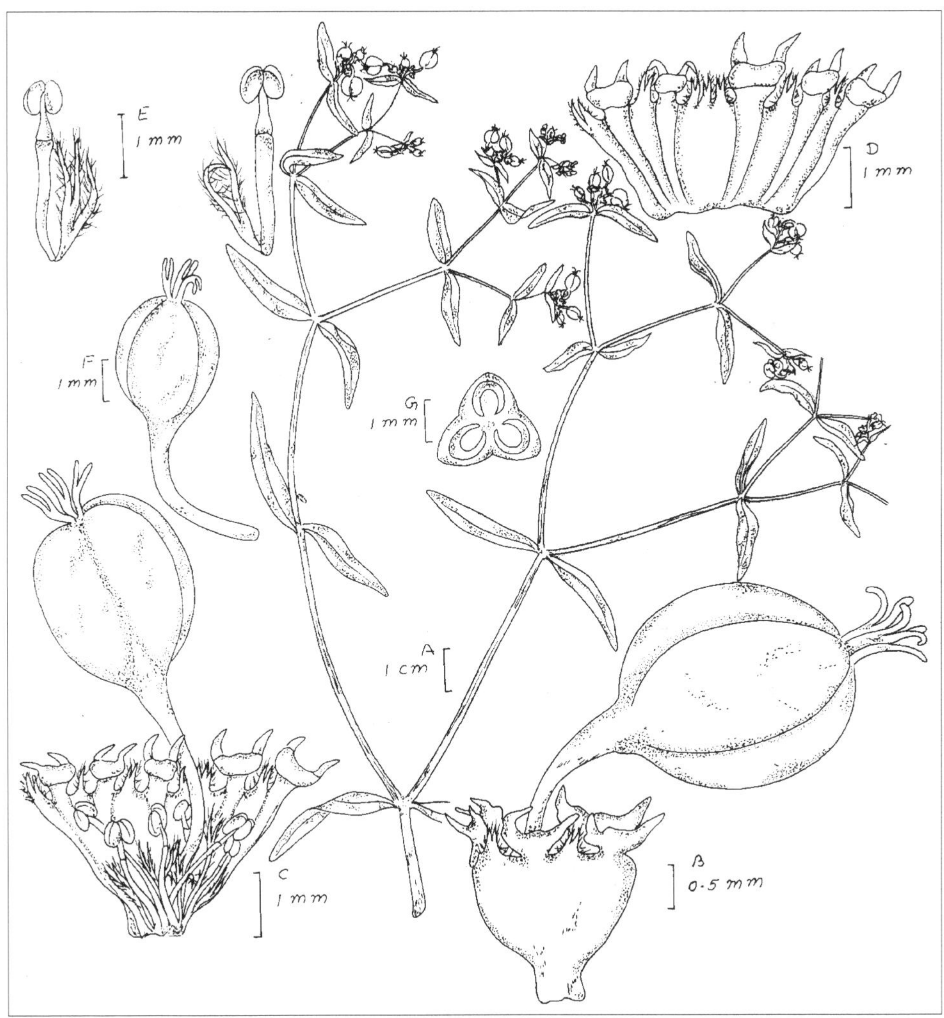

Figure 95: Euphorbia dracunculoides Lam.

A. Twig, B. Cyathium, C. Cyathium split open, D. Involucre split open, E. Stamens with bracteoles, F. Pistil, G. Ovary T.S.

Annual or sometimes perennial herb, to 50 cm high, much dichotomously branched from root stock; branchlets terete, striate. Leaves 1-5 × 0.2-0.5 cm, linear-linear lanceolate, base acute, margin entire, apex subacute, alternate below, opposite above, sessile, palminerved; floral leaves opposite, 2-3 × 0.3-0.6 cm. Cyathia solitary, terminal, thinly pubescent, subsessile; involucre 2.2 × 3.5 mm, broadly campanulate, thinly pubescent within; lobes short, triangular, ciliate; glands 5, semilunate, horned; appendages 0. Male florets 6-8; bracteole laciniate, ciliate; stalk 2.5 mm; anthers 0.5 × 0.6 mm. Female floret erect; ovary 2.3 mm, styles 3, each 2-fid, from above the middle. Capsule subglobose, roundly trigonous, 3.5 mm, keeled, smooth, glabrous; seeds 2.5 mm, ellipsoid, minutely tuberculate.

Common in waste lands and in cultivated fields in Nizamabad, Warangal, Khammam, Karimnagar and Mahabubnagar districts.

Fl. & Fr.: December-March.

Sriramsagar (NZB), *BR* 9004; Sirnapalli fields (NZB), *PSPB* 9164; Aklaspur (KRN), *GVS* 22515 (MH); Laxmipuram (WGL), *PVS* 76937 (BSID); Nellipakka RF (KMM), *RCS* 99026 (BSID).

Euphorbia elegans Spreng. in L., Syst. Veg. (ed. 16) 3: 794. 1826; Hook.f., Fl. Brit. India 5: 246. 1887; Gamble, Fl. Madras 2: 1274. 1925; Binojk. & N.P. Balakr., Genus *Euphorbia* in India 355. 2010 & in N.P.Balakr. & *al.*, Fl. India 23: 284. 2012. *Chamaesyce elegans* (Spreng.) Sojak, Cas. Nar. Muz. (Prague) 140 (3-4): 169. 1972.

Ascending herb, to 40 cm high, dichotomously branched; branchlets clothed with long scattered soft hairs. Leaves opposite, 1.5-2.8 × 0.7-1.5 cm, elliptic-oblong, base obliquely truncate-cordate, margin serrulate, apex obtuse, subsessile; floral leaves distichously imbricate, broadly elliptic or obovate, serrulate, hairy below, forming short spikes at the ends of the branchlets. Cyathia 1-3, axillary or subterminal within the imbricate leaves; involucre campanulate-turbinate, hairy; lobes forked, hairy; glands 4, equal, transversely oblong; appendages unequal, longer than the involucre, hairy within, near the base, sinuate at apex. Male florets 18-20; bracteole laciniate, ciliate; stalk 2.4 mm; anthers 0.4 × 0.6 mm. Female floret laterally pendulous; ovary 0.8 × 1 mm, subglobose, hairy; styles 3, each 2-fid, from near the base, capitate. Capsule 3.5 mm, hairy; seeds 1.5 mm, 4-angled, furrowed.

Occasional in Nalgonda and Mahabubnagar districts.

Fl. & Fr.: December- March.

Nagarjunasagar (NLG), *KMS* 9758 (CAL).

Euphorbia fusiformis Buch.-Ham. ex D. Don, Prodr. 62. 1825; Boiss. in DC., Prodr. 15: 93. 1866; Hook.f., Fl. Brit. India 5: 257. 1887; Binojk. & N.P. Balakr., Genus *Euphorbia* in India 364. 2010 & in N.P.Balakr. & *al.*, Fl. India 23: 332. 2012. *E. acaulis* Roxb., Fl. Ind. 2: 472. 1824; Gamble, Fl. Madras 2: 1276. 1925; Surya Prakash Babu in Pullaiah & Moulali, Fl. Andhra Pradesh 2: 856. 1997.

Vern.: *Barresepu Gadda.*

Herb, stem reduced to a stout underground rootstock, fusiform, 10-15 × 4-6 cm. Leaves in apical rosettes, obovate or oblanceolate, base cuneate, margin crenulate, apex obtuse-subacute, fleshy, subsessile. Cyathia in cymes, few, on long peduncles, appears after the leaves have fallen; bracts at the forks, 2 × 2.5 mm, ovate – lanceolate, scarious; involucre 3 × 3.5 mm, broadly cupular; lobes triangular, fringed; glands 4, equal; appendages minute. Male florets many, bracteolate; stalk 2 mm; anther 1 mm across. Female floret erect - subpendulous; ovary 1.5 mm, styles 3, connate at base, each 2-fid at above. Capsule 6 mm; seeds 2 mm, broadly ovoid, smooth.

Rare in the forest of Warangal and Mahabubnagar districts.

Fl. & Fr.: February – September.

Farahabad (MBNR), *TDCK* 15374; way to Gangaram (WGL), *PVS* 84003 (BSID).

Note: Boissier (1862) treated *E. acaulis* and *E. fusiformis* as distinct species and E. *humilis* Royle as a synonym of *E. fusiformis.* J. D. Hooker (1887), on the contrary, opined that *E. fusiformis* and *E. acaulis* were conspecific; he did not mention *E. humilis,* but also included *E. nana* in the synonymy. Cooke (1906) and Gamble (1925) also followed this treatment. Comparison of the original descriptions and type photographs confirms that *E. fusiformis, E. acaulis* and *E. humilis* are conspecific and that *E. fusiformis* Buch.-Ham. ex D.Don is the earliest validly published name for the species.

Euphorbia heterophylla L., Sp. Pl. 453. 1753; Binojk. & N.P. Balakr., Genus *Euphorbia* in India 355. 2010 & in N.P.Balakr. & *al.*, Fl. India 23: 330. 2012. *E. geniculata* Orteg. Nov. Rar. Pl. Matrit. 18. 1797; Hook.f., Fl. Brit. India 5: 239. 1887; Gamble, Fl. Madras 3: 1884. 1936.

Annual or biennial, erect herb, to 60 cm high, with white latex, dichotomously branched; branchlets terete, striate. Leaves 5-10 × 1.5-5 cm, obovate-rhomboid, base acute, margin serrate, apex acute, sparsely hairy, alternate below, opposite above; petiole to 2.5 cm. Cyathia 6-10, in terminal clusters; involucre 2.5 × 3 mm, campanulate; lobes laciniate, ciliate; gland 1, cupular; appendages 0. Male florets 12-15; bracteole ciliate; stalk 1-2 mm; anther 0.7 mm across. Female floret laterally pendulous; ovary 1.8 mm across; styles 3, each 2-fid, from above the middle. Capsule trigonous-subglobose, 5 mm; seeds 2 mm, 3-angled, tuberculate, brownish-black.

Common weed of wastelands and cultivated fields in all districts.

Fl. & Fr.: Throughout the year.

Janakapur near Asifabad (ADB), *PSPB* 9156; Neredukonda (ADB), *GO* & *DAM* 5107; Gowraram (NZB), *TP* & *BR* 6394; Chowtkoor (MDK), *TP* & *CP* 11920; Medak (MDK), *RG* 104115 (BSID); Mekavaripalli (RR), *MSM* & *PRSR* 11042; Mannanur (MBNR), *TDCK* 15372; Pakhal lake (WGL), *KMS* 13136 (CAL); Hussaisagar (HYD), *KMS* 8027 (CAL).

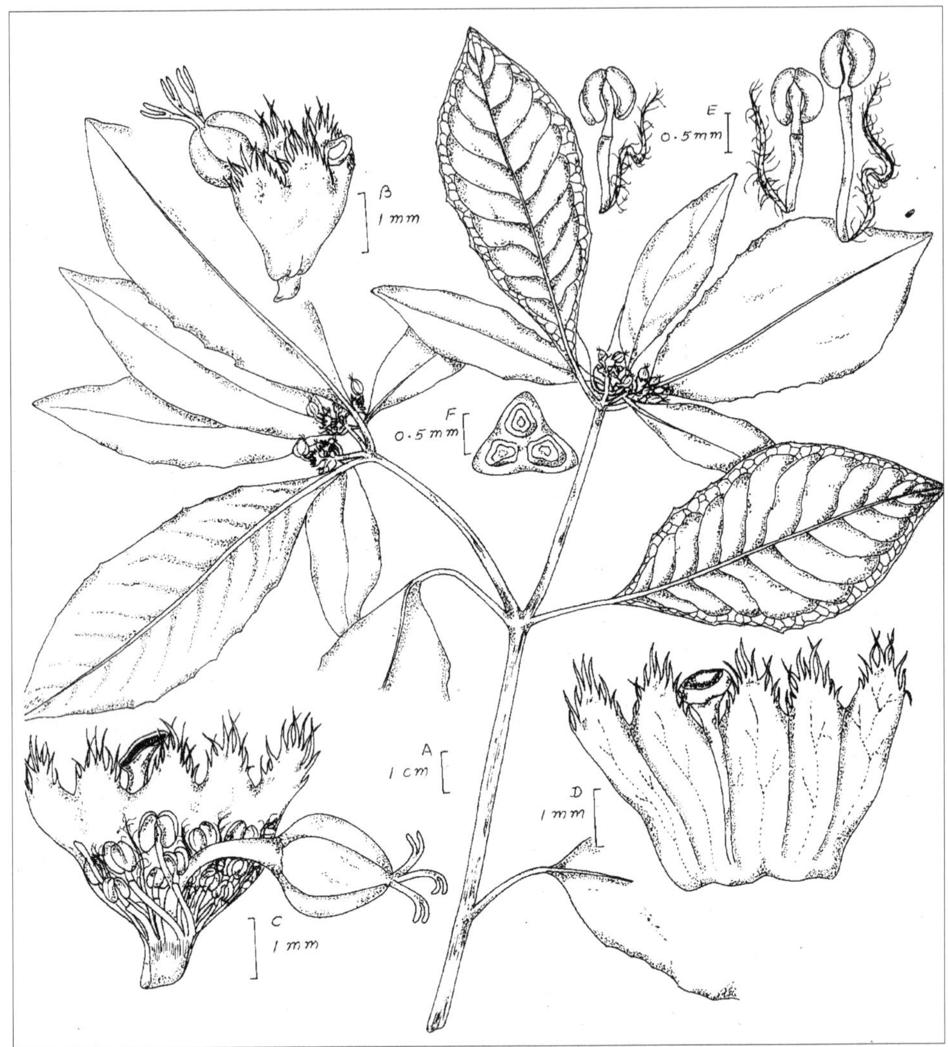

Figure 96: Euphorbia heterophylla L.

A. Twig, B. Cyathium, C. Cyathium split open, D. Involucre split open, E. Stamen with bracteole, F. Ovary T.S.

Euphorbia heyneana Spreng. in L., Syst. Veg. (ed.16) 3: 791. 1831. (as '*heyniana*') var. **heyneana;** Binojk. & N.P. Balakr., Genus *Euphorbia* in India 115. 2010 & in N.P.Balakr. & *al*., Fl. India 23: 286. 2012. *E. microphylla* Heyne ex Roth, Nov. Pl. Sp. 229. 1821, auct. non Lam. 1788; Hook.f., Fl. Brit. India 5: 252. 1887; Gamble, Fl. Madras 2: 1276. 1925. *E. orbiculata* Miq., Fl. Ind. Bot. 1: 421. 1859, (non Kunth 1817). *E. bombaiensis* Santapau, Bull. Bot. Soc. Bengal 8:17. 1955, excl. syn. *Chamaesyce heyneana* (Spreng.) Sojak in Cas. Nar. Muz. (Prague) 140: 169. 1972.

Vern.: *Alumu.*

Postrate herb, up to 15 cm long, stems many, spreading, slender, filiform, glabrous, internodes grey or reddish brown. Leaves opposite, subsessile, 3 × 1.5-3 mm, obovate-rounded or oblong, base obliquely truncate-cordate, margin entire, apex subacute-retuse, stipules 2-partite, laciniate. Cyathia solitary, axillary; involucre 0.9 x 1 mm, campanulate; lobes triangular, apiculate; glands 4; appendages minute, inconspicuous. Male florets 1-4, ebracteolate; stalk 1 mm; anthers 0.3 mm across. Female floret laterally pendulous; ovary 0.4 mm, styles 3, each 2-fid, from above the middle. Capsule 1.5 mm, obtusely keeled; seeds 1 mm, 4-angled, faintly furrowed.

Common weed in moist localities in most districts .

Fl. & Fr. December – February.

Kagaznagar (ADB), *TP* & *GO* 5502; Gannaram (NZB), *TP* & *BR* 6402; Yeddapally (NZB), *BR* & *KH* 6469.

Note: In 1821, Roth established this species basing on Heyne's collection from 'India Orientalis', as *E. microphylla.* As it happened to be a later homonym of *E. microphylla* Lam. (1788), Spreng. (1831) changed the specific epithet to *E.heyneana.* Santapau (1955) proposed a new name *E. bombaiensis* for this species observing that *E. heyneana* Spreng. doubtfully applies to Indian Plants. Rajagopal and Panigrahi (1968) suggested that *E. orbiculata* Kunth as the correct name for this species. However, *E. orbiculata* Kunth is a distinct species, restricted to the New World. Panigrahi (1974) cleared the confusion and recognized *E. heyneana* as the correct name for the Indian plants.

Euphorbia hirta L., Sp. Pl. 454. 1753; Gamble, Fl. Madras 2: 1275. 1925; Binojk. & N.P. Balakr., Genus *Euphorbia* in India 156. 2010 & in N.P.Balakr. & *al*., Fl. India 23: 284. 2012. *E. pilulifera* sensu Hook.f., Fl. Brit. India 5: 250. 1887, non L. 1753.

Vern.: *Reddyavari nanabalu.*

Ascending hispid herb, to 60 cm high; branchlets often 4-angled, densely patently strigose with multicellular yellowish or brownish yellow hairs and also minutely appressed-pubescent with whitish crisped hairs. Leaves opposite, 0.5-3 × 0.6-1.5 cm, elliptic-lanceolate, base oblilquely acute-rounded, margin serrulate-serrate, apex subacute-acute; petiole to 2 mm. Cyathia clustered in dense cymose glomerate heads, axillary, at upper nodes or terminal, often tinged reddish; heads with 20-30 cyathia, single or often paired on ca 1 cm long simple or dichotomous, leafless branches, subsessile or on 2-6 mm long leafless peduncles, male or bisexual; involucre 1 × 0.8

mm, campanulate; lobes triangular, laciniate; glands 4, minute; appendages small. Male florets 4-6, ebracteolate; stalk 1mm; anthers 0.3 mm across. Female floret laterally pendulous; ovary 1 × 0.9 mm; pubescent; styles 3, each 2-fid, from base. Capsule subglobose, sharply trigonous, 1.2-1.5 mm across, pubescent; seeds oblong, 1 mm, 4-angled, faintly furrowed, purplish or reddish grey, grey when mature.

Common weed along roadsides and in open forests in all districts.

Fl. & Fr: Throughout the year.

Jannaram (ADB), *TP* & *PVP* 4208; Kagaznagar (ADB), *GO* 4378; Manchippa RF (NZB), *TP* & *BR* 6026; Pillutla RF (MDK), *BR* & *CP* 11263; Gadirayal RF (RR), *MSM* & *KH* 10264; Nagarjunakonda (NLG), *KT* & *KMS* 9326 (CAL); Miryalaguda (NLG), *BVRS* 486 (NU); Pakhal (WGL), *ANH* 15969 (MH); Kodimial (KRN), *GVS* 20031 (MH); Khammam hills (KMM), *RR* 113808 (BSID); Lawal beat, Pasra (WGL), *RKP* 108164 (BSID); Garlapaadu RF (MBNR), *SRS* 109096 (BSID).

Note: Several floras treat this as either *E. hirta* L. or *E. pilulifera* L., or the latter as a synonym of the former. Boisser (1862), J.D.Hooker (1887), Prain (1903) and others called this species as *E.pilulifera*. Burman (1768), Duthie (1911), Haines (1925) and recently S.Carter (1984), Radcliffe-Smith (1986) and others treat this species as *E. hirta*. Study of the Linnean protologue of *E. pilulifera* reveals that it consists of more than one species, as the reference given by him, 'Burm. Zeyl. 224.t.105.f.1' points to *E. hirta*. Neverthless, the specimen No. 630.8 Linn (LINN) designated as type of *E. pilulifera* differs in several ways from *E. hirta* and from Burman's illustration.

E. hirta resembles *E. indica* and *E.hypericifolia* in the presence of glomerulate cyathia, but differs in its characteristic yellowish hairs.

Euphorbia hyssopifolia L., Syst. Nat. ed. 10. 1048. 1759; Binojk. & N.P. Balakr., Genus *Euphorbia* in India 166. 2010 & in N.P.Balakr. & *al.*, Fl. India 23: 289. 2012. *Chamaesyce hyssopifolia* (L.) Small, Bull. New York Bot. Gard. 3: 429. 1905; C.S.Reddy & V.S.Raju, J. Econ. Taxon. Bot. 26: 195. 2002.

Annual, erect, ascending or decumbent, glabrous herb; stems 2 or 3 branched and woody at base, node slightly thickened. Leaves stipulate, opposite, linear-lanceolate, 1.5-3.5 × 0.3-0.6 cm, glabrous, base inequilateral, serrate towards apex. Cyathia terminal and axillary in short stalked, leafy dichasia; glands sessile; limb of gland obsolete to prominent, white or pink. Male florets: 5-12; anthers red, visible outside. Female floret: gynophore glabrous; ovary glabrous, styles free. Capsules truncate, oblong-ovoid, 1.6 mm long, glabrous, lobed, splitting into 3, two-valved, one-seeded cocci; seeds oblong-ovoid, ca 1 mm long, ca 0.5 wide, 4-angled, faces with transverse ridges, brown or black.

Weed of gardens in Hanmakonda in Warangal district.

Fl. & Fr.: August – April.

Hanmakonda (WGL), *CSR* 1316 (KU), *VSR* 2950 (KU).

Note: C.S.Reddy and V.S.Raju (2002) reported this New World species from Hanmakonda, where it is supposed to be introduced and occur as a garden weed. *E. hyssopifolia* is closely related to *E. hypericifolia* L., both being of New World origin and now introduced to many Asian countries.

Euphorbia indica Lam., Encycl. 2: 423. 1786; V.S.Raju & P.N. Rao, Indian J. Bot. 2(2): 205. 1979; Binojk. & N.P. Balakr., Genus *Euphorbia* in India 168. 2010 & in N.P.Balakr. & *al.*, Fl. India 23: 290. 2012. *E. hypercifolia* auct. non L. 1753; Hook.f., Fl. Brit. India 5: 249. 1887; Gamble, Fl. Madras 2: 1275. 1925. *E. parviflora* auct. non L. 1759; Hook.f., Fl. Brit. India 5: 250. 1887; Surya Prakash Babu in Pullaiah & Moulali, Fl. Andhra Pradesh 2: 865. 1997.

Ascending or erect herb, to 45 cm high, sparsely pubescent; branches arising from rootstock, slender, filiform, fleshy, flexuous, branchlets often quadrangular. Leaves opposite, 1-2.8 × 0.5-1.5cm, elliptic-broadly oblong, obliquely truncate at base, serrulate, acute-obtuse, opposite, sparsely pubescent below; petiole to 4 mm; stipulate. Cyathia 13-15, aggregated in axillary and terminal clusters; involucre 1.3 × 1.2 mm, campanulate; lobes triangular, laciniate; glands 5, minute; appendages much larger than glands. Male floret bracteole laciniate or simple; stalk 1mm; anther 0.35 mm across. Female floret laterally pendulous; ovary 1.5 × 2 mm, hairy; styles 3, each 2-fid, near the base. Capsule 1.5 mm, subglobose, hairy; seeds 1 mm, 4-angled, minutely furrose or smooth.

Common along the forest paths, foothills and in cultivated fields in most districts.

Fl. & Fr.: July-March.

Bijjur RF (ADB), *GO* & *PVP* 4867; Boadh (ADB), *GO* & *PVP* 4638; Kowtal (ADB), *GO* & *PVP* 4830; Yancha (NZB), *PSPB* 9168; Ankole (NZB), *TP* & *BR* 6182; Gangapur (MDK), *TP* & *CP* 12083; Mohammadabad (RR), *MSM* & *KH* 10220; Mannanur (MBNR), *MVS* & *DV* 38638; Nagarjunasagar (NLG), *KMS* 19102 (MH).

Note: The identity of this species, since long, has been a taxonomic problem. Various authors have applied the names *E. hypericifolia, E. indica, E. bracteolaris* and *E. parviflora* to the specimens belonging to *E. indica*. Having failed to solve the problem J.D.Hooker (1887) brought these four names together under *E. hypericifolia* giving Boissier's diagnostic characters for each and left it to Indian Botanists to test their value and make the final decision. According to V.S.Raju and P.N.Rao (1979a), there is only one species in India and it is *E. indica. Euphorbia hypericifolia* L. is a plant of New World Tropics, characterized by glabrous capsules.

Euphorbia lacei Craib, Bull. Misc. Inf. Kew 1911: 456. 1911 & Contr. Fl. Siam. Dicot. 182. 1912; Esser & Chayamarit, Harvard Papers in Bot. 6(1): 262. 2001; Binojk. & N.P. Balakr. Genus *Euphorbia* in India 168. 2010 & in N.P.Balakr. & *al.*, Fl. India 23: 321. 2012. *E. barnhartii* Croizat, Euphorb. Antiq. 54. 1934; Surya Prakash Babu in Pullaiah & Moulali, Fl. Andhra Pradesh 2: 857. 1997. *E. trigona* Roxb., Fl. Ind. 2: 468. 1832 *nom. illeg.*; Hook.f., Fl. Brit. India 5: 256. 1887. *p.p;* Gamble, Fl. Madras 2: 1276. 1925.

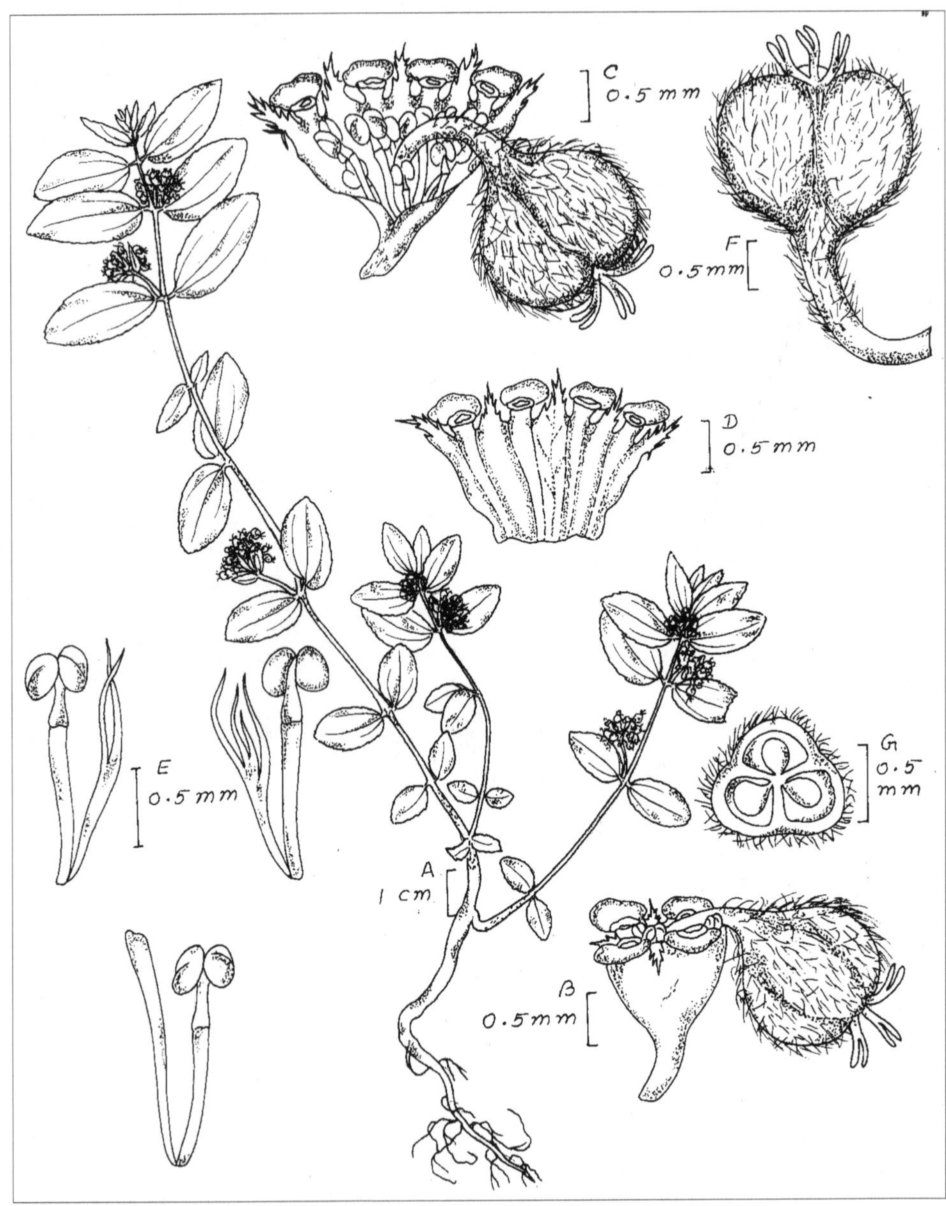

Figure 97: Euphorbia indica Lam.

A. Twig, B. Cyathium, C. Cyathium split open, D. Involucre split open, E. Stamens with bracteoles, F. Pistil, G. Ovary T.S.

Vern.: *Aavijamuku.*

Large succulent shrub or small tree, to 7 m high, with ascending branches, usually leafless; brachlets exclusively 3-winged, wings less broad in proportion to the length of the joints, with stipular divaricate persistent sharp spines. Leaves alternate, sessile, obovate-spathulate, base attenuate, margin entire, apex mucronate, deciduous. Cyathia in axillary, lax cymes; involucre cupular; glands 5, suborbicular, thick; appendages O; Male florets in 5 groups, each of 6-8, few sterile; bracteole laciniate, anthers subglobose, red. Female floret erect; ovary glabrous, styles 3, simple, slightly connate, stigmas emarginated. Capsule ovoid, trigonus, deeply sulcate and sharply keeled; seeds light brown to white, smooth.

Occasional in open plains usually in rocky places and along foothills in scrub jungles.

Fl. & Fr.: December-February.

Boddupalle (NLG), *BVRS* 639 (NU).

Note: Roxburgh in 1832 described this species as *E. trigona*. However, the specific epitet is a later homonym, preoccupied by *E. trigona* Mill. (1768) and Croziat (1934) gave the new name *'barnhartii'*. Eventhough allied to *E. trigona* Mill. in the size and shape of leaves, it differs by the longer slender bending and looping branches and the absence of whitish patches between the wings. Esser and Chayamarit (*loc.cit.*) found that *E. lacei* Craib is identical to *E. trigona* of Roxb. They suggested treating it as the correct name for this species.

Euphorbia lactea Haw., Syn. Pl. Succ. 127. 1812; Binojk. & N.P. Balakr. Genus *Euphorbia* in India 307. 2010 & in N.P.Balakr. & *al.*, Fl. India 23: 322. 2012.

Succulent, spiny shrubs or small trees, up to 10 m high; stems woody, up to 20 cm in diam. at base, older parts terete; branches many, crowded at apex, erect, bushy; young branches 3 (or 4) winged, ascending; wings ca 1 cm thick, wavy-margined, white-marbled or whitissh-greyish at the space between the wings; spines paired, ca 3 mm long, brownish black, persistent. Leaves rarely present, alternate, sessile, scalelike, ovate to orbicular, base rounded to cuneate, margin entire, apex obtuse, 3-5 x 2-3 mm, obscurely nerved, fleshy, caducous. Flowers and fruits not seen.

Usually cultivated in gardens as a fence, sometimes seen as an escape in Central University campus in Ranga Reddi district (Seshagiri Rao, 2012).

Euphorbia milii Des Moulins, Bull. Hist. Nat. Soc. Linn. Bordeaux 1 : 27-30, Pl. 1. 1826; Croizat, J. Arn. Arb. 21 : 506.1940; Binojk. & N.P. Balakr. Genus *Euphorbia* in India 338. 2010 & in N.P.Balakr. & *al.*, Fl. India 23: 328. 2012. *E. splendens* Bojer ex Hook., Bot. Mag. t. 2902; Gamble, Fl. Madras 2 : 1278. 1925.

Woody, somewhat scandent throny shrub, 25-50 cm high; stem terete, armed with stout, 1-2.5 cm long, sharp prickles. Leaves mostly on young shoots, crowded, obovate to oblong-spathulate, 12-15 x 4-5 cm, narrowed and decurrent at base, entire, undulate, obtuse and mucronate at apex, glabrous on both surfaces; petioles short or obscure, 2-3 mm long. Cyathia in long peduncled dichotomous cymes; peduncles 5 cm long, articulated, pinkish, each cyathia closely subtended by two, broadly ovate or suborbicular, bright red bracts, 1.5 x 3 cm across. Fruits not seen.

Grown in gardens, road dividers, lawns, houseyards as an hedge ornamental for its showy bracts

Fl. & Fr .: Almost round the year.

Euphorbia neriifolia L., Sp. Pl. 451.1753; Hook.f., Fl. Brit. India 5: 255. 1887; Gamble, Fl. Madras 2: 1277. 1925; Binojk. & N.P. Balakr., Genus *Euphorbia* in India 310. 2010 & in N.P.Balakr. & *al.*, Fl. India 23: 323. 2012. *E. ligularia* Roxb., Fl. Ind. 2: 465. 1832; Surya Prakash Babu in Pullaiah & Moulali, Fl. Andhra Pradesh 2: 863. 1997.

Vern.: *Aku-jemudu.*

Succulent shrub or small tree, to 6 m high; young branchlets 3-5-angled, with stipular divaricate persistent sharp spines, arranged in 5 spiral rows. Leaves alternate, sessile, 14-21 × 4-6 cm, obovate-oblong or subspathulately obovate, base cuneate, margin entire, apex acute-subacute, fleshy. Cyathia in axillary, subterminal, short pedunculate cymes, central sessile, usually male; involucre 6.5 × 10 mm, broadly cupular, thick; glands 5, transversely oblong, thick; appendages 0. Male florets in 5 groups each of 6-8, few sterile; bracteole laciniate; stalk 1.6 mm; anther 0.7 mm across. Female floret laterally pendulous; ovary 0.5 x 0.8 mm, styles 3, simple, capitate. Capsule 3 mm, 3-lobed; seeds globose, smooth.

Occasional in dry thorny scrubs, widely cultivated as a hedge plant.

Fl. & Fr.: December – February.

Surzapur (ADB), *TP* & *PVP* 4109; Alisagar (NZB), *BR* 7281; Sivanoor (MDK), *BR* & *CP* 11549; Cherukupally (NLG), *BVRS* 651 (NU).

Note: Roxburgh (1832) described *E. ligularia* based on the plants occurring in West Bengal. Boissier (1862), J.D.Hooker (1887) and others found *E.ligularia* Roxb. to be conspecific with *E. nerifolia* L.

Euphorbia nivulia Buch.-Ham., Trans. Linn. Soc. London 14: 286. 1825; Hook.f., Fl. Brit. India 5: 255; Gamble, Fl. Madras 2: 1277. 1925; Binojk. & N.P. Balakr., Genus *Euphorbia* in India 315. 2010 & in N.P.Balakr. & *al.*, Fl. India 23: 324. 2012.

Vern.: *Bontha jemudu, Aku-jemudu, Erra-kalli. Modu-jemudu, Jemudu.*

Armed, deciduous, succulent tree, up to 10 m high; trunk straight; branches whorled, branchlets cylindrical with stipular, well developed, divaricate, persistent, sharp spines on small corky bases arranged in regular rows; wood white, soft, even-grained. Leaves crowded at ends of branches, obovate to oblanceolate, 8-20 × 3-4 cm, subsessile, fleshy, subspathulate, glabrous, apex obtuse, margin entire, base cuneate, nerves not conspicuous. Cyathia in 2-4 cm long, terminal, lax cymes, involucre cup-shaped, transversely oblong, greenish yellow, bracts 2, broadly ovate, scarious, involucral lobes 5, broadly orbicular, glands 5. Male florets in 5 fascicles of 6 to 7 flowers each, anthers subglobose, longitudinally dehiscing. Female florets: ovary subglobose, styles connate up to middle, each bifid at tip, stigma papillose. Capsules trigonous, 6 × 8 mm, glabrous; seeds 4 mm, 4 angular, smooth.

Common along the rocky areas in open dry forests Adilabad, Nizamabad, Karimnagar, Ranga Reddi and Mahabubnagar districts.

Fl.: January-March. Fr.: April-May.

Tiryani RF (ADB), *PSPB* 9159; Laxmapur RF (NZB), *TP* & *BR* 6263; Kanmankalva RF (RR), *MSM* & *KH* 10302; Nilagiri viewpoint (MBNR), *SRS* 109722 (BSID).

Euphorbia perbracteata Gage, Bull. Misc. Inform. Kew 1914: 238. 1914; Binojk. & N.P. Balakr., Genus *Euphorbia* in India 230. 2010 & in N.P.Balakr. & *al.*, Fl. India 23: 313. 2012.

Vern.: *Vathani.*

Erect herb, up to 50 cm high, glabrous. Leaves alternate, sessile, linear-lanceolate or linear-oblong, 2-6 × 0.2-0.5 cm, pale green, base cuneate, margin entire, apex acute. Pseudoumbels 2-4-rayed, pseudoumbel leaves broader than stem leaves, base cuneate, margin entire, apex obtuse or subacute, 3-3.5 × 0.5-1 cm, rays 5-10 cm long, 2-4 times dichotomously branched; ray leaves opposite, connate at base, broadly orbicular, 1-2 × 1-1.5 cm. Cyathia solitary, terminal, sparsely hairy, bracts surrounding cyathia foliaceous, conspicuously veined and often connate at base, involucre turbinate, pubescent outside, involucral lobes 5, triangular, hairy along margin, glands 4. Male florets pedicels ca 2 mm long, glabrous, anthers subglobose. Female floret: gynophores ca 4 mm long, glabrous, ovary subglobose, trigonous, styles to 1 mm long, free, each bifid, stigma capitate. Capsule ovoid, subglobose; seeds subglobose, ovoid.

Weed in wastelands and moist areas in Adilabad and Karimnagar districts.

Fl. & Fr.: January-March.

Way to Chintakani, Mahadevpur (KRN), *NRR* & *TRS* 83199 (BSID); Kamareddy (NZB), *PSPB* 9161.

Euporbia prostrata Aiton, Hort. Kew. 2: 139. 1789; Hook.f., Fl. Brit. India 5: 266. 1887; Gamble, Fl. Madras 2: 1276. 1925; Binojk. & N.P. Balakr., Genus *Euphorbia* in India 121. 2010 & in N.P.Balakr. & *al.*, Fl. India 23: 295. 2012. *Chamaesyce prostrata* (Aiton) Small, Fl. SE U.S.A. 713. 1903. *Euphorbia chamaesyce* auct. plur. non L. 1753; Suryaprakash Babu in Pullaiah & Moulali, Fl. Andhra Pradesh 2: 858. 1997.

Vern.: *Pala alumu.*

Perennial, spreading, prostrate herb, up to 20 cm high, branchlets hispid-pubescent. Leaves opposite, distichous, elliptic to obovate, 3-9 × 1-4 mm, prominently 3-nerved, base obliquely truncate to rounded, margin crenulate, apex obtuse, always with small, axillary leaves, subsessile, stipulate. Cyathia axillary, solitary or 2 or 3 together in microphyllous branches, involucre tubular, 1 × 0.7 mm; glands 4, equal; appendage inconspicuous. Male florets 1-4, ebracteolate; stalk 0.5 mm; anther 0.1 mm, dehiscence transverse. Female floret: laterally pendulous, ovary subglobose, 0.5

mm across; styles 3, forked from base, 0.4 mm, stigma obtuse. Capsule prominently beaked, angles with stiff, spreading hairs; seeds obtusely 4-angled, 1 mm, faintly furrowed, reddish green, pale brown to pale grey.

A common weed of cultivated fields in black cotton soils in Karimnagar, Ranga Reddi, Hyderabad and Mahabubnagar districts.

Fl. & Fr.: July – February.

Note: In certain Indian Floras, the specific epithet used for this species is *E. chamaesyce,* with *E. prostrata* treated as a synonym. The two species are distinct and the former does not occur in India. *E. chamaesyce,* with green or pale green leaves and the cocci of fruits obtusely keeled and hairy throughout is a very common invasive weed found throughout North America and the Mediterranean parts of Europe and so far not been seen in Asia. *E. prostrata* differs from *E. chamaesyce* by the blackish green leaves and the acutely keeled fruits, which are hairy on the keels only. *E. prostrata* is also of American origin, but introduced to Asia long ago and is now widespread in tropical Asia (Binojkumar and Balakrishnan, 2010).

Euphorbia pulcherrima Willd. ex Klotzsch in Otto & Dietr., Neue Allgem. Deutsche Gartenz. 2: 27-28. 1834; Gamble, Fl. Madras 2: 1278. 1925; Binojk. & N.P. Balakr., Genus *Euphorbia* in India 360. 2010 & in N.P.Balakr. & *al.*, Fl. India 23: 331. 2012. *Poinsettia pulcherrima* (Willd.) Graham, Edinb. New Philos. J. 20: 412. 1836 & Bot. Mag. 63: t. 3493. Dec. 1836; Hook.f., Fl. Brit. India 5: 239. 1886.

Shrubs, up to 3 m high, somewhat woody, branchlets hollow. Lower leaves alternate, elliptic-ovate, 5-12 × 3-7 cm, base cuneate and subobtuse, margin entire, undulating, apex acute to acuminate, sparsely hairy; petioles 3-6 cm long, reddish green, canaliculated on upper side. Upper floral leaves alternate or whorled, elliptic, 3-11 × 1-5 cm, bright red or yellow or pale yellow, usually glabrous, petioles 1-3 cm long. Cyathia in leaf-opposed, condensed pseudodichasia; peduncles 2-5 mm long, involucres turbinate, glabrous outside, hairy inside; gland solitary, cupular, yellowish green. Male florets in 5 fascicles of 8-12 flowers each; pedicels 3-4 mm long; filaments red; bracteoles 4-6 mm long, laciniate. Female floret: gynophores 5-8 mm long, pale red; ovary subobvate, hidden within the cyathium, green; styles ca 4 mm long, connate up to middle, each bipartite, recurved. Fruit on deflexed red stalk, trilobite-subglobose, obtuse; seeds ovoid, smooth, pale grey.

Cultivated for its showy foliage as ornamental garden plant; rarely an escape along roadsides.

Fl. & Fr.: December – March.

Euphorbia rosea Retz. Observ. Bot. 4: 26. 1786; Hook.f., Fl. Brit. India 5: 251. 1887; Gamble, Fl. Madras 2: 1275. 1925; Binojk. & N.P. Balakr., Genus *Euphorbia* in India 174. 2010 & in N.P. Balakr. & *al.*, Fl. India 23: 292. 2012. *Chamaesyce rosea* (Retz.) Webster, J. Arnold Arbor. 48: 423. 1967.

Prostrate herb, dichotomously branched; branches brittle, wavy. Leaves 3-12 × 1.5-5 mm, oblong – obovate or linear spathulate, base obliquely cuneate – truncate, margin entire – serrulate, apex obtuse – retuse or rounded, opposite, subsessile, stipules triangular. Cyathia solitary or to 10 in axillary and subterminal lax clusters; involucre 2 × 1.8 mm, subcampanulate, puberulous; lobes triangular, ciliate; glands 4, transversely oblong; appendages of two upper glands obliquely oblong, crenulate; two lower glands suborbicular, short, crenulate. Male florets 4-6; bracteole laciniate; stalk 1 mm; anthers 0.45 mm across. Female floret laterally pendulous; ovary 1.2 mm, puberulous, styles 3, connate at base, each 2-fid, from above to quarter down. Capsule 2.5 mm, puberulous, bluntly keeled; seeds 1.5 mm, 4-angled, transversely furrowed.

Occasional in waste lands in Nizamabad district.

Fl. & Fr.: June – November.

Aṛsapally (NZB), *BR* & *KH* 6489.

Euphorbia rothiana Spreng., Syst. Veg. (ed. 16) 3: 796. 1826; Hook.f., Fl. Brit. India 5: 263. 1887; Gamble, Fl. Madras 2: 1278. 1925; Binojk. & N.P.Balakr., Genus *Euphorbia* in India 237. 2010 & in N.P.Balakr. & *al.*, Fl. India 23: 314. 2012. *E. laeta* Heyne ex Roth, Nov. Pl. Sp. 230. 1821, non Aiton 1789; Surya Prakash Babu in Pullaiah & Moulali, Fl. Andhra Pradesh 2: 863. 1997.

Annual erect herbs, up to 75 cm high, stem glaucous, much branched. Leaves oblong-oblanceolate, 2-4.5 × 0.4-0.7 cm, glabrous, base cuneate, margin entire, apex mucronate, sessile; floral leaves cordate, to 2.2 × 2.4 cm. Cyathia solitary or 2 – 3 in clusters, terminal and at the forks of branches; peduncle to 0.2 cm; involucres to 0.2 cm, 4-glandular. Male florets few; pedicels ca 2 mm long, bracteoles filiform; Female floret: gynophores ca 5 mm long, ovary subglobose, glabrous, styles 1-2 mm long, deeply bifid almost up to base, stigma capitate. Capsules oblong, to 0.4 × 0.3 cm, glabrous; seeds oblong, grooved laterally.

Occasional in cultivated fields in Adilabad and Nizamabad districts.

Fl. & Fr.: December – February.

Bhasar (ADB), *GO* & *PVP* 4641; Kamareddy (NZB), *BR* & *KH* 9628; Kandakurthi (NZB), *BR* 7264.

Note: Roth was the first to describe this species as *E. laeta* in 1821. However, the specific epithet '*laeta*' is pre-occupied in this genus, and hence, Sprengel (1826) changed it to '*rothiana*', although many Indian floras continued to call it *E. laeta* Roth. Raju (1985) clarified the nomenclature of this species.

Euphorbia sebastinei Binojk. & N.P.Balakr., Rheedea 3: 26-28, f.1. 1993 & Genus *Euphorbia* in India 125. 2010 & in N.P.Balakr. & *al.*, Fl. India 23: 297. 2012. *Chamaesyce sebastinei* (Binojk. & N.P.Balakr.) V.S.Raju, J. Econ. Taxon. Bot. 28: 92. 2004.

Annual decumbent herbs, 10-20 cm high; stems many from a thick root stock; branches 3-6 together on each node, almost on all nodes; nodes thickened; internodes 1-1.5 cm long, terete. Leaves subsessile, margin serrate, apex acute or apiculate, 5-10 × 3-5 mm, pubescent on both the surfaces; nerves obscure; stipules triangular, laciniate, ca 2 mm long, hairy; involucre turbinate, ca 1 × 1 mm, hairy; lobes triangular, toothed, exceeding the glands; glands 4, tiny, orbicular, rounded, limbs of glands obovoid, ca 0.5 × 0.8 mm, irregularly wavy at apical margin, white or pink. Male florets pedicel ca 0.5 mm long; anther subglobose, transversely dehiscing; bracteoles filiform, ca 0.7 mm long. Female floret: gynophore ca 1.5 mm long; ovary subglobose, ca 1 × 1 mm, hairy; styles 3, free from base, ca 0.4 mm long, each bifid at apex. Capsules ovoid, ca 1.5 ×1.5 mm, acutely keeled, hairy only on keels, maturing outside the cyathia; seeds oblong, obtusely tetragonous, ca 1 × 0.5 mm, smooth or obscurely 2-3-grooved on faces.

Rare along margins of lakes in Hyderabad.

Fl. & Fr.: July – February.

Hyderbad, Moosi river bank (HYD), *KMS* 5975 (Holotype, CAL; isotypes MH); Hyderabad, Hussain Sagar Jala vihar (HYD), *CSR* & *S.M.M.Javed* 3214 (WWF Herbarium, Hyderabad), 3322 (KU), *CSR* & *C.Pattnaik* 3408 (KU).

Euphorbia serpens Kunth, Nov. Gen. Sp. 2: 52. 1817; Mitra, J. Bombay Nat. Hist. Soc. 68: 854. 1971; Binojk. & N.P. Balakr., Genus *Euphorbia* in India 126. 2010 & in N.P.Balakr. & *al.*, Fl. India 23: 298. 2012. *E. microphylla* auct. non Heyne ex Roth 1821; Hook.f., Fl. Brit. India 5: 252. 1887 *p.p.*, *E. orbiculata* Kunth. var. *jawaharii* Rajagopal & Panigrahi, Taxon 17: 547. 1968.

Annual prostrate herb, many branches spreading from the root, dichotomously branched; branches striate, rooting at nodes. Leaves opposite, subsessile 3-5 × 1-4 mm, subquadrate, base obliquely truncate, margin entire, apex retuse; stipules 2-partite. Cyathia solitary, axillary, terminating in axillary shoots in upper leaf-axils; involucre 0.8 x 1 mm, broadly campanulate; lobes triangular, apiculate, one lobe serpent hooded; glands 4, equal; appendages minute. Male florets 1-4; bracteole laciniate, ciliate; stalk 0.6 mm; anther 0.25 mm across. Female floret laterally pendulous; ovary 0.7 × 0.6 mm; styles 3, each 2-fid, from above the middle. Capsule 1.5 mm; seeds 1 mm, 4-angled, faintly tuberculate.

Common weed in open places, along roadsides, waste lands and other moist areas in fields, dried ponds and low lying banks of rivers in Nalgonda amd Khammam districts.

Fl. & Fr.: June – December.

Billyguttalu (KMM), *RR* 106003 (BSID); Ramanagutta (KMM), *RR* 108003 (BSID); Rayagir (NLG), *BVRS* 19 (NU).

Note: *Euphorbia serpens* Kunth is often confused with *E. microphylla* Heyne but is distinguished by leaves subquadrate to ovate with retuse apex, stipules incised and branches rooting at nodes, while the latter has leaves obovate to oblanceolate with acute apex, stipules laciniate and branches not rooting at nodes.

Euphorbia thymifolia L., Sp. Pl. 454. 1753; Hook.f., Fl. Brit. India 5: 252. 1887; Gamble, Fl. Madras 2: 1276.1925; Binojk. & N.P. Balakr. Genus *Euphorbia* in India 345. 2010 & in N.P.Balakr. & *al.*, Fl. India 23: 292. 2012. *Chamaesyce thymifolia* (L.) Millsp., Pub. Field Columb. Bot. Mus. Ser. 2: 412. 1916.

Vern.: *Erra – chittipayalaku*.

Prostrate herb, forming dense tufts or mats of about 25 cm in diam., partially erect, up to 10 cm high, stems arising from a woody rootstock, slender, dichotomously branched, branches thinly pubescent. Leaves opposite, 3-7 × 1.5-4 mm, elliptic – oblong, base rounded – obliquely cordate, margin crenulate – serrate, apex rounded – obtuse, subsessile; stipules fimbriate. Cyathia solitary or 2 – 3 in axils; involucre 1 x 0.8 mm, campanulate; lobes triangular, laciniate; glands 4; appendages minute. Male florets 1-4, anthers oblong-obovte to subglobose, bracteoles setaceous. Female floret: Gynophore to 0.5 mm long, ebracteolate, stalk 1 mm, tomentose; styles 3, each 2-fid, from above the middle. Capsule 1.5 mm, obtusely keeled, keels tomentose; seeds 1 mm, 4-angled, minutely tuberculate.

Common in dry areas along roadsides, rocky areas and as a weed in gardens in most districts.

Fl. & Fr.: August – March.

Kadam dam (ADB), *GO* & *BR* 9417; Nizamsagar (NZB), *BR* & *CPR* 9246; Sirnapalli river (NZB), *BR* & *GO* 9082; Narsapur (MDK), *BR* & *CP* 11231; Srinivasapur (NLG), *BVRS* 408 (NU); Kodimial (KRN), *GVS* 20040 (MH).

Euphorbia tirucalli L., Sp. Pl. 452. 1753; Hook.f., Fl. Brit. India 5: 254. 1887; Gamble, Fl. Madras 2: 1276. 1925; Binojk. & N.P. Balakr., Genus *Euphorbia* in India 345. 2010 & in N.P.Balakr. & *al*., Fl. India 23: 335. 2012.

Vern.: *Tirukalli*.

Succulent shrub or small tree, to 5 m high; unarmed, branchlets terete glabrous, articulate. Leaves 0.4-1.5 × 0.1-0.2 cm, linear, cuneate at base, entire, obtuse – acute, caducous, subsessile, early deciduous. Cyathia subsessile, solitary or in small capitate glomerules, congested at the apices of the branchlets, yellow; bracts rounded, involucre 2 × 2.5 mm, cupular, coriaceous; lobes triangular, fringed and ciliate; glands 4-5, orbicular, thick; appendages 0. Male florets 6-8; bracteole laciniate; stalk 1.5 mm; anther 0.45 mm across. Female floret erect; perianth distinctly 3-lobed; ovary 2 × 1.7 mm, pubescent, styles 3, spreading, each 2-fid, from above the middle, recurved. Capsule 0.7 mm, globose, pubescent; seeds 3.5 mm, globose, smooth, buff, speckled brown and with dark brown ventral line.

Common in field hedges near villages in Adilabad, Nizamabad, Medak, Ranga Reddi, Hyderabad, Nalgonda and Mahabubnagar districts.

Fl. & Fr.; January – April.

Bheemaram (ADB), *GO* & *DAM* 5184; Akkannapet (MDK), *BR* & *CP* 11556; Lingampalle (RR), *MSM* & *KH* 10201; Rachakonda (NLG), *BVRS* 539 (NU).

Euphorbia tortilis Rottl. ex Ainslie, Mat. Ind. 2: 424. 1826; Hook.f., Fl. Brit. India 5: 256. 1887; Gamble, Fl. Madras 2: 1277. 1925; Binojk. & N.P. Balakr. Genus *Euphorbia* in India 328. 2010 & in N.P.Balakr. & *al*., Fl. India 23: 326. 2012.

Vern.: *Tirugu jemudu, Naga jemudu*.

Succulent shrub or small tree; up to 3 m high, branches jointed, branchlets 3-winged, twisted, with stipular divaricate persistent sharp spines. Leaves small, 4-6 × 4-5 mm, glabrous, fleshy, obovate – oblong, base cuneate, margin entire, apex subtruncate, Early deciduous, subsessile. Cyathia 3-4, in fascicles, axillary, pale yellowish green, bracts broadly triangular, obtuse at apex, involucres cup-shaped, 4 × 3 mm, lobes 5, laciniate along the margins, glands 5, fleshy, yellow. Male florets in 5 fascicles of 5-9 flowers each, anthers subglobose, transversely dehiscing. Female florets: ovary sharply 3-angled, styles 3, each 2-fid at above. Capsule trigonus; seeds 4-angled, smooth.

Occasional in the plains of scrub jungles in Ranga Reddi district (Seshagiri Rao, 2012)..

Fl. & Fr.: February – May.

Euphorbia trigona Mill., Gard. Dict., ed. 8: Euphorbia no. 3. 1768; Hook.f., Fl. Brit. India 5: 256. 1887 (as *E.trigona* Haw.); Binojk. & N.P.Balakr., Genus *Euphorbia* in India 331. 2010 & in N.P.Balakr. & *al*., Fl. India 23: 327. 2012; Gamble, Fl. Madras 2: 1276. 1925.

Shrubs, up to 2 m high, succulent, green or reddish green; main stems terete, up to 10 cm in diam.; branches many, ascending, appresed to stem, striate upwards, triangular, 3-winged, white-marbled in between the wings; podaria papillate, spines 2, divaricate, straight, 2-3 mm long, brown, persistent. Leaves elliptic to obovate-elliptic, base cuneate, margin entire, apex subacute to acute-apiculate, 1-4 x 0.5-2 cm, green to reddish beneath, persistent; petioles 1-3 mm long. Flowers and fruits not seen.

Usually seen as a garden plant (Binojkumar and N.P.Balakrishnan, *loc cit.*; Seshagiri Rao, 2012).

FALCONERIA Royle

Falconeria insigne Royle, Ill. Bot. Himal. Mts. 354. t. 98. f. 2. 1839; N.P.Balakr. & Chakrab., Fam. Euphorb. India 292. 2007 & in N.P.Balakr. & *al*., Fl. India 23: 347. 2012. *Sapium insigne* (Royle) Trimen, Cat. Pl. Ceylon 83. 1885; Hook.f., Fl. Brit. India 5: 471. 1888; Gamble, Fl. Madras 2 :1346. 1925; Surya Prakash Babu in Pullaiah & Moulali, Fl. Andhra Pradesh 2 : 885. 1997.

Vern.: *Devaswroopi, Garbhasula, Kadura*.

Deciduous tree, up to 15 m high; monoecious; bark grey, smooth, shining; wood white to whitish yellow, grey, straight- grained, medium textured, grey; branches thick, milky poisonous juice. Leaves crowded towards the ends of the branches, thick, elliptic-lanceolate or obovate, 6.5-22 × 4-8 cm, glabrous, bright green, base attenuate, margin crenate-serrate, often with glands on the margin, apex acuminate; secondary nerves 10-16 pairs, slender; petiole to 5 cm, with 2-conspicuous glands at the top. Flowers reddish in terminal, large (ca.12.5 cm long) spikes. Male flowers: above, sessile; calyx lobes 2, orbicular, concave; stamens 2, free, alternating with sepals, exserted. Female flowers: solitary, subsessile; calyx 2-lipped, with 2 or 3 ovate, long pointed sepals; ovary ovoid to obovoid, glabrous, styles 3, shortly recurved; stigma subsessile. Capsule globose, 1 cm dia., fleshy at first, dry and irregularly dehiscent when ripe, packed closely together along the rachis; seeds 3, ovoid, 0.6 cm in dia., smooth, pale.

Rare, in dry deciduous forests in Mahabubnagar district.

Fl.: January-February. Fr.: March.

Pullayapally (MBNR), *MVS* & *MB* 38661.

FLUEGGEA Willd.

1. Branchlets arrested, thorn like (armed); pistillodes 2, free .... **F. leucopyrus**
1. Branchlets unarmed; pistillodes 3, connate at base .... **F. virosa**

Flueggea leucopyrus Willd., Sp. Pl. 4: 757. 1805; Hook.f., Fl. Brit. India 5: 328. 1887; Gamble, Fl. Madras 2: 1296. 1925; N.P.Balakr. & Chakrab., Fam. Euphorb. India 342. 2007; Chakrab. & *al.* in N.P. Balakr. & *al.*, Fl. India 23: 412. 2012. *Securinega leucopyrus* (Willd.) Müll.Arg. in DC., Prodr. 15: 451. 1866; Surya Prakash Babu in Pullaiah & Moulali, Fl. Andhra Pradesh 2: 887. 1997.

Vern.: *Tella purugudu.*

Shrub, up to 5 m high, dioecious; branchlets terete, long thorn like, with leaves and flowers. Leaves 1-2 x 0.4-1.2 cm, obovate, base acute, margin entire, apex acute; petiole to 5 mm. Flowers in axillary clusters; pedicels to 5 mm, filiform; bracts minute. Male flowers 2 mm across; tepals 6, 1 × 0.6 mm, obovate, concave, obtuse; stamens 5, free, filaments 2 mm, anthers 0.5 mm across; disc of glands 5, free; pistillodes 2, 1.3 mm. Female flowers 2.7 mm across; tepals 6, 1.7 × 1 mm, free, obovate, concave, obtuse; ovary 1.8 × 1.5 mm, subglobose, styles 3, each 2-fid. Capsule 5 mm, globose; seeds plano-convex, smooth, brown.

Common along roadsides and also in open places in Hyderabad, Khammam, Nalgonda and Mahabubnagar districts.

Fl. & Fr.: June – October.

Rayalatipenta (MBNR), *TDCK* 16809; Ramanagutta (KMM), *RR* 107992 (BSID); Nagarjunasagar (NLG), *BVRS* 70 (NU).

Flueggea virosa (Roxb. ex Willd.) Voigt, Hort. Suburb. Calcutt. 152. 1845; Gamble, Fl. Madras 3: 1296. 1925; N.P.Balakr. & Chakrab., Fam. Euphorb. India 342. 2007; Chakrab. & *al.* in N.P. Balakr. & *al.*, Fl. India 23: 413. 2012. *Phyllanthus virosus* Roxb. ex Willd., Sp. Pl. 4: 578. 1805. *Securinega virosa* (Roxb. ex Willd.) Baill., Adansonia 6: 334. 1866, quod. syn. tantum. emend.; Surya Prakash Babu in Pullaiah & Moulali, Fl. Andhra Pradesh 2: 887. 1997. *Flueggea microcarpa* Blume, Bijdr. 580. 1826; Hook.f., Fl. Brit. India 5: 328. 1887 *p.p.*

Vern.: *Sulamunata.*

Shrub, up to 5 m high, dioecious; branchlets terete. Leaves 3-10 × 1.5-5.5 cm, variable, elliptic – obovate, cuneate – acute at base, entire, acute; petiole to 7 mm.

Flowers in axillary clusters; pedicels of male to 5 mm, filiform; female to 3 mm. Male flowers 2 mm across; tepals 5, 1.3 × 1 mm, ovate, obtuse, concave; stamens 5, filaments 1.5 mm, free; anthers 0.6 × 0.5 mm; pistillodes 3, 1.2 mm; disc of glands 5, free. Female flowers 1.7 mm across; tepals 5, 1.4 × 1 mm, ovate, obtuse, concave; ovary subglobose; styles 3, each 2-fid. Capsule 4 mm, globose; seeds plano-convex, faintly reticulate, brown.

Occasioinal in scrub and deciduous forests on bare slopes in Adilabad, Nizamabad and Warangal districts.

Fl. & Fr.: June – December.

Wankidi (ADB), *PSPB* 9155; Bheemaram (ADB), *GO* & *DAM* 5174; Mallipet (NZB), *BR* & *CPR* 9232; Pakhal (WGL), *KMS* 13126 (MH); Tekupalli (WGL), *PVS* 84082 (BSID).

GIVOTIA Griff.

Givotia rottleriformis Griff., Calcutta. J. Nat. Hist. 4: 388. 1844; Hook.f., Fl. Brit. India 5: 395. 1888; Gamble, Fl. Madras 2: 1342. 1925; N.P.Balakr. & Chakrab., Fam. Euphorb. India 232. 2007; Chakrab. & N.P.Balakr. in N.P.Balakr. & *al*., Fl. India 23: 264. 2012. *Givotia moluccana* (L.) Sreem., Taxon 24: 696. 1975; Surya Praksh Babu in Pullaiah & Moulali, Fl. Andhra Pradesh 2 : 885. 1997. *Croton moluccanum* L., Sp. Pl 1005. 1753.

Vern.: *Konda ponaku, Tella poliki, Yella dabba*.

Deciduous tree, up to 20 m high; bark smooth, peeling off in circular thick bosses; wood white; branchlets fulvous tomentose. Leaves chartaceous, broadly ovate or orbicular, 5-15 × 5-20 cm, glabrous above, yellowish tomentose beneath, base cordate, margin irregulary dentate, apex acuminate, glandular, 5-nerved; petiole to 10 cm with 2 glands. Flowers yellow in 10-15 cm long, axillary or sub terminal racemose or paniculate cymes. Male flowers: sepals oblong to obovate, cucullate, densely stellate pubescent; petals campanulate, 5-lobed with cuneate and truncate lobes, longer than sepals; disc of 5, free orbicular glands; stamens 8-20, outer ones ca 6 with free filaments and inner ones ca 10 with filaments connate at base and exserted, filaments unequal. Female flowers: sepals and petals as in male; ovary globose, stellate hairy; style 2-forked from base. Drupe oblong or subglobose, to 3 x 2.5 cm, brown, smooth; seed one, globose, hairy.

Occasional, in dry deciduous forests in most districts.

Fl.: May-July. Fr.: March.

Madharam (ADB), *TP* & *GO* 5467; Mothe RF (NZB), *BR* 9065; Narsapur RF (MDK), *TP* & CP 14026; Mohammadabad RF (RR), *MSM* & *KH* 10201; Appapur (MBNR), *VSR, DV* & *BR* 39279; Nagarjunakonda (NLG), *BVRS* 139 (NU); Mannanur (MBNR), *SSR* 15532.

Noite: Sreemadhavan (Taxon 24: 696. 1975) made the new combination *Givotia moluccana* (L.) Sreem. based on *Croton moluccanum* L., Sp. Pl. 1005. 1753, excl. ref. Burman. Radcliffe-Smith (Kew Bull. 22: 504. 1968) rejected the name *Croton*

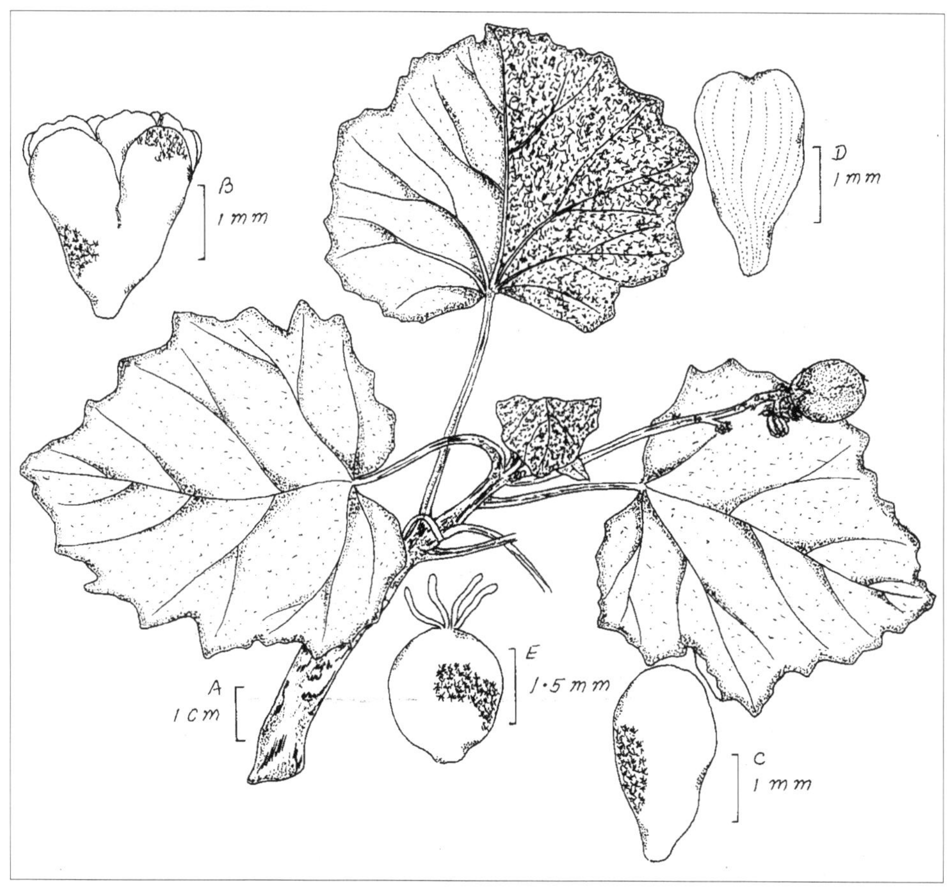

Figure 98: Givotia rottleriformis Griff., [Syn.: *Givotia moluccana* (L.) Sreem.]
B& A. Twig, B. Flower, C. Outer tepal, D. Inner tepal, E. Pistil.

moluccanum as *nomen confusum*, since the reference to Burman given by Linnaeus could not belong to this species. Hence earlier validly published name was adopted by N.P.Balakrishnan and Chakrabarty (2007) and the same is followed here also.

GLOCHIDION J.R.Forst. & G. Forst. *nom. cons.*

Glochidion zeylanicum (Gaertn.) Juss., Euphorb. Gen. Tent. 107. 1824; Hook.f., Fl. Brit. India 5: 310. 1887; Gamble, Fl. Madras 2: 1306. 1925; var. **zeylanicum;** N.P.Balakr. & Chakrab., Fam. Euphorb. India 353. 2007; Chakrab. & *al.*, in N.P.Balakr. & *al.*, Fl. India 23: 429. 2012. *Bridelia zeylanica* Gaertn., Fruct. 2 : 128. t. 109. 1790.

Vern.: *Pageri, Itepulla, Neeru mamidi.*

Tree, up to 8 m high; bark grey, thin, corky; branchlets zig-zag, coppery, glabrous. Leaves coriaceous, ovate-lanceolate or oblong, 13-17.5 × 5-7.5 cm, glabrous, glaucous above, paler beneath, base unequal, truncate-rounded, margin entire, apex sub-acuminate; secondary nerves 6-10 pairs; petiole 5 mm, thick. Flowers greenish, in axillary subumbellate clusters. Male flowers: pedicels 5-10 mm long; sepals 3 + 2-3, oblong, ovate, suborbicular or obovate, anther 3-8. Female flowers: sessile to pedicellate; pedicels 1-8 mm long; sepals 3 + 2-3, triangular, deltoid, oblong, ovate or suborbicular, occasionally subglobose; ovary subglobose, depressed, 4-7-locular, styles free, connivent to erect or united and columnar, conical to clavate. Capsule globose, 0.6 × 0.8 cm, obscurely lobed, glabrous, often beaked; seeds shining, wedge-shaped.

Occasional along the streams of dry deciduous forests in Karimnagar (Naqvi, 2001) and Nalgonda districts.

Fl. & Fr.: Throughout the year.

HOMONOIA Lour.

1. Leaves densely lepidote-scaly beneath; spikes 4-10 cm long .... **H. riparia**
1. Leaves not densely lepidote-scaly beneath; spikes not more than 2 cm long .... **H. retusa**

Homonoia retusa (Graham) Müll.Arg., Linnaea 34: 200. 1865; Hook.f., Fl. Brit. India 5: 456. 1887; Gamble, Fl. Madras 2: 1333. 1925; N.P.Balakr. & Chakrab., Fam. Euphorb. India 137. 2007; Susila Rani & N.P.Balakr. in Balakr. & *al.*, Fl. India 23: 125. 2012. *Adelia retusa* Graham, Cat. Pl. Bombay 185. 1839.

Shrub, up to 2 m high, dioecious; branchlets stout, woody. Leaves 1.5-3.5 × 1-2.5 cm, obovate or obovate – oblanceolate, acute – crenate at base, serrate – dentate, obtuse – retuse, coriaceous, sparsely glandular below; petiole to 2 mm. Flowers in short spikes, sessile; bracts ovate, acute, pubescent. Male flowers 3 mm across; tepals 3, 3 × 2.5 mm, broadly elliptic, concave; stamens many, anthers 0.4 × 0.3 mm. Female flowers 2 mm across; tepals 6, 1.8 × 0.7 mm, linear lanceolate, acute; ovary 1.5 mm, globose, styles 3, simple, papillose. Capsule 3.5 mm, globose; seeds rounded on the back, slightly angular on the inner face, smooth, yellow.

Rare in the river beds of Godavari in Adilabad, Nizamabad, Warangal and Karimnagar districts.

Fl. & Fr.: January – April.

Kadam dam (ADB), *GO* & *BR* 9419; Sriramsagar project (Godavari river bed) (NZB), *BR* 9001; Bhupatipur (Godavari river) (KRN), *GVS* 22481 (MH); Laxmipuram (WGL), *PVS* 76938 (BSID).

Homonoia riparia Lour., Fl. Cochinch 2: 637. 1790; Hook.f., Fl. Brit. India 5: 455. 1887; Gamble, Fl. Madras 2: 1333. 1925; N.P.Balakr. & Chakrab., Fam. Euphorb. India 137. 2007; Susila Rani & N.P.Balakr. in N.P.Balakr. & *al.*, Fl. India 23: 125. 2012.

Vern.: *Taniki chettu.*

Shrub, up to 3 m high, dioecious; branchlets pubescent. Leaves 13-18 × 1-2 cm, linear lanceolate, base acute – rounded, margin entire, apex acute, orbicular, scaleous and pubescent on veins below; petiole to 9 mm; stipules linear – lanceolate, acute. Flowers in axillary spikes, sessile, peduncle pubescent; bracts ovate, acuminate, pubescent. Male flowers 4 mm across; tepals 3, 3 × 2 mm, free, broadly elliptic, concave, acute, puberulous without; stamens many; anthers 0.2 mm across. Female flowers 3 mm across; tepals 5, 1.5 × 0.6 mm, free, ovate, acute – subacuminate, puberulous without; ovary 1 mm, globose, puberulous; styles 3, simple, papillose. Capsule 4 mm, puberulous; seeds rounded on the back, slightly angular on the inner face, smooth, yellow.

Occasional in the rocky crevices of streams and along riverbeds mixed with rocky boulders in Adilabad district.

Fl. & Fr.: November-January.

Potchera stream (ADB), *GO* & *DAM* 5045.

JATROPHA L.

1. Stipules well-developed, divided into narrow lobes or often modified into decurrent series of stalked glands:
    2. Lobes of leaves entire, glandular hairy along margins; flowers reddish or purplish red .... **J. gossypiifolia**
    2. Lobes of leaves serrate, each tooth ending in gland-tipped bristle; flowers green with pinkish tinge:
        3. Leaves shortly palmately lobed up to or above the middle; stipules short, with only a few gland-tipped segments; petals in female flowers connate to one-third the length; staminodes present in female flowers, almost stamen-like but much smaller and sterile .... **J. tanjorensis**
        3. Leaves deeply palmately lobed up to below the middle; stipules prominent, deeply divided into several gland-tipped segments, petals free or almost so to the base in female flowers; staminodes absent in female flower ...... **J. glandulifera**

1. Stipules absent or minute and subulate:
   4. Leaves peltate, 3-7-lobed .... **J. villosa**
   4. Leaves not peltate, 3-7-angular or lobed:
      5. Large branched shrubs; stem base not tuberous; leaves not glaucous beneath; petals united for half the length .... **J. curcas**
      5. Unbranched stunted shrubs; stem base tuberous; leaves glaucous beneath; petals free or connate at base only .... **J. heynei**

Jatropha curcas L., Sp. Pl. 1006. 1753; Hook.f., Fl. Brit. India 5 : 383. 1887; Gamble, Fl. Madras 2 : 1340. 1925; N.P.Balakr. & Chakrab., Fam. Euphorb. India 223. 2007; N.P.Balakr. in N.P.Balakr. & *al.*, Fl. India 23: 250. 2012.

Vern.: *Adavi amudam, Adavi nepalamu, Pedda-nepalamu, Kondamudam, Nepalamu.*

Small tree, up to 5 m high; monoecious; bark greenish-white, smooth, peeling off in thick flakes; wood white, very soft; branchlets glabrous. Leaves orbicular-cordate, 5-12 × 3.5-13 cm, glabrous, shallowly 3 to 5-lobed, lobes triangular, base cordate, margin entire, apex acuminate; petiole to 15 cm. Flowers greenish-white in axillary cymes. Male flowers: sepals ovate or elliptic, 3-4 x 1-2.5 mm; petals 5-10, united up to the middle, oblong or obovate, yellowish green; disc columnar, deeply 5-lobed; stamens 8-10, outer filaments free, inner 5 connate at base. Female flowers: sepals longer than in male flowers; petals absent or same as in male flowers, but free and smaller; ovary 3-locular, styles short and connate at base. Capsule globose, 3 cm across, rugose.

Common, semi-wild in forests and villages in all districts; and also grown as hedge plant along roadsides and out skirts of villages.

Fl.: April-June. Fr.: December-January.

Kadam (ADB), *GO* & *BR* 9411; Kamareddy (NZB), *PSPB* 9160; Maddigunta (NZB), *PSPB* 9162; Mingaram (NZB), *BR* 6315; Narsingi (MDK), *BR* & *CP* 11528; Lankalakunta (RR), *MSM* & *KH* 10273; Farahabad (MBNR), *TDCK* 15377; Yerraguntapalli (NLG), *BVRS* 634 (NU); Kodimial (KRN), *GVS* 20056 (MH).

Jatropha glandulifera Roxb., Fl. Ind. 3: 688. 1832; Hook.f., Fl. Brit. India 5: 382. 1887; Gamble, Fl. Madras 2: 1340. 1925. N.P.Balakr. & Chakrab., Fam. Euphorb. India 224. 2007; N.P.Balakr. in N.P.Balakr. & *al.*, Fl. India 23: 251. 2012.

Vern.: *Dundagaku.*

Shrub, up to 2.5 m high, monoecious; branchlets stout, short. Leaves 7-10 × 8-11 cm, palmately 3 to 5-lobed, posterior lobes smaller than anterior lobe, lobes obovate, base cordate, margin serrate, serratures gland tipped, apex acuminate; petiole to 5.5 cm, with glandular hairs; stipules divided into many glandular headed, filiform segments. Flowers yellow in axillary cymes; bracts linear – lanceolate, acute, glandular hairy. Male flowers: outer (sepals) 3.5 × 1.5 mm, slightly connate at base, lanceolate, obtuse; inner (petals) 4 × 2 mm, slightly connate at base, obovate, rounded, villous at base within; stamens 8, biseriate, 5 + 3; filaments 1.5 mm; anthers 1.1 mm across; disc of glands 5, free. Female flowers 5.5 mm across; tepals in two whorls, 5 + 5; outer

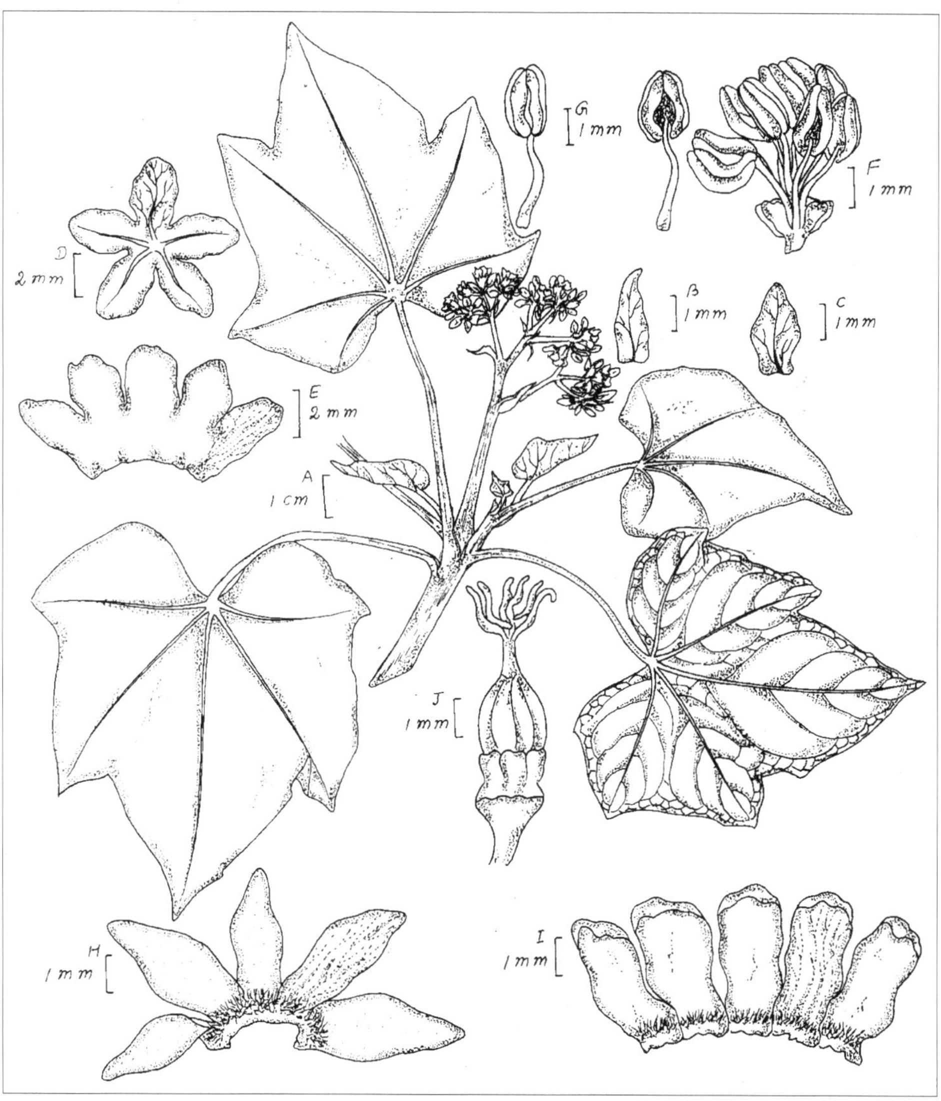

Figure 99: Jatropha curcas L.

A. Twig, B. Bract, C. Bracteole, D. Outer tepals, E. Inner tepals split open, F. Androecium, G. Stamens, H. Outer tepals split open, I. Inner tepals split open, J. Pistil.

(sepals) 2.5 × 1.2 mm, slightly connate at base, lanceolate, acute; inner (petals) 6 × 2.8 mm, slightly connate at base, obovate, rounded, villous at base within; ovary 4.5-3.5 mm; styles 3, each 2-fid, tuberculate; disc of glands 5, free. Capsule 2 cm, subglobse.

A weed of waste places especially in black cotton soils and along roadsides in Karimnagar, Medak, Ranga Reddi and Nalgonda districts.

Fl. & Fr.: August - January.

Central University campus (RR), *MSM* 19211; Medak (MDK), *RG* 104031 (BSID); Yerraguntapalli (NLG), *BVRS* 635 (NU).

Key to Varieties

Jatropha gossypiifolia L., Sp. Pl. 1006. 1753 ("*gossypifolia*")

1. Young leaves dark purple or bronze red, gradually becoming reddish green on maturity, glaucous, sparingly hairy at the divergence of primary veins; all parts including stalks of viscid glands purplish; styles connate at base; capsule smooth or sparingly hirsutulous .... var. **elegans**

1. Young leaves purplish green or greenish as they mature; densely hairy all over; all parts including stalks of viscid glands greenish; styles not forming a column, free; capsule densely hirsutulous .... var. **gossypifolia**

Jatropha gossypiifolia L. var. **elegans** (Pohl) Müll.Arg. in DC., Prodr. 15 (2): 1087. 1866. N.P.Balakr. & Chakrab., Fam. Euphorb. India 224. 2007; N.P.Balakr. in N.P.Balakr. & *al.*, Fl. India 23: 252. 2012. *Adenoropium elegans* Pohl, Pl. Bras. Icon. Descr. 1: 15. 1827; Hook.f., Fl. Brit. India 5: 383. 1887; Gamble, Fl. Madras 2: 1340. 1925.

Shrub, to 3 m high, monoecious; branchlets dark purple, glandular hairy. Leaves 4-10 × 4-11 cm, glaucous, cordiform, deeply 3 to 5-lobed, lobes obovate, base cordate, margin entire, glandular hairy, apex acute; petiole to 7 cm, with many fasicled and branched gland tipped hairs; stipules lanceolate, glandular; young leaves including petioles and glands dark purple, gradually becoming green. Flowers red in axillary cymes; pedicel to 5 mm; bracts lanceolate. Male flowers 6.5 mm across; tepals in two whorls, 5 + 5; outer (sepals) 3.5 × 1.1 mm, connate at base, lanceolate, acuminate, glandular hairy; inner (petals) 5 × 3 mm, free, campanulate, ovate, obtuse; stamens 8, biseriate, 5 + 3, filaments connate at base, anthers 0.9 mm across. Female flowers 7.5 mm across; tepals in two whorls, 5 + 5, outer (sepals) 6.5 × 2 mm, connate at base, lanceolate, acuminate, glandular hairy; inner (petals) 5.5 × 3.5 mm, free, campanulate, ovate, obtuse; ovary 2 mm, hairy, styles 3, stylar column short, each 2-fid; disc of glands 5, free. Capsule 2 cm, subglobose, sparingly hirsutulous; seeds oblong.

Common in waste lands along road sides in Adilabad, Nizamabad, Medak, Karimnagar, Ranga Reddi, Hyderabad and Nalgonda districts.

Fl. & Fr.: Throughout the year.

Bellampalli (ADB), *PSPB* 9151; Vempalli (ADB), *GO* & *PVP* 4972; Mamidipalli (NZB), *PSPB* 9166; Mosra (NZB), *TP* & *BR* 6061; Ponnala (MDK), *CP* 11363; Dharur (RR), *TP* & *MSM* 11099; Farahabad (MBNR), *TDCK* 16828; Bhongir fort (NLG), *BVRS* 4 (NU); Kodimial (KRN), *GVS* 20107 (MH); Tupakulagudem (WGL), *PVS* 76978 (BSID); Parnasala (KMM), *RCS* 104239 (BSID).

Jatropha gossypiifolia L. var. **gossypiifolia**. N.P.Balakr. in N.P.Balakr. & *al.*, Fl. India 23: 252. 2012. *Adenoropium gossypiifolium* (L.) Pohl, Pl. Bras. Icon. Descr. 1: 16. 1827. *Jatropha staphysagrifolia* Mill., Gard. Dict. ed. 8. 1768. *J. gossypiifolia* L. var. *staphysagriaefolia* (Mill.) Müll.Arg. in DC., Prodr. 15(2): 1087. 1866; Hook.f., Fl. Brit. India 5: 383. 1887; Gamble, Fl. Madras 2: 1340. 1925.

Vern.: *Seema nepalamu.*

Shrub, to 3 m high, monoecious; branchlets green. Leaves densely puberulous, cordiform, deeply 3 to 5-lobed, base cordate, margin entire, glandular hairy, apex acute; petiole with many fascicled and branched, gland tipped hairs; young leaves all parts including petioles and glands green. Flowers in axillary cymes; ovary densely hirsutulous; styles 3, independent. Capsule densely hirsutulous; seeds oblong.

Common in waste lands along road sides, near the tanks.

Fl. & Fr.: Throughout the year.

Amrabad (MBNR), *SKB, BSS* & *VSL* 32555; Hussainsagar banks (HYD), *VSR* 3000 (KUH).

Jatropha heynei N.P.Balakr., Bull. Bot. Surv. Ind. 3: 40. 1962; N.P.Balakr. & Chakrab., Fam. Euphorb. India 225. 2007; N.P.Balakr. in N.P.Balakr. & *al.*, Fl. India 23: 253. 2012. *J. heterophylla* Heyne ex Hook.f., Fl. Brit. India 5: 382. 1887, non Steud. 1840; Gamble, Fl. Madras 2: 1339. 1925.

Vern.: *Karuamudam dumpa.*

Shrub, up to 2 m high, monoecious; branchlets terete; Leaves 7-11 × 3.5-6.5 cm, oblanceolate, base cuneate, shortly decurrent, margin entire, apex acute; petiole to 3.5 cm, stipules linear – lanceolate. Flowers in terminal cymes; bracts lanceolate. Male flowers 4 mm across; tepals in two whorls, 5 + 5; outer (sepals) 3.3 × 1.1 mm, connate at base, lanceolate at base, oblong, obtuse; stamens 8-10; filaments mostly connate, to 2 mm; anthers 1.6 mm across. Female flowers 6 mm across; tepals in two whorls, 5 + 5; ovary subglose, glabrous; styles 3, connate at base, each 2-fid; disc of glands 5, free. Capsule subglobose; seeds ovoid, carunculate.

Occasional in dry deciduous forests in Adilabad, Karimnagar, Nalgonda, Hyderabad and Mahabubnagar districts.

Fl. & Fr.: Throughout the year.

Mallelatheertham (MBNR), *TDCK* 15392; Aklaspur (KRN), *GVS* 20218 (MH); Mudigonda (NLG), *BVRS* 148 (NU).

Note: *Jatropha heterophylla* Heyne ex Hook.f., is a later homonym of *Jatropha heterophylla* Steud. 1840 and hence illegitimate. Balakrishnan (*loc. cit.*) renamed it as *Jatropha heynei* N.P. Balakr.

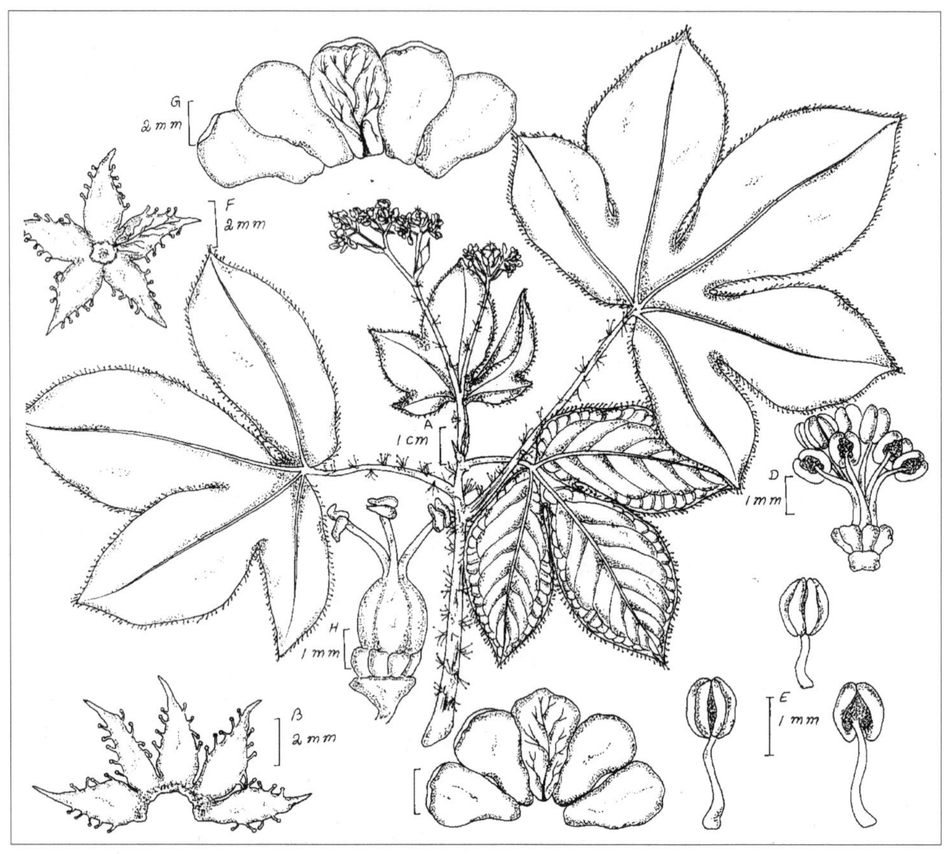

Figure 100: Jatropha gossypiifolia L. var. **elegans** (Pohl) Müll.Arg.

A. Twig, B. Outer tepals split open, C. Inner tepals split open, D. Androecium, E. Stamens, F. Outer tepals, G. Inner tepals split open, H. Pistil.

Jatropha tanjorensis Ellis & Saroja, J. Bombay Nat. Hist. Soc. 58: 834. 1961; Dehgan & Webster, Univ. Calif. Publ. 74: 64. 1979; N.P.Balakr. & Chakrab., Fam. Euphorb. India 227. 2007; N.P.Balakr. in N.P.Balakr. & *al.*, Fl. India 23: 254. 2012.

Shrub, up to 1 m high, polygamous; branchlets striate. Leaves 3-8 × 3.5-7 cm, cordiform, shortly 5-lobed, ovate – triangular, base cordate, margin serrate, serratures gland tipped, apex acuminate; sparsely pubescent; petiole to 6 cm, sparsely pubescent; stipules many, fasicled and branched, gland tipped hairs. Flowers in terminal cymes; pedicel to 3 mm, pubescent; bracts lanceolate, glandular hairy. Male flowers 4-5 mm across; tepals in two whorls, 5 + 5; outer (sepals) 4-4.5 × 1.8-2 mm, slightly connate at base, unequal, lanceolate, acuminate, hairy without, glandular at margins; inner (petals) 2.5 × 1.5 mm, slightly connate at base or free, obovate, obtuse, villous near the base within; stamens 8, free, filaments 0.5 mm, anthers 0.9 mm across, bearded. Bisexual flowers 9 mm across, tepals in two whorls, 5 + 5; outer (sepals) 7-9 × 2.5-3 mm, connate at base, unequal, lanceolate, acuminate, glandular and hairy; inner (petals) 5 × 2.5 mm, free, obovate, obtuse, villous near the base within; stamens 8, free, filaments to 1.1 mm, anthers 1 mm across, bearded; ovary 2 × 1.5 mm, subglobose, styles 3, connate at base each 2-fid; disc of glands 5, free. Capsule 3 × 2.5 mm, subglobose.

Occasional along the roadsides near the forest areas in Adilabad district.

Fl. & Fr.: October – April.

Wankidi (ADB), *PSPB* 11801.

Jatropha villosa Wight, Icon. Pl. Ind. Orient. 4(1): 1. 1848; N.P.Balakr. & Chakrab., Fam. Euphorb. India 227. 2007; N.P.Balakr. in N.P.Balakr. & *al.*, Fl. India 23: 256. 2012. *J. wightiana* Müll.Arg. in DC., Prodr. 15(2): 1080. 1866; Hook.f., Fl. Brit. India 5: 383. 1887; Gamble, Fl. Madras 2: 1340. 1925.

Undershrub, up to 2 m high, monoecious; branchlets striate, pubescent. Leaves 4-5.8 × 4-7 cm, cordiform, base cordate, margin entire, apex acute, pubescent along nerves; petiole to 18 cm; stipules deciduous. Flowers in terminal cymes; peduncle pubescent. Male flowers 5 mm across; tepals in two whorls, 5 + 5; outer (sepals) 4 × 2 mm, lanceolate, subacute, thinly pubescent; inner (petals) 3.5 – 1.6 mm, obovate, obtuse – subacute, villous at base within; stamens 5-8, filaments to 1.5 mm, anthers 1.5 × 4 mm; disc of glands 5, free. Female flowers 6 mm across; tepals as in the male flower; ovary 2 mm, globose, thinly pubescent, styles 3, connate at base, each 2-fid. Capsule globose, thinly pubescent.

Occasional in scrub jugles in Medak and Ranga Reddi districts.

Fl. & Fr.: October-December.

Narsapur (MDK), *CP* 11594; Tandur (RR), *MSM* 19244.

MACARANGA Du Petit-Thouars

Macaranga peltata (Roxb.) Müll.Arg. in DC., Prodr. 15: 1010. 1866; Gamble, Fl. Madras 2: 1326. 1925; N.P.Balakr. & Chakrab., Fam. Euphorb. India 142. 2007; Susila Rani & N.P.Balakr. in N.P.Balakr. & *al.*, Fl. India 23: 137. 2012. *Osyris peltata*

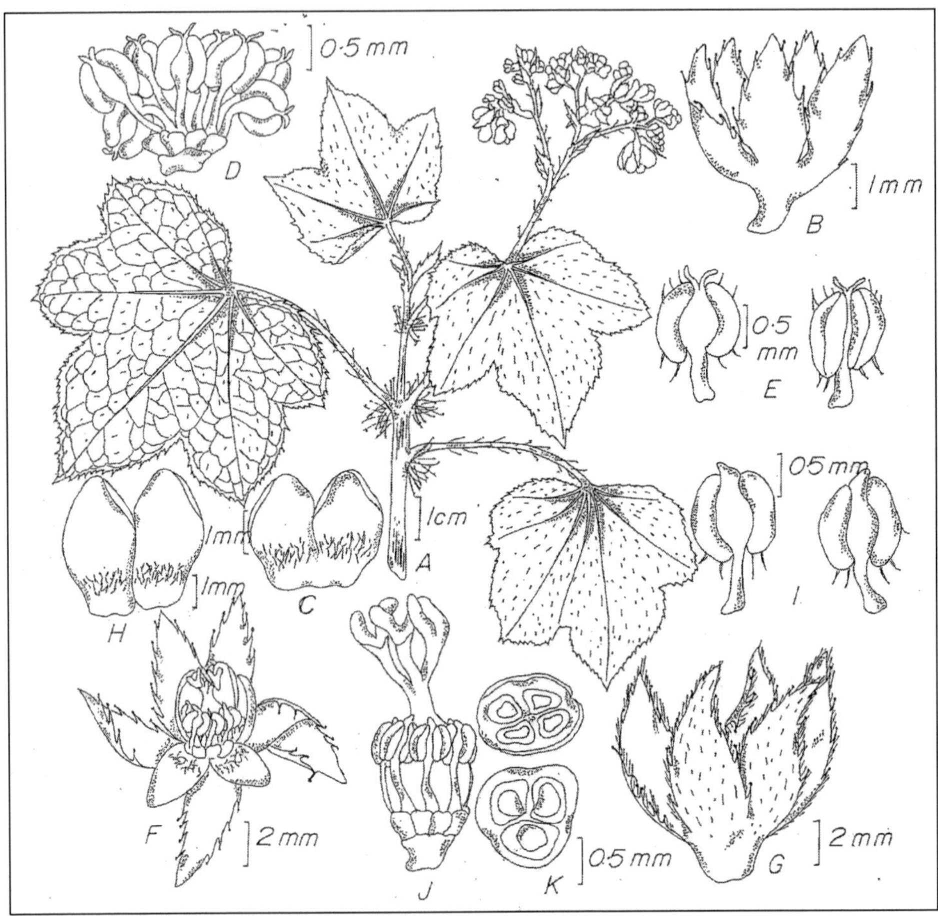

Figure 101: Jatropha tanjorensis Ellis & Saroja

A. Twig, B. Perianth outer lobes, C. Perianth inner lobes (part), D. Androecium, E. Stamens, F. Flower, G. Perianth outer lobes, H. Perianth inner lobes (part), I. Stamens, J. Stamens with pistil, K. Ovary T.S.

Roxb., Fl. Ind. 3:755. 1932. *Macaranga roxburghii* Wight, Icon. Pl. Ind. Orient. t. 5:23. 1852 & t. 1949 f.4.1853; Hook.f., Fl. Brit. India 5: 448. 1887.

Vern.: *Boddi, Kondajaphera, Konda tamara.*

Small or moderate-sized, resinous tree; branchlets glaucous; bark pale, with lenticels; tender parts tomentose. Leaves large, peltate, orbicular-ovate, 10-25 × 8-19 cm, margin usually entire, apex acuminate, more or less hairy on the nerves beneath and glandular, somewhat rusty above or the mature ones almost glabrous, basal nerves 3 strong principal and 6 others, secondary nerves 7-10, ending usually in hirsute glands; petiole usually 7.5-20 cm; stipules ovate to oblong or ovate-lanceolate. Panicles tomentose, mostly from the axils of fallen leaves, male spikes 7.5-10 cm long; bracts large, tomentose, foliaceous, at the base of the branches, orbicular or transversely oblong, denticulate, concealing the flower-clusters, 3.7-7.5 mm broad according to the position; stamens 2-5, usually 3, anthers mostly 4-valved. Female flowers in nearly simple racemes or branched at the base, 3 cm long; bracts large, rusty tomentose, deciduous; pedicels 6.2-15 mm long; tepals very minute, rounded; ovary densely covered with glands, glabrous or only slightly puberulous; stigma lateral, sessile, broadly peltate, papillose. Capsule globose, 6-7.5 mm diam., often grooved, with deciduous yellow glands; seed 1, globose, black.

Occasional in forests in Mahabubnagar district.

Fl.: March, Fr.: April- June.

Saleswaram (MBNR), *KP, VSR* & *BR* 39208.

MALLOTUS Lour.

1. Leaves narrowly oblong-ovate to lanceolate, mostly twice as long as broad, with red granules beneath; stamens 15-30; styles 3; fruits 3-loculed, densely clothed with crimson-red resinous powdery glands .... **M. philippensis**

1. Leaves broadly ovate-deltoid, mostly as long as broad, yellow tomentose and glandular beneath; stamens 40-80; styles 2; fruits 2-loculed, densely fulvous tomentose, not with red resinous powdery glands .... **M. repandus**

Mallotus phillippensis (Lam.) Müll.Arg., Linnaea 34: 1196, 1865; Hook.f., Fl. Brit. India 5: 44. 1887; Gamble, Fl. Madras 2: 1322. 1925; Beddome, Fl. Sylv. S. India t. 289.1872 ('*philippinensis*'); N.P.Balakr. & Chakrab., Fam. Euphorb. India 150. 2007; Susila Rani & N.P.Balakr. in N.P.Balakr. & *al.*, Fl. India 23: 159. 2012. *Croton philippensis* Lam., Encycl. 2: 206. 1786.

Vern.: *Kumkuma chettu, Sindhuramu, Vasantha gandhamu, Chandirmu, Hanumanthuni bottu, Naga-rakta.*

Evergreen tree, up to 12 m high; bark grey, nearly smooth, exfoliating in small woody plates; wood grey to light red, smooth, fairly straight-grained, medium fine and even-textured; branchlets striate, densely grey-tomentose. Leaves oblong-ovate to lancecolate, 4-19 × 2-10 cm, glabrous above, pubescent and numerous red orbicular glands beneath, base cuneate, margin entire, apex acuminate, strongly 3-nerved, secondary nerves 4-7 pairs above basal ones; petiole to 5.5 cm, with 2 small sessile glands one on each side of the summit. Flowers pale yellow in axillary or terminal spikes, male spikes long; female spikes short. Male flowers: bracts triangular, ca 1 mm long; sepals 3 or 4, ovate to elliptic-lanceolate; stamens 15-30, filaments free. Female flowers: bracts triangular; sepals 3-5, connate at base, ovate-lnceolate; ovary 2-3 mm across, densely crimson-red glandular; styles 3, free above, 2-5 mm long. Capsule 3-lobed, red glandular, loculicidally 3-valved, 1 cm across, smooth; seeds globose, black, 4 mm across with a moderately thick brittle testa.

Common, on hills of deciduous forests and open scrub jungles in Mahabubnagar and Warangal districts.

Fl.: October-December. Fr.: February-March.

Appapur (MBNR), *TDCK* 13644; Pakhal (WGL), *KMS* 11636 (MH), *RKP* 108264 (BSID).

Mallotus repandus (Willd.) Müll.Arg., Linnaea 34:197. 1865; Hook.f., Fl. Brit. India 5:442. 1887; Gamble, Fl. Madras 2: 1321. 1925; N.P.Balakr. & Chakrab., Fam. Euphorb. India 150. 2007; Susila Rani & N.P.Balakr. in N.P.Balakr. & *al.*, Fl. India 23: 161. 2012. *Croton repandus* Willd., Ges. Naturf. Freunde Berlin Neue Schriften 4: 206. 1803.

Vern.: *Kanda veltu.*

Large shrub, sometimes scandent or subscandent with thorns on the trunk; branchlets stellate-pubescent or tomentose. Leaves ovate or ovate-deltoid, 5-10 × 4.3-8.7 cm, base cordate, subtruncate, retuse, rarely obtuse, margin entire or somewhat sinuate-toothed, apex acute or acuminate, glabrescent above, densely stellate-pubescent and closely covered with small glands beneath, secondary nerves 2-3, petiole 2.5-6 cm long; stipules minute, tomentose. Inflorescence densely pubescent. Male flowers in terminal panicled racemes; tepals 3-5, yellow, suborbicular or lanceolate, 3 mm long, concave, tomentose, glandular within; anthers with 2 oblong cells, slightly produced beyond the connective. Female flowers green, in simple racemes; tepals linear, caducous; ovary tomentose, 2-lobed and celled; stigma plumose, sessile. Capsule 2-valved, 1.2 cm diam., densely fulvous-tomentose, glandular; seeds subglobose, black.

Occasional in hedges and also in scrub jungles in Warangal district.

Fl.: November-February, Fr.: March-April.

Narasampet (WGL), *KMS* 11588 (MH).

MANIHOT Mill.

1. Shrubs 1.5-2.7 m high, rarely up to 4 m high; leaves epeltate or slightly peltate for less than 5 mm from margin; lobes narrowly oblanceolate, elliptic; flowers in lax racemes; calyx 7-12 mm long, hairy inside; fruits 6-winged .... **M. esculenta**

1. Trees, 5-10 m high; leaves peltate for atleast 4-5 mm from margin; lobes broadly obovate to elliptic; flowers in panicles; calyx 15-18 mm long, glabrous inside; fruits not winged .... **M. glaziovii**

Manihot esculenta Crantz, Inst. Rei. Herb. 1:167. 1766; N.P.Balakr. & Chakrab., Fam. Euphorb. India 230. 2007; Chakrab. & N.P.Balakr. in N.P.Balakr. & *al.*, Fl. India 23: 261. 2012. *M. utilissima* Pohl, Pl. Bras. Ic. 1:32.t.24. 1827; Hook.f., Fl. Brit. India 5: 239. 188; Gamble, Fl. Madras 2: 1346. 1925.

Vern.: *Karra pendalamu.*

Shrub, 1.5-2.7 m high, with large tuberous roots; stems very nodose. Leaves 3-7-palmatipartite; petiole to 12 cm long; segments oblanceolate, 7-12 × 2.5-4 cm, base cuneate, margin entire or sparsely denate, apex acute to acuminate. Flowers in terminal and pseudoaxillary, laxly paniculate or branched cymes, bracts caducous. Male flowers: pedicel 4-7 mm long; calyx 7-12 mm long, campanulate, 5-lobed up to the middle; disc central; stamens 10, free. Female flowers: pedicel 1-2 cm long; calyx as in male, up to 12 mm long; disc pulveniform; ovary with 6 longitudinal ridges, stigma warty. Fruits globose or ellipsoid, muricate-rugulose.

Frequently cultivated.

Fl. & Fr.: February – March.

Note: Variegated type with yellow-green ornamental foliage is often cultivated in the gardens.

Manihot glaziovii Müll.Arg. in C. Martius, Fl. Bras. 11(2): 446. 1874; Gamble, Fl. Madras 2: 1346. 1925; N.P.Balakr. & Chakrab., Fam. Euphorb. India 230. 2007; Chakrab. & N.P.Balakr. in N.P.Balakr. & *al.*, Fl. India 23: 262. 2012.

Vern.: *Siyara-rabbaru, Sima-pittati, Badama.*

Tree, 5-10 m high. Leaf-lobes 3 to 6, elliptic to obovate, 8-18 × 4.5-8 cm, glabrous, dark green above, glaucous beneath, margin entire, apex obtuse with narrow apicule; secondary nerves regular, 12-15 pairs, midvein raised and prominent above; petiole peltate, 5-20 cm. Flowers greenish-yellow in racemose, lax panicles. Male flowers: pedicels ca 7 mm long; calyx 15 - 18 mm long; divided for half way down. Female flowers: pedicels 10-12 mm long; calyx as in male; ovary smooth, stigma sub-penicillate. Capsule green, globose, 3 cm across, pendulous.

Introduced and cultivated, planted in gardens.

Fl.: & Fr.: Throughout the year.

MICROSTACHYS A. Juss.

Microstachys chamaelea (L.) Müll.Arg., Linnaea 32: 95. 1863; Esser, Blumea 44: 176. 1999; N.P.Balakr. & Chakrab., Fam. Euphorb. India 293. 2007; Chakrab. & N.P.Balakr. in N.P.Balakr. & *al.*, Fl. India 23: 348. 2012. *Tragia chamaelea* L., Sp. Pl. 1391. 1753. *Sebastiania chamaelea* (L.) Müll.Arg. in DC., Prodr. 15: 1175. 1866; Hook.f., Fl. Brit. India 5: 475. 1888; Gamble, Fl. Madras 2: 1344. 1925; Surya Prakash Babu in Pullaiah & Moulali, Fl. Andhra Pradesh 2: 886. 1997.

Herb, up to 60 cm high, monoecious; branchlets terete, striate, puberulous. Leaves 2.5-6.5 × 0.3-0.5 cm, linear – lanceolate, cuneate at base, minutely dentate, apiculate; petiole to 0.3 mm, thinly hairy; stipules minute, lanceolate or triangular, ciliate. Flowers in axillary spikes; male flowers many; female flowers solitary at base; bracts lanceolate. Male flowers 1.6 mm across; tepals 5, unequal, ovate – lanceolate, acute, ciliate; stamens 3, free, filaments 0.3 mm, anthers 0.3 mm across. Female flowers 1.9 mm across; tepals 3, ovate or obovate, acute, ciliate; ovary 1.2 mm, echinate, styles 3, simple. Capsule 5 mm, subglobose, 3 cocci, each with 2-vertical rows of spinules; seeds oblong, smooth, yellow.

Common weed along forest paths in most districts.

Fl. & Fr.: July – February.

Penganga river (ADB), *GO* & *PVP* 4281; Vempalli RF (ADB), *TP* & *GO* 5496; Mosra forest nursery (NZB), *TP* & *BR* 6071; Pocharam tank (MDK), *TP* & *CP* 12047; Lingampalle (RR), *MSM* 12766; Kambalpally (NLG), *BVRS* 247; Aklaspur (KRN), *GVS* 22495 (MH); Murugampadu RF (KMM), *RCS* 102439 (BSID); Pakhal (WGL), *RKP* 108269 (BSID).

PEDILANTHUS Neck. *nom. cons.*

Pedilanthus tithymaloides (L.) Poit., Ann. Nat. Hist. 19: 390.t.19.1812; Hook.f., Fl. Brit. India 5:239. 1887; Gamble, Fl. Madras 2: 1346. 1925; N.P.Balakr. & Chakrab., Fam. Euphorb. India 281. 2007; Binojk. & N.P. Balakr. in N.P.Balakr. & *al.*, Fl. India 23: 337. 2012. *Euphorbia tithymaloidea* L., Sp. Pl. 453. 1753.

Vern.: *Kantapala, Seema citramulam*.

Scandent undershrub, up to 1 m high, monoecious; branchlets zigzag, succulent. Leaves 4-11 × 2.5-5.5 cm, brittle, base attenuate, margin entire, apex acute, subsessile. Cyathia terminal; bracts 4, unequal, thin – fleshy, scarlet red, posterior 2 shorter, anterior 2 longer, obtuse, ciliate, caducous; involucre 1.5 cm, obliquely bilateral; glands 4, unequal; appendages 0. Male florets 22-25, ebracteolate; filaments to 1 mm, each 2-fid, above. Female floret: solitary, raised by an erect pedicel, ovary 3-locular, styles stout, cylindrical, ca 7 mm long, stigmas 3, each bifid. Fruits not seen.

Common in plains, in gardens and often in hedges.

Fl.: March, Fr.: not seen.

Mamidipalli (NZB), *BR* & *KH* 9012.

PHYLLANTHUS L.

1. Perennial undershrubs, shrubs, or trees:
    2. Shrubs:
        3. Fruit capsular; filaments free:
            4. Stamens 5 or 6:
                5. Sepals 5; stamens 5 .... **P. tenellus**
                5. Sepals 6; stamens 6:
                    6. Male pedicels up to 9 mm long; seeds hollowed out at hilum .... **P. pinnatus**
                    6. Male pedicels up to 1.5 mm long; seeds not hollowed at hilum: .... **P. leschenaultii**

4. Stamens 3 .... **P. lawii**

3. Fruit baccate; filaments free or in two whorls with inner 3 connate .... **P. reticulatus**

2. Trees:

7. Leaves broadly elliptic to ovate, 5-9 × 2.5-4.5 cm; female disc small or absent; fruit drupaceous, indehiscent .... **P. acidus**

7. Leaves linear-oblong to narrowly elliptic-oblong, 0.8-2.2 × 0.2-0.7 cm; female disc urceolate; fruits fleshy, but fundamentally capsular and at length dehiscent .... **P. emblica**

1. Annual herbs or sub-shrubs:

8. Capsule verrucose:

9. Leaf margin serrulate; stipules lanceolate; filaments connate; seeds transversely furrowed .... **P. urinaria**

9. Leaf margin entire; stipules sagittate-peltate; filaments free; seeds minutely tuberculate .... **P. simplex**

8. Capsule smooth or other than verrucose:

10. Tepals 5; disc glands 5 .... **P. amarus**

10. Tepals 6; disc glands 6:

11. Stamens 3:

12. Proximal leaves of leaf-bearing branches double the size of distal leaves .... **P. airy-shawii**

12. All leaves of branchlets more or less of the same size:

13. Prostrate or slightly ascending herb .... **P. rotundifolius**

13. Erect herbs:

14. Branches winged and hispidulous .... **P. scabrifolius**

14. Branches not winged, glabrous:

15. Stipules peltate, cordate or sagittate at base .... **P. maderaspatensis**

15. Stipules neither peltate nor cordate or sagittate at base .... **P. fraternus**

11. Stamens 5:

16. Male flowers 1-3 .... **P. debilis**

16. Male flowers 5-6 .... **P. rheedii**

Phyllanthus acidus (L.) Skeels, U.S.D.A. Bur. Pl. Industr. Bull. 148: 7.1909; Chakrab. & *al.* in N.P.Balakr. & *al.*, Fl. India 23: 445. 2012. *Averrhoa acida* L., Sp. Pl. 428. 1753. *Cicca acida* (L.) Merr., Interpr. Herb. Amb.314. 1917. *Cicca disticha* L., Mant. 1: 124. 1767. *Phyllanthus distichus* Müll.Arg. in DC., Prodr. 15(2): 413. 1866; Hook. f., Fl. Brit. India 5: 304.

Vern.: *Racha usiri, Chinna usiri.* English: Star gooseberry.

Small to moderate-sized tree; branches thick, tuberculate and closely marked with the scars of the deciduous, leafy, slender branchlets. Leaves ovate or elliptic to obovate 5-7.5 × 2-3 cm, glabrous, pale-glaucous beneath, base rounded; petiole to 3 mm; stipules minute, toothed. Flowers in slender racemes on leafless brancelets. Male flowers red, minute; pedicels 1-2 mm long; sepals 2 + 2, ovate to suborbicular; disc glands 4, suborbicular; stamens 4, free, 2-3 mm long. Female green, larger; pedicels 2-4 mm long; sepals 2 + 2, ovate to suborbicular; disc lobulate; ovary subglobose, styles free, 1.5-2 mm long, deeply bifid, reflexed. Fruit yellowish, depressed-globose, 6-8-lobed, 1.6 cm dia., pericarp fleshy, acidic.

Cultivated in gadens, house compounds etc.

Fl.: May. Fr.: June-July.

Uses: Fruit is sour, edile, cooked or pickled or is made into jam or jelly.

Phyllanthus airy-shawii Brunel & Roux, Nordic J. Bot. 4: 470. 1984; Mitra & Nayar, Taxon 37:470. 1988; N.P.Balakr. & Chakrab., Fam. Euphorb. India 385. 2007; Chakrab. & *al.* in N.P.Balakr. & *al.*, Fl. India 23: 481. 2012. *P. debilis* 'Herb'. Ham. ex Hook.f., Fl. Brit. India 5: 299. 1887, *p.p.* maj. non Klein ex Willd. 1805; Surya Prakash Babu in Pullaiah & Moulali, Fl. Andhra Pradesh 2: 879. 1997.

Herb, 20-60 cm high; stem subcylindric, smooth, first producing 4-5 basal leaves then triangular cataphylls. Cataphylls slightly tinged with anthocyanin, 1.8-2.8 mm long with stipules 1.9-3.1 mm; branchlets glabrous, strong, 15-30 cm long, hardly compressed. Leaves early caducous when dry, ovate to elliptic, 1.2-3 × 0.8-1.5 cm, acute or obtuse, yellowish green; petiole 0.6-1.1 mm; stipules 2.2-3.1 mm. Two to four basal axils of the branchlets each with a male multiflorous cymule, the distal axils with a single female flower. Male flower early caducous after anthesis; pedicel 1-1.3 mm long; tepals 0.6-1 × 0.5-0.6 mm, in 2 unequal whorls; disc of 6 papillose glands alternating with the tepals. Filaments fused into column but distally free; anthers transversely dehiscing. Female flowers: pedicels elongating to 3.1-3.8 mm; tepals 6, 1.5-2.8 × 0.9-1.6 mm, in 2 unequal cycles; disc glandular, annular with denticulate margin; ovary smooth, glabrous; styles bifid. Capsule 2.4-2.8 mm diam; seeds brownish, ungula-shaped, with 6-7 longitudinal ribs on the back.

Common weed in cultivated fields, waste places and road sides.

Fl. & Fr. May-January.

Neelwai RF (ADB), *TP* & *GO* 5425.

Phyllanthus amarus Schumach., & Thonn., Beskr. Guin. Pl. 421 & Kongl. Danske Vidensk. – Selsk. Skr. 4: 195. 1829; N.P.Balakr. & Chakrab., Fam. Euphorb. India 386. 2007; Chakrab. & *al.* in N.P.Balakr. & *al.*, Fl. India 23: 482. 2012. *P. niruri* auct. non L. 1753; Hook.f., Fl. Brit. India 5: 298. 1887 p.p; Gamble, Fl. Madras 2: 1290. 1925. *P. nanus* Hook.f., Fl. Brit. India 5: 298. 1887.

Vern.: *Nela usiri.*

Herb, up to 50 cm high, monoecious; branchlets slender, spreading. Leaves 4-9 × 2-5 mm, elliptic-obovate or oblong, base rounded – obtuse, margin entire, apex obtuse, glaucous below; petiole 0.5 mm, very short; stipules lanceolate, acute. Flowers axillary; male flowers few, 1-3 in in upper axils; female flowers many, solitary in the remaining lower axils. Male flowers 1.2 mm across; tepals 5, 0.8 × 0.4 mm, free, ovate, acute; stamens 3, filaments 0.3 mm, connate, anthers 0.3 mm across; disc of glands 5, minute. Female flowers 1.6 mm across; tepals 5, 0.9 × 0.4 mm, free, ovate-oblong, acute – obtuse, with white central margin; ovary 0.5 mm, 6-lobed, styles 3, each 2-fid, at above; disc of 5 glands. Capsule 2.5 mm, depressed globose, scarcely lobed; seeds trigonus, vertically muricate.

Common weed of open waste lands and in cultivated fields in all districts.

Fl. & Fr.: Throughout the year.

Jannaram (ADB), *TP* & *PVP* 4203; Nadipalli (NZB), *BR* & *KH* 6447; Narsapur (MDK), *BR* 11230; Gadirayal (RR), *MSM* & *KH* 10267; Errapenta (MBNR) *TDCK* 1683; Mannanur RF (MBNR), *SRS* 107447 (BSID); Khammam hills (KMM), *RR* 113806; Hanmakonda (WGL), *RKP* 105366 (BSID).

Phyllanthus debilis Klein ex Willd., Sp. Pl. 4: 582. 1805; Hook.f., Fl. Brit. India 5: 299. 1887; Gamble, Fl. Madras 2: 1290. 1925; N.P.Balakr. & Chakrab., Fam. Euphorb. India 386. 2007. *P. debilis* Wall. ex Hook.f., Fl. Brit. India 5: 299. 1887. *P. niruri* auct. non L. 1753: Hook. f., Fl. Brit. India 5: 298. 1887, *p.p.*

Herb, up to 1 m high, monoecious; branchlets terete. Leaves 1-2 × 0.8-1.2 cm, elliptic-obovate, base rounded-acute, margin entire, apex acute-broadly acute, glaucous below; petioles to 1.5 mm; stipules narrowly triangular, acuminate, scarious. Flowers axillary; male flowers few, 1-3 in the lower axils; female flowers many, solitary in the upper axils. Male flowers 3 mm across; tepals 6, 1 × 0.8 mm, obovate – oblong, subacute, with scarious margin; stamens 3, filaments 0.5 mm, connate, anthers 0.25 mm across; disc of glands 6, stellately lobed. Female flowers 4 mm across; tepals 6, 1.8 × 1 mm, obovate-oblong, acute-subacute, with broad scarious margin; ovary 0.8 mm, globose, 6-lobed; styles 3, each 2-fid, from above the middle; disc of glands 6, stellately lobed. Capsule 4 mm, depressed globose, smooth; seeds trigonus, globose, vertically striate.

Common in shady places in the hill forest and near the foothills in all districts.

Fl. & Fr.: August -January.

Bhimnagar RF (NZB), *TP* & *BR* 6210; Arsapally (NZB), *BR* & *KH* 6488; Narsapur RF (MDK), *TP* & *CP* 14025; Anantagiri RF (RR), *TP* & *MSM* 11058; Kodimial (KRN), *GVS* 20028 (MH); Gangaram (WGL), *PVS* 84015 B.

Note: *Phyllanthus debilis* described by Hook.f. in Fl. Brit. India is *P. airyshawii* Brunel & Roux (Mitra and Nair in Taxon 37: 470. 1988).

Phyllanthus emblica L., Sp. Pl. 982. 1753; Hook. f., Fl. Brit. India 5 : 289. 1887; N.P.Balakr. & Chakrab., Fam. Euphorb. India 369. 2007; Chakrab. & *al.* in N.P.Balakr. & *al.*, Fl. India 23: 446. 2012. *Emblica officinalis* Gaertn., Fruct. 2: 122. t. 108. f. 2. 1790; Gamble, Fl. Madras 2: 1295. 1925.

Vern.: *Pedda usiri chettu, Triphalamu, Pulla usirika,Usiri.* Eng.: Indian gooseberry

Deciduous, monoecious tree, up to 15 m high; branches spreading; bark greyish-white, smooth, peeling off in flakes; wood red or reddish-brown, more or less shallowly and irregularly interlocked grained, coarse and uneven textured; branchlets glabrous and finely pubescent. Leaves oblong, 1.3-2.3 × 0.6-1 cm, glabrous, light green, base truncate, margin entire, apex obtuse, apiculate; petiolule 1 mm. Flowers greenish-yellow in axillary fascicles on the leaf bearing branchlets, often on the naked portion below the leaves. Male flowers: pedicels 1-4.5 mm long; sepals 3 + 3, oblong-obovate; flabellate-obovate to spathulate-oblanceolate, subequal; disc glands 6, clavate or subtriangular; stamens 3, filments connate into a column, anthers connate into a squarish or globose mass. Female flowers: subsessile or pedicels up to 0.5 mm long; sepals 3 + 3, oblong-obovate to oblanceolate; disc shortly cupular-annular, deeply 6-lobed; ovary ovoid, styles 3, 2-3 mm long, shortly connate below into a column, bifid to quadrifid above, recurved. Drupe pale yellow, depressed globose, 1-2 cm in dia., 3-celled, 6-seeded; seeds reniform, 3-gonous, shining, reddish-brown, small, 6 × 7 mm.

Very common, in dry deciduous forests in all districts

Fl.: March-May. Fr.: October-February.

Sirpur RF (ADB), *GO* & *PVP* 4911; Bhimnagar (NZB), *TP* & *BR* 6207; Narsapur RF (MDK), *BR* & *CP* 14104; Anantagiri RF (RR), *MSM* & *PRSR* 12783; Farahabad (MBNR), *TDCK* 13660; Kambalpally (NLG), *BVRS* 621 (NU); Aklaspur (KRN), *GVS* 22514 (MH); Pakhal (WGL), *KMS* 13165 (MH); Bhadrachalam (KMM), *RCS* 104312 (BSID).

Phyllanthus fraternus G.L.Webster, Contrib. Gray Herb. 176: 53. 1955 & J. Arn. Arb. 38: 309, f. 19G & H. 1957; N.P.Balakr. & Chakrab., Fam. Euphorb. India 387. 2007; Chakrab. & *al.* in N.P.Balakr. & *al.*, Fl. India 23: 485. 2012. *P. niruri* auct. plur. non L. 1753: Hook.f., Fl. Brit. India 5: 298. 1887, *p.p.*

Slender, erect, annual herb, 30-40 cm high. Leaves elliptic-oblong to elliptic-oblanceolate, 5-13 × 1.5-5 mm, obtuse or rounded at apex and base, lateral nerves 4-7 pairs, indistinct; petioles 0.5 mm long. Flowers in cymule. Male flowers: pedicels 1 mm long; sepals 6, suborbicular-obovate, disc glands 6, lobulate; stamens 3, filaments united into a short column. Female flowers: pedicels 1.5-2 mm long; sepals 6, oblong-oblanceolate, disc thin, flat, irregularly deeply lobed into 6-10 segments, ovary sessile, subglobose, styles minute, free. Fruits trilobite-subglobose, 1.7-2 mm diam.; seeds longitudinally 7-8-ridged on the back, concentrically 7-8-ridged on the sides.

Weed in waste places and cultivated fields in the plains at low altitudes in Hyderabad and Mahabubnagar districts.

Fl. & Fr.: August – October.

Phyllanthus lawii Graham ex Müll.Arg. in DC., Prodr. 15: 353. 1866; Hook.f., Fl. Brit. India 5: 290. 1887; Gamble, Fl. Madras 2: 1289. 1925; Chakrab. & *al.* in N.P.Balakr. & *al.*, Fl. India 23: 465. 2012.

Shrub, up to 1.5 m high, monoecious; branchlets terete, striate. Leaves 0.3-0.5 × 0.2-0.3 cm, elliptic – oblong, base rounded or subcordate, margin entire, rounded – apiculate; petiole 0.4 mm; stipules lanceolate – subulate. Flowers axillary; male flowers solitary or few; female flowers solitary. Male flowers: 3 mm across; pedicels filiform, ca 5 mm long; sepals 3+3, linear – oblong, obtuse; stamens 3, filaments connate. Female flowers similar to male flowers; ovary globose, styles 3, short, each 2-fid at above, recurved. Capsule globose, slightly depressed; seeds trigonus, smooth.

Occasional along the river banks in Khammam district.

Fl. & Fr.: September – January.

Rathamhutta hills (KMM), *RCS* 98974 (BSID).

Phyllanthus leschenaultii Müll.Arg., Linnaea 32: 37. 1863 & in DC. Prodr. 15(2): 398. 1866. *P. missionis* Hook.f., Fl. Brit. India 5: 297. 1887; Gamble, Fl. Madras 2: 1290. 1925.

Slender, glabrous shrub, 2-2.5 m high. Leaves alternate, very thin, petioled, elliptic-oblong, base acute, apex obtuse. Flowers very small, pedicels much shorter than leaves, nerves 6-8 pairs, distinct, stipules small, ovate-lanceolate. Flowers very small, pedicels much shorter than the leaves; calyx lobes with obscure scarious margins. Male flowers 3-nate; sepals slightly enlarged, but not reflexed in fruit; stamens 3, filaments free; ovary 3-celled, style short, reflexed. Capsule smooth, coriaceous; seeds ribbed.

This taxon was reported from Narsapur town, Medak district by Sebastine & *al.* (1960).

Fl. & Fr.: August – December.

Narsapur town (MDK), *KMS* 7964 (MH).

Phyllanthus maderaspatensis L., Sp. Pl. 982. 1753; Hook.f., Fl. Brit. India 5: 292. 1887; Gamble, Fl. Madras 2: 1289. 1925; N.P.Balakr. & Chakrab., Fam. Euphorb. India 378. 2007; Chakrab. & *al.* in N.P.Balakr. & *al.*, Fl. India 23: 469. 2012.

Vern.: *Nalla usirika.*

Herb, up to 60 cm high, monoecious; branchlets subterete. Leaves 0.2-0.7 × 0.2-0.9 cm, obovate, base cuneate, margin entire, apex retuse – obtuse or rounded – truncate, glaucous below; petiole to 3 mm; stipules peltate, lanceolate, acute. Flowers axillary;

male flowers few, minute, 1-3 in upper axils; female flowers many, solitary, in remaining lower axils, subsessile; bracts 2, narrowly triangular, acute. Male flowers 1.8 mm across; tepals 6, 0.7 × 1 mm, obovate, obtuse, with scarious margin; stamens 3, filaments 0.3 mm, connate, anthers 0.5 mm acorss; disc of glands 6, minute. Female flowers 2.5 mm across; tepals 6, 1.5 × 0.6-1.2 mm, obovate – broadly ovate, acute, unequal; ovary 0.8 mm, globose, styles 3, small, each 2-fid, at above. Capsule 4 mm, depressed globose, 3-lobed; seeds trigonus, vertically muricate.

Common in wastelands near moist shady localities in most districts .

Fl. & Fr.: June – April.

Boadh (ADB), *GO* & *PVP* 4629; Dichpalli (NZB), *BR* & *KH* 6439; Toopran (MDK), *BR* & *CP* 11282; Mekavaripalli (RR), *TP* & *MSM* 11040; Boddupalle (NLG), *BVRS* 643 (NU); Kodimial (KRN), *GVS* 20077 (MH); Kanthikonda (WGL), *RKP* 111538 (BSID); Billyguttalu (KMM), *RR* 106001 (BSID).

Phyllanthus pinnatus (Wight) Webster, J. Arnold Arbor. 38: 52. 1957. in adnot. Chakrab. & *al.* in N.P.Balakr. & *al.*, Fl. India 23: 478. 2012. *Chorisandra pinnata* Wight, Icon. Pl. Ind. Orient. t. 1994. 1853; N.P.Balakr. & Chakrab., Fam. Euphorb. India 382. 2007.

Vern.: *Nalla pooli.*

Large shrub, up to 5 m high, monoecius; branchlets subterete. Leaves variable, elliptic – sub orbicular or obovate, base cuneate, margin entire, apex acute, glaucous below; petiole to 4 mm. Flowers in axillary clusters; male flowers many; female flowers few; pedicels of male to 5 mm, filiform, female to 2 mm. Male flowers 3 mm across; tepals 6, free, ovate, fimbriate, subacute; stamens 6, free, filaments 2 mm, anthers 0.7 mm across; disc of glands 6, free. Female flowers 2 mm across; tepals 6, free, ovate, fimbriate, subacute; ovary 1 mm, globose, 3-lobed, styles 3, each 2-fid, at above; disc of glands 6, connate. Capsule 2.8 mm, depressed globose.

Common in dry deciduous forests and also in open rocky places in Mahabubnagar district.

Fl. & Fr.: June – September.

Phyllanthus reticulatus Poir. in Lam., Encycl. 5: 298. 1804; Hook.f., Fl. Brit. India 5: 288. 1887; N.P.Balakr. & Chakrab., Fam. Euphorb. India 382. 2007; Chakrab. & *al.* in N.P.Balakr. & *al.*, Fl. India 23: 479. 2012. *Kirganelia reticulata* (Poir.) Bail., Etud. Gen. Euphorb. 613. 1858; Gamble, Fl. Madras 2: 1294. 1925.

Vern.: *Nalla purugudu.*

Large shrub, up to 6 m high, monoecious; branchlets subterete, pubescent. Leaves 0.8-1.2 × 0.4-1 cm, elliptic – oblong, base acute, margin entire, apex acute – rounded; petiole 2 mm, pubescent; stipules lanceolate, acute. Flowers axillary; male flowers many, 3-6 in fascicles in upper axils; female flowers few, solitary, in lower axils; pedicels to 5 mm, pubescent; bracts small, lanceolate. Male flowers 2.5 mm across; tepals 5 or 6, obovate-oblong, obtuse, unequal; stamens 5, biseriate, 2 + 3; outer shorter, free, 1.1 mm, inner longer, connate, 1.9 mm; anthers 0.5 mm across; disc of glands 5.

Female flowers 2 mm across; tepals 5 or 6, 1.5 × 1 mm, connate at base, obovate-broadly ovate, obtuse, unequal; ovary 1.5 mm, globose, styles 3, minute, each 2-fid; disc of glands 5. Drupe 5 mm, globose, smooth; seeds trigonus, crustaceous.

Common in hedges and at the foot hills of forests in most districts.

Fl. & Fr.: July – March.

Indanapalli (ADB), *GO* & *DAM* 5409; Wankidi (ADB), *GO* 4470; Banjapalli (NZB), *BR* 9549; Manchippa (NZB), *TP* & *BR* 6006; Marrikunta (MDK), *CP* 11387; Medak (MDK), *BR* & *CP* 11407; Gadirayal RF (RR), *MSM* & *KH* 10249; Farahabad (MBNR), *TDCK* 15360; Subedari (WGL), *RKP* 105254 (BSID); Rollapadu (KMM), *RR* 108079 (BSID); Yadgir (NLG), *BVRS* 548 (NU).

Phyllanthus rheedii Wight, Icon. Pl. Ind. Orient. t. 1895. f.l. 1852; Hook.f., Fl. Brit. India 5: 293. 1887; Gamble, Fl. Madras 2: 1289. 1925; N.P.Balakr. & Chakrab., Fam. Euphorb. India 379. 2007; Chakrab. & *al.* in N.P.Balakr. & *al.*, Fl. India 23: 474. 2012. *P. kozhikodianus* Sivar. & Manilal, J. Indian Bot. Soc. 56: 165-168. ff. 1-6. 1977; C.S.Reddy & *al.*, Suppl. Fl. Andhra Pradesh 107. 2008.

Undershrub, up to 50 cm high, monoecious; branchlets terete, angular, pubescent. Leaves 1-2.3 × 0.4-0.9 cm, elliptic – elliptic oblong, base rounded – acute, margin entire, apex apiculate, thinly pubescent; petiole to 4 mm, pubescent; stipules lanceolate. Flowers axillary; male flowers few, 5-6 in upper axils; female flowers many, solitary, in the remaining lower axils. Male flowers 2.5 mm across; tepals 6, 2 × 1.1 mm, obovate – oblong, subacute; stamens 5, biseriate, 2 + 3; outer shorter, free, inner longer, connate, anthers 0.5 m across; disc of glands 6, free. Female flowers 4 mm across; tepals 6, 2.4 × 1.4 and 2.2 × 1 mm, unequal, obovate – broadly oblong, subacute; ovary 0.5 mm, globose, styles 3, each 2-fid, from above the middle; disc of glands 6, free. Capsule 4-5 mm, globose; seeds striate.

Occasional in the hills of Khammam district.

Fl. & Fr.: October – December.

Phyllanthus rotundifolius Klein ex Willd., Sp. Pl. 4: 584. 1805; Hook.f., Fl. Brit. India 5: 299. 1887; Gamble, Fl. Madras 2: 1290. 1925; N.P.Balakr. & Chakrab., Fam. Euphorb. India 389. 2007; Chakrab. & *al.* in N.P.Balakr. & *al.*, Fl. India 23: 489. 2012.

Prostrate or slightly ascending herb; branches trailing from a stout root stock. Leaves 3-6 × 2-5 mm, ovate – orbicular, margin entire, apex obtuse – acute or rounded; petiole 0.5 mm; stipules lanceolate, acute. Flowers axillary, solitary; male flowers few in the upper axils; female flowers many in the remaining axils; pedicels of female longer than male, to 3 mm; bracts 0.5 mm, lanceolate. Male flowers 1.5 mm across; tepals 6, 1 × 0.6 mm, ovate – oblong, obtuse; stamens 3; filaments connate at base; anthers 0.3 mm across; disc of glands 6, free. Female flowers 2.5 mm across; tepals 6, 1.5 × 1mm, oblong – broadly ovate, obtuse, with scarious margins; ovary 0.6 mm, globose; styles 3, each 2-fid at above, recurved; disc annular, cushion-shaped. Capsule 2.5 mm, smooth; seeds with vertically arranged minute tubercles.

Wastelands mostly in sandy areas in Medak and Nalgonda districts.

Fl. & Fr.: August-January.

Medak (MDK), *RG* 104022 (BSID); Nagarjunasagar (NLG), *KT* 9741 (MH).

Phyllanthus scabrifolius Hook.f., Fl. Brit. India 5: 299. 1887; Viswanatha & *al.*, J. Econ. Taxon. Bot.35: 208.2011.

Annual, erect herb, 20-35 cm high, stem and branches winged, dentate or toothed, laciniate. Cataphylls 1.5-2 mm long, narrowly triangular, lanceolate; stipules triangular-lanceolate. Leaf blade 4-15 × 2-10 mm, thick, obovate, obovate-elliptic, broadly elliptic or rounded, base cuneate, margin entire or serrate to dentate, apex acuminate to apiculate, obtuse or occasionally mucronate, densely scaberulous below, sparsely scaberulous above; petiole 1-1.5 mm long. Cymules unisexual with solitary female flower in upper axils and 1-3 male flowers in lower axils. Male flowers: minute, filiform; calyx lobes 6, membranous, biseriate, unequal, outer lanceolate, inner ovate-obovate; stamens 3, filaments connate below, free and spreading above, disc segments 6, saucer-shaped with tuberculate surface. Female flowers: pedicels 1.5-2 mm long; calyx lobes 6, subequal, greenish, outer linear-obovate, lanceolate, inner linear-obovate, disc rounded with irregularly lobed margin, styles 3, free, recurved from base. Capsules 3-4 mm across, depressed-globose, 3-lobed, smooth or minutely puberulous; seeds trigonous, brownish.

Rare on dry lateritic soil and dry gravelly soil on hilly slope in Nalgonda district (Viswanatha & *al.*, 2011).

Fl. & Fr.: July – December.

Nedigonda (NLG), *R.R.Rao, R.N.Kulkarni, R.Murugan & Viswanathan* 484 (CIMAP. Bangalore).

Phyllanthus simplex Retz., Observ. Bot. 5: 29. 1788; Hook.f., Fl. Brit. India 5: 295. 1887; Gamble, Fl. Madras 2: 1289. 1925; N.P.Balakr. & Chakrab., Fam. Euphorb. India 380. 2007; Chakrab. & *al.* in N.P.Balakr. & *al.*, Fl. India 23: 476. 2012. *P. virgatus* Forst. f., Insul. Austr. 65. 1786; Surya Prakash Babu in Pullaiah & Moulali, Fl. Andhra Pradesh 2: 883. 1997.

Vern.: *Uchchi usirika.*

Herb, up to 50 cm high, monoecious; branchlets winged. Leaves 0.6-2 × 2-0.4 cm, linear – oblong or lanceolate, entire, acute – apiculate, subsessile; stipules sagittate – peltate, acuminate, scarious. Flowers solitary in axils; male flowers few, in lower axils; female flowers many, in upper axils; pedicels of male to 3 mm, female to 11 mm. Male flowers 1.1 mm across; tepals 6, 0.8 × 0.3 mm, oblong, obtuse; stamens 3, free, filaments 0.35 mm, anthers 0.35 mm across; disc of glands 6, free. Female flowers 2 mm across; tepals 6, 1 × 0.4 mm, oblong, obtuse, mucronate; ovary 0.5 mm, tuberculate, styles 3, horizontally spreading, each 2-fid, from above the middle; disc annular, crenulate. Capsule 3 mm, globose, 3-lobed, minutely tuberculate; seeds trigonus, minutely tuberculate.

Common in moist localities and along the streams in all districts.

Fl. & Fr.: August – May.

Neelwai (ADB), *TP* & *GO* 5432; Manchippa (NZB), *TP* & *BR* 6044, *BR* & *PSPB* 9621; Pocharam RF (MDK), *TP* & *CP* 12059; Tandur (RR), *TP* & *MSM* 12703; Dhararam (MBNR), *TDCK* 13613; Rechapalli (KRN), *GVS* 20162 (MH); Aklaspur (KRN), *GVS* 20209 (MH); Pakhal (WGL), *KMS* 13146 (MH); Ramavaram (KMM), *RR* 112573 (BSID); Gangaram (WGL), *PVS* 84014 (BSID).

Phyllanthus tenellus Roxb., Fl. India 3: 668. 1832; Webster, Brittonia 22: 63. 1970; Mitra, Bull. Bot. Surv. India 27: 154. 1985; N.P.Balakr. & Chakrab., Fam. Euphorb. India 382. 2007; Chakrab. & *al.* in N.P.Balakr. & *al.*, Fl. India 23: 480. 2012.

Annual herb, up to 60 cm high; main stem smooth, subterete with spirally arranged cataphylls. Lateral branchlets 5-15 cm long, with 10-25 leaves; leaf blades membranous, elliptic to obovate, 6-25 × 4-11 mm, pale below, base acute to rounded, apex acute or obtuse, stipules lanceolate, acuminate. Monoecious, cymules axillary on lateral branchlets; proximal cymules with 2 or 3 male flowers and 1 or 2 female flowers, distal cymules with 1-2 female flowers, male flowers with pedicels 0.5-0.7 mm long; disc segments 5; stamens 5, filaments free, distally bent, anthers dehiscing horizontally. Female flowers with smooth flexuous capillary pedicels becoming 3-8 mm long, basally dilated into a pulvinus, calyx lobes 5, ovate, acute or obtuse, whitish except for the unbranched midrib, disc platelliform, ovary smooth, styles free, spreading, bifid ½ to their length. Capsules greenish, 1.7-1.9 mm broad; seeds pale brown, densely papillose.

Weed of gardens in Warangal district (C.S. Reddy and V.S. Raju, 2002).

Fl. & Fr.: Throughout the year.

Hanmakonda (WGL), *VSR* 3010 (KU), *CSR* 1350 (KU).

Phyllanthus urinaria L., Sp. Pl. 982. 1753; Hook.f., Fl. Brit. India 5: 293. 1887; Gamble, Fl. Madras 2: 1289. 1925; N.P.Balakr. & Chakrab., Fam. Euphorb. India 389. 2007; Chakrab. & *al.* in N.P.Balakr. & *al.*, Fl. India 23: 490. 2012.

Vern.: *Erra usirika.*

Herb, to 70 cm high, monoecious, slightly branched; branchlets flattened or slightly winged, thinly puberulous. Leaves 0.5-1.5 × 0.2-0.4 cm, oblong – linear oblong, base obliquely rounded, margin serrulate, apex apiculate; petiole very short; stipules lanceolate, acuminate. Flowers minute, solitary in axils; male flowers few in upper axils; female flowers many in remaining lower axils, subsessile; bracts 1 mm, linear lanceolate, acute, scarious. Male flowers 1.4 mm across; tepals 6, 0.8 × 0.35 mm, ovate – oblong, subacute; stamens 3, connate, subsessile; anthers 0.5 mm across; disc of glands 6. Female flowers 2 mm across; tepals 6, 1 × 0.5 mm, oblong, subacute; ovary 0.8 mm, globose, verrucose, styles 3, each 2-fid, adpressed to ovary. Capsule 3 mm, globose, verrucose; seeds trigonus, transversely furrowed.

Occasional in shady, moist localities in the forests as a forest undergrowth in Adilabad, Khammam and Ranga Reddi district.

Fl. & Fr.: December – March.

Ankusapuram (ADB), *GO* 4326; Parnasala RF (KMM), *RCS* 102464 (BSID); Vikarabad (RR), *MSM* 19245.

PUTRANJIVA Wall.

Putranjiva roxburghii Wall., Tent. Fl. Nap. 61. 1826; Hook.f., Fl. Brit. India 5. 1336. 1887; Gamble, Fl. Madras 2:1311.1925; N.P.Balakr. & Chakrab., Fam. Euphorb. India 335. 2007; Chakrab. & N.P.Balakr. in N.P.Balakr. & *al.*, Fl. India 23: 405. 2012. *Drypetes roxburghii* (Wall.) Hurursawa, J. Fac. Sci. Univ. Tokyo Sect. 3. Bot. 6: 337. 1954; Suryprakash Babu in Pullaiah & Moulali, Fl. Andhra Pradesh 2: 854. 1997.

Vern.: *Kudrajuvi, Kaduru, Putrajivika, Veku, Mahaputta-jivi.*

Moderate-sized, dioecious, evergreen tree; bark dark grey with horizontal lenticels; wood greyish-white, close-grained; branchlets puberulous, lenticellate. Leaves coriaceious, elliptic-oblong, 5-12.5 × 1-8 cm, glabrous, base obtuse, some what oblique, margin distantly serrulate, apex shortly acuminate, secondary nerves ca. 11 pairs, not prominent; petiole to 1 cm. Flowers greenish-yellow, dense, sessile in axillary clusters around branchlets. Male flowers: pedicelss 1.5-2 mm long; sepals 3-5, suborbicular; stamens 2-4, 1.5-2 mm long. Female flowers: pedicels 6-15 mm long; sepals 5 or 6, ovate to elliptic; ovary ovoid, styles 2 or 3, 3-7 mm long, often connate below; stigmas flabellate. Drupe yellowish, globose or ellipsoid, 2 × 1.5 cm, pubescent; seed 1, crustaceous.

Ocasional, usually grows on alluvial soil along the rivers or in swamps or evergreen forests in Adilabad district; sometimes cultivated in gardens for hedges and on roadsides. Occurs also in Mahabubnagar district.

Fl.: March-April, Fr.: October-December.

Jodevagu – Bheemaram nursery (ADB), *GO* & *DAM* 5178; Amrabad (MBNR), *BR, SKB* & *BSS* 32573.

Uses: Nirmal toys are made with the wood of this species.

RICINUS L.

Ricinus communis L., Sp. Pl. 1007. 1753; Hook.f., Fl. Brit. India 5: 457. 1887; Gamble, Fl. Madras 2: 1335. 1925; Purseglove, Trop. Crops, Dicots 180. 1977; N.P.Balakr. & Chakrab., Fam. Euphorb. India 154. 2007; Susila Rani & N.P.Balakr. in N.P.Balakr. & *al.*, Fl. India 23: 166. 2012.

Vern.: *Amudamu*. English: Castor.

Monoecious, evergreen shrub, up to 4 m high. Leaves alternate, palmatifid, 6-10-lobed, each 1-nerved with many, lateral nerves, peltate; lobes lanceolate, 10-18 × 3-7 cm, thinly pubescent below, margin serrate, apex acuminate; petiole to 20 cm. Paniculate racemes terminal, to 10-20 cm. Male flowers below, 1.5 cm across, female ones above, 6 mm across. Male flowers: perianth cupular, splitting into 3-5 lobes, lanceolate, valvate, 4 mm, margin inrolled, acuminate; stamens numerous; filaments connate and repeatedly branched; anther cells divergent, 4 mm. Female flowers: tepals

5, sub-equal, lanceolate, valvate, 5 mm, acute, ovary globose, echinate, 3 mm, 3-locular, ovules 3, pendulous, styles 3, stout, 4 mm, papillose, stigmatiferous, disc 0. Capsule 3-lobed, 2 cm across, prickly; seeds oblong; testa smooth, marbled, shiny, carunculate.

Cultivated for its oil seeds.

Fl. & Fr.: Through the year.

Kadthala (ADB), *GO* & *DAM* 5014; Tadwai (WGL), *RKP* 110810 (BSID); Khammam to Wyra (KMM), *RR* 108083 (BSID).

SAUROPUS Blume

Sauropus quadrangularis (Willd.) Müll.Arg., Linnaea 32: 72. 1863; Hook.f., Fl. Brit. India 5: 335. 1887; Gamble, Fl. Madras 2: 1303. 1925; N.P.Balakr. & Chakrab., Fam. Euphorb. India 395. 2007; Chakrab. & *al.* in N.P.Balakr. & *al.*, Fl. India 23: 497. 2012. *Phyllanthus quadrangularis* Willd., Sp. Pl. 4: 585. 1805.

Shrub, to 1 m high, monoecious; branchlets quadrangular. Leaves 1-2 × 0.8-1.4 cm, elliptic – oblong, base acute – obtuse, margin entire, apex apiculate; petiole to 2 mm; stipules minute, lanceolate – subulate. Flowers in axillary cymes; male 2-3, female solitary; pedicels to 6 mm, filiform. Male flowers 3 mm across; tepals 6, 1 × 1.8 mm, connate at base, triangular, rounded – emarginated; stamens 3, connate at base, anthers 3.5 mm across. Female flowers 4 mm across; tepals 6, 2.2 × 1.8 mm, free, obovate, rounded; ovary 1.5 mm, globose, styles 3, each 2-fid, recurved. Drupe 7 mm, depressed globose.

Rare, in sandy, damper parts of forests as undergrowth in Warangal district (C.S.Reddy, 2001).

Fl. & Fr.: September – November.

Lavvala (WGL), *CSR* 1559 (KU).

SYNADENIUM Boissier

Synadenium grantii Hook. f., Curtis Bot. Mag. t. 5633. 1867; N.P.Balakr. & Chakrab., Fam. Euphorb. India 281. 2007& in N.P.Balakr. & *al.*, Fl. India 23: 338. 2012.

Succulent shrub, up to 5 m high. Leaves 10-16 × 4-6 cm, broadly lanceolate cuneate at base, decurrent, entire, obtuse – subacute, alternate; petiole to 1 cm. Cyathia in axillary or subterminal cymes; rim cushion like, recurved, papillose; involucres 5.5 mm across, broadly cupular; bracts 5, concave, fimbriate. Male florets 25-30, in 5 groups; stalk 2.5 mm; bracteolate, laciniate; anther 0.6 x 0.8 mm. Female floret laterlly pendulous; ovary 2 mm, 3-lobed; styles 3, each 2-fid, at above. Capsule 3-lobed, thinly pubescent; seeds oblong.

Occasional as a hedge plant (C.S. Reddy, 2001; Seshagiri Rao, 2012)

Fl. & Fr.: November – January.

Hanamkonda (WGL), *CSR* 495 (KU).

TRAGIA L.

1. Twining hispid herb; leaves simple, base rounded – cuneate; styles circinate revolute .... **T. involucrata**
1. Hispid herb or under shrub; leaves simple or palmately 3-partite, base truncate – cordate; styles entire .... **T. plukenetii**

Tragia involucrata L., Sp. Pl. 980. 1753; Hook.f., Fl. Brit. India 5: 465. 1888; Gamble, Fl. Madras 2: 1332. 1925; N.P.Balakr. & Chakrab., Fam. Euphorb. India. 181. 2007; N.P.Balakr. in N.P.Balakr. & *al.*, Fl. India 23: 196. 2012.

Vern.: *Dulagondi.*

Twinning hispid berb, with stinging hairs, monoecious; branchlets striate. Leaves 3-6.5 × 1-3.5 cm, elliptic ovate – lanceolate or oblong-lanceolate, base rounded or cuneate, margin serrate, apex acuminate, sparsely hispid; petiole to 1.5 cm, hispidate; stipules ovate, acuminate. Flowers in axillary and terminal leaf-opposed racemes, subsessile; male flowers many, minute, from above the middle; female flowers few, near the base; bracts 3 mm, lanceolate, acute. Male flowers 2.5 mm across; tepals 3-4, 1.4 × 1 mm, slightly connate at base, broadly ovate or orbicular, concave, subacute; stamens 3-4, anthers 0.45 mm across; pistillode small, 3-fid. Female flowers 5.8 mm across; tepals 6, 5 × 2 mm, free, accresent, pinnatifid, lanceolate, laciniate, hispid without; ovary 2 mm, globose, hispid, styles 3, connate at base, circinately revolute. Capsule 5 × 9 mm, hispid; seeds globose, smooth.

Frequent along forest outskrits and in scrub jungles in Adilabad, Karimnagar, Mahabubnagr and Nalgonda districts.

Fl. & Fr.: December – March.

Kunthala (ADB), *GO & DAM* 5126; Mallelatheertham (MBNR), *TDCK* 15357, *SRS* 108987 (BSID); Kambalpally (NLG), *BVRS* 322 (NU).

Tragia plukenetii Radcl.- Sm., Kew Bull. 37: 688. 1983; N.P.Balakr. & Chakrab., Fam. Euphorb. India. 182. 2007; N.P.Balakr. in N.P.Balakr. & *al.*, Fl. India 23: 199. 2012. *T. cannabina* L. f., Suppl. Pl. 415. 1781, *nom. illeg.;* Gamble, Fl. Madras 2; 1333. 1925. *T. involucrata* L. var. *cannabina* (L.f.) Müll.Arg. in DC., Prodr. 15: 944. 1866; Hook.f., Fl. Brit. India 5: 465. 1888.

Vern.: Pedda dulagondi.

Herb or undershrub, to 1 m high, with sparsely hispid stinging hairs, monoecious; branchlets striate. Leaves 1.5-7 × 1.5-4 cm, lanceolate or plamately 3-lobed, middle longer than laterals, base truncate – cordate, margin serrate – crenate dentate, apex acuminate, sparsely hispid; petiole 1.5 cm, hispidate; stipules ovate, acuminate. Flowers in axillary racemes; male flowers many, at above; female flowers few at base; bracts 1.4 mm, lanceolate. Male flowers 2 mm across; tepals 3, 1.5 × 1 mm, free or slightly connate at base, broadly ovate – broadly elliptic, concave, subacute; stamens 3, filaments 0.3 mm, anthers 0.4 mm across. Female flowers 6 mm across; tepals 4-6, 4.2 × 2.4 mm, free, ovate, pinnatifid, densely hispid without and near the margin

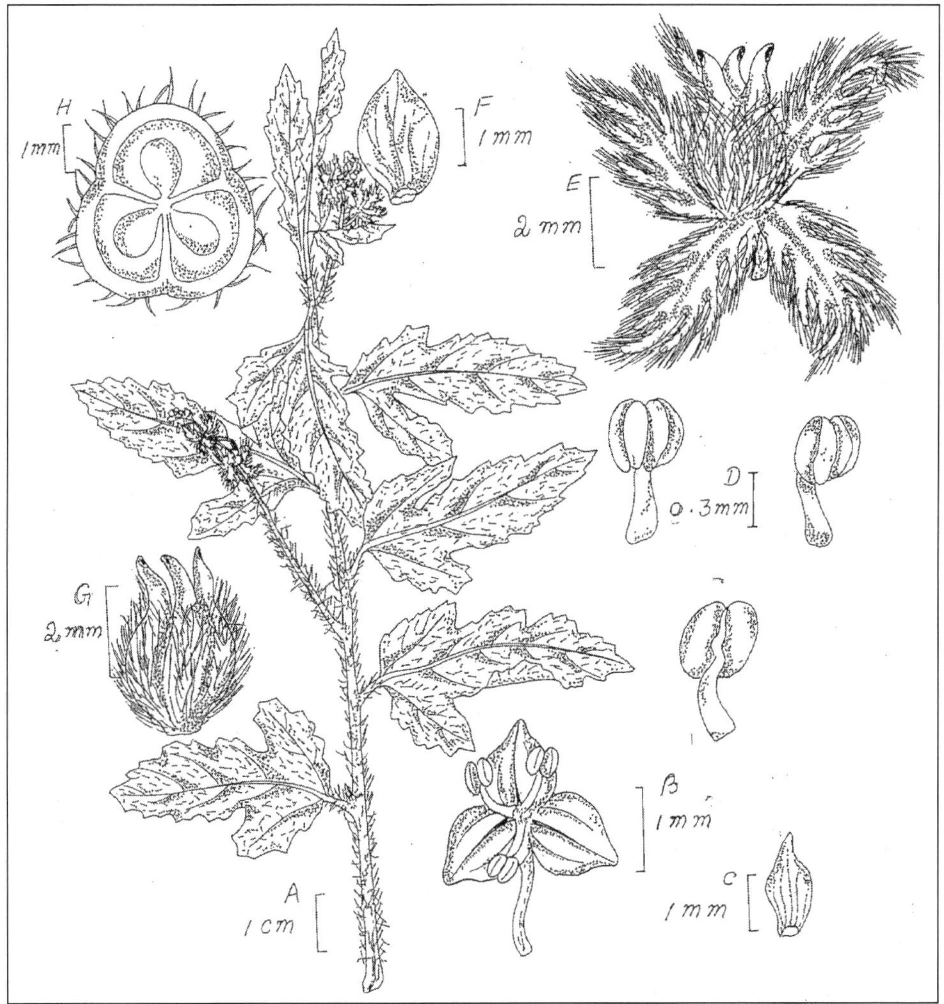

Figure 102: Tragia plukenetii Radcl.- Sm.

A. Twig, B. Flower, C. Bract, D. Anthers, E. Flower, F. Bract, G. Pistil, H. Ovary T.S.

within; ovary 2.5 mm, subglobose, hispid, styles 3, slightly connate at base, not revolute. Capsule 4 × 7 mm, hispid; seeds globose, smooth.

Common in hedges and in forest outskrits in Nizamabad, Medak, Ranga Reddi, Karimnagar and Nalgonda districts.

Fl.& Fr.: December – March.

Sirnapalli (NZB), *PSPB* 9163; Uppalwai (NZB), *BR* & *GO* 9097; Medak (MDK), *CP* 11794; Naskal (RR), *MSM* & *PRSR* 15121; Mudigonda (NLG), BVRS 613; Vempalli (KRN), *GVS* 20183 (MH); Jallada – Kothaguem (KMM), *RR* 106099 (BSID); Jakaram (WGL), *RKP* 105329 (BSID).

Note: Linnaeus f. when uniting Linnaeus' *Croton hastatus* and *C. urens* (Sp. Pl. ed. 1: 1005. 1753) and transferring them to *Tragia*, failed to choose between one of these epithets. Hence, his name *Tragia cannabina* becomes superfluous and illegitimate. As the epithets *hastatus* and *urens* have since been preoccuipied A. Radcliffe-Smith has named it after Leonard Plukenet (1642-1706).

ULMACEAE

1. Leaves penninerved; stamens twice the number of perianth lobes; fruit samara; plants deciduous .... **Holoptelia**
1. Leaves 3-nerved from base; stamens as many as perianth lobes; fruit a drupe; plants evergreen:
   2. Plants dioecious, leaf scabrous above, glaucous beneath .... **Trema**
   2. Plants with polygamous flowers, leaves glabrous on both surfaces, green beneath .... **Celtis**

CELTIS L.

Celtis timorensis Span., Linnaea 15: 343. 1841. *C. cinnamomea* Lindl. ex Planch., Ann. Sci. Nat. Bot. (Ser.3) 10: 303. 1848; Hook.f., Fl. Brit. India 5: 482. 1888; Fischer, Fl. Madras, 3: 1349. 1925.

Ver.: *Koti-Bira, Chipuru, Kaki-mushti*.

Evergreen tree, up to 10 m high; bark pale grey, smooth; wood dull white; young branches pubescent. Leaves membranous, ovate-elliptic, 5.5-11 × 2.2-5.2 cm, glabrous, blackish when dry, base obtuse or rounded, margin entire, apex acute-acuminate, 3-ribbed, basal lateral nerves nearly reaching the apex; petiole to 1.4 cm. Flowers small, whitish, pedicellate on 2 cm long, axillary, dichotomous racemes. Drupe ovoid, globose, 0.6 cm across, orange when ripe.

A common tree in dry deciduous forests of Nalgonda district

Fl.: Mar.-April, Fr.: May.

Kambalpally (NLG), *BVRS* 319 (NU).

HOLOPTELEA Planch. (*'Holoptelaea'*)

Holoptelea integrifolia (Roxb.) Planch., Ann. Sci. Nat. Bot. (Ser.3) 10: 259. 1848; Hook.f., Fl. Brit. India 5: 481. 1888; Fischer, Fl. Madras 3: 1348. 1925. *Ulmus integrifolia* Roxb., Pl. Coromandel t. 78. 1798 & Fl. Ind. 2: 68. 1832.

Vern.: *Nauli, Pulari, Pedda nevili, Tapasi, Tappachi, Nevili, Namlinara*.

Polygamous, deciduous tree, up to 15 m high; bark whitish-grey, pustular, smooth, exfoliating in quadrilateral scales; branchlets terete, tender parts pubescent. Leaves chartaceous, elliptic-ovate, 8–10 × 4–8 cm, coriaceous, appressed-pubescent below, puberulous along nerves above when young, base rounded, subcordate, margin entire, apex abruptly acuminate; secondary veins 5–6 pairs; petiole to 1.3 cm long; stipules scarious. Flowers yellow in axillary fascicles or racemes, at the scars of the old shoots; bracts scaly; male flowers and bisexual flowers mixed in raceme; tepals 4, free, concave, 1.5 mm, imbricate, unequal, pubescent, ciliate, obtuse; stamens 7–9, biseriate, filament 1.7 mm; anthers slightly curved, 1.3 mm, hairy, introrse; ovary compressed-ovate, 2 × 1.5 mm, stalk elongating in fruit; style-arms 2, stigmatose. Fruit a dry, winged, compressed samara, 1 × 0.7 cm, wings membranous, 3 cm across, yellow, reticulate; seed flat, exalbuminous.

Common, in deciduous forests and in villages along roadsides in most districts.

Fl.: January-February, Fr.: April - May.

Mudheli (NZB), *TP* & *BR* 6366; Narsingi (MDK), *BR* & *CP* 11529; Kusumasamudram RF (RR), *MSM* 10334; Yerlapenta (MBNR), *TDCK* 13690; Chityal (NLG), *BVRS* 412 (NU); Pakhal (WGL), *ANH* 15942 (MH); Bhadrachalam RF (KMM), *RCS* 99037 (BSID).

TREMA Lour.

Trema orientalis (L.) Blume, Mus. Bot. 2: 62. 1856; Hook.f., Fl. Brit. India 5: 484. 1888; Fischer, Fl. Madras 3 : 1350. 1925. *Celtis orientalis* L., Sp. Pl. 1044. 1753.

Vern.: *Morali, Boggu chettu, Priyalu, Aava, Konda jonna, Ada nelli*. Eng.: Charcoal tree.

Evergreen, dioecious tree, up to 12 m high; trunk straight with spreading branches; bark greyish brown, smooth, with numerous lenticels; wood light reddish-grey; branchlets terete, sparsely pubescent. Leaves chartaceous, ovate or ovate-lanceolate, 7-18 × 2-7 cm, somewhat rough above, whitish beneath, clothed with soft tomentum, base unequal, rounded or sub-cordate, margin serrulate, apex acuminate, secondary nerves 3-5 pairs, basal pairs reaching to the half the length of the leaf; petiole to 1.5 cm, pubescent. Flowers greenish in 2 cm long, axillary, pubescent cymes; the male cymes rather more compact than the female and usually on different trees. Drupe ovoid, 0.3 cm long, black when ripe, glabrous.

Occasional in deciduous forests, along perennial streams.

Fl. & Fr.: Throughout the year.

Kambalpally (NLG), *BVRS* 315 (NU).

CANNABACEAE

CANNABIS L.

Cannabis sativa L., Sp. Pl. 1027. 1753; Hook.f., Fl. Brit. India 5: 487. 1888; Fischer, Fl. Madras 3: 1350. 1928.

Vern.: *Ganjai-chettu, Ganja-chettu, Bangi-aku.*

Large, erect, aromatic, resinous herb or sub-shrub; branchlets angled, sparsely pubescent. Leaves 3.5-6.5 × 0.5-0.8 cm, palmately compound, leaflets sub-sessile, 3-5-fid above, lateral ones smaller, lanceolate, alternate above, opposite below, base cuneate, margin serrate, apex acuminate, chartaceous, thinly scabrid above, appressed pubescent, with sessile glands below. Flowers dioecious, axillary, males cymose or panicled, female in small cymes or heads, with large bracts. Fruit an achene.

Occasionally cultivated clandestinely.

Fl. & Fr.: January – April.

Armoor (NZB), *BR* 7296; Farahabad (MBNR), *TDCK* 13684; Irlapenta (MBNR), *SKB* & *BSS* 32376.

MORACEAE

1. Flowers in a hollow receptacle; flowers hidden .... **Ficus**
1. Flowers not in a hollow receptacle; flowers patent:
    2. Plants with a milky sap:
        3. Flowers monandrous .... **Artocarpus**
        3. Flowers tetrandrous .... **Streblus**
    2. Plants with watery sap .... **Morus**

ARTOCARPUS J.R. Forst. & J.G.A. Forst. *nom. cons.*

1. Leaves elliptic or obovate; base cuneate; fruits 30-100 cm long, 25-30 cm in diameter .... **A. heterophyllus**
1. Leaves ovate or rhomboid, base rounded; fruits 5-15 cm long, 4-10 cm in diameter .... **A. hirsutus**

Artocarpus heterophyllus Lam., Encycl. 3: 209. 1789. *A. integrifolius auct. non* L.f. 1781; Fischer, Fl. Madras 3: 1369.1928; Hook.f., Fl. Brit. India 5: 541. 1888.

Vern.: *Panasa.* Eng.: Jack fruit tree.

Large, evergreen tree; crown dense, round; bark black, mottled green, smooth; wood bright yellow, darkening on exposure; branchlets glabrous. Leaves elliptic or obovate, coriaceous, 10-17 × 5.6-10.1 cm, dark green and shining, base cuneate, margin entire, apex obtuse with short acumen, secondary nerves ca. 7 pairs, prominent, 2 cm apart, curved upwards; petiole to 3 cm. Flowers greenish, crowded on globose, oblong or cylindrical, solitary, axillary and cauliflorus receptacles. Fruit oblong-globose, greenish-yellow, large, clothed with fleshy and enlarged perianth, carpels spiny; seeds reniform, greyish, and oily.

It is grown by farmers for its fruits.

Fl.: November-December. Fr.: July-August.

Artocarpus hirsutus Lam., Encycl. 3 : 21.1789.

Large evergreen tree, up to 70 m high with grey bark; young shoots hairy. Leaves rhomboid or ovate, base rounded, apex subacute, margin entire; petioles stout, 1.3-3 cm long; stipules nearly 2.5 cm long, lanceolate. Flowers greenish yellow in solitary or paired, narrowly cylindrical, smooth, axillary receptacles. Male receptacle elongated, to 14 cm long. Fruit syncarp, 5-15 x 4-10 cm. cylindrical to ellipsoid, yellow to orange yellow when ripe, echinate with long process; seeds ovoid, 1.3-2 cm long.

Planted in houseyards, in gardens for its evergreen foliage and edible fruits.

Fl. & Fr .: January-May.

FICUS L.

Note: Chaudhary & *al.* (2012) gave an excellent Synopsis fo the genus *Ficus* in India.

1. Plant a climber .... **F. pumila**
1. Plants shrubs or trees (male plants climbing in *F. heterophylla*):
   2. Juvenile leaves often pinnately lobed .... **F. heterophylla**
   2. All leaves entire:
      3. Receptacles in dense clusters or fascicles on tubercles or on leafless racemes of panicled branches from the trunk or larger branches:
         4. Leaves glabrous on both surfaces, margin entire .... **F. racemosa**
         4. Leaves scabrid-hispid above, pubescent or tomentose beneath, margin toothed or serrate .... **F. hispida**
      3. Receptacles solitary, paired or 2-6 together in the axils of leaves:
         5. Leaves with very many fine lateral nerves-cultivated .... **F. elastica**
         5. Leaves not as above:
            6. Receptacle distinctly peduncled:
               7. Leaves lobed-cultivated .... **F. carica**
               7. Leaves entire or not lobed .... **F. tinctoria**
            6. Receptacles sessile:
               8. Apices of leaves caudate:
                  9. Caudate-acumen short, leaf base narrowed .... **F. talboti**
                  9. Caudate-acumen long, leaf base broad:
                     10. Leaf base cordate .... **F. arnottiana**
                     10. Leaf base truncate or subcordate .... **F. religiosa**

8. Apices of leaves obtuse, blunt or acute-shortly bluntly acuminate:
    11. Leaves tomentose or pubescent beneath even at maturity:
        12. Midrib usually with an elongated gland at base; receptacle scarlet red, puberulous .... **F. benghalensis**
        12. Midrib egandular; receptacles green tomentose .... **F. mollis**
    11. Leaves glabrous:
        13. Lateral nerves above 20 pairs, inter secondaries as prominent as secondaries:
            14. Branches drooping; aerial roots numerous; leaves shining, base indistinctly 3-veined at base .... **F.benjamina**
            14. Branches not drooping; aerial roots few; leaves not shining, base distinctly 3-veined at base .... **F. microcarpa**
        13. Lateral nerves less than 20 pairs, inter secondaries not prominent as secondaries:
            15. Receptacles very shortly peduncled; aerial roots few .... **F.virens**
            15. Raceptacles sessile; aerial roots absent:
                16. Raceptacles 2-6 together on tubercles from leafless axils .... **F. tsjahela**
                16. Raceptacles in pairs in leafy axils .... **F. amplissima**

Ficus amplissima J.E. Sm. in Rees, Cyclop. 14 : Ficus no. 68. 1810; Chaudhary & *al.*, Taiwania 57: 207. 2012. *F. tsiela* Roxb. ex Buch.-Ham., Trans. Linn. Soc. London. 15 : 149. 1827; King in Hook.f., Fl. Brit. India 5: 515.1888; Fischer, Fl. Madras 3: 1362. 1928.

Vern.: *Pedda juvi, Pitta juvvi.*

Spreading, deciduous tree, up to 20 m high; trunk cylindrical, bark greenish-grey, smooth, latex milky; wood whitish. Leaves alternate, ovate-oblong, 7–12 × 5–8 cm, glabrous, coriaceous, base truncate to rounded, acute, margin entire to undulate, apex acuminate; lateral veins 8–12 pairs, flattened above, slightly raised below, inconspicuous intercostal veins; petiole 3.5–5 cm, gland at apex on lower side or at base of lamina; stipule lanceolate, to 2.5 cm long, covering leaf at young stage, 2–2.5 × 0.5–0.8 cm. Figs monoecious, axillary, subsessile, 1–1.8 cm diameter, in leaf axils, depressed globose or pyriform, apex truncate to broadly rounded, glabrous, smooth, purple when ripe; basal bracts 3, broadly ovate, 3 mm across, obtuse; orifice circular, closed by 3, flat, apical scales; internal bristles 0; tepals 3, broadly lanceolate, 1 mm long, brownish, margin scarious. Male flowers distinct; stamen 1, included; filament 0.4 mm long; anther oblong, parallel, unequal, to 0.5 mm across. Female flowers sessile; ovary bean-shaped, white, 1 mm across; style filiform, 2 mm long. Gall flowers similar, pedicellate. Achenes smooth, 2 mm long.

Common, in deciduous forests in Adilabad, Medak, Warangal, Karimnagar and Mahabubnagar districts.

Syconia: January-March.

Mancheryal (ADB), *GO* & *DAM* 5189; Gangapur (MDK), *TP* & *CP* 12075; Farahabad (MBNR), *TDCK* 15396; Way to Appapur (MBNR), *SSR* & *KJ* 15541; Aklaspur (KRN), *GVS* 20231 (MH); Eturnagaram (WGL), *RKP* 110842 (BSID).

Ficus arnottiana (Miq.) Miq., Ann. Mus. Bot. Ludg.-Bat. 3: 287. 1867; King in Hook.f., Fl. Brit. India 5: 513. 1888; Fischer, Fl. Madras 3: 1363. 1928; Chaudhary & *al.*, Taiwania 57: 20. 2012. *Urostigma arnottianum* Miq. in Hook. London J. Bot. 6: 564. 1847.

Vern.: *Kalla-ravi, Banda ravi.*

Tree, up to 7 m high, shining, glabrous; bark pale, smooth; wood white. Leaves alternate, ovate or elliptic-ovate, 7–15 × 5–12 cm, subcoriaceous, glabrous, shiny, base cordate, margin entire to slightly undulate, apex acuminate or sub caudate-cuspidate, 1–1.5 cm; veins 5 from base, laterally 6–8 pairs, intercostals transverse, zigzag in the middle, flattened above, raised below; petiole 7–12 cm, glands below the base of lamina; stipules 4–6 cm long. Figs in axillary pairs or crowded near apex, subsessile or pedunculate, to 2 mm long, globose, 5–7 mm across, reddish when ripe; orifice plane; bracts 3, ovate, concave, 1 × 2.5 mm, obtuse, caducous; internal bristles 0; tepals gamophyllous, 3–4-lobed, brownish red, glabrous. Male flowers around the orifice, sparsely scattered in the interior of fig; stamen 1, subsessile; anthers 2-celled, ovate-oblong, 0.3 mm across, dehiscing longitudinally. Female flowers creamy, some what reddish on stylar side; ovary depressed-globose, 0.4 mm; style filiform, 1.5 mm long. Gall flowers stalked. Achenes smooth.

Common, in all districts in rocky areas in the hils in Nalgonda and Mahabubnagar districts.

Syconia: August onwards.

Appapur (MBNR), *SKB* & *BR* 30726; Saleswaram (MBNR), *BR* & *SKB* 30741; Kambalpally (NLG), *BVRS* 320 (NU).

Ficus benghalensis L., Sp. Pl. 1059. 1753; King in Hook.f., Fl. Brit. India 5: 499. 1888; Fischer, Fl. Madras 3: 1361.1928; Chaudhary & *al.*, Taiwania 57: 210. 2012. *F. indica* L., Sp. Pl. 1060. 1753. *Urostigma benghalensis* (L.) Gasp. Nov. Gen. Fic. 7. 1844.

Vern.: *Marri, Pedda marri, Rohi, Vatairilsja*; Eng.: The Banyan, India Fig.

Evergreen tree, up to 25 m high, 50 m canopy, branches spreading, aerial roots form as pillars from branches, bark greyish white, smooth; wood grey-white; young branches pubescent, latex milky. Leaves alternate, elliptic-ovate, 10–17 × 7–12 cm, coriaceous, glabrous above, pubescent below, base rounded, margin entire, apex obtuse, to subacute; laterally 5 or 6 veins arise from base, lateral veins 3–6 pairs, flattened above, raised below, intercostals transverse, zigzag in the middle; petiole to 4 cm long, glabrous, glands below the base of lamina; stipules deciduous, leaving annular scars, 2 × 1.5 cm. Figs monoecious, 1 or 2, axillary, depressed-globose, 1.5–2 cm across, puberulous without, sessile, dark pink or red when ripe; bracts 4–5, cupular, 6 mm across, shortly connate, obtuse, persistent; orifice subumbonate; internal bristles 0; tepals 3–5, shortly connate, 2 mm across, brownish, glabrous. Male flowers dispersed with female; stamen 1; filament 0.8–1 mm long; anther oblong, parallel, unequal, to 1 mm across, shortly mucronate; connective brown. Female flowers sessile; ovary obovoid, glabrous, 1.5 × 1 mm, dark brown on stylar side; style erect or curved, 2 mm long, tapering. Gall flowers similar to female flowers, pedicellate. Achenes globose-ellipsoid, 2 × 1.5 mm, dark brown.

Very common, on roadsides, wild in village surroundings and old walls in all districts. It is sacred tree of Hindus and most popular shade providing tree.

Syconia: September-March.

Sirpur road (ADB), *GO* & *PVP* 4931; Nasarullabad (NZB), *TP* & *BR* 6179; Medak (MDK), *BR* & *CP* 11413; Mohammadbad RF (RR), *MSM* & *KH* 10208; Peddavuru (NLG), *BVRS* 515; Bhupatipur (KRN), *GVS* 20175 (MH); Madikonakutta, Narsapur (MDK), *KMS* 6726 (MH); Koriguttalu (KMM), *RR* 106044 (BSID); Vatavarlapalli (MBNR), *VBH* 84961 (BSID).

F. benghalensis L. var. **krishnae** (C.DC.) Corner, Gard. Bull. Singapore 21(1): 14. 1965; Chaudhary & *al.*, Taiwania 57: 210. 2012. *F. krishnae* C.DC., Bull. Herb. Boiss., ser. 2, 2: 760. 1902

The unique feature of the tree is that the leaves have a pocket-like fold at the base.

Cultivated in gardens. and considered as Horticultural variety of *F. benghalensis.*

Ficus benjamina L., Mant. Pl. 129. 1767; King, Ann. Roy. Bot. Gard. (Calcutta) 1. t. 52A. 1887 & in Hook.f., Fl. Brit. India 5: 508. 1888; Fischer, Fl. Madras 3: 1367. 1928; Chaudhary & *al.*, Taiwania 57: 210. 2012.

Vern.: *Konda-juvvi., Putta-juvvi, Konda golugu.*

An evergreen tree, up to 20 m high; crown spreading, aerial roots numerous, trunk cylindric; bark whitish-grey, smooth; branches drooping, branchlets glabrous. Leaves subcoriaceous, elliptic-obovate, 5.5-8.5 × 2.8-4.8 cm, glabrous and shining, drying grey-green to olivaceous brownish, base rounded or acute, margin entire, apex abruptly acuminate, 3-nerved, lateral nerves 6-14 pairs, secondary nerves almost as prominent, raised on both sides, connected by an intramarginal nerves from base; petiole to 2.2 cm. Receptacles monoecious, obovoid, ellipsoid or subglobose, sessile or stalked, pyriform, axillary, paired, ripening yellow to orange, basal bracts 2-3, crescentic; orifice slightly raised, circular, closed by 3 small flat apical bracts.

Often planted for shade in avenues and as an ornamental tree

Syconia: March- May.

Ficus carica L., Sp. Pl. 1753; Bailey, Man. Cult. Pl. 339. 1949; Chaudhary & *al.*, Taiwania 57: 194. 2012.

Vern.: *Anjur, Anjuram.*

Much branched, small deciduous tree, branching from the base. Leaves thick, broad-ovate nearly orbicular, 3-5-lobed, 10-20 cm long, rough above, pubescent beneath, base cordate, margin entire, apex acuminate, 3-5-nerved; secondary nerves 6-8 pairs above basal pairs; petiole to 6 cm. Receptacles solitary or paired, axillary and subterminal, pear-shaped, variable in size, yellowish- purple when ripe.

Introduced and cultivated; commerically grown.

Syconia: January-June.

Ficus drupacea Thunb., Diss. Ficus 6, 11. 1786, var. **pubescens** (Roth) Corner, Gard. Bull. Singapore 17: 381. 1960; Chaudhary & *al.*, Taiwania 57: 211. 2012. *Ficus mysorensis* Heyne ex Roth var. *pubescens* Roth, Roem & Schult. Syst. Veg. 1: 508. 1817; King, Ann. Roy. Bot. Gard. (Calcutta) 1: t. 14. 1887&in Hook.f., Fl. Brit. India 5: 500. 1888; Fischer, Fl. Madras 3: 1361. 1928.

Tree, up to 25 m high; branchlets fulvous-tomentose. Leaves alternate, broadly elliptic-ovate, 10-22 × 6-10 cm, subcoriaceous, 3-nerved from base, sometimes with an additional pair, lateral nerves ca. 14 pairs, flattened above, raised below, intercostals transverse, ± zigzag, prominent on either side, glabrous above and below, tender leaves puberulous below, base broadly acute to cordate, margin entire, apex abruptly acuminate, cusp 1-1.5 cm; petiole to 3 cm, glandular at apex below; stipules broadly lanceolate, 2.5 cm, tomentose. Fig monoecious, axillary, solitary, sessile, ellipsoid-globose, 1.5-2 cm across; fig-wall thick, ripening orange, tomentose without; basal bracts orbicular, concave, 6 mm, brown-pilose without, obtuse; orifice umbonate, closed by 3-4, apical bracts; internal bristles a few, linear-lanceolate, to 4 mm. Tepals 3-4, free, ovate-concave, 1 mm, brown, acute. Male: disperse, pedicellate; stalk to 4 mm, stamen 1, exserted, filament 0.5 mm, anther oblong, parallel, 1 mm. Female: sessile, ovary obovoid, 0.7 mm, brown; style filiform, 2 mm. Achenes smooth.

Seshagiri Rao (2012) reported from central University campus.

Ficus elastica Roxb. ex Hornem., Hort. Bot. Hafn. Suppl. 7: 1819; King in Hook.f., Fl. Brit. India 5: 508. 1888; Chaudhary & *al.*, Taiwania 57: 212. 2012.

Vern.: *Rubber chettu.* Eng.: Indian Rubber tree.

Small to large, evergreen tree with numerous aerial roots; trunk much buttressed and fluted. Leaves thick, coriaceous, oblong or elliptic, 15-25 cm long, glossy green, base rounded or somewhat narrowed, margin entire, apex with sharp point, lateral nerves numerous, fine parallel, petiole to 4 cm; stipules pink, about half as long as leaf lamina. Receptacles yellow, paired, sessile on old leaf axils.

Grown as an ornamental tree in gardens for its glossy ornamental leaves or for shade.

Ficus heterophylla L.f., Suppl. Pl. 442. 1781; King in Hook.f., Fl. Brit. India 5:518. 1888; Fischer, Fl. Madras 3: 1366. 1928; Corner in Dassan. & Fosberg, Rev. Handb. Fl. Ceylon 3: 272. 1981; Chaudhary & *al.*, Taiwania 57: 201. 2012.

Very variable shrub, creeping, sarmentose or sometimes erect; branchlets pithy and more or less hollow, brown hairy, hispid or with dense spreading pubescence. Leaves very variable, linear, linear-lanceolate, ovate, elliptic to broadly ovate-oblong, unlobed or shallowly or deeply pinnatifidly 2-many-lobed, 2.5-12.5 cm long, base obtuse, rounded, subcordate or cordate, margin shallowly serrate or dentate, apex acute or shortly acuminate, more or less hispid or scabrid above, softly pubescent, glabrescent or scabrellous beneath, petiole 0.4-7.5 cm long, usually hispid; stipules oblong-lanceolate, 3.7-7.5 mm. Figs axillary, solitary, young more or less pyriform or ellipsoid and hispid, mature globose or subglobose, 1.8-2.5 cm long and 1.2 cm across, mouth broad; peduncle 5-10 mm long. Gall flower pedicelled, tepals 4, linear, 2.5-3 mm. Male flowers: Tepals 3, shorter than in gall flowers, stamen 1.

Occasional on old walls, in crevices of stones in Karimnagar district (Naqvi, 2001).

Figs: January-May.

Ficus hispida L.f., Suppl. Pl. 442. 1781; King in Hook.f., Fl. Brit. India 5:522. 1888; Fischer, Fl. Madras 3:1367. 1928; Chaudhary & *al.*, Taiwania 57: 205. 2012.

Vern.: *Kakimedi, Beddu marri, Kukka bodda, Brahmamedi, Bomma medi*.

Small tree, up to 5 m high, bushy; branchlets hispid, bark grey, rough; wood dark grey, latex milky to reddish. Leaves alternate, obovate-elliptic, oblong, 10–18 × 5–10 cm, coriaceous, glaucescent above, scabrid on either side, hispid below, base truncate, cordate; margin entire, apex acute or obtuse; basal veins 4–5, lateral veins 6 pairs, intercostals transverse, prominent below; petiole with a sub nodal gland, to 4 cm long; stipules lanceolate, 1.5 cm long. Figs dioecious on special shoots and on branchlets, depressed-globose, base narrowed, 1.5–2.5 cm across, ripening yellow, faintly ribbed, hispid to sticky-pubescent without; peduncle stout, 0.5 cm long; bracts triangular, to 2 mm long; orifice slightly raised, closed by 5–6 apical bracts; internal bristles 0; perianth cupular, truncate, to 2 mm across; tepals 0, glabrous. Male flowers ostiolar, 2-seriate; stamen 1, subsessile or stalked; ovary depressed-globose, 0.7 mm across; style 1.5 mm long, hairy, tip clavate. Gall flowers similar, but larger, distinctly stalked. Achenes lenticular, 1.5 mm, keeled, with prominent hilum.

Common, in hedges, thickets, vacant plots in all districts.

Syconia: December- March.

Jannaram (ADB), *TP* & *PVP* 4190; Bheemaram (ADB), *GO* & *DAM* 5183; Gowraram (NZB), *TP* & *BR* 6390; Narsapur RF (MDK), *TP* & *CP* 12015; Rudraram RF (RR), *TP* & *MSM* 12725; Appapur (MBNR), *TDCK* 13646; Farahabad (MBNR), *TDCK* 15365; Kambalpally (NLG), *BVRS* 317; Kodimial (KRN), *GVS* 20133 (MH); Pakhal (WGL), *ANH* 15946 (MH); Madisagutta temple (KMM), *RR* 106057 (BSID).

Ficus microcarpa L.f., Suppl. Pl. 442. 1782; Chaudhary & *al.*, Taiwania 57: 212. 2012. *F. retusa auct. non* L.; King, Ann. Roy. Bot. Gard. Cal. 1:50 1887 & in Hook.f., Fl. Brit. India 5:511.1888; Fischer, Fl. Madras 3: 1362. 1928.

Vern.: *Juvvi, Konda juvvi, Yerra-juvvi, Konda pillara, Nandireka, Billa juvi, Hama vudaga.*

Evergreen tree, up to 15 m high; aerial roots few; bark brown, nearly smooth; wood light reddish-grey; branchlets glabrous. Leaves coriaceous, elliptic or obovate, 4–8 × 2.5–4.5 cm, glabrous, shining, base cuneate-acute, margin entire, apex rounded to retuse; basal veins 3, lateral veins 12–19 pairs, closely pinnate, raised on either side, intercostals 0, secondary lateral veins as prominent as primary veins; petiole to 1.5 cm long, glandular at apex below, canaliculated, glabrous; stipules ovate-lanceolate, to 1 cm long. Figs monoecious, 6–8 mm across, sessile, axillary, paired or solitary, globose, glabrous, yellowish-red when ripe; bracts 3, ovate, elliptic, boat-shaped, 2 mm across, obtuse, persistent; orifice plane or slightly raised, closed by 3, flat, apical bracts; internal bristles minute, sparse; tepals 3–4, free, ovate-lanceolate, 1.5 mm long, brownish, glabrous, Male flowers dispersed; stamen 1, hardly exserted; filament 0.3 mm; anther unequal, ovate-oblong, 0.7 mm, mucronate. Female flowers sessile; ovary ovoid-globose, red-brown, 1 mm across; style 1.5 mm long, tapering. Gall flowers similar to female flowers, pedicellate. Fig wall fleshy and smooth. Achenes smooth.

Common in Adilabad, Nizamabad, Khammam and Mahabubnagar districts, often planted in avenues

Syconia ripening by October.

Bhadrachalam (KMM), *JSG* 15809 (MH).

Ficus mollis Vahl, Symb. Bot. 1:82. 1790; Chaudhary & *al.*, Taiwania 57: 212. 2012. *F. tomentosa* Roxb. ex Willd., Sp. Pl. 4: 1136. 1806; King in Hook.f., Fl. Brit. India 5: 501.1888; Fischer, Fl. Madras 3: 1361. 1928.

Vern.: *Jivi, Juvvi, Kal juvi.*

Evergreen tree, up to 15 m high; bark greenish-white, smooth, scaly; wood white; young branches densely, yellowish tomentose; aerial roots several from spreading branches, not used for support. Leaves subopposite, often in unequal pairs, broadly oblong to elliptic-ovate to pandurate, 10–15 × 6–10 cm; coriaceous, glaucescent above, scabrid on either side, hispid below or densely rust-grey tomentose beneath, base cordate with a gland at the base of the midrib beneath, margin entire, apex acute or obtuse; basal veins 3, lateral veins 6 pairs, impressed above transverse raised, prominent below; petiole with a subnodal gland, to 4 cm long; stipules ovate lanceolate, 1 cm long, hispid. Figs monoecious on branchlets, sessile, axillary in pairs, rust tomentose, depressed-globose, base narrowed, 1.5–2 cm across, ripening yellow, faintly ribbed; bracts 3, orbicular, concave 4 mm across, velvety-tomentose; orifice circular, slightly raised, closed by 3 apical bracts; internal bristles sparse to abundant. Perianth free, truncate, to 2 mm across; tepals 3–4, ovate-lanceolate, 1 mm long, brown, glabrous. Male flowers dispersed subsessile; stamen 1, filament 0.3 mm long; anther oblong, 0.7 mm across. Female flower sessile; ovary globose, 1 mm across, brown; style 1.5 mm, tapering. Achenes lenticular, 1.5 mm smooth.

Common, in forests and villages in dry localities in most districts.

Fl. & Fr.: Almost throughout the year.

Potchera water falls (ADB), *GO & DAM* 5043; Velutla RF (NZB), 9035; Manchippa RF (NZB), *BR & KH* 9033; Anantasagar RF (RR), *MSM & NV* 10409; Farahabad (MBNR), *TDCK* 16830; Mannanur RF (MBNR), *TDCK* 13639; Ramappa temple area, Palempet (WGL), *RKP* 105346 (BSID).

Ficus pumila L., Sp. Pl. 1060. 1753; Bailey, Man. Cult. Pl. 339. 1949; Fischer, Fl. Madras 3 : 1367. 1928.

Climber with clinging roots; young stems hairy. Leaves dimorphic, ovate-oblong, 4-10 x 2.5-6 cm, cordate at base, margins entire, apex obtuse to subacute, finely white-villous beneath; petioles 8-25 mm long; stipules appresed hairy. Hypanthodia on erect fruiting branches, solitary, axillary, sessile to shortly pedunculate, subglobose to pyriform 2.5-5 cm long, greenish yellow, subtended by ovate-lanceolate basal bracts, apical orifice umbonate, internal bristles copious, short. Male flowers large, numerous, in several whorls. Female flowers sessile to long-pedicelled. Figs obovoid, turbinate, 3.5-7 cm long, purple-black, glabrous to glaucous.

Cultivated as a cover on walls of the buildings, flyover bridges etc. as an ornamental.

Syconia: July – October.

Ficus racemosa L., Sp. 1060. 1753. *F. glomerata* Roxb., Pl. Coromandel 2: 13. t. 123. 1799; King in Hook.f., Fl. Brit. India 5: 535. 1888; Fischer, Fl. Madras 3: 1364. 1928; Chaudhary & *al.*, Taiwania 57: 203. 2012.

Vern.: *Medi, Atti, Moida, Bodda, Paidu.*

Evergreen tree, up to 20 m high; latex milky; bark greenish smooth, brownish-red, scaly, peeling off; wood soft, smooth, branchlets white-pilose. Leaves alternate, oblanceolate, 9–20 × 3–6 cm, subcoriaceous, glaucous above, glabrous below, base rounded to acute, margin entire, apex acute; lateral veins 8–10 pairs, some what flattened and prominent on both sides, intercostals obscure; petiole not articulate, glandular at the basal nerve axils, 3–5 cm long; stipules lanceoalte, 1–1.5 cm long. Figs monoecious on separate branches, cauliflorous, subglobose, 1.5–2.5 cm across, white-pilose, walls thick, soft, ripening purplish red, ribbed, base tapering, apex truncate or broadly rounded; peduncle to 7 mm long; basal bracts ovate-triangular, to 2 mm across, obtuse, persistent; orifice subumbonate or slightly sunken, closed by 5–6 bracts; internal bristles 0; perianth shortly cupular. Tepals 3–5, ovate-lanceolate, 1.5 mm across, scarious, lanceolate, acute. Male flowers ostiolar; peduncle to 3 mm long; stamens 2, exserted; filaments 1 mm, connate below; anthers oblong, parallel, 0.8 mm across. Female flowers stalked; ovary stalked, brown, 1.5 mm long; style to 3 mm long. Gall flowers similar to female, long-stalked; style short. Achenes lenticular, 1 mm, obscurely keeled.

Common, along river banks, roadsides, outskirts of villages in all districts.

Syconia: Throughout the year.

Neredugunda (ADB), *GO* & *DAM* 5105; Yellareddy (NZB), *TP* & *BR* 6328; Narsapur (MDK), *CP* 11575; Pakhal (WGL), *KMS* 13168 (MH).

Ficus religiosa L., Sp. Pl. 1059. 1753; King in Hook.f., Fl. Brit. India 5: 513. 1888; Fischer, Fl. Madras 3; 1363. 1928; Chaudhary & *al.*, Taiwania 57: 208. 2012.

Vern.: *Ravi, Ragi, Raigu, Ashwarthamu*; Eng.: The peepal tree.

Deciduous, fast growing tree, up to 20 m high; often epiphytic when young; bark greenish, smooth; wood greyish-white, fibrous; branchlets yellowsh grey, glabrous. Leaves alternate, broadly ovate, cordate, 5–13 × 4.5–12 cm, coriaceous, glabrous, base truncate or somewhat cordate, margin entire, apex abruptly acuminate, cuspidate to 8 cm long, linear-lanceolate; basal nerves 3, lateral nerves 8–10 pairs, raised on either side, inter costals zigzag; petiole to 12 cm long, articulate, glandular at the apex below; stipules ovate-lanceolate, to 1.5 cm long. Figs monoecious, axillary, paired, sessile, obovoid or globose, 4–6 mm across, ripening pink, purple or black; fig-wall thick; basal bracts cupular, to 4 mm long; margin truncate, puberulous, persistent; orifice umbonate, closed by 3 apical bracts; internal bristles 0. Male flowers ostiolar, sessile; tepals 2, free, ovate-lanceolate, 0.7 mm across; stamens 1; filament 0.2 mm long; anthers oblong, parallel, 0.7 mm long. Female flower sessile; tepals 3–4, free, linear-lanceolate, 1 mm long, brownish, glabrous; ovary ovoid-oblong, 1 mm across, red-brown; style 1.5 mm long, dilated above. Gall flowers similar to female. Achenes smooth.

Common, along roadsides, near temples, wild in village surrounding, river banks in all districts; it is a sacred tree of Hindus, and Buddhists. It is also planted for shade.

Fl. & Fr.: May-Aug.

Nirmal (ABD), *GO* & *BR* 9403; Gowraram (NZB), *TP* & *BR* 6389; Medak (MDK), *TP* & *CP* 14081; Central University campus (RR), *TP* & *MSM* 15139; Puranapul (HYD), *KMS* 5994 (MH); Aklaspur (KRN), *GVS* 20232 (MH); Umamaheswaram (MBNR), *SRS* 109158 (BSID); Mallepally (NLG), *BVRS* 538 (NU).

Ficus talbotii King, Ann. Roy. Bot. Gard. (Calcutta), 1: t. 63. 84Q. 1887 (*talboti*) & in Hook.f., Fl. Brit. India 5:512. 1888; Fischer, Fl. Madras 3: 1363. 1928; Chaudhary & *al.*, Taiwania 57: 213. 2012.

Evergreen tree, 10-20 m high; aerial roots absent; bark green, smooth; wood very white, turning yellow. Leaves subcoriaceous, narrow elliptic to broad ovate, 7-11 × 4-6 cm, glabrous, base rounded, margin entire, apex shortly, acutely caudate, 3-nerved; secondary nerves 8-10 pairs, raised on both surfaces; petiole to 3 cm, glandular at apex below. Receptacles obovoid, sessile, ripening yellowish, spotted with red colour dots, in axillary pairs; achenes smooth.

Occasional in forests in Mahabubnagar district.

Fls. & Frs.: December-February.

Borapur (MBNR), *BR, SKB* & *BSS* 32513.

Ficus tinctoria Forster f., Fl. Ins. Austr. 76. 1786 subsp. **gibbosa** (Blume) Corner, Gard. Bull. Singapore 17: 476. 1960; Chaudhary & *al.*, Taiwania 57: 203. 2012. *F. gibbosa* Blume, Bijdr. 9: 466. 1825; King, Ann. Roy. Bot. Gard. (Calcutta) 1: 4. t. 2. 1887 & in Hook.f., Fl. Brit. India 5: 497. 1888; Fischer, Fl. Madras 3: 1365. 1928. *Ficus tinctoria* Forster f. subsp. *parasitica* (K.D.Koenig ex Willd.) Corner, Gard. Bull. Singapore 17: 476. 1961.

Vern.: *Konda juvvi, Pakki.*

Tree, up to 10 m high; often epiphytic, enclosing the trunks of tree in a perfect network of branches and creeping along the walls and on the side of wells, without aerial roots, bark greyish, smooth; wood brownish grey. Leaves thinly coriaceous, rhomboid or sub-rhomboid or elliptic ovate, 4-12 × 3-6 cm, hispid and dark green above, scabrid and minutely hispid and paler beneath; apex acute, margin entire, base oblique, 5-7 mm across, paired, globose, peduncled, hairy; male receptacles yellow; female receptacles red when ripe.

Common in open places in Adialabad, Karimnagar and Nalgonda districts.

Syconia: February-April.

Kambalpally (NLG), *BVRS* 311 (NU).

Ficus tsjahela Rheede ex Burm. f., Fl. Ind. 227. 1768; King, Ann. Roy Bot. Gard. Calcutta, 1:57, t. 70 & 84. 1887(*tsjakela*) & in Hook.f., Fl. Brit. India 5; 14. 1888; Fischer, Fl. Madras 3: 1862.1928; Corner in Gard. Bull. Singapore 21: 7. 1965; Chaudhary & *al.*, Taiwania 57: 20. 2012.

Deciduous tree, up to 18 m high, aerial roots absent; bark coloured, rough; wood brown; branchlets glabrous, dark brown. Leaves coriaceous, oblong-elliptic, or elliptic-ovate, 6-19 × 3-10.5 cm, glabrous, dark green above, pale beneath, ashy grey when dry, base truncate or rounded, margin entire, apex shortly acuminate, 3-ribbed; 7-12 pairs above basal ones, conspicuous, to 2.5 cm apart from basal ones to just above lateral nerves; petiole to 8 cm. Receptacle monoecious, depressed-globose, 6 mm across, greyish-purple, sessile, 2-6 together on tubercles in fallen leaf branches, basal bracts 3, bifid, orifice plane, circular; apical bracts 3.

In moist regions of moist deciduous forests above 600 m.

Syconia: Ripen by May.

Borapuram (MBNR), *BR, SKB* & *BSS* 32375.

Ficus virens Aiton, Hort, Kew. 3: 451. 1789; Chaudhary & *al.*, Taiwania 57: 209. 2012. *F. infectoria* Roxb., Fl. Ind. 3: 551. 1832 (non Willd); 1806; King in Hook.f., Fl. Brit. India 5; 515. 1888; Fischer, Fl. Madras 3: 1362. 1928. *F. infectoria* Roxb. var. *lambertiana* (Miq.) King, Ann. Roy. Bot. Gard. (Calcutta) 1: 63. t. 76. 1887; Fischer,

Fl. Madras 3: 1363. 1928. *F. virens* Aiton var. *lambertiana* (Miq.) Raizada, Indian For. 94: 443. 1968; Sudhakar Reddy & *al*., Suppl. Fl. Andhra Pradesh 109. 2008. *Urostigma lambertiana* Miq., London J. Bot. 6: 565. 1847.

Vern.: *Banda Juvvi, Jati Juvvi*.

Large, spreading, deciduous tree, epiphytic in early stages, with as few aerial roots; bark greenish-grey, smooth; wood grey. Leaves subcoriaceous, 7-18 × 2.65.7 cm, glabrous, apex abruptly acuminate, margin entire to slightly undulate, base acute, truncate or subcordate; secondary nerves 7-12 pairs, basal pair shorter, intercostals zig zag; petiole 2.5-7 cm, articulate, glandular at the apex below. Receptacles axillary, paired, shortly peduncled, globose, whitish with red dots.

Common, in plains to hills in most districts.

Fl. & Fr.: October-May.

Gandhari RF (NZB), *TP* & *BR* 6333; Narsapur (MDK), *SSR* 16194; Anantagiri (RR), *SSR* 18517; Vempali (KRN), *GVS* 20188 (MH); Pakhal (WGL), *ANH* 15943 (MH), *RKP* 108256 (BSID); Borapur (MBNR), *BR* & *SKB* 32514; Uskavagu, Mannanur (MBNR), *SRS* 108901, 107441 (BSID).

MORUS L.

Morus alba L., Sp. Pl. 986. 1753; Hook.f., Fl. Brit. India 5: 492. 1888; Fischer, Fl. Madras 3: 1370. 1928.

Vern.: *Resham chettu*; Eng.: Mulberry.

Deciduous, medium-sized tree; crown round; branches grey or greyish yellow; young shoots pubescent. Leaves thin, sub-coriaceous, broadly ovate, 7-12 × 4-7 cm, glabrous and shining above, slightly pubescent along nerves beneath, base subcordate, margin dentate or coarsely toothed, apex acute, basal nerves 3-5; secondary nerves 3-5 pairs; petiole 1-2.5 cm. Flowers monoecious, greenish, aggregated into spike. Fruit red, then purple, sweet.

Usually cultivated, for rearing the silk worms.

Fl. & Fr.: Throughout the year.

Nizamabad outskirts (NZB), *BR* & *KH* 6490.

STREBLUS Lour.

Streblus asper Lour., Fl. Cochinch. 1: 615. 1790; Hook.f., Fl. Brit. India 5: 489.1888; Fischer, Fl. Madras 3: 1353. 1928.

Vern.: *Stianike, Barranki, Barniki, Pakki, Bamigi.*

Evergreen, scraggy or bushy tree, up to 7 m high; latex milky, bark smooth, grey; wood yellowish or brownwish-white; branchlets hairy. Leaves thick-coriaceous, elliptic-obovate or rhomboid, 1.7-6.8 × 1-3 cm, rough on both surfaces, base acute or cuneate, margin irregularly dentate, apex acute, lateral nerves 4-6 pairs, raised on under surface; petiole to 2 mm. Stipules linear, caducous. Flowers dioecious, male flowers in globose heads, cream-coloured, axillary, peduncle to 5 mm; female flowers solitary on axillary, usually fascicled peduncles. Fruit globose, 6 mm across, yellow,

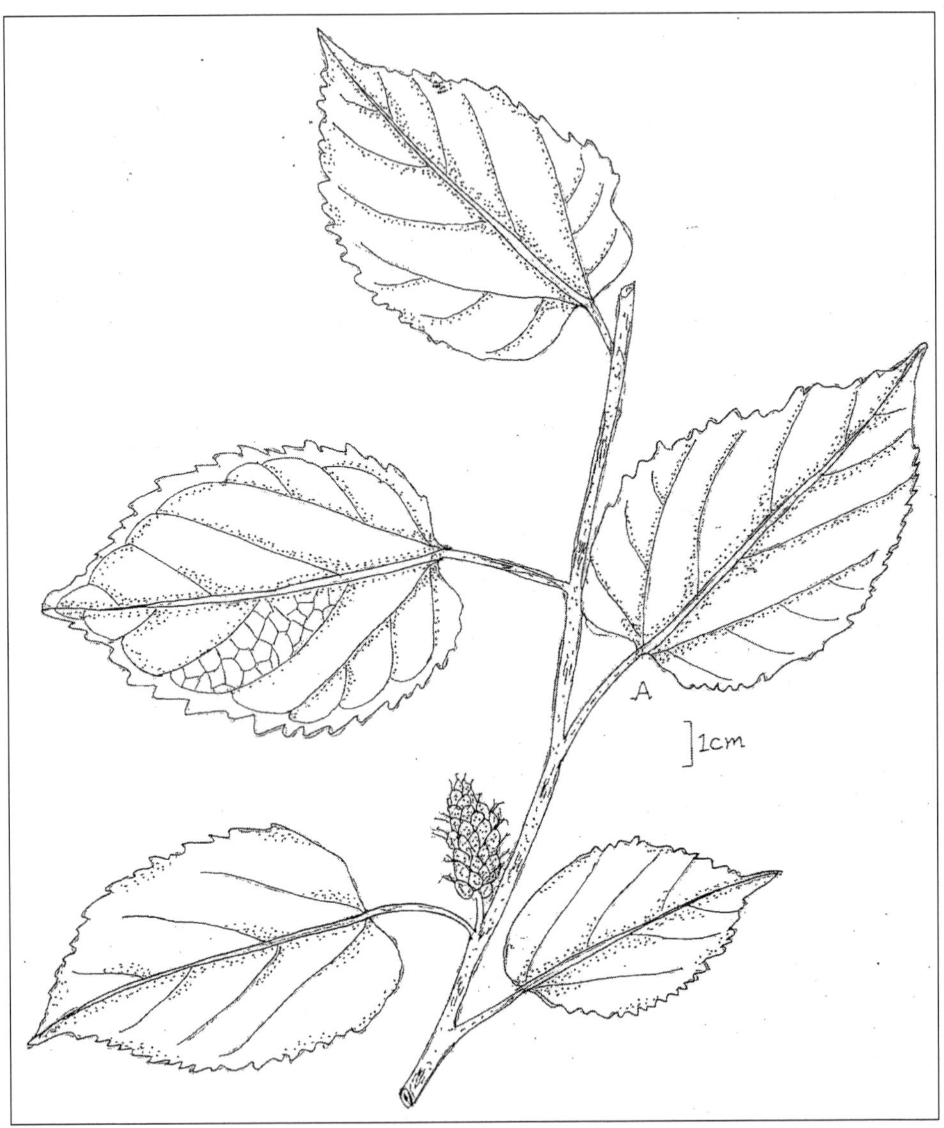

Figure 103: Morus alba L.

A. Twig.

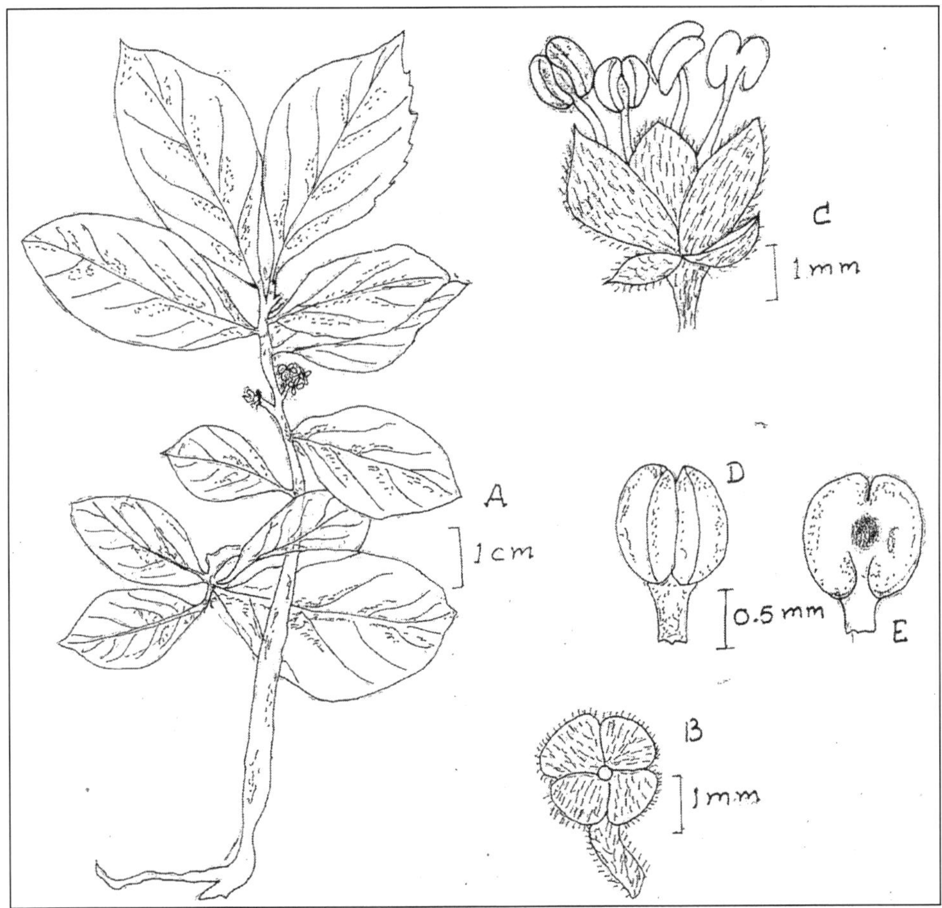

Figure 104: Streblus asper Lour. Male plant
A. Twig, B. Flower bud, C. Flower, D & F. Anthers.

enclosed in the enlarged fleshy sepals, style persistent; seed 1, globose, smooth, 3 mm across.

Common, in shrubberies, hedges in dry deciduous forests in all districts.

Fl.: January- March, Fr.: April.

Jodevagu (ADB), *GO* & *DAM* 3182; Neelwai (ADB), *GO* & *DAM* 5159; Ibrahimpet (NZB), *BR* & *KH* 9045; Kotgir (NZB), *BR* & *KH* 9645; Pocharam Sanctuary (MDK), *BR* & *CPR* 4613; Gadirayal RF (RR), *MSM* & *KH* 10333; Gowrammasari RF (MBNR), *TDCK* 13691; Yadagirigutta (NLG), *BVRS* 549 (NU); Kudurpalli (KRN), *SLK* 70723 (NBG); Near Mahadevpur (KRN), *SLK* 70793 (NBG); Pakhal (WGL), *RKP* 108271 (BSID); Dummagudem RF (KMM), *RCS* 99062 (BSID).

URTICACEAE

1. Plants with stinging hairs .... **Girardinia**
1. Plants without stinging hairs:
    2. Stigmas penicillate .... **Pilea**
    2. Stigmas filiform (not penicillate) .... **Pouzolzia**

GIRARDINIA Gaudich.

Girardinia diversifolia (Link) Friis, Kew Bull. 36: 145. 1981. *Urtica diversifolia* Link, Enum. Hort. Berol. Alt. 2: 385. 1822 (non. Blume 1825). *Girardinia leschenaultiana* Decne in Jacq. Voy. Inde 4: 152. 1844; Fischer, Fl. Madras 2: 1372. 1928. *G. zeylanica* Decne, *loc.cit.*; Fisher, Fl. Madras 2: 1373. 1928.

Tall perennial herbs, 1-1.5 m high, closely armed with slender stinging hairs; stems and branches furrowed. Leaves 7.5-12 × 9-12.5 cm, pedately 3-5-lobed, base truncate to subcordate, alternate, margin coarsely serrate, the serratures triangular, apex acuminate, chartaceous, sparsely scabrous above, densely so along the nerves below. Male flowers on simple heads armed with numerous stinging hairs. Female flowers in capitate heads, clustered along the rachis, axillary. Achenes compressed.

Occasional in hills of Mahabubnagar district.

Fl. & Fr.: February – March.

Borapuram (MBNR), *BR* & *BSS* 32387.

PILEA Lindl. nom. cons.

Pilea microphylla (L.) Liebm., Mexic. Meldeagt. Pl. V, 2: 302. 1851; Hook.f., Fl. Brit. India 5: 551. 1888; Fischer, Fl. Madras 3: 1379. 1928. *Parietaria microphylla* L., Syst. Nat. ed. 10. 2: 1308. 1759. *Pilea muscosa* Lindl., Coll. Bot. t. 4. 1821; Hook.f., Fl. Brit. India 5: 551. 1888.

Prostrate herb, to 20 cm long. Leaves decussate, elliptic ovate in unequal pairs, to 4 × 2 mm, subsucculent, 1-nerved, base acute, margin entire, apex subacute or obtuse; stipules connate, intrapetiolar, raphides transverse; petiole to 1 mm. Flowers in stalked umbellate clusters. Male: 1.5 mm across, tepals 4, connate, minutely 2-4-toothed, valvate, stamens 4, filaments 0.7 mm, anthers 0.3 mm. Female: 0.4 mm across, ovary ellipsoid, 0.5 mm. Achenes ellipsoid, 1 mm, 1-locular, ovules erect. Stigma shortly penicillate, subsessile, albumen scanty, cotyledons broad.

Grows on moist walls; weed of nurseries and pots (C.S. Reddy and V.S. Raju, 2002).

Fl. & Fr.: July – January.

Central University campus (RR), *TP* & *MSM* 15133.

POUZOLZIA Gaudich.

1. Male flowers 4-merous; achenes winged .... **P. zeylanica**
1. Male flowers 5-merous; achenes not winged .... **P. auriculata**

Pouzolzia auriculata Wight, Icon. Pl. Ind. Orient. 6: 42. 1853; Hook.f., Fl. Brit. India 5: 582. 1888; Fischer, Fl. Madras *3: 1383. 1928.*

Erect scabrid herbs to under shrubs, stems woody, young branches angled, pubescent. Leaves 2-7 × 1-2 cm, elliptic-lanceolate, alternate, base acute to rounded, margin entire; apex acuminate, appressed hairy. Flowers in axillary, sessile clusters, greenish. Achenes white.

Occasional along the streams in forests in Mahabubnagar district.

Fl. & Fr.: January – November.

Pouzolzia zeylanica (L.) Benn. & R.Br., Pl. Jav. Rar. 66. 1838. *Parietaria zeylanica* L., Sp. Pl. 1052. 1753. *P. indica* L., Mant. Pl. 128. 1767. *Pouzolzia indica* (L.) G. Benn., Pl. Jav. Rar. 67. 1838; Hook.f., Fl. Brit. India 5: 581. 1888; Fischer, Fl. Madras 3: 1382. 1928.

Perennial rooted, weak stemmed herbs or shrubby, very variable in size and habit, stems grooved, hirsute. Leaves 2-4.5 × 0.8-2 cm, very variable, gradually reducing in size, elliptic-ovate, alternate, base rounded, margin entire, apex acute, pilose below, less so above. Flowers white in subsessile clusters. Achenes 1.3 x 1 mm, broadly ovoid, acute, smooth and shining, white with a red coloured tip.

Common in Ranga Reddi and Mahabubnagar districts as a weed.

Fl. & Fr.: January – April.

Kusumasamudram RF (RR), *MSM* 10334.

CASUARINACEAE

CASUARINA Adans.

Casuarina equisetifolia L., Amoen. Acad. 4: 143. 1759; Hook.f., Fl. Brit. India 5: 598. 1888; Fischer, Fl. Madras 3: 1389. 1928.

Vern.: *Sarugudu.*

Tall leafless trees, up to 15 m high; monoecious; branchlets arising from the axils of scales; bark greyish brown or dark brown. Leaves scaly, alternating with ribs of the upper node. Male flowers in terminal catkins; bracts arranged in 10-25 whorls and each whorl having 6-7 bracts with one flower in each bract; bracteole membranous, ovate; tepals 2, membranous; filaments exserted. Female in globose-ovoid cone like heads; bracts similar to those of male inflorescence; perianth 0; pistil bicarpellary, ovary unilocular, stigmas 2-3 mm long. Nutlets compressed, winged.

Widely planted.

Fl. & Fr.: February – August.

Bhadrachalam (KMM), *RCS* 98796 (BSID).

CERATOPHYLLACEAE

CERATOPHYLLUM L.

Ceratophyllum demersum L., Sp. Pl. 992. 1753; Hook.f., Fl. Brit. India 5: 639. 1888; Fischer, Fl. Madras 3: 1391. 1928.

Submerged, root-less aquatic herbs, to 3 m; branchlets brittle, green, reddish, glabrous. Leaves whorled, 1.5-2.5 cm long, bright green, sometimes sparsely sericeous, margin denticulate, apex acute. Flowers monoecious, axillary, solitary. Fruit 4 cm long with 3 spines.

Occasional in ponds, ditches and pools in Adilabad, Medak, Ranga Reddi, Warangal and Nalgonda districts.

Fl. & Fr.: September – December.

Mallampet RF (ADB), *TP* & *GO* 5452; Medak tank (MDK), *BR* & *CP* 11402; Malkalcheruvu (RR), *MSM* & *KH* 10567; Near Dindi project (NLG), *BVRS* 675 (NU); Pakhal (WGL), *KMS* 11682 (MH).